Ocular Infection

Second Edition

David Seal
Applied Vision Research Centre, City University, London, England, UK

Uwe Pleyer
Charité, Humboldt University, Berlin, Germany

With Contributions from

Gregory Booton
The Ohio State University, Columbus, Ohio, USA

Consuelo Ferrer
Instituto Oftalmológico de Alicante, Alicante, Spain

Susanne Gardner
Atlanta, Georgia, USA

Regis Kowalski
University of Pittsburgh, Pittsburgh, Pennsylvania, USA

Philip Thomas
Joseph Eye Hospital, Tiruchirrapalli, Tamil Nadu, India

informa
healthcare

New York   London

Informa Healthcare USA, Inc.
52 Vanderbilt Avenue
New York, NY 10017

© 2007 by Informa Healthcare USA, Inc.
Informa Healthcare is an Informa business

No claim to original U.S. Government works
Printed in the United States of America on acid-free paper
10 9 8 7 6 5 4 3 2 1

International Standard Book Number-10: 0-8493-9093-1 (hb : alk. paper)
International Standard Book Number-13: 978-0-8493-9093-7 (hb : alk. paper)

Library of Congress Cataloging-in-Publication Data

Seal, David
   Ocular infection / by David Seal, Uwe Pleyer. – 2nd ed.
      p. ; cm.
Includes bibliographical references and index.
ISBN-13: 978-0-8493-9093-7 (hb : alk. paper)
ISBN-10: 0-8493-9093-1 (hb : alk. paper)
1. Eye–Infections. I. Pleyer, Uwe, 1957-II. Title.
   [DNLM: 1. Eye Infections–diagnosis. 2. Eye Infections–pathology.
3. Eye Infections–therapy. WW 160 S438o 2007]
   RE96.S43 2007
   617.7–dc22   2007028097

Visit the Informa web site at
www.informa.com

and the Informa Healthcare Web site at
www.informahealthcare.com

Ocular Infection

Preface to the Second Edition

Many changes have taken place in ophthalmic practice over the last 10 years, and these are reflected in this completely revised and expanded second edition. Likewise, there has been a change in the primary authorship of the book, with Professor Uwe Pleyer taking over for Anthony Bron and John Hay. In this edition, we have rearranged the chapters and expanded the basic science but still kept the book to a reasonable length. We have supplemented the clinical text with numerous flow diagrams, which we hope will please those readers who wish to have "one page" summaries for quick reference. However, the book also retains in-depth discussions of key topics related to the science, diagnosis, and treatment of ocular infection, for the benefit of those who wish to study the subject in more detail.

David Seal
Uwe Pleyer
London and Berlin

Preface to the First Edition

This book has been written to further the understanding of infective eye disease. We have reviewed the literature extensively in an effort to offer a solid foundation for the diagnosis and management of ocular infection. To this end tables have been constructed to highlight important practical points. We hope that this book will prove useful to ophthalmologists, optometrists, microbiologists, and hospital pharmacists concerned with good practice in medicine and the treatment of ocular infection. Our aim is the avoidance of unnecessary blindness due to lack of understanding of the disease process and its chemotherapy.

We are very grateful to all our colleagues, both in the U.K. and overseas, for their advice and assistance with this book, and we especially thank Professor Harold Lambert for his encouragement. In particular, we would like to thank the following, including colleagues in the Tennent Institute, who have supplied clinical and laboratory photographs: Thomas Byers, John Dart, Gordon Dutton, Linda Ficker, Colin Kirkness, Barbara Leppard, Susan Lightman, Ian Mackie, Ron Marsh, Keith Rodgers, Elizabeth Wright, and Peter Wright. We are also particularly grateful to John McCormick for clinical and laboratory photography in the Tennent Institute and to Dorothy Aitken and Sam Cameron for technical electron microscopy.

David Seal
Anthony J. Bron
John Hay
Glasgow and Oxford

Acknowledgments

Alan Tomlinson, Department of Optometry, Glasgow Caledonian University, Glasgow, U.K. for advice on pathophysiology of contact lens wear over many years (Chapter 6 and Appendix D) and *Acanthamoeba*.

Consuelo Ferrer ,Vissum, Alicante, Spain for assistance with molecular biology (Chapter 3) and advice with Appendix D and for photomicrographs of *Fusarium*. Jorge Alio, Chief Ophthalmologist, Vissum, Alicante, Spain for supplying photos of fungal keratitis and for much useful discussion on causes and prevention of post-operative endopthalmitis.

David Lloyd, Renaissance Healthcare, Epsom, U.K. and Klaus Geldsetzer, Santen gmbh Munich, Germany for pharmaceutical advice and for assistance with Chapter 4 and Appendices C and D.

Gholam Peyman, Department of Ophthalmology, University of Tucson, Arizona, U.S.A.; Kirk Wilhelmus, Cullen Eye Institute, Baylor College of Medicine, Houston, Texas, U.S.A.; Steven Barrett, Charing Cross Hospital, London U.K.; Crawford Revie and George Gettinby, Department of Modelling Science, University of Strathclyde, Glasgow U.K.; Peter Barry, Department of Ophthalmology, St. Vincent's Hospital, Dublin, Eire; Per Montan, Department of Ophthalmology, St. Erik's Hospital, Gothenburg, Sweden; Gunther Grabner, Department of Ophthalmology, University Hospital, Salzburg, Austria; Luis Cordoves, Department of Ophthalmology, University Hospital, Tenerife, Canary Islands, Spain; Christopher Lowman and Udo Reischl, Departments of Ophthalmology and Molecular Biology, University of Regensburg, Regensburg, Germany; Marie-Jose Tassignon, Department of Ophthalmology, University Hospital, Antwerp, Belgium; Hugo Verbraeken, Department of Ophthalmology, University Hospital, Ghent, Belgium; Suleyman Kaynak, Department of Ophthalmology, University Hospital, Izmir, Turkey and colleagues in the ESCRS (European Society of Cataract & Refractive Surgeons) for their help, discussions, and advice with prophylaxis and management of post-operative endophthalmitis (Chapter 8).

John Hay, Glasgow, U.K. for his original construction of Table 1 in Appendix B and for drug information.

Magnus Peterson, Department of Modelling Science, University of Strathclyde, Glasgow U.K. for Tables 1–5 in Chapter 10 and good advice.

Philip Thomas, Joseph Eye Hospital, Tiruchirrapalli, India for assistance with mycological methods (Chapter 2) and drugs (Appendix A) and *Fusarium* (Appendix D).

Tara Beattie, Department of Civil Engineering (Public Health), University of Strathclyde, Glasgow U.K.; Andrew Rogerson, Marshall University, Huntington, West Virginia, U.S.A.; Megan Shoff, Department of Evolution, Ecology, and Organismal Biology, The Ohio State University, Columbus, Ohio U.S.A. and Scott Schatz, Department of Optometry, Nova SE University, Ft. Lauderdale, U.S.A. for many useful discussions on *Acanthamoeba* pathogenicity (Chapter 6).

We are also grateful for assistance by Jill Bloom, Senior Drug Information Pharmacist, Moorfields Eye Hospital, London, U.K. and by members of Bausch & Lomb, Kingston, London U.K. for helpful discussions with Appendix C.

Finally, we are most grateful to Alan Burgess, Informa U.K. and for the general assistance of publishing staff of Informa U.S.A.

Contents

Contributors

Gregory C. Booton   Department of Molecular Genetics, Ohio State University, Columbus, Ohio, U.S.A.

Consuelo Ferrer   Molecular Biology Laboratory, Vissum Instituto Oftalmológico de Alicante, Alicante, Spain

Suzanne Gardner   Atlanta, Georgia, U.S.A.

Regis P. Kowalski   Charles T. Campbell Ophthalmic Microbiology Laboratory, University of Pittsburgh, Pittsburgh, Pennsylvania, U.S.A

Philip Thomas   Joseph Eye Hospital, Tiruchirrapalli, Tamil Nadu, India

ONE

Pathogenesis of Infection and the Ocular Immune Response

SURFACE PROTECTION OF THE EYE

The normal conjunctiva and cornea are protected by a triple-layered tear film comprising an outer oily layer from the meibomian glands, an aqueous layer from lacrimal glands, and an inner layer of mucus, derived chiefly from conjunctival goblet cells. Blinking maintains the integrity of this protective layer. The goblet cells and subsurface vesicles of the conjunctiva create the deep mucus layer of the tear film, which is anchored to the ocular surface glycocalyx.

Tears produced by the lacrimal gland have a high concentration of immunoglobulin A (IgA) (0.6 G/L), lysozyme (1 G/L), and lactoferrin (1.2 G/L), which have an antimicrobial action. This occurs from the coating of bacteria by tear IgA, growth inhibition by lactoferrin iron chelation, and the action of lysozyme. Lysozyme (acetylmuramidase) splits the bond between acetylmuramic acid and acetylglucosamine in the peptidoglycan of the bacterial cell wall with a direct lytic action on *Micrococcus lysodeikticus*. Synergy with immunoglobulin and complement is required for lysis of other bacteria. Immunoglobulin G (IgG) (0.05 G/L) and complement are found in low concentration in tears but leak through the conjunctiva from inflamed vessels and are important for causing cell wall lysis by the complement cascade.

From an age of about 40 years, tear volume and the concentration of lacrimal proteins (lysozyme, lactoferrin, and IgA) decrease due to reduced lacrimal gland function, but the eye does not become infected more frequently unless there is concomitant development of a severely dry eye. This emphasizes the beneficial effect of constant tear flow, removing bacteria and other debris from the ocular surface, although the moderately dry eye develops a compensatory antibacterial mechanism based on leakage of IgG and complement.

The corneal epithelium consists of five layers, is transparent, and rests on Bowman's membrane. It is generated by the limbal stem cells. The epithelium can resurface an epithelial defect within 24–48 hours. Its surface cells are linked by tight junctions and it provides an important barrier to the invasion of the corneal stroma by microbes. This is well demonstrated by the increased frequency of microbial keratitis in certain soft contact lens wearers with a compromised, hypoxic epithelium, especially those using extended wear lenses. Similarly, while bacterial keratitis is rarely found in an eye with a normal epithelium, it is more common in patients with recurrent epithelial defects, such as those due to chronic herpetic disease.

INVASION OF THE EYE BY PATHOGENIC MICROBES

In order to cause infection, the organism has to penetrate into the eye. This can be facilitated by an epithelial defect or be due to a contaminated contact lens (as noted) or direct corneal trauma. Penetration of the posterior segment by a contaminated, high-velocity fragment, from a hammer-and-chisel injury or shrapnel, can cause endophthalmitis, for instance by *Bacillus cereus* (considered in detail in Chapter 8). The blood-borne route must always be considered for endophthalmitis when there is no history of accidental or surgical trauma, particularly the possibility of endocarditis, meningococcal bacteremia, or a previous, possibly contaminated, blood transfusion. Intravenous drug addicts are also more likely to develop endophthalmitis, particularly with *Candida albicans*.

For invasion of the corneal stroma to occur, the organism must first adhere to the epithelium. *Pseudomonas aeruginosa* is particularly adept at invading the compromised epithelium from a superficial source, often from contaminated contact lenses or eye drops. In rabbit experiments, *P. aeruginosa* has been shown by us to be able to adhere to damaged epithelium and to invade the stroma within 1 hr, due to the release of toxins and proteases. For the contact lens wearer, the initial problem involves bacterial, amebal, or polymicrobial contamination of the storage case with adherence of bacteria—or adsorption of *Acanthamoeba* (1)—to the lens, which is often coated with a biofilm protecting the organisms. The contact lens then acts as a mechanical vector, transferring the microbes to the corneal epithelium. The situation is exacerbated by the sequestration of tear fluid behind the lens.

Penetration alone by an organism is not, in most instances, sufficient to cause infection. The organism must proliferate to establish sufficient numbers of cells to overcome host defenses. The importance of bacterial load is emphasized by the finding that while up to 25% of intraoperative aqueous samples at cataract surgery contain bacteria, the rate of postoperative endophthalmitis following cataract surgery is low, at 0.1% to 0.7%. In most cases the bacterial inoculum is insufficient to result in multiplication, the cells being inactivated by antibacterial substances present in the anterior chamber (AC) or by antibiotics given intraoperatively. In a few patients, bacterial proliferation does occur, along with adhesion to the intraocular lens, and a devastating endophthalmitis results (Chapter 8).

The virulence of the organism also dictates the outcome for the patient. This usually relates to the production of lethal toxins, by the bacterium, which are quickly effective at causing tissue necrosis, such as the exotoxins and proteases liberated by *P. aeruginosa*. *Streptococcus pyogenes* is highly virulent for the eye, producing exotoxin A and needing only a small inoculum, possibly as low as 10 cells, to cause necrotizing fasciitis of the lids or a fulminant endophthalmitis within 24 hours of cataract surgery. While this situation applies to all tissues of the body, it is different for *P. aeruginosa*, which is highly pathogenic for the cornea but not for the skin or other organs of the immunocompetent host, except for the brain. The cornea is unique in that it is both avascular and, in youth at least, lacks professional antigen-processing cells except in the peripheral epithelium. Its ability to mount an immediate immunological reaction to trauma is therefore limited and, hence, *P. aeruginosa* can establish a fulminant necrotic infection in the first 24 hours. This is unusual in other, vascularized sites in the body. The posterior segment also possesses immune-privileged compartments and provides an excellent environment for bacterial multiplication.

IMMUNITY AGAINST INFECTION

Factors determining the immunity of the cornea are illustrated in Figure 1 using *Staphylococcus aureus* protein A as an example of a bacterial antigen.

LOCAL CORNEAL SURFACE IMMUNITY

This depends on IgM, C1 complement, and Langerhans cells (LCs), all of which are found in the peripheral, but not the central, cornea. IgG diffuses into the stroma from the limbus, achieving 50% of the serum concentration. Inflammation of the cornea, as in herpetic keratitis, can stimulate centripetal LC migration and this occurs to some extent with age. LCs are professional antigen-processing cells and are the only corneal cells to express the major histocompatibility (MHC) class II antigens without prior inducement by cytokines.

LCs take up antigens by pinocytosis or endocytosis. Upon exposure to antigen, LCs and other antigen-presenting cells (APCs), such as corneal epithelial cells and macrophages, undergo functional maturation and gain the ability to present the antigen to CD4 T helper (Th) cells attracted to the area by cytokines. Antigens taken up into APCs are enzymatically degraded to oligopeptides with an unfolded secondary structure, which are expressed at their surface as antigenic peptides bound to MHC molecules.

Macrophages can be transformed into APCs by interleukin-1 (IL-1) released from corneal epithelial cells, which induces the expression of MHC class II molecules at their surface. These APCs then process antigenic peptides to form a binary complex with MHC class II molecules. Macrophages can also process particulate antigens, including whole bacteria such as staphylococci and free-living amoebae such as *Acanthamoeba*, but can more effectively process soluble antigens such as the protein A of *Staphylococcus aureus*, internalized in endocytic vesicles. This is in contrast to Langerhans cells, which only process soluble antigen.

T lymphocytes exert their local effector function by secreting cytokines in tissues that act directly on target cells (2). Interferon (IFN-g) induces expression of MHC class II molecules in keratinocytes, epithelial cells, endothelial cells, and fibroblasts, which can all, to a variable degree, act as APCs, processing and presenting immunogenic peptides complexed with MHC class II molecules. However, they differ in their capacity to produce costimulatory signals and do not stimulate resting T cells, which require IL-2 to become activated.

LOCAL DELAYED-TYPE HYPERSENSITIVITY

Delayed hypersensitivity (DH) is induced by organisms, such as *Onchocerca volvulus* and *Staphylococcus aureus*, which can excite a cell-mediated immune reaction, expressed by the Th1 lymphocyte cells and mediated by cytokines (3). It has been proposed as a mechanism to explain episodes of marginal ulceration consequent to recurrent blepharitis caused by *Staphylococcus aureus* (Chapter 5) (4). This is discussed below, and illustrated in Figure 1, as an example of immune mechanisms operating in the peripheral cornea.

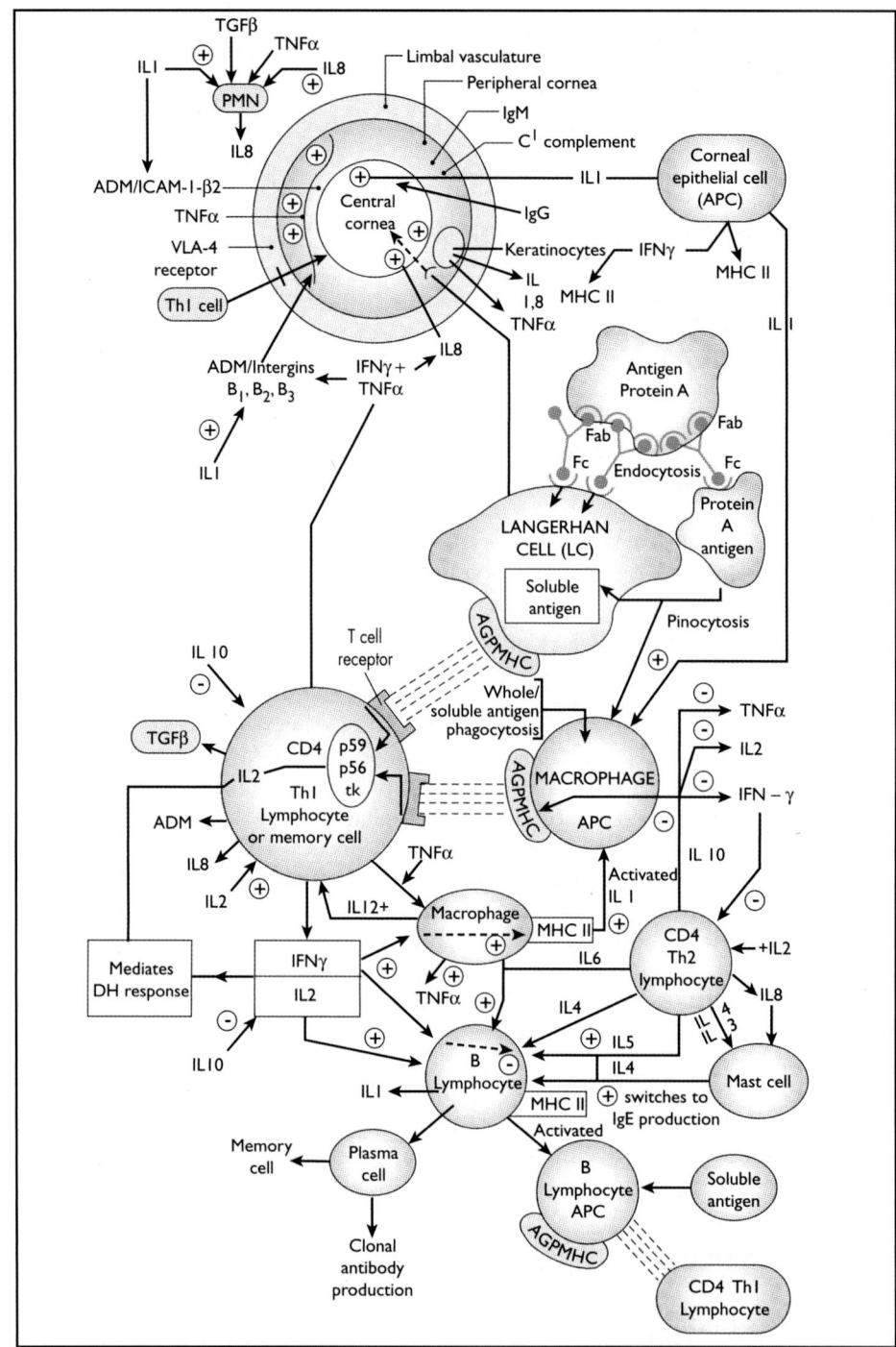

Figure 1 Immunity against infection in the cornea, with *Staphylococcus aureus* protein A as the example antigen. *Abbreviations*: ADM, adhesion molecules; AGPMHC, antigenic (oligo)peptide MHC group; APC, antigen-presenting cell;CD, cluster determinant antigen; IFN, interferon; IL, interleukin; MHC, major histocompatibility complex (II); PMN, polymorphonuclear cell; Th, T helper lymphocyte; tk, tyrosine kinase; TNF, tumor necrosis factor; VLA, very late antigen (super IgG) receptor.

Staphylococcus aureus repeatedly colonizes the lid margins of both normals (approximately 10%) and atopes (approximately 50%). It is likely that diffusible cell wall antigens from *Staphylococcus aureus*, especially protein A but also ribitol teichoic acid, reach the APC of the corneal epithelium across the tear film. In the peripheral cornea and also the conjunctiva, these antigens are processed by dendritic LCs. The LCs have receptors for the Fc portion of the antibody molecule, which assists endocytosis of antigens such as protein A by the binding of specific antibody. In the latter case, there can also be cross-linking between specific (Fab) and non-specific (Fc) sites both found on protein A.

The CD4 Th1 lymphocyte is responsible for most DH reactions. The Th1 subtype (as opposed to the Th2 subtype), mediates DH by a cascade of cytokines produced by T-cell clones. Th1 cells produce IL-2 and IFN-g, responsible in part for the induration response of the DH reaction in the skin. Th1 and Th2 cells regulate the actions of each other on antibody production; i.e., IFN-g produced by the Th1 cell inhibits the stimulatory effect of IL-4, produced by the Th-2 cell, on antibody production by B lymphocytes. This may be the mechanism for the clinical observation that a strong DH response is associated with a weak antibody response and vice versa.

Activated CD4 Th1 cells secrete IFN-g at the site of antigen entry and induce MHC class II expression on "non-professional" APCs, thereby activating them. The recognition of the MHC–peptide complex of the APC by a T-cell receptor initiates an intracellular signal transduction pathway via the p56 and p59 tyrosine kinases. The molecules that are then produced increase binding of Th cells to APCs and are called adhesion molecules (ADMs); for CD4 cells this includes the 55 kd monomeric transmembrane glycoprotein belonging to the Ig gene superfamily and for CD8 T cytotoxic cells (CTC), two 34 kd alpha chain molecules.

The cellular response is regulated by expression of ADMs on inflammatory cells and vascular endothelium, which in turn is controlled by cytokines, which can also act as chemotactic factors for polymorphonuclear cells (PMNs). These ADMs cause adhesion of PMNs and lymphocytes to the local limbal vascular endothelium, when they bind to it and migrate to the site of activated lymphocytes in the peripheral cornea.

Memory Th lymphocytes will also migrate to this site if the patient has undergone systemic enhancement of CMI to the particular antigen, in this example protein A of *Staphylococcus aureus*. The expression of integrins (b1, b2, b3) attracts these activated Th1 cells to the inflammatory site. The expression of b1 is upregulated on the surface of activated memory cells, which bind to the counter receptor on vascular endothelium [very late antigen-4 (VLA-4), a member of the Ig superfamily] and results in extravasation into the site (peripheral cornea) of the processed antigen (in this example, protein A).

IMMEDIATE HYPERSENSITIVITY (IgE) REACTION

In sites such as the conjunctiva, Th2 lymphocytes play a key role in immediate hypersensitivity reactions. They produce IL-4 and IL-5, which stimulate B lymphocytes to switch to IgE production and to express IgE receptors. Furthermore, IL-4 induces mast cell proliferation. Th2 cells are found in the conjunctiva of children with vernal disease. The finding that adult patients with severe allergic keratoconjunctivitis, with very high IgE antibodies, have lids colonized by *Staphylococcus aureus* but lack DH to its protein A antigen is explained by the exaggerated Th2 response suppressing the expected Th1 response.

OCULAR IMMUNE PRIVILEGE

Ocular immune privilege has five primary features (5):

1. Blood-ocular barriers, considered in Chapter 4 ("Pharmacology and Pharmacokinetics")
2. Absence of lymphatic drainage
3. Soluble immunomodulatory factors in the aqueous humor
4. Immunomodulatory ligands on the surface of ocular parenchymal cells
5. Tolerogenic APCs in the anterior and posterior chambers

Immune-privileged sites such as the eye have evolved with the mechanisms above to prevent the induction of inflammation. Only recently have neuropeptides, constitutively present in ocular tissue, been recognized as part of the immune privilege mechanism (6).

Local corneal surface immunity consists of IgM, C1 complement, and Langerhans (dendritic) cells; until recently, these cells were thought to be found only in the peripheral corneal epithelial sheet in the normal non-inflamed eye. Irritation of the central cornea, such as with herpetic infection, can result in centripetal migration of these dendritic cells to promote an immuno-inflammatory response. Dekaris *et al.* (7) suggested that IL-1 induction of the dendritic cell migration was mediated by TNF receptor function rather than TNF-α, which had an independent stimulatory role. They also found that p55 and p75 signaling pathways were important in mediating dendritic cell migration.

The normal cornea has large numbers of dendritic cells in the epithelium, which are important for antigen recognition and processing, with monocytes (APCs) in the stroma (8). In addition, conjunctival macrophages can function as APCs and have been shown important to control *Acanthamoeba* invasion by phagocytosis (9), as well as to have a role in corneal graft rejection (10,11). The dendrite is the only corneal cell to express the major histocompatibility class (MHC) II molecules without prior inducement by cytokines. Class II MHC molecules are responsible for presenting processed peptides from exogenous antigens to receptors on CD4+ T helper lymphocytes. A recent advance has been the discovery of the CIITA protein, a non-DNA binding activator of transcription that is a master control gene for class II gene expression (12); current research is focused on understanding situations where class II gene expression occurs in a CIITA-independent pathway and the molecular basis for this expression.

Hamrah *et al.* (13,14) have studied the phenotype and distribution of dendritic cells in the periphery and central zone of normal and inflamed corneas. They have shown that in the periphery nearly one half expressed MHC class II positive (CD80+CD86+) molecules, and found that the central cornea contained immature and precursor dendritic cells that were uniformly MHC class II negative (CD80−CD86−). They have also found immature monocyte APCs in the stroma of the central cornea. After 24 hours of inflammation, they found an increased number of dendritic cells in the central cornea with upregulation of expression of MHC class II receptors with costimulatory molecules CD80 and CD86; in addition, they found a CD11c-CD11b positive population of macrophages (monocytes) exclusively in the posterior stroma. Hazlett *et al.* (15) have investigated the B7/CD28 costimulatory pathway for LC migration and activation in the central cornea. This new data means that the concept of the cornea being immune privileged from a lack of resident lymphoreticular cells needs revision, as the cornea is capable of diverse cellular mechanisms for antigen presentation (14).

Corneal epithelial cells or non-activated macrophages in the anterior and posterior chambers can produce IL-1, so activating macrophages which then express MHC class II molecules and become APCs. They will then process antigenic peptides in a binary complex with MHC class II molecules similar to dendritic cells. Macrophages can process particulate antigens including whole bacteria such as staphylococci and propionibacteria, but can more effectively process soluble antigens viz. protein A, internalized in endocytic vesicles.

T lymphocytes exert their local effector function by secreting cytokines in tissue for direct interaction with target cells. IFN-γ induces expression of MHC class II molecules by keratinocytes, epithelial cells, endothelial cells, and fibroblasts, which can all vary in their capacity to serve as APCs. They can all process and present immunogenic peptides complexed with MHC class II molecules. However, they differ in capacity to produce costimulatory signals and do not stimulate resting T cells, which require IL-2 to become activated.

Diamond *et al.* (16) demonstrated an antimicrobial effect in rabbit aqueous humor. They found that it gave a 90% drop ($1 \log_{10}$) in viable colony count for *Staphylococcus aureus* and *P. aeruginosa* compared to controls of rabbit serum. They found that rabbit aqueous humor could be bactericidal for *Staphylococcus aureus* and *P. aeruginosa* but not for micrococcus, *Escherichia coli*, or *S. pneumoniae*. Defensins are naturally occurring antimicrobial peptides (17). Reverse transcription-polymerase chain reaction (RT-PCR) was performed on human post-mortem ciliary body samples for beta defensins-1 (HBD-1) and beta defensins-2 (HBD-2), and alpha defensins 5 and 6. This study demonstrated that HBD-1 is constitutively present in the aqueous and vitreous, probably at sub-bactericidal concentrations. HBD-2 was absent from aqueous, but cytokine stimulation studies suggested that it may be generated in response to inflammatory cytokines during infections. HBD-2 has a wider antibacterial spectrum, is 10-fold more potent, and may play a more significant role in antimicrobial defense than HBD-1 (17). Caution is required for the use of defensins therapeutically because they also promote cell proliferation and fibrin formation, which are two key elements in the ocular scarring processes such as proliferative vitreoretinopathy.

Immune complex activation occurs by stimulation of either the C1 complement component, with a resulting cascade to C5a that acts chemotactically for polymorphonuclear leukocytes (PMNs) and causes release of cytokines, or activation of the alternative pathway (Fig. 2). Aqueous humor contains three complement regulatory proteins: membrane cofactor protein (MCP), decay-accelerating factor (DAF), and CD59, and a cell surface regulator of complement (Crry) that inhibits complement activation (18–20). These proteins have been found in normal human aqueous humor and vitreous as well as in Lewis rats, in which experimental results suggested that the complement system is continuously active at a low level in the normal eye but is tightly regulated. Severe inflammation developed if these regulatory proteins were blocked with monoclonal antibody or otherwise inactivated. The cornea, choroid, and retina are particularly susceptible to autoimmune mediated, immune-complex disease; however, the vitreous and subretinal space are not. The expression of ICAM-1-$\beta 2$ integrin is needed to attract neutrophils

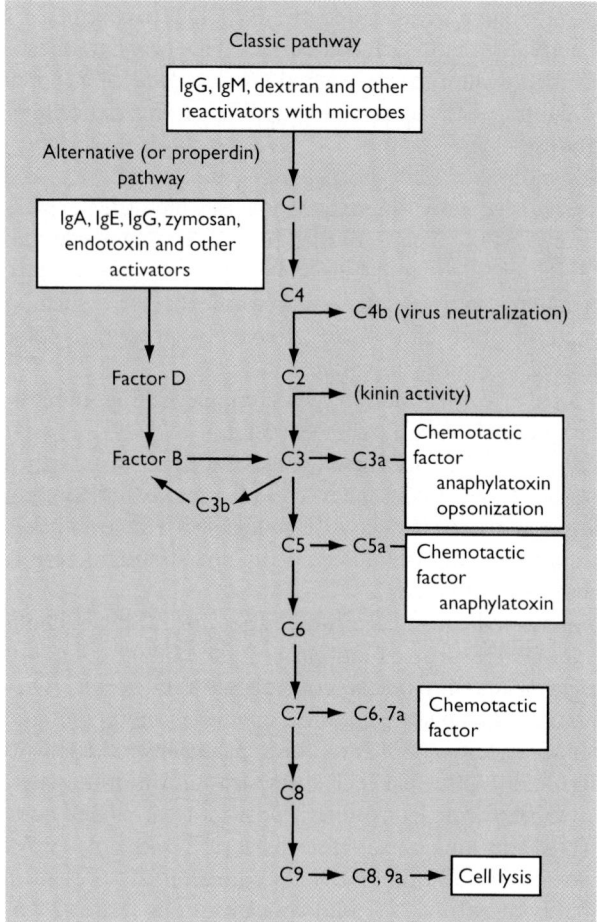

Figure 2 Classic and alternate complement pathways.

to the site, as part of the inflammatory process, and will attract activated Th1 lymphocytes as well.

DELAYED-TYPE HYPERSENSITIVITY FROM CELL-MEDIATED IMMUNITY

Delayed-type hypersensitivity (DTH) from cell-mediated immunity (CMI) has been proposed in humans (21) for recurrent episodes of chronic blepharitis due to *Staphylococcus aureus*, leading to local conjunctival enhancement of CMI, expressed as DTH, to cell wall antigens of *Staphylococcus aureus*, especially protein A. *Staphylococcus aureus* repeatedly colonizes the lids of both normals (approximately 10%) and atopes (approximately 50%) from other sites of human carriage, as well as causing chronic folliculitis in patients with blepharitis. DTH is also the mechanism that is suppressed in anterior chamber-associated immune deviation (ACAID) and posterior chamber-associated immune deviation (POCAID) (Fig. 3). Johnson *et al.* (22) first showed in rabbits, which had been immunized subcutaneously with a whole

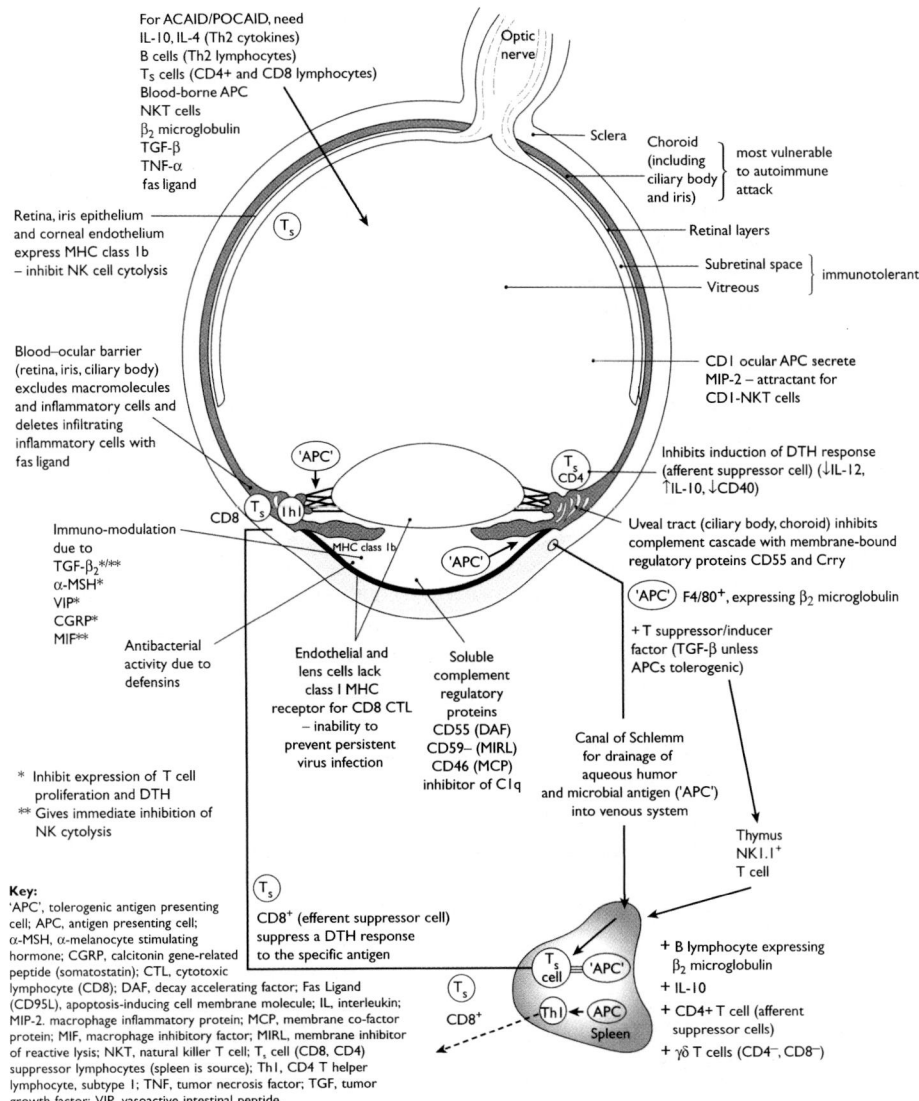

For ACAID/POCAID, need
IL-10, IL-4 (Th2 cytokines)
B cells (Th2 lymphocytes)
T_S cells (CD4+ and CD8 lymphocytes)
Blood-borne APC
NKT cells
β_2 microglobulin
TGF-β
TNF-α
fas ligand

Retina, iris epithelium
and corneal endothelium
express MHC class Ib
– inhibit NK cell cytolysis

Blood–ocular barrier
(retina, iris, ciliary body)
excludes macromolecules
and inflammatory cells and
deletes infiltrating
inflammatory cells with
fas ligand

Immuno-modulation
due to
TGF-β_2*/**
α-MSH*
VIP*
CGRP*
MIF**

Antibacterial
activity due to
defensins

* Inhibit expression of T cell
  proliferation and DTH
** Gives immediate inhibition of
   NK cytolysis

Optic
nerve

Sclera

Choroid
(including
ciliary body
and iris)

most vulnerable
to autoimmune
attack

Retinal layers

Subretinal space
Vitreous

immunotolerant

CD1 ocular APC secrete
MIP-2 – attractant for
CD1-NKT cells

Inhibits induction of DTH response
(afferent suppressor cell) ($\downarrow$IL-12,
$\uparrow$IL-10, $\downarrow$CD40)

Uveal tract (ciliary body, choroid) inhibits
complement cascade with membrane-bound
regulatory proteins CD55 and Crry

'APC' F4/80+, expressing β_2 microglobulin

+ T suppressor/inducer
factor (TGF-β unless
APCs tolerogenic)

Endothelial and
lens cells lack
class I MHC
receptor for CD8 CTL
– inability to
prevent persistent
virus infection

Soluble
complement
regulatory
proteins
CD55 (DAF)
CD59– (MIRL)
CD46 (MCP)
inhibitor of C1q

Canal of Schlemm
for drainage of
aqueous humor
and microbial antigen ('APC')
into venous system

MHC class Ib

'APC'

CD8

Thymus
NK1.1+
T cell

Key:
'APC', tolerogenic antigen presenting
cell; APC, antigen presenting cell;
α-MSH, α-melanocyte stimulating
hormone; CGRP, calcitonin gene-related
peptide (somatostatin); CTL, cytotoxic
lymphocyte (CD8); DAF, decay accelerating factor; Fas Ligand
(CD95L), apoptosis-inducing cell membrane molecule; IL, interleukin;
MIP-2. macrophage inflammatory protein; MCP, membrane co-factor
protein; MIF, macrophage inhibitory factor; MIRL, membrane inhibitor
of reactive lysis; NKT, natural killer T cell; T_s cell (CD8, CD4)
suppressor lymphocytes (spleen is source); Th1, CD4 T helper
lymphocyte, subtype 1; TNF, tumor necrosis factor; TGF, tumor
growth factor; VIP, vasoactive intestinal peptide.

CD8+ (efferent suppressor cell)
suppress a DTH response
to the specific antigen

Spleen

+ B lymphocyte expressing
β_2 microglobulin
+ IL-10
+ CD4+ T cell (afferent
suppressor cells)
+ $\gamma\delta$ T cells (CD4–, CD8–)

Figure 3 Anterior (ACAID)- and posterior (POCAID)-associated immune deviation.

cell vaccine to develop systemic enhancement (CMI/DTH) to *Staphylococcus aureus*, that there was tolerance instead of DTH in the AC. This situation applies similarly to most other organisms able to induce a systemic DTH reaction.

The CD4 T helper cell (Th) is responsible for most DTH reactions (23). The Th1 subtype mediates DTH, not given by the Th2 subtype, based on an array of cytokines produced by T-cell clones including IL-2, IFN-γ, and TNF-α (Fig. 1). The Th2 subtype stimulates B lymphocytes with IL-4, IL-5, IL-10, and IL-13 cytokines to produce an antibody response (Fig. 1). Other cytokines not characterized as either Th1 or Th2 include IL-12, which promotes IFN-γ and TNF-α production and hence a Th1 response, and IL-18. Polymorphonuclear cells, macrophages, and dendritic cells can produce both Th1 and Th2 cytokines, but ocular APCs cannot produce the Th1-inducing cytokine IL-12. DTH is thought responsible for the underlying mechanisms for the pathogenesis of trachoma, HSV stromal keratitis, and some forms of idiopathic uveitis. DTH is recognized as the primary cause of experimental autoimmune uveitis (EAU) in rodents. ACAID (see below) is believed to be a mechanism that prevents trivial activation of the DTH arm of the immune response as a method for sustaining immuno-homeostasis in the eye.

Macrophages produce IL-1, which stimulates other macrophages to become APCs and to present antigen to Th1 lymphocytes. IL-1 also regulates macrophage inflammatory protein-2 (MIP-2) (24), which acts as a chemo-attractant for PMN influx and may be susceptible to downregulation to reduce the inflammatory response (25).

Th1 produces IL-2 and interferon (IFN)-γ, responsible in part for the induration response of the DTH reaction. Th1 and Th2 cross-regulate each other's activities via cytokine production, with one inhibiting production by the other, e.g., IFN-γ inhibits the effect of IL-4 on B lymphocytes. Their role in cross-regulating each other and thereby maintaining a balance in the immune response is responsible for the adaptive immunity to various pathogens, e.g., Th1 protects against intracellular pathogens while Th2 protects against helminths. Several novel therapeutic strategies hold promise for modulating the alloimmune response to corneal allografts by either promoting antigen-specific tolerance or redirecting the host's response from a Th1 pathway toward a Th2 pathway to reduce immune rejection (11).

The cell–cell interaction between the APC and the Th lymphocyte combination of CD3-TCR, recognizing the MHC/oligopeptide antigen expressed, and the complex of CD4 and CD5 receptors, activates p59 and p56 tyrosine kinases by dephosphorylation as an "activity cascade." This in turn induces synthesis of IL-2, and its receptor expression, to give proliferation of selected antigenic specificity, resulting in the inflammatory DTH reaction.

ANTERIOR CHAMBER AND POSTERIOR CHAMBER ASSOCIATED IMMUNE DEVIATION

ACAID and POCAID (Fig. 3) exist to protect the single cell–layered corneal endothelium, the AC, and the subretinal space and vitreous from immune-mediated damage by cell-mediated immunity or DTH (26,27). This POCAID effect is not present in the retina or choroid. Immune deviation is achieved by avoidance of CD8 cytotoxic T lymphocyte (CTL)-mediated cytolysis and inhibition of CD4 Th1 lymphocyte response in the AC (10,26). These aims are achieved by an absence of class I MHC expression on endothelial and lens cells, so avoiding lysis by class I restricted CTLs. The problem with this mechanism for the AC endothelium is the inability to prevent persistent viral infection. This also applies to the immune-privileged retinal pigment epithelium (RPE), in which cytomegalovirus (CMV) can persist. Recent evidence (28) has found that CMV immediate early (IE) gene expression in RPE cells deviates ocular antiviral inflammation via Fas ligand (FasL) (Fig. 3). TNF-α and IFN-γ, found elevated in AIDS patients with retinitis, sensitize RPE cells to FasL-mediated cytolysis, contributing to retinal destruction rather than inflammation. One explanation proposed for the normally-reduced immune response is engagement of the pro-apoptotic molecule Fas by its ligand (FasL), which leads to apoptosis and consequently limits an inflammatory response (29).

The prevention of a DTH (cell-mediated) immune reaction is gained by the camerosplenic axis (Fig. 3). Microbial antigen in the AC is discharged in the aqueous to the venous blood flow, often phagocytosed within "tolerogenic" APCs, and so to the spleen instead of to the lymphatics. However, there is experimental evidence that up to 25% of antigen in the AC reaches T cells in the submandibular lymph nodes of AC-injected mice via the uveal/scleral pathway but because of the inability of ocular APC to produce IL-12, and hence to stimulate a Th1 reaction, the responding T cells fail to differentiate into functioning Th1 cells (10). Within the spleen there is production of T suppressor(T_s) lymphocytes that migrate back to the AC to inhibit Th1 lymphocytes. These T_s cells also suppress a DTH response, acting systemically to the specific antigen. This systemic protection is lost if the spleen, which is the source of the T_s cells, is removed. A functioning spleen has been shown essential for developing and maintaining ACAID.

In order to achieve this suppression or immune tolerance, there is an afferent (CD4 T_s) and efferent (CD8 T_s) mechanism. The afferent CD4 T_s lymphocyte, generated in the spleen, inhibits induction of the DTH response within the AC and subretinal space. The CD4 T_s cells go to the spleen via the aqueous humor/venous system with the tolerogenic APC that has captured eye-derived antigens. Within the spleen there is multicellular interaction with natural killer (NK) T cells, splenic B lymphocytes expressing β_2 microglobulin, IL10, and $\gamma\Delta T$ cells (CD4-, CD8-), which creates an antigen-presentation environment that leads to CD4+ and CD8+ alpha/beta T cells, which suppress induction of Th1 and Th2 immune expression (5). The spleen releases CD8 T_s efferent suppressor lymphocytes, which can suppress a DTH response to a specific antigen and which return to the eye via the arterial blood flow. Splenic production of T_s lymphocytes can also be observed with intravenous (IV) injection of antigens, but while the classic model of intravenous antigen-induced immune deviation shares many characteristics with ACAID, including down-regulation of DTH, there are also major differences (10). In particular, IL-4 is required for IV-induced immune deviation but not for ACAID, while IL-10, B cells, blood-borne APCs, IV NKT cells, and $\beta_2$microglobulin are not required for IV-induced immune deviation but are for ACAID (10).

There is need for TNF-α and the Th2 cytokine IL-10 within the AC to induce an ACAID response to an antigen. The AC also contains TGF (tumor or transforming growth factor)-β, which renders local APCs tolerogenic and is another necessary part of ACAID, particularly if the host has systemically enhanced DTH to the specific antigen. Not all antigens produce ACAID or POCAID, with notable exceptions being purified protein derivative of *Mycobacterium tuberculosis*, large simian virus-40 T antigens, and some experimental tumor allografts (26). The reasons and mechanism for such antigen selection remain unclear.

Macrophages within the AC are not always able to kill bacteria such as *Propionibacterium acnes* and coagulase-negative staphylococci (CNS). This is well illustrated in chronic 'saccular' or granulomatous endophthalmitis due to *P. acnes* and, less frequently, CNS (30–32). Macrophages phagocytose *P. acnes* and CNS within the capsule fragment (Chapter 8) but are unable to kill the bacteria due to the lack of a functioning cell-mediated immune system with cytokine expression, so that these bacteria are able to multiply within them. The macrophages within the AC function as tolerogenic APCs rather than as scavenger cells. There is a chronic inflammatory reaction recognized clinically as "saccular," plaque, or granulomatous endophthalmitis with white plaque in the capsular bag (Chapter 8). The AC tap is often culture-negative because the bacteria are intracellular only, but a PCR test can be positive (Chapter 3). The bacteria may be killed when released from the macrophages by the antibacterial effect within the aqueous humor.

Our hypothesis suggests that the continued intracellular replication of the bacteria may allow the macrophage to express MHC class II molecules and to give a limited Th1 response susceptible to suppression with corticosteroids. This explains why the inflammation returns when corticosteroid therapy is withdrawn and why the capsule fragment often has to be removed surgically, after which there is no recurrence of chronic inflammation. The most useful antibiotic to use is clarithromycin, by the oral or intravitreal route, which penetrates into the AC and is concentrated up to 200 times intracellularly within macrophages to kill intracellular bacteria (Chapter 4)—therapeutic success has been reported with oral therapy at 500 mg bd (33–35) without the need for surgical removal of the capsule and intra-ocular lens. Clarithromycin can be injected intravitreally with a nontoxic dose of up to, but not more than, 1.0 mg in 0.1 mL (36).

Specific production of antibodies within the vitreous and posterior chamber during endophthalmitis has been investigated by Ravindranath _et al._ (37). Using a rat model of _Staphylococcus epidermidis_ endophthalmitis, they found IgG and IgM but not IgA antibodies against glycerol teichoic acid (GTA) of the staphylococcus cell wall were present in the vitreous from day 1 onwards, and declined by day 7. Plasma cells were seen in the vitreous between days 1 and 3 and B lymphocytes (CD45+, CD3−) were present in pooled vitreous humor. In contrast, serum anti-GTA IgM antibodies were raised for one week. These vitreous antibodies may be involved in neutrophil-mediated opsonophagocytosis, leading to spontaneous sterility within their rat model. Other recent work by Meek _et al._ (38) has investigated the specificity of intraocular produced antibodies. They demonstrated that antibodies recognizing the same antigen in both serum and intraocular fluid differ in the epitopes that are recognized, demonstrating that the intraocular compartment determines its own antibody profile against intraocular pathogens. The authors presented several models for how an exclusive intraocular B cell lymphocyte repertoire may function.

In the conjunctiva, Th2 lymphocytes play a key role in immediate hypersensitivity reactions, producing IL-4 and IL-5 to stimulate B cells and expression of IgE receptors; IL-4 induces mast cell proliferation. This explains why adult patients with severe allergic keratoconjunctivitis, who have very high systemic IgE antibodies associated with IL-5 induction of eosinophils, have lids colonized by _Staphylococcus aureus_ but lack DTH to _Staphylococcus aureus_ protein A (39). Th2 cells are found in the conjunctiva of children with vernal disease. Recent experimental models have demonstrated that inhibition of Th2 cells and their secreted cytokines could be a therapeutic target for inhibiting chronic allergic inflammation on the ocular surface (40).

IMMUNOPATHOGENESIS

Considerable work has been conducted by Hazlett _et al._ (41–43) exploring the role of Langerhans dendritic cells (DCs), cytokines, and subsets of T lymphocytes in the pathogenesis of _Pseudomonas aeruginosa_ infection of the cornea in a mouse model. Firstly, they have shown that DCs are crucial in the innate immune response (41).

In B6 mice, which produce a Th1-type response to a challenge of *P. aeruginosa* and ulcerate (44), the early induction of DCs made no difference, but in BALB/c mice, which give a Th2 response, early induction of DC was disastrous—instead of the infection resolving and healing, the cornea perforated (41). They compare this model with extended-wear contact lens wearers (EWCLWs) in humans, in whom chronic irritation leads to centripetal migration of DCs into the central cornea (45); such EWCLWs may therefore be at greater risk of corneal ulceration due to *P. aeruginosa* (Chapter 6 and Appendix D).

Secondly, they explored the differing pathogenesis between a Th1 and a Th2 response in their mouse model (42). They found that the presence of IL-12 is important to stimulate production of IFN-γ and TNF-α and therefore to give a Th1 response (39). If IL-12 was absent, there was unchecked bacterial growth; if present, the bacterial growth was stopped, although the cornea ulcerated. They found that differing genetic strains of mice would either produce a Th1 response (B6, BL10) or a Th2 response (BALB/c) (46).

This important work begins to explain how the immune response to a bacterial challenge, such as contamination of the AC with phacoemulsification surgery, influences the outcome in regard to numbers and virulence of the organism. The combination of the type of immune response (Th1 or Th2 with relevant cytokines) together with the bacterial virulence factors and toxins, if any, will determine whether there is an acute or chronic inflammatory response or, when the immune response is satisfactory, there is no inflammatory effect at all. A high bacterial count can also influence the outcome overcoming the host immune response to lead to infection.

Others have made important contributions studying the role of DTH, Th1 cells, and latency for the immunopathogenesis of herpes virus infections (47). Khanna *et al.* (48) have challenged the concept that herpes simplex virus type 1 (HSV-1) latency represents a silent infection that is ignored by the host immune system, and suggested that antigen-directed retention of memory CD8(+) T cells specific for the immunodominant gB(498-505) HSV-1 epitope are selectively retained in the ophthalmic branch of the latently infected trigeminal ganglion. They conclude that CD8(+) T cells provide active surveillance of HSV-1 gene expression in latently infected sensory neurons.

IMMUNE SUPPRESSION

Immune modulation can be obtained with the drug cyclosporine, which suppresses the activity of the p56 and p59 tyrosine kinases in the Th1 lymphocytes, when the DTH activity cascade is blocked. This drug also acts on calcineurin-blocking T-cell activation. Cyclosporin acts specifically and reversibly on lymphocytes. It does not depress hemopoiesis and has no effect on the function of PMNs. It is a potent immunosuppressive drug and prolongs survival of allogeneic transplants involving heart, kidney, pancreas, bone marrow, lung, and cornea. Cyclosporin prevents and treats rejection and graft-versus-host disease. It has also been found beneficial in psoriasis, atopic dermatitis, rheumatoid arthritis, uveitis, and Sjogren's syndrome, all conditions considered to have an immunological pathogenesis. Cyclosporin inhibits the development of cell-mediated reactions. It blocks the lymphocytes in the early phase of their cycle, as described above, and inhibits lymphokine production and release including IL-2 (T cell growth factor).

Fucidin (fusidic acid) is an anti-staphylococcal antibiotic, for which there is a topical preparation (Fucithalmic), and also has a similar immunosuppressive action to cyclosporin. This may explain its beneficial effect for treating the blepharitis of rosacea, usually associated with both *Staphylococcus aureus* colonization of the lids and systemically enhanced DTH to its antigens (49). In the future there may be potential for therapeutic application of ACAID mechanisms to prevent progression of ocular immuno-inflammatory disease (27).

Corticosteroids act by affecting the functions of cell populations and their distribution. They have a specific role inhibiting phospholipase conversion of membrane phospholipids to arachidonic acid (Fig. 4).

Corticosteroids have both anti-inflammatory and immunosuppressive effects. The anti-inflammatory effects are non-specific to the cause of the disease process and will inhibit the inflammatory reaction to nearly any type of stimulus. The effects include the following:

Anti-inflammatory

- Inhibition of increased capillary permeability due to the acute inflammatory reaction and suppression of vasodilation
- Inhibition of degranulation of neutrophils, mast cells, and basophils with monocytopenia and eosinopenia
- Reduction of prostaglandin synthesis by inhibiting phospholipase (Fig. 4)

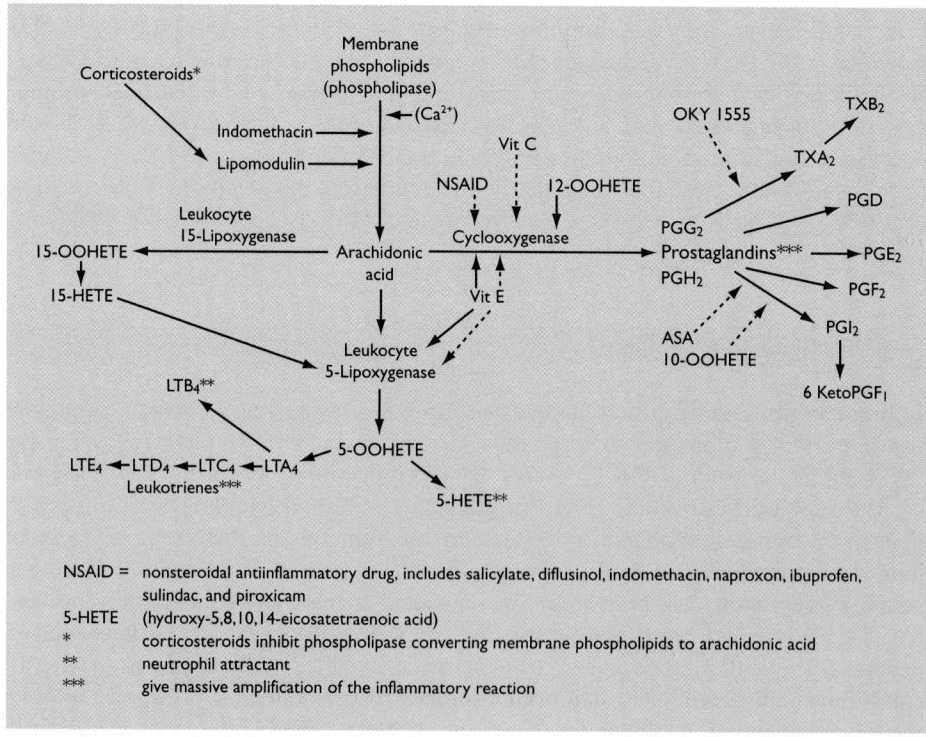

NSAID = nonsteroidal antiinflammatory drug, includes salicylate, diflusinol, indomethacin, naproxon, ibuprofen, sulindac, and piroxicam
5-HETE   (hydroxy-5,8,10,14-eicosatetraenoic acid)
*   corticosteroids inhibit phospholipase converting membrane phospholipids to arachidonic acid
**   neutrophil attractant
***   give massive amplification of the inflammatory reaction

Figure 4 Arachidonic acid metabolic pathways.

▪ Temporary neutrophil leukocytosis four to six hours after administration but with reduced adherence to vascular endothelium

Immunosuppressive

▪ Induction of lymphocytopenia—T lymphocytes are more susceptible than B lymphocytes so antibody production is not reduced while DTH/CMI reactions are modified at low concentrations
▪ Inhibition of migration of macrophages and neutrophils
▪ Suppression of lymphocyte proliferation
▪ Suppression of action of cytokines
▪ Inhibition of DTH/CMI by decreasing recruitment of macrophages
▪ Alteration of antigen processing by macrophages
▪ Depression of bactericidal activity of monocytes and macrophages

Corticosteroids reduce the activity of macrophages but not necessarily their activation by cytokines. They do inhibit cytokine release with additional suppression of PMN activity, explaining progression of fungal and protozoal or parasitic infections in their presence, since it is these latter infections especially that require macrophages as scavenger cells.

The use of corticosteroid therapy in managing endophthalmitis has continued to be a necessary albeit controversial treatment (Chapter 8). Our findings generally correspond to those provided by others in that intravitreal corticosteroid therapy should be used for bacterial, and considered for fungal, endophthalmitis and should be injected at the same time as the antibiotics.

Nonsteroidal anti-inflammatory drugs (NSAIDs) act by inhibiting the conversion of arachidonic acid to prostaglandins by cyclo-oxygenase (Fig. 4). NSAIDs such as flurbiprofen (Froben)® may in the future supersede conventional steroid preparations, such as dexamethasone or prednisolone, by removing unwanted side effects. This is particularly so when corticosteroids are introduced systemically for treatment of severe ocular inflammatory disease. Flurbiprofen, given orally, has been found particularly effective for suppressing the acute inflammation of ocular infection, especially keratitis, episcleritis, and limbitis, as well as providing analgesia without interfering with satisfactory chemotherapy. It is also a mydriatic.

LOCAL IMMUNE COMPLEX DISEASE

This occurs in the peripheral cornea by the activation of the C1 component of complement and the formation of C5a, which is chemotactic for PMNs. This also causes release of IL-8. The marginal ulceration that occurs in the autoimmune connective tissue diseases, such as rheumatoid arthritis, is thought to be immune complex–mediated, involving rheumatoid factor (IgM) directed against self-IgG. The integrin ICAM-1-β2 is involved in neutrophil invasion as part of this inflammatory process.

A similar mechanism has been proposed for marginal ulceration associated with blepharitis due to *Staphylococcus aureus*, as an alternative explanation to the DH theory by Mondino (50). It is suggested that toxins of *Staphylococcus aureus* cause punctate epithelial breakdown that allows staphylococcal antigens, especially protein A, to penetrate the epithelium and complex with IgM found in the peripheral cornea. This is conceived to activate C1 and its cascade to C5a, causing tissue damage. It is also a possibility that the marginal ulceration that occasionally occurs in chlamydial infection is due to immune complex disease.

IMMUNE DEPOSITION IN THE CORNEA

The cornea is unique in that insoluble complexes, formed by soluble antigens and antibodies when each is at equal concentration, can be visualized as localized rings in the corneal stroma. When either the antigen or the antibody is present in excess concentration, the complexes remain soluble. The reaction occurs between a microbial antigen in the central cornea and an antibody (IgG) that diffuses across the cornea from the limbal vessels. Centripetally migrating PMNs degranulate within the ring. Such an example is illustrated in Figure 5A, when there was an *Acanthamoeba* infection under partial host control at 3 months. The depositions produced are called Wessley rings, which appear as double circles, and, while they have often been associated with herpetic keratitis, they have also been recognized with bacterial, fungal, and viral infections (Fig. 6).

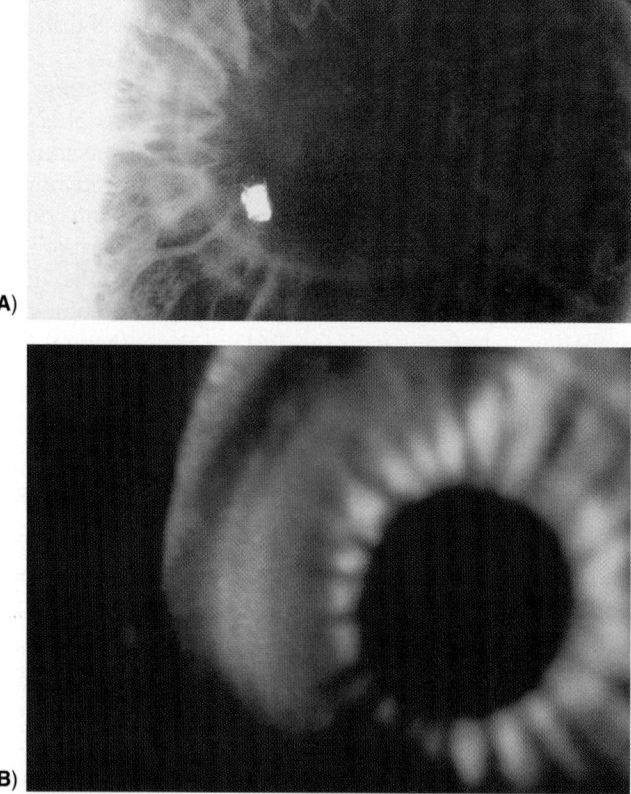

(A)

(B)

Figure 5 **(A)** Wessley rings in cornea associated with on *Aconthamoeba* infection. **(B)** Immune ring in an eight-year-old boy.

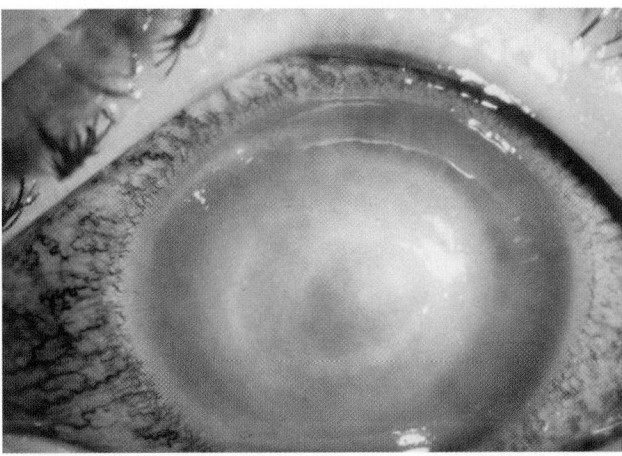

Figure 6 Wessley ring in CL-associated *Pseudomonas aeruginosa* keratitis.

MUCOSA-ASSOCIATED LYMPHOID TISSUE

The immune protection at mucosal surfaces, including the ocular surface, is maintained in part by mucosa-associated lymphoid tissue (MALT). One of the main functions of MALT is to establish a balance between immunity and tolerance in order to prevent destruction of the delicate mucosal tissues by constant inflammatory reactions, which applies in particular to the eye. This is maintained by an anti-inflammatory cytokine milieu in mucosal tissues and is most likely regulated by antigen-presenting dendritic cells as key regulators of the immune system. They normally favor anti-inflammatory T- or B-cell responses in mucosal locations. A major defense mechanism of MALT is the production of secretory immunoglobulins, mainly of the IgA and partly of the IgM isotype, by differentiated B cells (plasma cells). In contrast to the IgG isotype that prevails in the blood, IgA has very little complement-binding activity and therefore does not initiate inflammatory reactions during host defense. IgA attaches to bacteria to 'coat' them, reducing their adhesion to epithelial cells.

The lymphoid cells of MALT migrate in a regulated fashion, guided by specialized vessels, adhesion molecules, and soluble chemotactic factors. They migrate between the different mucosal organs, which are hence assumed to constitute a functionally inter-related mucosal immune system. By these migration pathways, MALT is also connected to the central immune system. MALT is a regular component of the normal human ocular surface and termed eye-associated lymphoid tissue (EALT) (Fig. 7).

MALT consists of a diffuse lymphoid tissue (A) and of an organized follicular tissue (B), shown here in different enlargement. Mucosal tissues in general are composed of two sheets, the luminal epithelium (e) with its basement membrane (bm) and an underlying lamina propria (lp), that both contain lymphocytes. The lamina propria is composed of loose connective tissue with small blood vessels (b), afferent lymph vessels (l), and numerous cells including lymphoid cells [T lymphocytes (black), B-lymphocytes (blue), plasma cells (p)].

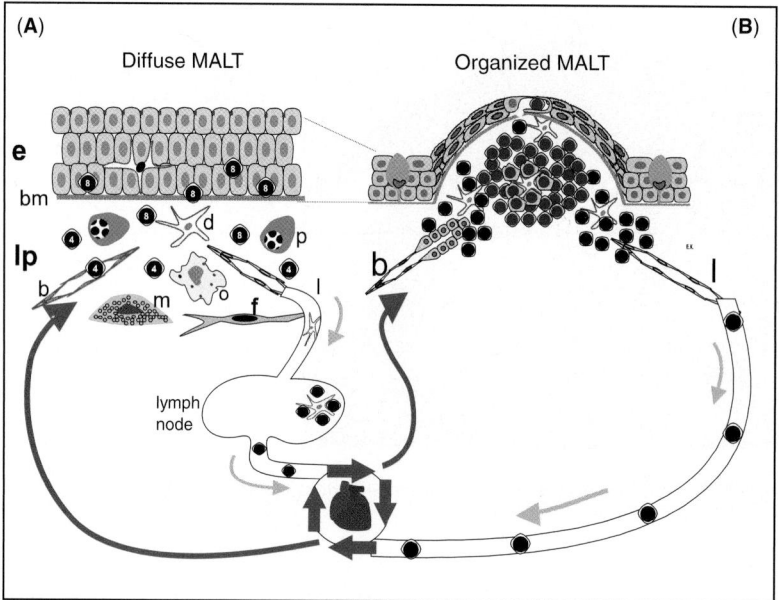

Figure 7   Structure and function of the mucosal immune system. *Source*: From Ref. 51.

Accessory cells occur like fibroblasts (f), macrophages (m), mast cells (mc), or dendritic cells (dc). Intraepithelial lymphocytes are mainly CD8+ suppressor/cytotoxic T cells whereas in the lamina propria of the diffuse tissue (A) they occur in roughly equal numbers together with CD4+ T helper cells. Follicular lymphoid tissue (B) is formed by accumulations of B lymphocytes with parafollicular T-cell zones, vessels, and an overlying specialized follicle-associated epithelium for antigen transport towards the follicle. Naïve lymphocytes enter follicular regions via blood vessels (b) and get in contact to antigens; antigen-specific lymphocytes proliferate, differentiate, and leave via lymphatics (l). They finally reach the blood circulation and may later emigrate to populate the same or other mucosal tissues as effector cells (T cells and plasma cells).

EYE-ASSOCIATED LYMPHOID TISSUE

Whole mounts of complete normal human ocular tissues have shown that lymphoid cells are a normal tissue constituent and form continuous mucosa-associated lymphoid tissue in the lacrimal gland, conjunctiva, and lacrimal drainage system (Fig. 8).

The ocular surface is an integral part of the mucosal immune system of the body. The diffuse lymphoid tissue with an effector function by lymphocytes and plasma cells is continuous from the lacrimal gland along the excretory ducts into the conjunctiva as conjunctiva-associated lymphoid tissue (CALT) and continues through the lacrimal canaliculi inside the lacrimal drainage system as lacrimal drainage-associated lymphoid tissue (LDALT). The lymphoid tissue of these three organs together constitutes EALT. Follicular tissue for the detection of ocular antigens occurs in CALT and LDALT. Effector cells that are primed in follicular

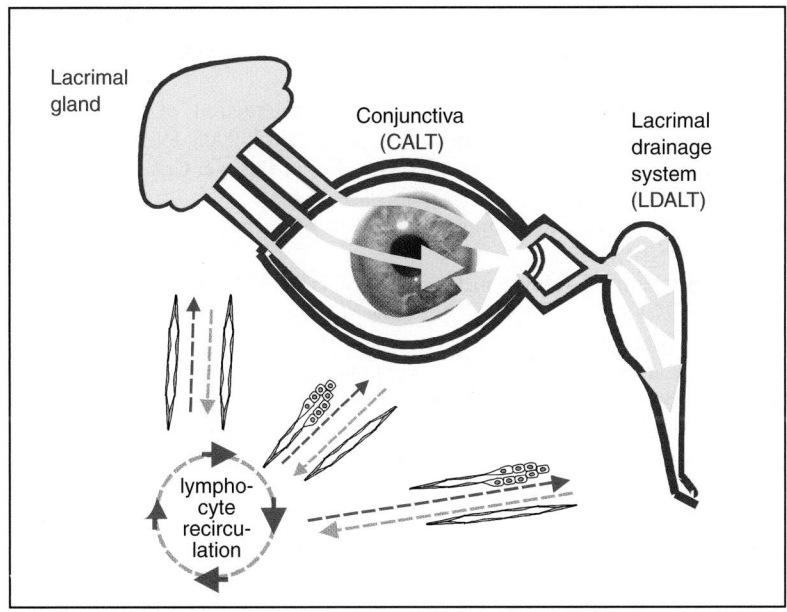

Figure 8   Eye-associated lymphoid tissue (EALT). *Source*: From Ref. 51.

tissue against ocular surface antigens can migrate in a regulated fashion via specialized vessels between the organs of EALT and the other parts of the mucosal immune system and can hence provide them with effector cells that are specifically directed against antigens that occur at the ocular surface.

REFERENCES

1.   Simmons P, Tomlinson A, Connor R, Hay J, Seal DV. Effect of patient wear and extent of protein deposition on adsorption of *Acanthamoeba* to five types of hydrogel contact lenses. Optom and Vis Sci 1996; 73:362–8.
2.   Cousins SW, Rouse BT. Chemical mediators of ocular inflammation. In: JS Pepose, GN Holland, KR Wilhelmus (eds), Ocular Infection and Immunity. Mosby Year Book, Chicago:1996; 50–70.
3.   Hendricks RL, Tang Q. Cellular immunity and the eye. In: JS Pepose, GN Holland, KR Wilhelmus (eds), Ocular Infection and Immunity. Mosby Year Book, Chicago:1996; 71–95.
4.   Seal DV, Ficker L. Immunology and therapy of marginal ulceration as a complication of chronic blepharitis due to *S. aureus*. In: J Lass (ed.). Advances in Corneal Research. Selected Abstracts of the World Congress on the Cornea IV. New York: Plenum Press, 1997, pp 19–25.
5.   Streilein JW. Ocular immune privilege: the eye takes a dim but practical view of immunity and inflammation. J Leukoc Biol 2003; 74:179–185.
6.   Taylor AW. Neuroimmunomodulation and immune privilege; the role of neuropeptides in ocular immunosuppression. Neuroimmunomodulation 2002–2003; 10:189–198.
7.   Dekaris I, Zhu SN, Dana MR. TNF-alpha regulates corneal Langerhans cell migration. J Immunol 1999; 162:4235–4239.

8. Brissette-Storkus CS, Reynolds SM, Lepisto AJ, Hendricks RL. Identification of a novel macrophage population in the normal mouse corneal stroma. Invest Ophthalmol Vis Sci. 2002; 43:2264–71.

9. Hurt M, Proy V, Niederkorn JY, Alizadeh H. The interaction of *Acanthamoeba castellanii* cysts with macrophages and neutrophils. J Parasitol. 2003; 89:565–72.

10. Niederkorn J. Immune privelege in the anterior chamber of the eye. Crit Rev Immunol 2002a; 22:13–46.

11. Niederkorn JY. Immunology and immunomodulation of corneal transplantation. Int Rev Immunol. 2002b; 21:173–96.

12. Radosevich M, Ono-Santa J. Novel mechanisms of class II MHC gene regulation. Immunol Res 2003; 27:85–106.

13. Hamrah P, Liu Y, Zhang Q, Dana MR. Alterations in corneal stromal dendritic cell phenotype and distribution in inflammation. Arch Ophthalmol 2003a; 121:1132–1140.

14. Hamrah P, Huq-Syed O, Liu Y, Zhang Q, Dana MR. Corneal immunity is mediated by heterogenous population of antigen presenting cells. J Leukoc Biol 2003b; 74:172–178.

15. Hazlett L, McClellan S, Barrett RP, Rudner X. B7/CD28 costimulation is critical in susceptibility to *Pseudomonas aeruginosa* corneal infection: a comparative study using monoclonal antibody blockade. J Immunol 2001; 166:1292–1299.

16. Diamond JP, Moule K, Leeming JP, Tavare J, Easty DL. Purification of an antimicrobial peptide from rabbit aqueous humour. Curr Eye Res. 1998; 17:783–7.

17. Haynes RJ, McElveen JE, Dua HS, Tighe PJ, Liversidge J. Expression of human beta-defensins in intraocular tissues. Invest Ophthalmol Vis Sci. 2000; 41:3026–31.

18. Sohn JH, Kaplan HJ, Suk HJ, Bora PS, Bora NS. Chronic low level complement activation within the eye is controlled by intraocular complement regulatory proteins. Invest Ophthalmol Vis Sci. 2000a; 41:3492–502.

19. Sohn JH, Kaplan HJ, Suk HJ, Bora PS, Bora NS. Complement regulatory activity of normal human intraocular fluid is mediated by MCP, DAF, and CD59. Invest Ophthalmol Vis Sci. 2000b; 41:4195–202.

20. Bardenstein DS, Cheyer CJ, Lee C, *et al.* Blockage of complement regulators in the conjunctiva and within the eye leads to massive inflammation and iritis. Immunology. 2001; 104:423–30.

21. Ficker L, Ramakrishnan M, Seal D, Wright P. Role of cell-mediated immunity to staphylococci in blepharitis. Am J Ophth 1991; 111:473–479.

22. Johnson JE, Cluff LE, Goshi K. Studies on the pathogenesis of Staphylococcal infection. J Exp Med 1960; 113:235–247.

23. Hendricks RL, Tang Q (1996) Cellular immunity and the eye. In: Pepose GS, Holland G, Wilhelmus K. (eds), Ocular Infection and Immunity. Mosby Year Book, Chicago: 71–95.

24. Rudner XL, Kernacki KA, Barrett RP, Hazlett LD. Prolonged elevation of IL-1 in *Pseudomonas aeruginosa* ocular infection regulates macrophage-inflammatory protein-2 production, polymorphonuclear neutrophil persistence, and corneal perforation. J Immunol 2000; 164:6575–6582.

25. Kernacki KA, Barrett RP, Hobden JA, Hazlett LD. Macrophage inflammatory protein-2 is a mediator of PMN influx in ocular bacterial infection. J Immunol 2000; 164:1037–1045.

26. Niederkorn JY, Ferguson TA (1996) Anterior chamber associated immune deviation. In: Pepose GS, Holland G, Wilhelmus K. (eds). Ocular Infection and Immunity. Mosby Year Book, Chicago: 96–103.

27. Streilein J, Streilein JW. Anterior Chamber Associated Immune Deviation (ACAID): regulation, biological relevance and implications for therapy. Int Rev Immunol 2002; 21: 123–152.

28. Scholz M, Doew HW, Cinatt J. Human cytomegalovirus retinitis: pathogenicity, immune evasion and persistence. Trends Microbiol 2003; 11:171–178.

29. Green DR, Ferguson TA. The role of Fas ligand in immune privilege. Nat Rev Mol Cell Biol. 2001; 2:917–24.

30. Abreu JA, Cordoves L, Mesa CG, Mendez R, Dorta A, De la Rosa, MA. Chronic pseudophakic endophthalmitis versus saccular endophthalmitis. J Cat Refract Surg 1997; 23:1122–1125.
31. Warheker PT, Gupta SR, Mansfield DC, Seal DV, Lee WR. Post-operative saccular endophthalmitis caused by macrophage-associated staphylococci. Eye 1998a; 12: 1019–1021.
32. Kresloff MS, Castellarin AA, Zarbin MA. Endophthalmitis. Surv Ophthalmol 1998; 43: 193–224.
33. Warheker PT, Gupta SR, Mansfield DC, Seal DV. Successful treatment of saccular endophthalmitis with clarithromycin. Eye 1998b; 12:1017–1019.
34. Okhravi N, Guest S, Mathesom M, et al. Assessment of the effect of oral clarithromycin on visual outcome following presumed bacterial endophthalmitis. Curr Eye Res 2000; 21: 691–702.
35. Karia N, Aylward GW. Postoperative Propionibacterium acnes endophthalmitis. Ophthalmol 2001; 108:634.
36. Unal M, Peyman GA, Liang C, Hegazy H, Molinari LC, Chen J, Brun S, Tarcha PJ. Ocular toxicity of intravitreal clarithromycin. Retina 1999; 19:442–446.
37. Ravindranath RM, Hasan SA, Mondino BJ. Immunopathologic features of Staphylococcus epidermidis-induced endophthalmitis in the rat. Curr Eye Res 1997; 16: 1036–1043.
38. Meek B, Speyer D, de Jong PTVM, de Smet MD, Peek R. The ocular humoral immune response in health and disease. Prog Retin Eye Res 2003; 22:391–415.
39. Tuft S, Kemeney M, Buckley R, Ramakrishnan M, Seal D. Role of S.aureus in chronic allergic conjunctivitis. Ophthalmology 1992; 99:180–184.
40. Calonge M, Siemasko KF, Stern ME. Animal models of ocular allergy and their clinical correlations. Curr Allergy Asthma Rep 2003; 3:345–351.
41. Hazlett L, McClellan S, Rudner X, Barrett RP. The role of Langerhan cells in Pseudomonas aeruginosa infection. Investig Ophthalmol Vis Sci 2002a: 43:189–197.
42. Hazlett L, Rudner XL, McClellan SA, Barrett RP, Lighvani S. Role of IL-12 and IFN-γ in Pseudomonas aeruginosa corneal infection. Investig Ophthalmol Vis Sci 2002b; 43: 419–424.
43. Hazlett LD, Houang X, McClellan SA, Barrett RP. Further studies on the role of IL-12 in Pseudomonas aeruginosa corneal infection. Eye 2003; 17:863–871.
44. Houang X, Hazlett L. Analysis of Pseudomonas aeruginosa corneal infection using an oligonucleotide array. Investig Ophthalmol Vis Sci 2003; 44:3409–3416.
45. Hazlett LD, McClellan SM, Hume EB, Dajcs JJ, O'Callaghan RJ, Willcox MD. Extended wear contact lens usage induces Langerhans cell migration into cornea. Exp Eye Res 1999; 69:575–577.
46. Hazlett LD, McClellan SM, Kwon B, Barrett R. Increased severity of Pseudomonas aeruginosa corneal infection in strains of mice designated as Th1 versus Th2 responsive. Investig Ophthalmol Vis Sci 2000; 41:805–810.
47. Hendricks RL. Immunopathogenesis of viral ocular infections. Chem Immunol. 1999; 73:120–36.
48. Khanna KM, Bonneau RH, Kinchington PR, Hendricks RL. Herpes simplex virus-specific memory CD8+ T cells are selectively activated and retained in latently infected sensory ganglia. Immunity. 2003; 18:593–603.
49. Seal D, Wright P, Ficker L, Hagan K, Troski M, Menday P. Placebo-controlled trial of Fusidic acid gel and Oxytetracycline for recurrent blepharitis and rosacea. Brit J Ophthalmol 1995; 79:42–45.
50. Mondino BL. Inflammatory diseases of the peripheral cornea. Ophthalmology 1988; 95: 463–68.
51. Knop E, Knop N, Pleyer U. Clinical aspects of MALT. In Pleyer U, Mondino BJ (eds), Essentials in Ophthalmology. Uveitis and immunological disorders. 2004 Springer, Berlin, Heidelberg, 2005; pp 65–89.

TWO

Microbiology

BACTERIOLOGY

Cell Wall

Bacteria are divided into Gram-positive and Gram-negative groups based on their cell wall structure. The bacterial cell wall is a rigid structure surrounding a flexible cell membrane. The cell wall maintains the shape of the cell: its rigid wall compensates for the innate flexibility of the phospholipid membrane and also maintains the cell's integrity when the intracellular osmotic gradient is unfavorable. The wall is also responsible for attachment: teichoicacids attached to the outer surface of the wall serve as attachment sites for bacteriophages. Flagella, fimbriae, and pili all emanate from the wall.

The composition of the Gram-positive cell wall is 90% peptidoglycan polymer made of alternating sequences of N-acetylglucosamine (NAG) and N-acetyl-muraminic acid (NAMA) with each layer cross-linked by an amino acid bridge. The peptidoglycan polymer imparts thickness to the Gram-positive cell wall. In contrast, peptidoglycan makes up only 20% of the Gram-negative cell wall. Periplasmic space and an outer membrane also diffe]rentiate the Gram-negative organism and contain proteins that destroy potentially dangerous foreign matter. The outer membrane, composed of lipid, protein, and lipopolysaccharide (LPS), is porous because porin proteins allow free passage of small molecules. The lipid portion of LPS also contains lipid A, a toxic substance, which imparts the pathogenic virulence associated with some Gram-negative bacteria.

Gram stain was the innovation of Hans Christian Joaquim Gram, a Danish physicist, who sought to distinguish bacterial organisms based on their different cell wall structures. The crystal violet primary stain in Gram stain preferentially binds peptidoglycan. Because the cell wall of Gram-negative bacteria is low in peptidoglycan content and high in lipid content, the primary crystal violet stain is washed out when the decolorizer (acetone) is added. Instead, the Gram-negative organism incorporates the safranin counterstain, which stains red.

Methylene blue is a simple stain that is taken up by all bacteria and can be used to show if bacteria are present or not. Other useful stains include acridine orange (all bacteria, fungi, and protozoa), modified Ziehl-Neelsen (weak acid, no alcohol) (*Nocardia* and *M. leprae*), full Ziehl-Neelsen (strong acid and alcohol) (*Mycobacteria*), lactophenol blue (wet preparation for fungi), immunofluorescent stains, and fluorescein-labeled molecular probe stains. Use of molecular biology techniques with polymerase chain reaction (PCR) for detecting bacterial ribonucleic acid (RNA) and deoxyribonucleic acid (DNA) is described in detail in Chapter 3.

Organisms

Even within the broad rubric of Gram reaction, many different taxonomy schemes are used to group various organisms. For instance, the organisms are further subdivided by their physical morphology (cocci or rod), metabolic pathway (aerobe, anaerobe, or facultative anaerobe), and survival function (spore-forming). In this section, these organisms were grouped in the simplest way possible (Table 1).

The Gram-positive cocci are grouped together based on their Gram stain reaction, thick cell wall structure, and spherical shape. *Staphylococcus, Streptococcus,* and *Enterococcus* species are members of this group with medical importance. *Enterococcus* organisms are a part of the group D streptococci that were designated a new genus. Various biochemical tests are employed to differentiate these species. For instance, staphylococci are catalase positive, where as streptococci and *Enterococcus* species are catalase negative.

Staphylococcus.   There are 23 recognized staphylococcal species; three of them, *S. aureus, S. epidermidis,* and *S. saprophyticus* are recognized as being most clinically significant. The *Staphylococcus* genus appears as Gram-positive cocci in grape-like clusters (Fig. 1). Staphylococci are classified into *S. aureus* and non-aureus organisms by the coagulase test. The former is coagulase positive while the latter are all coagulase negative, hence known collectively as CNS.

The pathogenic effects of staphylococci are mainly produced by a variety of toxins. For instance, enterotoxin causes food poisoning and exfoliative toxins cause scalded skin syndrome. *S. aureus,* the pyogenic and virulent organism, is almost always resistant to penicillin because of beta-lactamase production, but is usually susceptible to the beta-lactamase–stable penicillins (e.g., methicillin, nafcillin, oxacillin, [di]cloxacillin), but may become resistant on the basis of altered penicillin-binding proteins. Methicillin-resistant *S. aureus* (MRSA) has become a major concern within hospitals, as part of hospital acquired infection. *S. aureus* has a propensity for invasion of the vascular system, leading to bacteremia, and is distinguished from other bacteria by its ability to establish metastatic sites of pyogenic infection throughout the body, including endophthalmitis.

Systemically, *S. aureus* is the most pathogenic of this group of organisms and is also associated with endophthalmitis, pneumonia, meningitis, boils, arthritis, and osteomyelitis. *S. epidermidis* is a coagulase-negative staphylococcus that is part of normal skin flora. Infection with *S. epidermidis* usually occurs in the presence of a foreign body such as an intraocular lens, catheter, prosthetic device, surgery, or trauma. *S. saprophyticus* is a common cause of infection of the urinary tract, both upper (i.e., kidney—pyelonephritis) and lower (i.e., bladder—cystitis) segments, in sexually active young women. This bacterium may possess an adhesin for urothelial cells.

Intraocularly, involvement of staphylococcal organisms with postoperative endophthalmitis has been well documented. *S. epidermidis* was the most frequent organism isolated in the Endophthalmitis Vitrectomy Study (EVS), although bias in recruitment explains this (Chapter 8). *S. aureus* is associated with a severe form of acute postoperative endophthalmitis while CNS causes both subacute and chronic postoperative endophthalmitis.

Streptococcus.   The *Streptococcus* genus consists of Gram-positive cocci in chains or clusters (Fig. 2). Several classifications have been proposed for this genus. Hemolytic activity has been used as a preliminary criterion for classifying some streptococci.

Table 1	Gram-positive and Gram-negative Bacteria

Gram-positive bacteria

- Cocci
    - *Staphylococcus* species
    - *Streptococcus* species
- Rods
    - *Bacillus* species
    - *Listeria* species
    - *Lactobacillus* species
    - *Erysipelothrix* species
    - *Corynebacterium* species
    - *Actinomyces* species
    - *Propionibacterium* species
    - *Clostridium* species

Gram-negative bacteria

- Rods
    - *Enterobacteriaceae*
- *Escherichia coli*
- *Klebsiella* species
- *Proteus* species
- *Serratia* species
- *Providencia* species
- *Salmonella* species
- *Shigella* species
- *Citrobacter* species
- *Morganella* species
    - Pleomorphic group
- *Haemophilus* species
- *Brucella* species
- *Pasteurella* species
- *Legionella* species
    - Nonfermenters
- *Pseudomonas* species
    - Others
- *Vibrio* species
- *Campylobacter* species
- *Helicobacter* species
- Coccobacilli
    - *Haemophilus influenzae*
    - *Neisseria*

- α-hemolysis - incomplete lysis causing "greening" around colonies on blood agar; viridans group
- β-hemolysis - complete lysis; contains most of major human pathogens
- Non-hemolytic

Rebecca Lancefield devised a scheme of classification based on the antigenic characteristics of a cell-wall carbohydrate called the C substance. Group A was named *S. pyogenes*; more than 90% of streptococcal disease in humans is caused by beta-hemolytic group A streptococci, which is extremely virulent although it also colonizes the throats of 10% of the population. Group A streptococcal diseases are divided into suppurative (primary) and non-suppurative (sequelae) diseases. Suppurative diseases include impetigo, erysipelas, puerperal fever, cellulitis, and

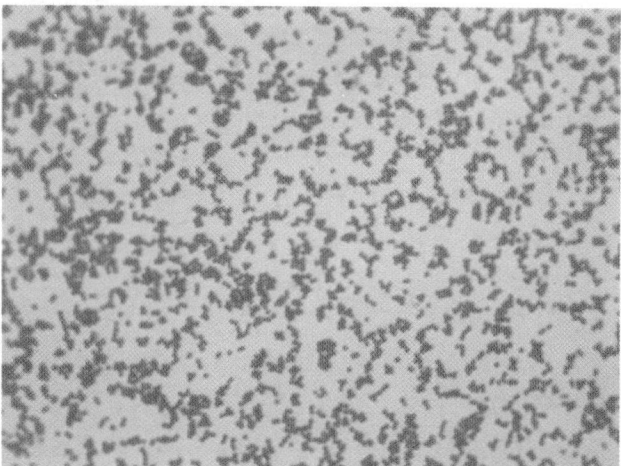

Figure 1 Gram-positive staphylococci in clusters.

necrotizing fasciitis. It also causes a highly virulent and purulent acute postoperative endophthalmitis. Subsequent non-suppurative complications include scarlet fever, rheumatic fever, glomerulonephritis, and erythema nodosum. Streptococci perpetuate their pathology via elaborated toxins including erythrogenic toxin, cardiohepatic toxins, nephrotoxins, hemolysins, and spreading factors.

Lancefield Group B streptococcus is called *S. agalactiae* and is found in the oral cavity, intestinal tract, and vagina. It is not beta-hemolytic. In the United States, it is the most frequent cause of life-threatening disease in newborns including meningitis and the occasional case of endophthalmitis. Adult diseases include meningitis, endocarditis, and pneumonia.

Lancefield Groups C and G streptococci are beta-hemolytic and cause pyogenic infections in the elderly and compromised individuals including endophthalmitis; they also cause animal infections, which may be the source of the human infection.

Lancefield Group D is divided into enterococcal and nonenterococcal organisms. Enterococcal members such as *S. faecalis*, *S. faecium*, and *S. durans* have

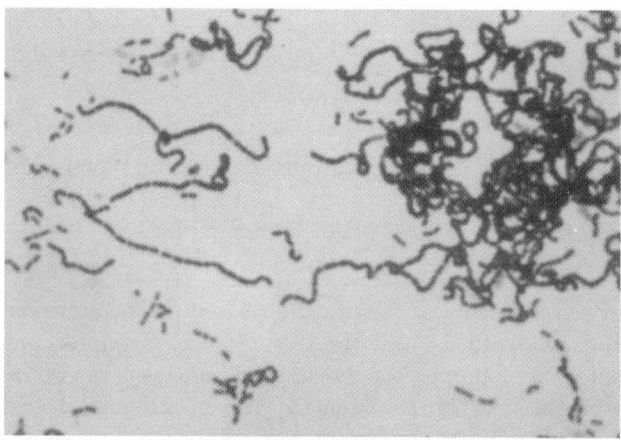

Figure 2 Microaerophilic streptococci in long chains (Gram stain).

been assigned the genus name *Enterococcus* and thus are referred to as *E. faecalis*, *E.faecium*, and *E. durans*, respectively. Enterococcal organismshave long been known to cause infective endocarditis, but have also been recognized to cause endophthalmitis (1), nosocomial infection, and "superinfection" in patients receiving antimicrobial agents. Furthermore, *Enterococcus* species are now receiving increased attention because of their resistance to multiple antimicrobial drugs, probably contributing to their prominence in nosocomial infections. The most common enterococci-associated nosocomial infections are those of the urinary tract, followed by surgical wound infections and bacteremia.

Viridans streptococci are mostly alpha-hemolytic bacteria that do not fit into the Lancefield rubric; they are responsible for subacute bacterial endocarditis but occasionally possess a selection of substance antigens. Patients with a history of rheumatic fever or congenital heart disease are susceptible to seeding of these displaced organisms on the heart valves or other vital systems. They have also been found as an increasingly important cause of postoperative endophthalmitis in the United Kingdom, following phakoemulsipication cataract surgery. They can be highly virulent within the vitreous cavity, causing a severe endophthalmitis with loss of all vision and even the globe (Chapter 8). The source is not clear but they may be derived from the respiratory tree of the chronic bronchitic patient.

Gram-positive Rods (Spore Forming)

The Gram-positive rods are ubiquitous organisms divided into three groups based upon their ability to produce endospores and their morphological appearance.

Bacillus. Bacilli are ubiquitous organisms found in soil, water, airborne dust, and even the human intestines. *Bacillus* (aerobic) (Fig. 3) and *Clostridium* (anaerobic) bacteria are the only spore-forming organisms. On blood agar plates, bacilli produce large, spreading, gray-white colonies with irregular margins. In the spore state, these organisms are resistant to heat and chemicals; however, autoclaving at 120°C and 15 lbs of pressure will destroy them. Although most *Bacillus* species are harmless saprophytes, *B. anthracis* and *B. cereus* cause significant pathology.

B. anthracis. *B. anthracis*, the organism that causes anthrax, primarily affects herbivorous animals. The bacterium is transmitted to humans through inhalation of endospores or direct contact with animal products. In order to be fully potent, *B. anthracis* must produce a capsule and a protein exotoxin. Under the microscope, these organisms are nonmotile and appear to have square ends and to be attached by a joint to other cells (Fig. 4). Sources of infection are usually industrial or agricultural. Unfortunately, recent events have added terrorism as another cause. The infection is classified as one of three types:

- Cutaneous infection (95% of human cases)
- Inhalation anthrax (rare)
- Gastrointestinal anthrax (very rare)

Although extraocular and eye surface complications have been reported (2), no intraocular complications have been noted. Cutaneous anthrax leading to corneal scarring from cicatricial ectropion (3) has been reported, as well as preseptal cellulitis (4).

B. cereus. Unlike *B. anthracis*, *B. cereus* is a motile bacterium that can cause toxin-mediated food poisoning. Its association with traumatic endophthalmitis has been documented in detail in the literature (Chapter 8). *B. cereus* has also been

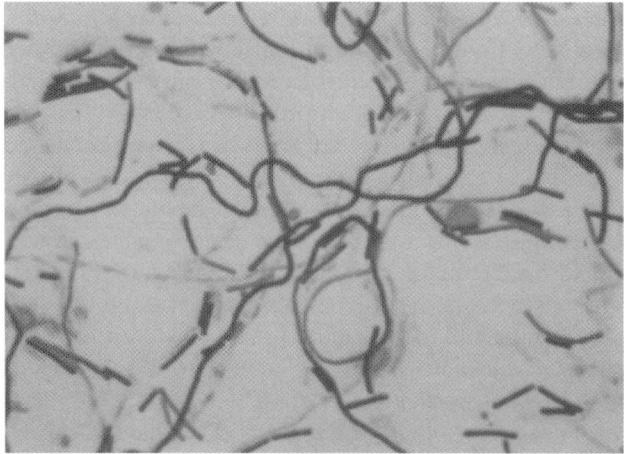

Figure 3 Long chains of *Bacillus* sp. in culture; may form spores in human tissue, which does not occur with *Clostridium perfringens*.

associated with endogenous endophthalmitis, especially in immunocompromised hosts. *B. cereus* produces beta-lactamases and is resistant to beta-lactams, including third-generation cephalosporins. It is also found in different foods, including stew, cereal, fried rice, and milk. The toxin is produced after the infected foods are cooked and causes symptoms similar to those of staphylococcal food poisoning.

Clostridium. *Clostridium* species are Gram-positive, anaerobic spore-forming, motile rods found ubiquitously in organisms inhabiting water, soil, vegetation, and the human bowel. These organisms are especially fond of soil. Under the microscope, they resemble long drumsticks with a bulge located at their terminal ends.

Extremely durable *Clostridium* spores are produced when the environment becomes unfavorable. In the active form, potent exotoxins are produced that are responsible for tetanus, botulism, and gas gangrene.

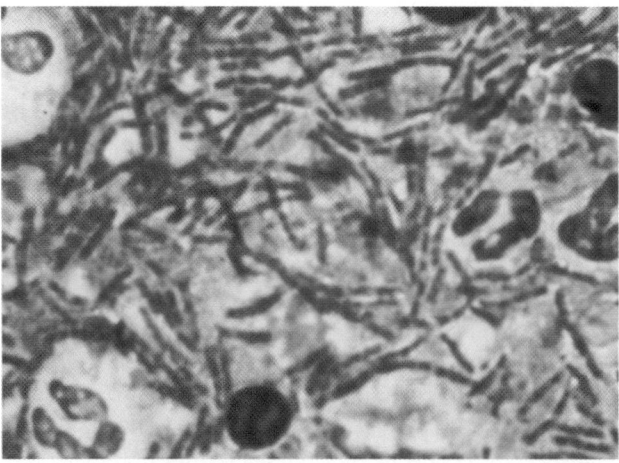

Figure 4 Methylene blue stain of *Bacillus anthracis*.

Endophthalmitis cases related to *Clostridium* organisms have been reported following trauma (Chapter 8) and in one case, after cataract surgery (5). Coinfection with *C. tetani* and *Bacillus* species was noted in one particular traumatic case (6), the source being contamination with mud. Endogenous endophthalmitis secondary to metastasis of *Clostridium* organisms from the biliary tract has also been reported (7). In one case of panophthalmitis secondary to *Clostridium septicum*, the patient presented with spontaneous gas gangrene panophthalmitis and a gas bubble was present in the anterior chamber (8). Gram stain of clostridia from cases of gas gangrene shows Gram-positive bacilli with no spores, as the bacterium does not produce spores when causing active infection; it does so when exposed to air or if the laboratory cultures are a few days old.

Gram-positive Bacilli (non–spore forming)

Listeria monocytogenes causes listerosis. Neonates experience granulomatosis infantisepticum. It can also be spread by undercooked meat. Seventeen cases of endophthalmitis associated with *Listeria* species have been reported in the literature since 1977. Most occur as endogenous endophthalmitis in immunocompromised hosts, typically with elevated intraocular pressure, pigment dispersion, and dark hypopyon (9).

Erysipelothrix rhusiopathiae is usually found in decomposing nitrogenous material; swine are major reservoirs. Most human infections are due to occupational exposure. Systemic infection can cause endocarditis. No cases of endophthalmitis have yet been reported with this organism.

Corynebacteria diphtheriae is found in upper respiratory lesions and skin. Humans are the only natural host of this bacterium, which causes respiratory and cutaneous diphtheria. Along with coagulase-negative staphylococci, *Corynebacteria* species are a frequent cause of subacute endophthalmitis following cataract surgery (10,11). They also comprise the majority of positive cultures after cataract surgery in patients without evidence of endophthalmitis, implying contamination at the time of surgery (10,12).

Lactobacillus organisms are normal flora in the vagina. One case of posttraumatic endophthalmitis caused by *Lactobacillus* species has been reported in the literature (13).

Gram-negative Organisms

Gram-negative bacteria are deficient in peptidoglycans. They loosely bind the crystal violet primary stain, which is removed on exposure to acetone for five seconds (Gram-positive bacteria retain the stain). Therefore, they are stained with the secondary safranin stain, which imparts its red hue. The LPS layer is primarily responsible for endotoxin production. They are found in the gastrointestinal tract and are a common cause of urinary tract infection.

Gram-negative Diplococci

Neisseria Organisms. *Neisseria* species are pathogens as well as normal flora in humans. They are aerobic and facultative anaerobic, Gram-negative diplococci requiring a moist environment and warm temperatures for optimum growth. Where the diplococci abut, they are flat sided, so each member of the pair presents a bean-shaped configuration. They are found within polymorphs. These organisms grow well on chocolate agar containing antibiotics that inhibit growth of Gram-negative bacteria (Fig. 5). The two most clinically significant *Neisseria* species are *N. gonorrhoeae* and *N. meningitidis*.

N. meningitidis (meningococcus). Humans are the only natural hosts for *N. meningitidis*, and transmission is through nasal droplets. Meningococcal diseases can be divided into nasopharyngitis, septicemia, meningitis, arthritis, conjunctivitis, and, occasionally, bilateral endogenous endophthalmitis. Different strains of *N. meningitidis* are classified by their capsular polysaccharides. This bacterium is the second leading cause of meningitis in the United States. Both exogenous and endogenous endophthalmitis cases related to *N. meningitidis* have been reported (14, 15). Meningococcal endophthalmitis has been reported as both a preceding event and sequelae of septicemia (Chapter 8) (16,17).

N. gonorrhoeae. Among the *Neisseria* species, only *N. gonorrhoeae* strains are always pathogenic. This organism colonizes mucosal surfaces of the cervix, urethra, rectum, oropharynx, and nasopharynx, causing the correlating symptoms. Ocular surface sequelae of gonococcal ophthalmia have been well documented but intraocular manifestations have not been reported.

Enterobacteriaceae. The *Enterobacteriaceae* are a family of large Gram-negative rods that are usually associated with intestinal infections, but are found in almost all natural habitats. These are among the most pathogenic and frequently encountered organisms in clinical microbiology. Systemically, they have also been known to cause meningitis, pneumonia, bacillary dysentery, typhoid, and food poisoning.

In general, Gram-negative organisms account for 16% to 18.5% of all cases of culture-proven post-surgical endophthalmitis and up to 30% of posttraumatic cases in rural areas. Among *Enterobacteriaceae* organisms are the *Proteus* species, *Serratia marcescensm, Morganella morganii, Citrobacter* species, *Escherichia coli*, and *Klebsiella pneumoniae*. This list includes non-*Enterobacteriaceae* organisms such as *Pseudomonas* species and *Haemophilus influenzae*. *Enterobacteriaceae* organisms have been associated with both endogenous and exogenous endophthalmitis, especially in immunocompromised hosts. However, no cases of endophthalmitis due to *Shigella*, *Edwardsiella*, or *Providencia* species have been reported.

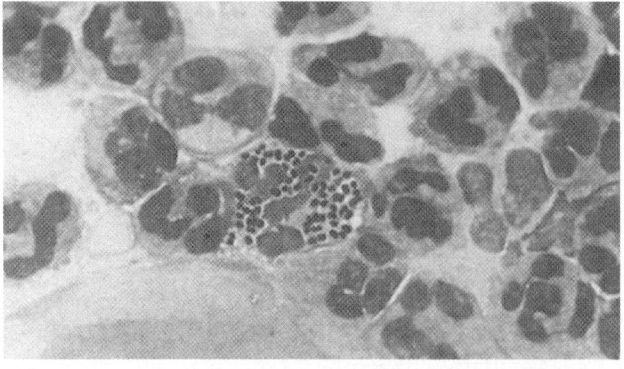

Figure 5   Gram stain of *Neisseria meningitidis* in PMNs.

Ocular sequelae can include central artery occlusion resulting from septic thrombi. *Klebsiella* and *Serratia* species appear to be among the organisms most likely within this group to cause endophthalmitis. Sixty-one percent of the 44 patients with *Klebsiella* endophthalmitis reported in the literature from 1981 to 1994 had diabetes, 68% had suppurative liver disease, and 16% had a urinary tract infection (18). Although most cases of endogenous endophthalmitis result from metastasis in immunocompromised hosts, a case of exogenous endophthalmitis is caused by *Serratia* organisms after an injury with a wood fragment has also been reported (19). Among the less frequent causes of endophthalmitis, *Proteus* organisms were found in a diabetic patient who underwent a triple procedure (penetrating keratoplasty, cataract extraction, and posterior chamber intraocular lens implantation) as a result of contamination from the donor cornea (20). *Morganella* organisms have also been isolated from a postoperative cataract patient (21). In a case of endophthalmitis induced by *Salmonella* organisms, theuse of snake powder as a food seasoning was suspected as the cause. Salmonella has also been isolated from a case of endophthalmitis following pneumopexy.

Other Gram-negative Rods

Haemophilus. *Haemophilus* organisms comprise a significant portion of the indigenous flora of the upper respiratory tract. Given their ubiquity, most of the population is a carrier for one ormore *Haemophilus* species. It represents a largegroup of Gram-negative rods with an affinity for growing in blood. The blood medium provides two factors that *Haemophilus* species require for growth: X factor (hemin) and V factor (nicotinamide dinucleotide). *Haemophilus* colonies will usually form satellites around *Staphylococcus* colonies because staphylococci can provide the necessary factors required for optimum growth. Morphologically, *Haemophilus* bacteria usually appear as tiny coccobacilli under the microscope, but can also assume many other shapes. Some classification systems have grouped them into a pleomorphic category. Using methylene blue stain on a smear may aid in identification.

Haemophilus species are classified by their capsular properties into six different serological groups, a–f. Species that possess the type b capsule are the most clinically significant because of their virulent properties.

H. influenzae. Infection with *H. influenzae* occurs following inhalation of infected droplets. Most invasive upper respiratory infections are caused by type b encapsulated strains (HIB). Systemic sequelae can occur after hematogenous spread and can cause meningitis, epiglottitis, cellulitis, septic arthritis, pneumonia, and endophthalmitis. HIB is the major cause of meningitis in children under four years of age who have yet to form protective antibodies. This species may exist with or without a pathogenic polysaccharide capsule. Strains that lack the capsule usually cause mild localized infections (otitis media, sinusitis), as opposed to HIB, which can cause several serious infections. An HIB vaccine has been found highly protective for infants.

H. influenzae has been well documented asa major cause of bleb-associated endophthalmitis. There have also been cases following strabismus surgery (22) and trauma (23). *H. aegyptius* and *H. paraphrophilus* have also been reported as causing endophthalmitis. *H. ducreyi* causes chancroid but has not been associated with endophthalmitis.

Brucella organisms only cause incidental infections in humans; the organism is thought to have been eradicated in many countries, except in agrarian environments.

It colonizes the genital tracts of animals and infects the milk. *Brucella abortus* infects cattle and *B. melitensis* infects goats and was the original cause of Malta fever. A case of endophthalmitis due to *B. melitensis* has been reported in a 17-year-old woman in Saudi Arabia, probably due to drinking raw goat's milk (24). Populations avoid brucellosis by pasteurizing milk and producing farm animals from which *Brucella* has been eradicated.

Francisella tularensis causes tularemia from tick bites. This highly infectious disease is carried by rodents, deer, pets, and many other animals. Oculoglandular tularemia has been reported but endophthalmitis has not been noted as a sequelae.

Pasteurella multocida infects humans from dog or cat bites. Patients tend to exhibit swelling, cellulitis, and some bloody drainage at the wound site. Infection may also move to nearby joints where it can cause swelling and arthritis. No ocular complications have been noted in the literature.

Legionella pneumophila. *L. pneumophila* causes legionnaires' disease and Pontiac fever. The first discovery of bacteria from genus *Legionella* came in 1976 when an outbreak of pneumonia at an American Legion convention led to 29 deaths. Transmission is through respiratory droplets and causes a gradual onset of flu-like symptoms. Severe pneumonia can develop that is not responsive to penicillins or aminoglycosides but does respond to the quinolones and azithromycin. Legionnaires' disease also has the potential to spread into other organ systems of the body such as the gastrointestinal tract and the central nervous system. No ocular complications have been reported.

Miscellaneous Gram-negative Rods

Vibrio. *Vibrio* organisms are obligate aerobes with a recognizable curved shape and a single polar flagellum (Fig. 6). These are waterborne organisms transmitted to humans, especially fishermen, via infected water or through fecal transmission. In hot countries with poor sewerage or water treatment, cholera can occur in epidemic proportions due to *V. Cholerae*.

Ocular involvement usually occurs as a result of trauma involving infected warm seawater, such as the Gulf of Mexico or the South China Sea and Hong Kong. In a

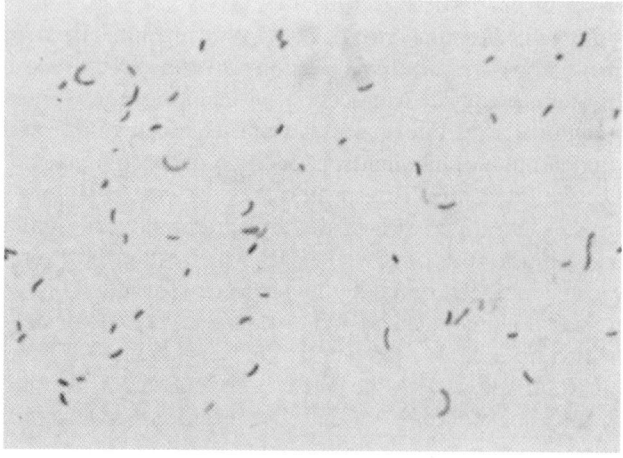

Figure 6   Gram-negative rod: *Vibrio vulnificus.*

recent review, Penland et al. describe 17 cases of eye infections involving *Vibrio* species, mainly due to *V. vulnificus* in which 11 (65%) involved exposure to seawater or shellfish. Three cases of exogenous endophthalmitis were included in this survey (25).

Campylobacter. *Campylobacter* organisms are microaerophiles that can survive in a low oxygen environment and prefer a relatively high concentration of carbon dioxide, making a suitable growth environment a challenge. Cell motility is achieved through polar flagella that emanate from a curved rod-shaped cell. The unique shape of the cell and flagella are extremely useful in Gram stain identification. *C. jejuni* is an organism that causes gastrointestinal infection. Humans acquire the organisms by eating undercooked chicken or drinking contaminated milk and water.

Helicobacter. These are also microaerophilic organisms with curved cell bodies. Aside from the well-known association of *H. pylori* with stomach ulcers, *Helicobacter* species have been linked to ocular manifestations of acne rosacea (26). *H. pylori* has been reported to be found frequently in glaucoma patients (27).

Pseudomonas. *Pseudomonas* species are motile, Gram-negative, aerobic rods that, along with *Xanthomonas* species, comprise a group of bacteria called pseudomonads. These organisms are found commonly in soil and water, as well as on plant and animal surfaces. These bacteria demonstrate significant antibiotic resistance, including innate resistance to many antibiotics, and are capable of surviving in harsh conditions. They also produce a slime layer that is resistant to phagocytosis. *Pseudomonas aeruginosa* is the most common human pathogen and the most frequently isolated non-fermenter in the laboratory. It has several features that distinguish it from other *Pseudomonas* species:

- Able to grow at 42°C
- Produces a bluish pigment (pyocyanin), a greenish pigment (pyoverdin), and a red pigment (pyorubin)
- Characteristic fruity odor

Upon infection, the LPS layer helps the cell adhere to host tissues and prevents leukocytes from ingesting and lysing the organism. Lipases and exotoxins destroy host cell tissue, leading to the complications associated with infection. *P. aeruginosa* prefers moist environments but can survive in a medium as nutritionally deficient as distilled water. It will also grow on most laboratory media.

Pseudomonas species are often encountered in hospital and clinical settings and are a major cause of nosocomial infections. Immunocompromised patients, burn victims, cystic fibrosis patients, and patients on respirators or with indwelling catheters are most susceptible. Infection can occur at many sites and can lead to urinary tract infection, sepsis, endocarditis, pneumonia, pharyngitis, pyoderma, and dermatitis.

Ocular infectious complications can cause devastating vision loss with potential for infection at any location. It is the commonest cause of contact lens–associated keratitis (Chapter 6 and Appendix D). Both endogenous and exogenous forms of endophthalmitis and panophthalmitis caused by *Pseudomonas* spp. have been described. The etiologies range from keratitis to trauma and surgery, especially but not necessarily with an immunocompromised state. *P. aeruginosa* is highly virulent and therefore it must be managed aggressively with vitrectomy and intravitreal antibiotics (gentamicin, amikiacin or ceftazidime) or the eye will be lost within 48 hours.

Other *Pseudomonas* species include *P. cepacia*, an opportunistic pathogen of cystic fibrosis patients. *Stenotrophomonas maltophila* (formerly known as

Xanthomonas maltophila) is very similar to the pseudomonads. This motile bacterium is a cause of nosocomial infections in immunocompromised patients. *S. maltophila* also harbors significant resistance to many antibiotics considered effective for treating infections caused by *Pseudomonas* organisms. However, most strains of the bacterium are susceptible to trimethoprim or sulfamethoxazole.

Acinetobacter. *Acinetobacter* species are oxidase-negative, non-motile bacteria that appear as Gram-negative pairs under the microscope. Identifying the different species of this genus is possible with the use of fluorescence-lactose-denitrification medium that determines the amount of acid produced from metabolizing glucose. Although many *Acinetobacter* species can cause infection, *A. baumannii* is the most frequently encountered species in the clinical laboratory, causing hospital-acquired infections that include skin and wound infections, pneumonia, and meningitis.

Five cases of endophthalmitis caused by *Acinetobacter* species have been reported in the literature; all were exogenous, resulting from either trauma or surgery. Some of the postoperative cases presented in a similar fashion to *Propionibacterium acnes* (28).

Anaerobes

Anaerobic organisms require an oxygen-free environment to grow; strict anaerobes are completely inhibited from growing in the presence of oxygen. Anaerobes represent 5–10% of all clinical infections. Anaerobic infections usually have the following characteristics:

- Foul-smelling discharge
- Proximity to mucosal membrane
- Necrotic tissue
- Gas formation in tissues or discharge
- Infection following a bite

The significant anaerobes are listed below.

Actinomyces species are Gram-positive, non-spore forming anaerobic bacilli. They have been described as bacteria that form filamentous branches. They are difficult to culture, and thioglycolate fluid medium should be used, in which they grow as "bread crumb" colonies at a depth reflecting their ideal microaerophilic environment. Infection usually occurs through trauma with abscess formation at the implantation site. Six cases of postoperative endophthalmitis have been reported (29). They also cause canaliculitis (Chapter 5).

Propionibacterium species are among the most commonly isolated Gram-positive anaerobes. They are found as saprophytes in humans, animals, and dairy products. *P. acnes,* among the most commonly isolated non–spore-forming anaerobic clinical specimens, is found as normal flora on the skin, nasal pharynx, oral cavity, and gastrointestinal and genitourinary tracts. Systemically, *P. acnes* has been associated with acne vulgaris, endocarditis, wound infections, central nervous system shunt infections, and abscesses. Chronic postoperative endophthalmitis, of a granular type involving the capsular bag, is the main ocular complication (Chapter 8). Here, the organisms are found within macrophages and may not be isolated from an anterior chamber tap. They can be identified by a Gram stain of the capsule fragment (fixed histology) (Fig. 7) and be seen within macrophages by electron microscopy (Fig. 8). Microscopically, *Propionibacterium* organisms clump and show

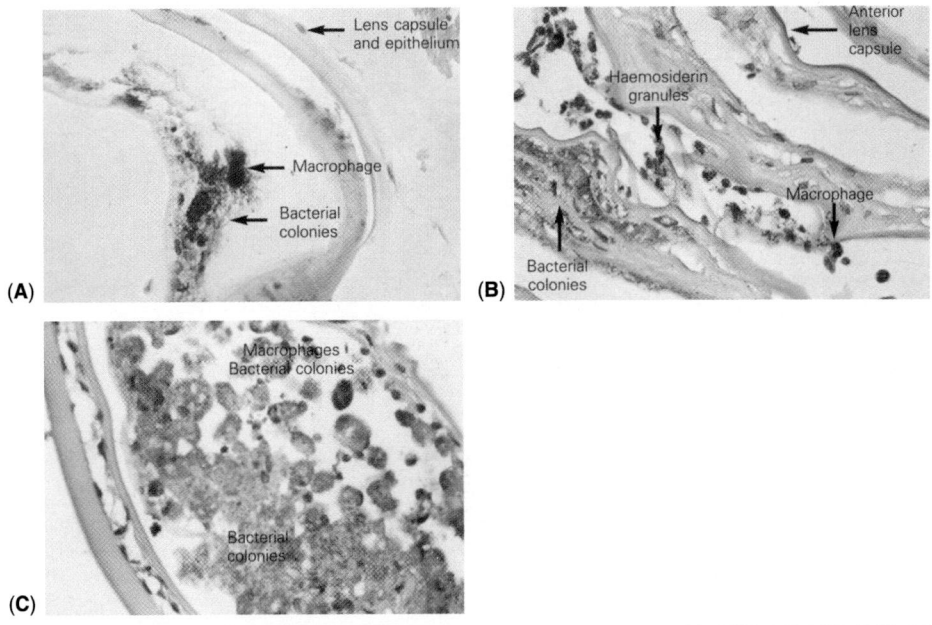

Figure 7 (**A–C**) Hematoxylin and eosin stain of the capsule fragment removed surgically from three patients with chronic "saccular" endophthalmitis showing macrophages and bacterial colonies.

a tendency to branch. Colonies grow best in an anaerobic or microaerophilic environment using blood agar. Incubation should occur for a minimum of 14 days as they grow slowly.

Culturing Media

There are three types of culturing medium: non-selective (viz. blood, chocolate, or nutrient agars), selective, and differential. A selective medium promotes the growth of desired organisms while inhibiting the undesirable organisms. A differential medium provides a visible indication of a physiological property, such as the fermentation of a carbohydrate. Routine culture media can be stored in a refrigerator, although only fresh plates of media shouldbe used. If the media appear dry or the edges of the Petri dish are recessed, then these should not be used. Inoculation should occur at room temperature. For most organisms encountered in ophthalmology, blood and chocolate agars (for bacteria), Sabouraud's agar (for fungi), and thioglycolate and brain heart infusion broth medias (for bacteria and fungi) should be sufficient. Other cultures can be helpful, such as for mycobacteria, depending on specific need and availability.

Blood Agar Plate

Blood agar consists of an agar base, derived from seaweed, with a peptic digest of animal tissue, dextrose, and yeast extract. It is an enriched, non-selective, general purpose medium able to grow most aerobes (except *Neisseria*, *Haemophilus*, and *Moraxella* species), anaerobes (with vitamin K, cysteine, and hemin supplementation), and fungi. Using this differentiating medium, the degree of hemolysiscaused by hemolysins is assessed to differentiate among Gram-positive cocci.

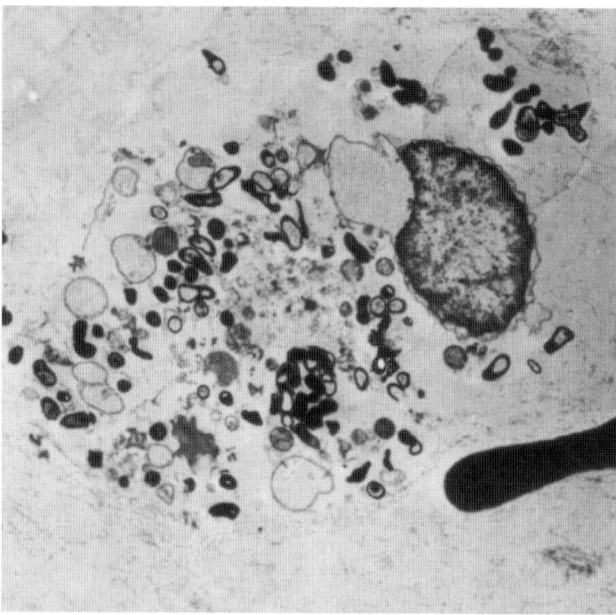

Figure 8   Electron micrograph shows *Propionibacterium acnes* in a macrophage. *P. acnes* was cultured from an eye with late-onset chronic plaque-type endophthalmitis.

- Beta hemolysins completely lyse the red blood cells and hemoglobin; this results in complete clearing around colonies.
- Alpha hemolysis refers to the partial lysis of red blood cells and hemoglobin and produces a greenish discoloration of the blood agar around the colonies.
- No hemolysis, sometimes called gamma hemolysis, results in no change of the medium.
- Hemolysis must be determined before two days of incubation, as later reactions are non-specific.

Thioglycolate Medium

This is a semisolid, nutrient medium that supports the growth of organisms of all oxygen requirements: anaerobes, aerobes, and facultative anaerobes. Thioglycolate reaction causes a gradation of oxygen and establishes an anaerobic environment at the bottom of the medium. An oxygen indicator turns the medium pink or blue at the top of the tube. This medium is boiled before use to drive off oxygen, which is less soluble at hot temperatures. The specimen is inoculated after cooling. Obligate aerobes will grow only at the top of the tube of medium, microaerophiles in the middle, while anaerobes grow only at the bottom.

Sabouraud's Agar

Sabouraud's agar is a selective medium with low pH that inhibits the growth of most bacteria, making it selective for yeasts and most but not all fungi. It is a rich medium, good for yeasts, but vegetative plant fungi may not grow, preferring a low nutritious medium such as cornmeal agar.

Chocolate Agar

Chocolate agar is comprised of sheep blood that provides factors X (hemin) and V (nicotinamide adenine dinucleotide) necessary for growth of *Haemophilus* species, *N. gonorrhoeae*, *N. meningitidis*, and *Moraxella* organisms. However, Thayer-Martin medium (Oxoid Ltd, Cambridge U.K.) containing antimicrobials (vancomycin 3.0 mg, colistin 7.5 mg, and nystatin 12.5 units per mL of agar) is used to supplement the chocolate agar when *N. gonorrhoeae* is suspected. These incorporated antimicrobials will suppress the growth of other organisms that inhibit the growth of *N. gonorrhoeae*. Thayer-Martin plates are incubated in an atmosphere containing 3% to 10% CO_2.

MacConkey Agar

MacConkey agar is mainly used to differentiate between various Gram-negative rod-shaped organisms. It is also used to inhibit growth of Gram-positive organisms; thus it is both differential and selective. Many facultative anaerobes in the intestine are lactose fermenters (*E. coli*). Several well-known pathogens (viz. *Shigella* and *Salmonella* species) are unable to ferment lactose. Bacterial colonies that can ferment lactose turn the medium red as the result of the response of the pH indicators to the acidic environment created by the fermentation. Organisms that do not ferment lactose do not cause a color change.

Eosin-Methylene Blue (EMB) Agar

EMB agar, a differential medium that inhibits the growth of Gram-positive organisms, is used to identify Gram-negative enteric rods. Gram-positive bacteria are inhibited by the dyes eosin and methylene blue, which are added to the agar. The medium also allows distinction between lactose-fermenting and non-lactose fermenting organisms. Two types of coliform colonies are noted:

- Coli-type: very dark colonies but with a green hue in reflected light caused by the precipitation of methylene blue in the medium from the very high amount of acid produced from fermentation. Colonies are composed of methyl red-positive lactose-fermenters such as most strains of *E. coli* and some *Citrobacter* strain.
- Aerogenes-type: less dark colonies with a dark center surrounded by a wide, light-colored, mucoid rim. This results in a "fish-eye" appearance. Those organisms that form this type of colony are methyl red-negative lactose-fermenters, including most strains of *Klebsiella* and *Enterobacter*.

Mannitol Salts Agar (MSA)

MSA is a common medium used for the isolation of *S. aureus*. This medium contains 7.5% salt, which is inhibitory to most bacteria other than staphylococci. While most *Staphylococcus* spp. are capable of growing in this high salt concentration, *S. aureus* is also capable of fermenting the carbohydrate mannitol with production of an acid that turns the pH indicator from red to yellow. Nonpathogenic staphylococci can grow on the medium but produce no acid from it.

Bile-Esculin Agar

Bile-esculin agar is used to identify group D streptococci, which are able to grow in the presence of bile, an emulsifying agent produced in the liver. Group D streptococci also have the ability to hydrolyze esculin, which turns the medium black and denotes a positive test. Other bacteria capable of growing in the presence of bile do not turn the medium black. A variation of this medium uses sodium azide to inhibit the growth of all other Gram-positive and Gram-negative bacteria.

Brain-Heart Infusion Broth

This is a highly nutritious and buffered liquid used as an adjunct to the solid media. Because the medium is a liquid, the material picked up by the swab but not released onto the solid agar thus has an opportunity to grow. Any antibiotics or other inhibitors of bacterial growth will be diluted and, therefore, have less effect. Inoculation of broth also allows the use of antimicrobial removal devices. However, a fluid medium does not permit confirmation that growth is occurring along an inoculum streak nor allow quantification of the amount of growth. It is a useful qualitative culture medium for enhancing the growth of a small number of bacteria.

Löwenstein-Jensen Medium

Löwenstein-Jensen medium, commonly used to identify *Mycobacterium* species, is used on initial culture, as it will support growth of a very small inoculum. This is an egg/potato-based medium containing the dye malachite green to inhibit slow-growing contaminants. Fluid media viz. Kirschner's compatible with or supplied by the Bactec Automated Detection (Beckton-Dickinson, Franklin Lakes, New Jersey, U.S.A.) system may also be used.

Incubation

Agar plates should be incubated for a minimum of 48 hours at 37°C for bacteria and 32°C for fungi. Cultures for bacteria should be plated out on two agar plates for both aerobic and anaerobic conditions. If a microaerophilic bacterium is suspected, then that growth condition should also be included. Both bacteria and fungi grow better in an atmosphere of 5% carbon dioxide than in air alone. Anaerobic plates should be incubated in an anaerobic cabinet for 7 days or for 14 days if *P. acnes* is suspected. Fluid enrichment media should be incubated for 14 days at 37°C and then subcultured.

Plating and Staining

The streak plate technique is the most widely used method of obtaining isolated colonies from a mix of cultures. This technique dilutes the number of organisms by decreasing the density, allowing individual colonies to be isolated from other colonies. Each colony is considered "pure," because theoretically it began with an individual cell.

1. Inoculation is begun in the first, or primary, quadrant of the agar plate. A light touch is necessary to avoid penetrating or scraping the agar surface. The plate is covered with a lid.
2. The loop is flamed and then allowed to cool.
3. A growth from quadrant one is chosen and is streaked onto quadrant two.

4. The loop is again flamed and allowed to cool. The procedure is repeated for quadrants three and four.

Gram Staining Procedure

Gram staining is a method for classifying bacteria based on their cell wall characteristics. Before a Gram stain procedure is begun, the specimen must be properly spread onto the glass slide. A smear can be prepared from a clinical sample, such as aqueous humor or vitreous, or a solid agar or broth laboratory medium. The specimen should be fixed to the slide before starting the procedure. For specimens other than vitreous, the smear can be fixed with heat in a flame. Vitreous specimens must *not* be heat fixed but allowed to slowly air dry. This can take up to 45 minutes!

The required reagents are:

- Crystal violet (the primary stain)
- Iodine solution (the mordant)
- Decolorizer (ethanol is a good choice)
- Safranin (the counterstain)
- Water (preferably in a squirt bottle)

STEP 1: The entire slide is flooded with crystal violet. The stain is left on the slide for 60 seconds, then gently washed with water for 5 seconds, taking care not to wash away the specimen, which now appears blue-violet.

STEP 2: The slide is flooded with iodine solution, which is allowed to remain for 60 seconds before gentle rinsing with water for 5 seconds. The specimen should still be blue-violet. Step 3 follows immediately.

STEP 3: A decolorizer, ethanol, is added. However, excess application of ethanol will lead to a false Gram-negative result and an insufficient amount will lead to a false Gram-positive result. The ethanol is added by drops until the blue-violet color is no longer emitted from the specimen. This takes up to but not longer than 5 seconds. A 5-second gentle rinse with water follows.

STEP 4: The counterstain, safranin, is added. The slide is flooded with safranin, which is allowed to remain for 60 seconds to permit the bacteria to incorporate the safranin. Gram-positive cells will incorporate little or no counter-stain and will remain blue-violet in appearance. Gram-negative bacteria, however, will take on a pink color and are easily distinguishable from the Gram-positive cells. Excess dye is removed with a 5-second water rinse.

The slide is blotted gently or allowed to air dry before viewing under the microscope with a x10 objective and a x100 oil immersion lens. The smear should never be rubbed. Record bacteria as Gram-positive (black), or Gram-negative (red), cocci or rods, and whether in pairs, tetrads, chains, clusters, palisades, or diffusely spread. Yeasts stain brown/black and fungal mycelium stains weakly red.

Acridine Orange

Stain for 10 min with 0.1% dissolved in buffer at pH 3. Wash with tap water. View under ultraviolet light in a dark room; bacteria, fungi, and protozoa that have taken up the stain fluoresce yellow. This is a useful non-specific stain for bacteria including *Nocardia*, but it does not stain mycobacteria or fungi. It also stains *Acanthamoeba* and other protozoa.

Modified Ziehl-Neelsen (for Nocardia spp. and Mycobacterium spp.):

Add concentrated carbol fuchsin and heat slide with a taper, until the stain begins to boil, for 20 min. Wash with tap water. Add 5% acetic acid for 2 min. Wash with tap water and counterstain with malachite green for 5 min. Wash with tapwater. View with light microscopy, and observe for long, possibly branching, bacilli (*Nocardia*) or slightly curved shorter bacilli (*Mycobacterium*), both of which stain red against a green background. If present, either of these two bacteria will be found in large numbers. Differentiate between them by decolorizing the slide with acetone for 10 seconds, and treat the slide with concentrated acid-95% alcohol for 15 seconds; wash and counterstain with malachite green (full Ziehl-Neelsen stain). *Nocardia* will promptly decolorize, and hence disappear, while *Mycobacterium* will retain the carbol fuchsin stain, being acid-alcohol-fast bacilli; the species that usually infects the cornea is the saprophyte *M. chelonae*.

Corneal Smears

Collect two smears, keeping one for additional stains. Stain with Gram stain. If no bacteria or fungi are seen and keratitis persists, destain with acetone and perform the acridine orange stain. If negative, destain again and perform the modified Ziehl-Neelsen stain as given above. If positive, repeat with the full Ziehl-Neelsen stain to differentiate between *Nocardia* and *Mycobacteria* as described above. Further specialized stains include immunofluorescent stains raised against particular microbes or end-stains such as the methenamine silver stain for fungi and parasites.

MYCOLOGY

Philip Thomas
Joseph Eye Hospital, Tiruchirrapalli, Tamil Nadu, India

The major causes of ocular mycoses are species of *Candida*, especially *C. albicans*, *Aspergillus* (especially *A. fumigatus*), *Fusarium* (especially *F. solani*), *Scedosporium apiospermum* (*Pseudallescheria boydii*), *Blastomyces dermatitidis*, *Coccidioides immitis*, *Cryptococcus neoformans*, *Histoplasma capsulatum*, and *Sporothrix schenckii* (30). Most of these organisms produce endophthalmitis but may also cause focal (localized) chorioretinitis or a granulomatous lesion in the iris or ciliary body (31). Fungal causes of keratitis are considered in Chapters 6 and 9 and Appendix D.

Types of specimens required from the various ocular tissues to make a diagnosis of fungal infection are listed in Table 2 (30). Techniques for detection of fungi by direct microscopy are given in Table 3 and a guide to the recognition of fungi is given in Table 4. Figure 9 is a flow chart for identification of ocular fungi in culture. Use of molecular biology techniques with PCR for detecting fungal infection is described in detail in Chapter 3.

Examples are shown in Figures 10–12 for *Candida albicans*, Figures 13–15 for *Aspergillus* spp., Figures 16 for *Histoplasma capsulatum*, Figures 17–19 for *Scedosporium apiospermum*, Figures 20–23 for *Fusarium* sp, and Figures 24–25 for *Paecilomyces lilacinus*.

(Text continues on page 46.)

Table 2 Diagnosis of Fungal Infection

Clinical entity and samples to be collected	Histopathology[a]				Smears[b]			Culture[c]
	HPE	GMS	PAS	Other	Gram	Giemsa	Other	
Orbital mucormycosis								
Biopsies from necrotic tissue (also nose, paranasal sinuses, oropharynx)	+	+	+	CFW	–	–	–	Usually negative
Aspirated purulent material	–	–	–	–	–	–	GMS, PAS	Diagnosis by H&E
Chronic orbital cellulitis								
Material drained from lesion by sterile syringe and needle	–	–	–	–	+	+	Fungus stains	SBA, SDA
Blepharitis								
Biopsied tissue	+	+	+	CFW	–	–	–	BHIB, Thio
Cotton/calcium alginate swab (moistened with BHIB) used to scrub anterior lid margins and any ulcerated area	–	–	–	–	+	+	Fungus stains	SBA, SDA, BHIB, Thio
Lid biopsy for certain lesions (e.g. *Blastomyces*)	–	+	+	Gram	–	–	–	–
Dacryocanaliculitis								
Sterile spatula used to transfer purulent material expressed by compression of lid and canaliculus	–	–	–	–	+	+	Fungus stains	SBA, SDA, BHIB, Thio
Sterile spud or small chalazion curette used to scrape canaliculus if concretions are present (concretions crushed on slide and stained)-	–	–	–	–	+	–	AFB	SBA, Thio
Dacryocystitis								
Swab used to collect material released by spontaneous external fistulization	–	–	–	–	+	+	Fungus stains	SBA, SDA, BHIB, Thio
Swab used for conjunctival cultures (see below)	–	–	–	–	+	+	Fungus stains	SBA, SDA, BHIB, Thio
Sterile syringe and needle to drain material from lacrimal sac	–	–	–	–	+	+	Fungus stains	SBA, SDA, BHIB, Thio
Dacryoliths removed and bisected	+	+	+	CFW	–	–	–	SBA, SDA, BHIB, Thio

(Continued)

Table 2 Diagnosis of Fungal Infection (*Continued*)

Clinical entity and samples to be collected	Histopathology[a]				Smears[b]			Culture[c]
	HPE	GMS	PAS	Other	Gram	Giemsa	Other	
Dacryoadenitis								
Lacrimal gland surgically removed and bisected:								
(a) One part as such for HPE[d]	−	+	+	+	−	−	Fungus stains	−
(b) Other part aseptically ground and suspended in sterile buffered saline	−	−	−	−	+	+	−	SBA, SDA, BHIB, Thio
Conjunctivitis								
Rhinosporidiosis–surgical excision of lesions for HPE	+	−	−	−	−	−	−	No culture; diagnose by H&E
Other infections:								
(a) Calcium alginate/cotton-tipped swabs (moistened in BHIB if lesions are dry) used to scrub inferior tarsal conjunctiva and fornix–*do not use anaesthetic*	−	−	−	−	−	−	−	SBA, SDA, BHIB, Thio
(b) Spatula or sterile swab gently scraped across inferior tarsal conjunctiva; scrapings spread thinly on slides for staining–use *local anaesthetic for this*	−	−	−	−	+	+	−	−
(c) Conjunctival biopsy	+	+	+	+	−	−	−	SBA, SDA, BHIB, Thio
Scleritis								
Material obtained from conjunctiva (see above) and cornea (see below)	−	−	−	−	+	+	Fungus stains	SBA, SDA, BHIB, Thio
Scrapings and swabs from scleral abscess	−	−	−	−	+	+	Fungus stains	SBA, SDA, BHIB, Thio
Scleral biopsy or corneal biopsy (in nodular, diffuse or necrotising scleritis)	+	+	+	Gram, AFB, CFW	−	−	−	SBA, SDA, BHIB, Thio

Keratitis

							Culture media
Use sterile swabs to collect ipsilateral and contralateral lid and conjunctiva material	−	−	−	+	+	Fungus stains	SBA, SDA, BHIB, Thio
Use Kimura spatula, Bard-Parker knife, sterile razor or surgical blade or hypodermic injection needle to collect corneal scrapes	−	−	−	+	+	Fungus stains	SBA, SDA, BHB, Thio Snyder, CTA, SDA
Epithelial biopsy, superficial keratectomy, penetrating keratoplasty to obtain corneal tissue for HPE and culture	+	+	CFW	−	−	−	SBA, SDA, BHB, Thio Snyder, CTA, SDA

Endophthalmitis

							Culture media
Conjunctival swabs only if leaking bleb or wound	−	−	−	+	+	Fungus stains	SBA, SDA, BHIB, Thio
Sterile syringe used to collect aqueous and vitreous aspirate	−	−	−	+	+	Fungus stains	CTA
Vitreous wash material passed through membrane filter; filter removed and cut into pieces for culture	−	−	−	−	−	−	SBA, SDA, CTA
Vitreous biopsy	+	+	CFW, KOH	−	−	−	SBA, SDA, BHIB, Thio

[a] H&E, Haematoxylin-eosin; PAS, periodic acid-Schiff; GMS, Gomori methenamine silver; CFW, calcofluor white; KOH, potassium hydroxide.

[b] Fungus stains: KOH, potassium hydroxide, ink-KOH, LCB, lactophenol cotton blue, CFW, FCL, fluorescein conjugated lectins; AFB, Kinyoun acid-fast stain; for *Nocardia* PAS, GMS.

[c] SBA, 5% sheep blood agar incubated at 25°C and 37°C (also anaerobically for Actinomyces); SDA, Sabouraud glucose-neopeptone agar (Emmons' modification) incubated at 25°C (with chloramphenicol/gentamicin); BHIB, brain heart infusion broth incubated at 25°C (with chloramphenicol/gentamicin); Thio, thioglycolate broth (may be supplemented with vitamin K and Haemin); CTA, cystine tryptone agar; Snyder, Snyder agar pH 4.8.

Abbreviation: HPE, histopathological examination. *Key*: + = use this test; − = not appropriate for test.

Table 3   Techniques for Detection of Fungi by Direct Microscopy

Method	Advantages	Disadvantages
A. Potassium hydroxide (KOH) mount i. Sensitivity: 33–94% in keratitis ii. Used for endophthalmitis and other ocular infections	a. Single-step, inexpensive simple method b. 86% positivity in a series of *Fusarium* keratitis. c. Oil immersion magnification not required	1. Artifacts are common 2. Corneal cells do not swell to produce transparent preparations 3. *Candida* and other yeasts may escape detection
B. Ink-KOH mount i. Used in keratitis ii. Useful for corneal scrapings and biopsy specimens iii. Ink has fine particles which attach to fungal cell walls	a. Rapid, simple, inexpensive single-step method b. *Fusarium, Aspergillus* and *Candida* can be seen and identified c. Comparable to calcofluor in sensitivity d. Better sensitivity than KOH	1. Optimal viewing time is at 12-18 h 2. Wet mount cannot be kept for long 3. Stain has a short shelf-life since ink tends to precipitate out
C. KOH-dimethyl sulphoxide For digestion with PAS[a] or acridine orange counterstain	a. Important ocular fungi can be seen b. Combines digestion of tissue with good counterstaining	1. UV microscope needed if acridine orange is used 2. Both counterstains require careful reagent preparation and expertise in staining
D. Gram staining i. Sensitivity 54–73% in cases of keratitis ii. Used for detection of fungi in all types of ocular specimens	a. Stains *Candida* blastoconidia and pseudohyphae b. Stains hyphae of important ocular fungi c. Bacteria stained and differentiated d. Preparations can be re-stained with GMS[c] e. Takes 5 min to perform	1. May stain fungal hyphal cytoplasm irregularly or not at all in some cases 2. Stains fibrous protein, causing opacity 3. False-positive artifacts common 4. Crystal violet precipitation may obscure detail and cause confusion
E. Giemsa staining i. Sensitivity of 54–75% in cases of keratitis ii. Used for detection of fungi in all types of ocular specimens	a. Permits differential staining of tissue and cellular elements b. Stains yeast cells and pseudohyphae and hyphae of ocular fungi c. Stains chlamydiae, viral inclusions and protozoa	1. Has some disadvantages similar to those of Gram stain 2. Tissue cells also stain, forming opaque area where smear is thick 3. False-positive artifacts common 4. Buffer and working solutions need careful preparation 5. Staining time of 60 min

(*Continued*)

Method	Advantages	Disadvantages
F. Lactophenol cotton blue (LCB) staining i. Sensitivity: 78% in a series of keratitis ii. Can also be used for fluid ocular samples (AC tap, vitreous aspirates)	a. Stain easily available with shelf-life of 1 year b. Rapid, simple, inexpensive one-step method c. Important ocular fungi can be seen and identified d. Wet mount is a semi-permanent preparation which can be kept for years e. *Acanthamoeba* can be detected	i. No digestion of tissue, hence optimal results may not be obtained with this preparation ii. Unusual fungi may escape detection iii. Contrast between fungi and background material may sometimes be insufficient iv. Impurities in cotton blue stain may cause problems in interpretation
G. Methenamine silver staining (modified) i. Sensitivity of 89% in a series of keratitis ii. Can be used for other ocular samples in addition to corneal scrapes	a. Stains fungal cell wall clearly; no interference from back-ground b. Positives more frequent and reliable than in methods A–F c. Negatives more reliable	1. Excessive deposit of silver may obscure cell wall and septa 2. Stains cellular debris and melanin 3. Gelatin-coated slides needed 4. Controls needed; reagents and procedure need standardization 5. Takes 60 min to perform
H. Calcofluor white (CFW) i. Fluorescent dye with high affinity for polysaccharides, e.g. chitin of fungal cell walls ii. Used to detect fungi in corneal scrapings and fluid ocular samples (aqueous, vitreous) iii. Sensitivity of 94-95% reported in Indian studies	a. Excellent sensitivity and good specificity b. Detects yeast blastoconidia and also fungal hyphae of important ocular fungi (*Aspergillus, Fusarium*) c. Material from KOH mounts can be used for subsequent CFW staining	1. Not all fungi are adequately stained 2. UV microscope needed 3. Corneal collagen stained; may interfere with detection of fungi 4. Reagents and procedure require standardization and expertise
I. Fluorescein-conjugated lectins (FCLs) i. Lectins (ubiquitous proteins) bind specifically to carbohydrates present in fungal cell wall ii. Used for staining fungi in corneal scrapings and tissue sections; provides	a. Selective staining with a panel of FCL (e.g. concanavalin, wheat germ agglutinin, lens culinaris agglutinin) allows identification of fungal species b. Important ocular fungi (*Fusarium, Aspergillus,*	1. More clinical studies needed to establish efficacy 2. UV microscope needed 3. Reagents and procedures require standardization and expertise

(Continued)

Table 3 Techniques for Detection of Fungi by Direct Microscopy (*Continued*)

Method	Advantages	Disadvantages
consistently bright staining of fungal structures	Candida) can be identified c. Found useful in clinical cases of mycotic keratitis	
J. Immunofluorescence (IF) Direct IF, indirect IF and anticomplement staining methods used	a. Permits selective staining of fungi in ocular specimens b. Can be used for different ocular specimens	1. Expensive 2. Antibodies of good quality necessary 3. Reagents and procedures require standardization and expertise 4. More clinical studies needed

[a] PAS, periodic-acid Schiff.
[b] UV, ultraviolet.
[c] GMS, Gomori methenamine silver staining.

Samples for fungus isolation should generally be plated out onto Sabouraud's medium or brain heart infusion agar. Thioglycolate or brain heart infusion fluid medium should also be inoculated. Cornmeal agar is not generally recommended as a primary isolation medium for filamentous fungi but is used to induce sporulation of a filamentous fungus to facilitate identification, once the fungus has been isolated in the primary isolation medium. Samples should be incubated at 32°C for four weeks in 5% carbon dioxide. The plates should be sealed in a plastic box for the agar to remain moist.

Contents of Media

Sabouraud's Agar (Emmons' Modification)
Dextrose 20 g; mycological peptone 10 g; agar 20 g; deionized water 1 L; final pH 6.8 to 7.0; chloramphenicol added to a final concentration of 40 mg/L or gentamicin added to a final concentration of 50 mg/L to prevent bacterial contamination.

Cornmeal Agar
Infusion from cornmeal (ground yellow corn) 50 g; agar: 15 g; deionized water 1 L; final pH 6.0.

Thioglycolate Broth
Yeast extract 5 g; casitone 15 g; sodium chloride 2.5 g; *l*-cystine 0.25 g; thioglycolic acid 0.3 mL; agar 0.75 g; methylene blue 0.002 g; deionized water 1 L; final pH 7.2.

Table 4 Guide to Recognition of Fungi and Actinomycetes in Ocular Specimens (Corneal Scrapings and Fluid Specimens)

Fungal genus or species	Direct microscopic appearance
Aspergillus[a] *A. fumigatus* *A. flavus*, etc.	Septate, dichotomously branched, hyaline hyphae of uniform width (3–6 μm) often in clusters, occasionally conidial heads
Fusarium[a] *F. solani*	Septate, branched hyaline hyphae, 2–4 μm width, occasional irregular and bulged
Dematiaceous filamentous fungi[a] *Curvularia* *Exserohilum* *Bipolaris* *Altemaha*	Septate, branched, regular, brownish hyphae, 2–6 μm width, occasionally bulged and occasionally with conidia, often constricted at septations
Other hyaline filamentous fungi[a] *Acremonium* *Penicillium* *Paedlomyces* *Pseudallescheha* *VertJcillium*	Septate, regular, branched hyaline hyphae, 2–4 μm width with occasional intercalary chlamydospores (hyphae of *Pseudallescheha* are very thin and may be light brown in colour)
Zygomycetes[b] *Mucor* *Rhizopus* *Absidia*	Broad, thin-walled, infrequently septate or aseptate hyaline hyphae, 6–25 μm wide with non-parallel contours and random branches; often appear bizarre and vacuolated
Lasiodiplodia theobromoe[a]	Septate, branched (sometimes dichotomous) brownish hyphae, 3–5 μm width, highly bulged; cyst-like structures often present
Candida[c] *C albicans and others*	Oval, budding, yeast-like cells (blastoconidia) 2–6 μm diameter and septate hyphae or pseudohyphae
Cryptococcus	Pleomorphic yeast-like cell (2–20 μm diameter); mucin-positive capsules and single or multiple narrow-based buds
Rhinosporidiosis[d]	Large sporangia (100–350 μm diameter) with thin walls that enclose numerous endospores (6–8 μm diameter); connective tissue shows many blood vessels
Dimorphic fungi (histomorphic features)[d] *Histoplasmosis* (*H. capsulatum*) *Blastomycosis* *Paracoccidioidomycosis* *Sporotrichosis*	Spherical to oval, budding, yeast-like cells (2–4 μm diameter), found within mononuclear phagocytes Spherical to oval multinucleate, large (8–15 μm diameter) yeast-like cells, with thick walls and single broad-based bud
Actinomycetes[c] *Actinomyces, Nocardia*	Large (10–60 μm) spherical yeast-like cells with multiple buds attached by narrow necks ('steering wheel' forms) Pleomorphic, medium-sized (2–10 μm), spherical to oval, 'cigar shaped' yeast forms, that produce single and, rarely, multiple buds Gram-positive, thin branching filaments 1 μm diameter (some *Nocardia* species are partially acid-fast; stain with modified Ziehl-Neelsen)

[a] Best seen in wet preparations (potassium hydroxide, lactophenol cotton blue, etc.).
[b] Best seen in haematoxylin and eosin (H&E) Gomori methenamine silver or periodic acid-Schiff-stained preparations.
[c] Best seen in Gram-stained smears.
[d] Seen in tissue sections (H&E).

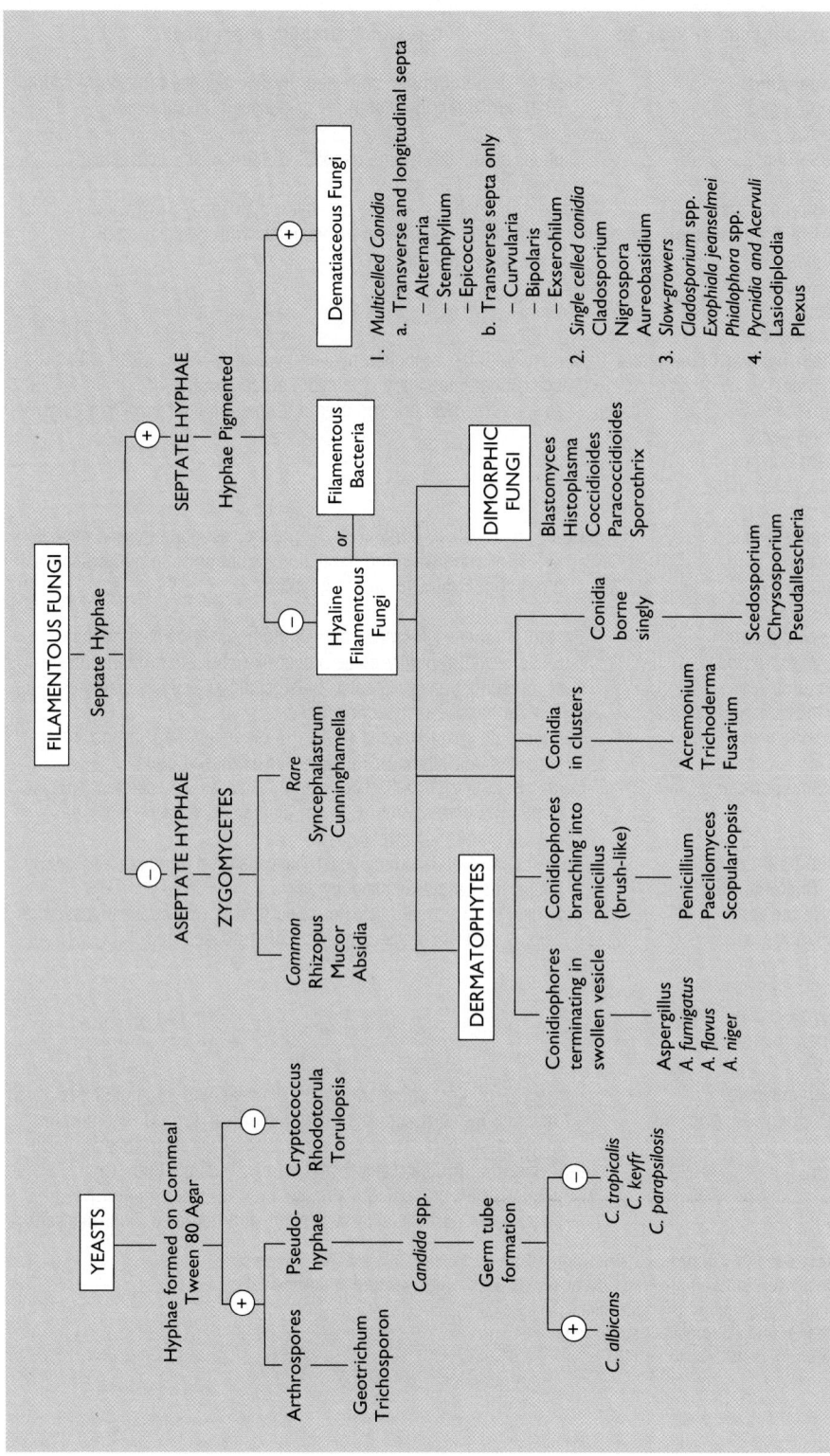

Figure 9 Flow chart for identification of ocular fungi in culture.

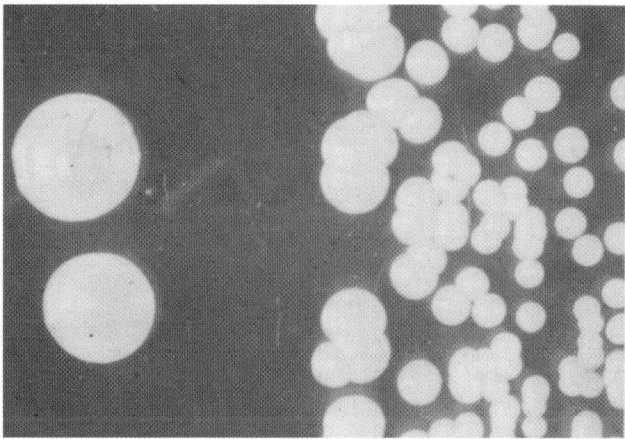

Figure 10   White colonies of *Candida albicans* on agar.

Brain Heart Infusion

Infusion from calf brains 200 g; infusion from beef heart 200 g; proteose peptone 10 g; dextrose 2 g; sodium chloride 5 g; disodium phosphate 2.5 g.

Antifungal Susceptibility Testing

This is currently performed for yeasts and mycelial fungi by broth or agar dilution methods, where minimum inhibitory concentrations (MICs) are determined in drug-impregnated broth or agar media. The measurement of zones of inhibition of growth against impregnated discs (disc diffusion method) on agar is now widely used for sensitivity testing of yeasts.

The selection of a medium depends on the drug being tested.

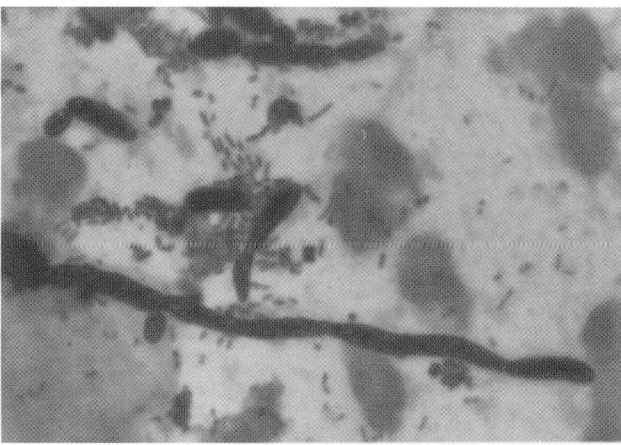

Figure 11   Pseudomycelium of *Candida albicans* (Gram stain).

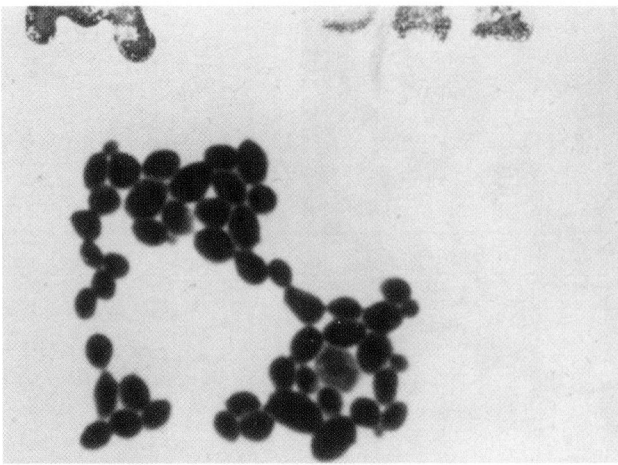

Figure 12 Typical budding yeast cells of *Candida albicans* (Gram stain).

Broth media used: unbuffered yeast-nitrogen base (YNB) for flucytosine (5FC), Antibiotic Medium 3 FDA for polyenes, and casein-yeast extract-glucose (CYG) broth or YNB for imidazoles.

Agar media used: unbuffered yeast morphology agar (YMA) for 5FC, Antibiotic Medium 12, buffered YMA or Kimmig agar for polyenes, and CYG or Kimmig agar for azoles.

Buffered YNB (for broth tests) and buffered YMA (for agar tests) are useful for all three groups of compounds. Inocula are prepared from 24 to 48 hour-old cultures in YNB and adjusted to contain 10^6CFU/mL. Conidial suspensions or culture blocks containing hyphae and spores are used as the inocula for mycelial fungi.

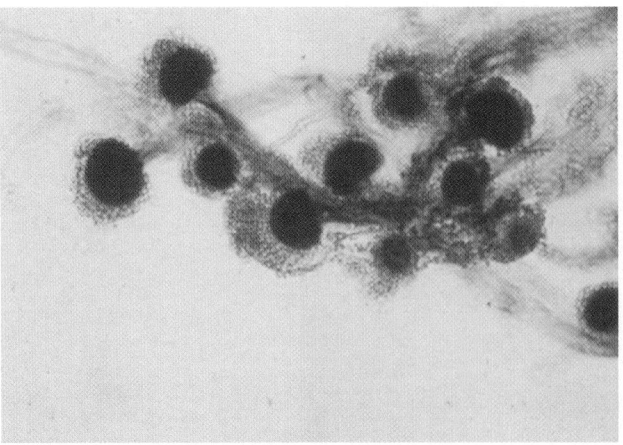

Figure 13 *Aspergillus* conidiosphore seen in cultures.

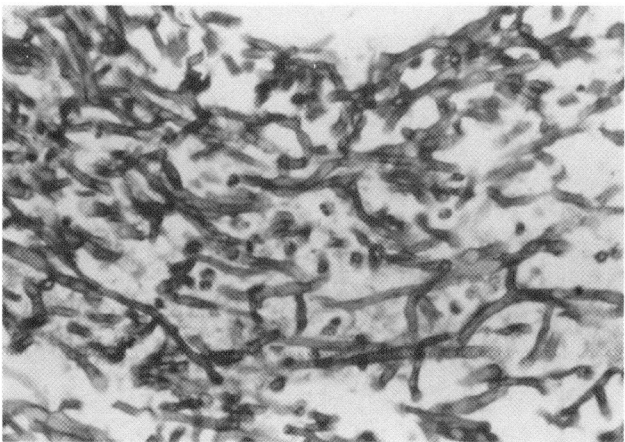

Figure 14   Periodic acid-Schiff stain of branching mycelium *Aspergillus* species.

Drugs commonly tested include 5FC, ketoconazole, miconazole, fluconazole, itraconazole, and amphotericin B; expected MICs for *Candida albicans* to be susceptible are 0.5, 0.1, 0.1, 0.3, 0.1, and 0.5 mg/L respectively. The MIC for each imidazole can vary widely, especially for *Candida* spp. and other yeasts, so that sensitivity tests are useful.

The reference microdilution method of the Clinical Laboratory Standards Institute [formerly the National Committee for Clinical Laboratory Standards (NCCLS)] of the United States, namely M-27 A (NCCLS, 2002), provides a standardized guideline for in vitro testing of yeast isolates, which has shown 85% to 90% intra- and inter-laboratory reproducibility. Fluconazole resistance, due to active efflux from the cell common for all imidazoles, is well recognized in AIDS patients. There is also a broth microdilution protocol (NCCLS M38-P, 1998) for the

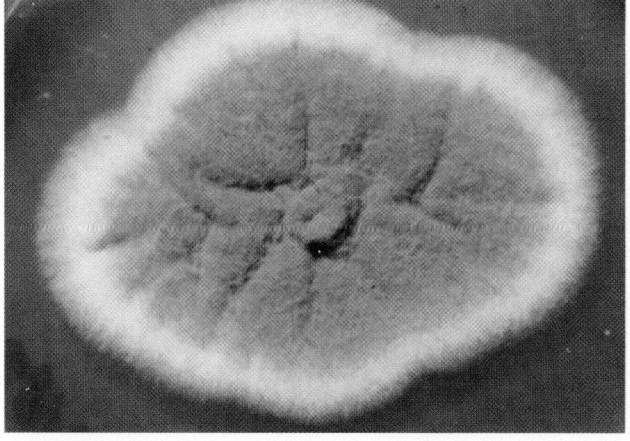

Figure 15   *Aspergillus fumigatus* (green pigment).

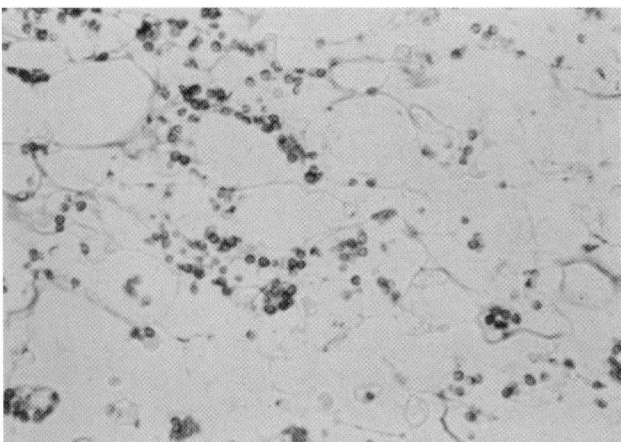

Figure 16 *Histoplasma capsulatum* in tissue.

susceptibility testing of conidium-forming fungi, including *A. fumigatus*, using RPMI 1640 as the test medium to determine the MIC.

Synergy testing can be performed, but satisfactory in vitro results may not give clinical success. In particular, amphotericin + 5 FC has been successful for cryptococcal infection. Amphotericin + fluconazole has been shown additive for candidiasis in murine studies, but amphotericin + ketoconazole or itraconazole has been shown to be antagonistic in animal models of aspergillosis.

Satisfactory new methods for testing MICs include application of E test drug-impregnated strip technology for *Candida* spp. versus posaconazole (32) and *Aspergillus* spp. and of the Bio-Cell Tracer method for *Aspergillus niger*. In in vitro susceptibility testing of *Aspergillus* spp., the MICs obtained by the E test have been found to correlate with those obtained by the NCCLS broth microdilution method (M38-A), the degree of agreement being 92.2% when E test MICs are read at

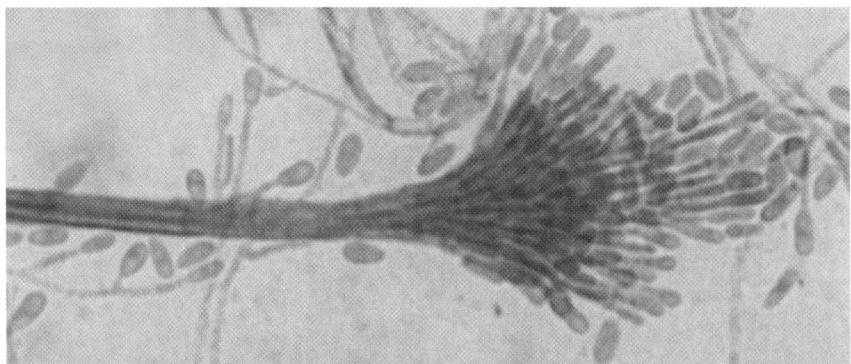

Figure 17 Microscopy of *Scedosporium apiospermum* (teleomorph *Pseudoallescheria boydii*). Conidia are abundant, yellow to pale-brown in color, oval in shape (6–12 × 3.5–6 μm) with a scar at the base, and are formed from single or branched, long, slender annellides that are sometimes aggregated into bundles of tree-like synnemata (Graphium state).

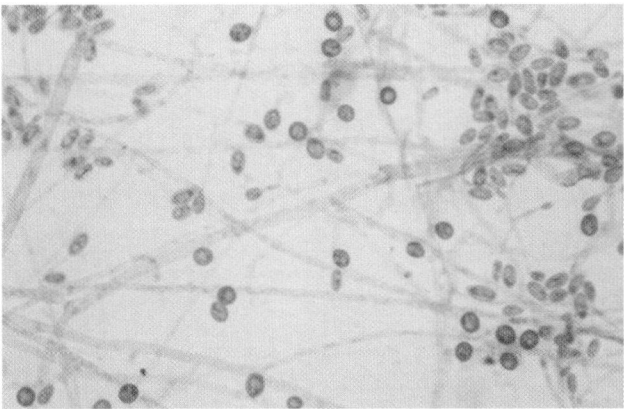

Figure 18 In slide preparations the annellidic tip usually has only a single conidium remaining.

24 hours and 83.1% agreement when both methods are read at 48 hours (33). When posaconazole is used as the test antifungal agent, overall agreement between E-test and broth microdilution MICs are 84% for *Aspergillus* spp. and 100% for the less common opportunistic molds, except *Penicillium* spp. (67%) (34).

Determination of in vitro Susceptibility of Ocular Filamentous Fungal Isolates to Antifungal Agents

This can be performed by an agar dilution method or a broth dilution method (NCCLS M38 A).

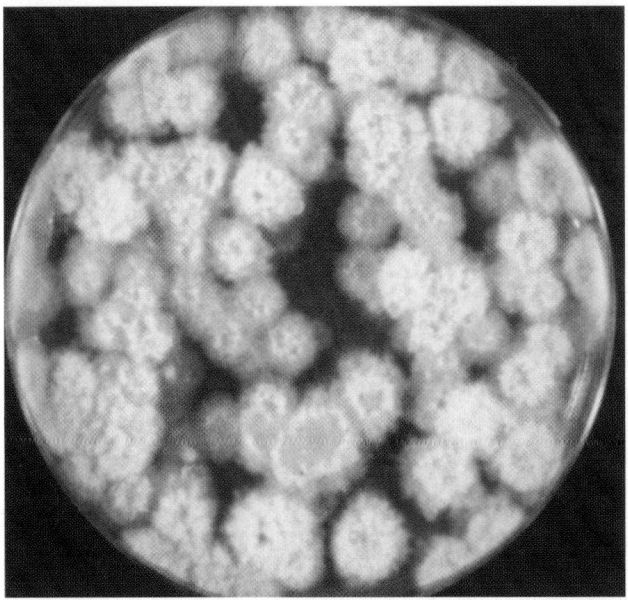

Figure 19 *Scedosporium apiospermum* plate culture. Colonies are high, flat to dome-shaped and floccose in texture, initially white and later pale smoky-brown in color, and grow rapidly (40 mm/ 7 days).

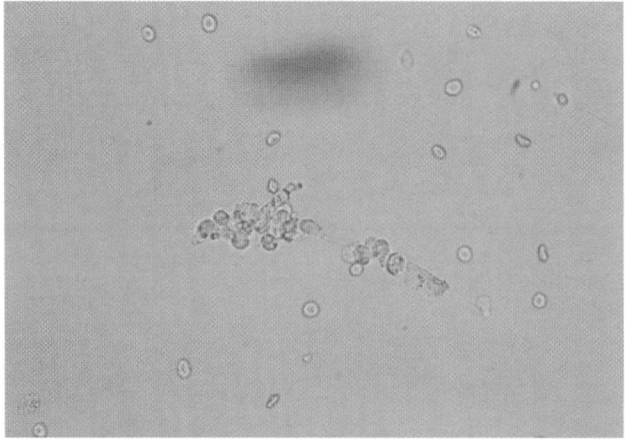

Figure 20 Wet preparation of aqueous showing branching hyphae and conidia of *Fusarium oxysporum*. *Source*: Courtesy of C. Ferrer.

Preparation of fungal inoculum: Fungi are grown for 24 to 48 hours (or 7 days for *Curvularia* spp.) on slants of neutral Sabouraud's dextrose agar (NSDA). After adding 1 drop of Tween 20, the colonies are brushed with a sterile cotton swab and the mixture of conidia and hyphal fragments are suspended in 1 mL of sterile water. Heavy particles are allowed to settle after vortexing, and the homogenous suspension adjusted to achieve the required concentration. The Carshalton strain of *Candida kefyr* may be used as a positive control for intertest reproducibility.

Agar Dilution Method

Test medium: Kimmig agar containing peptone from casein 4.3 g, peptone from meat 9.3 g, dextrose 10 g, sodium chloride 11.4 g, agar 15 g, deionized water 1 L, pH 6.5 is

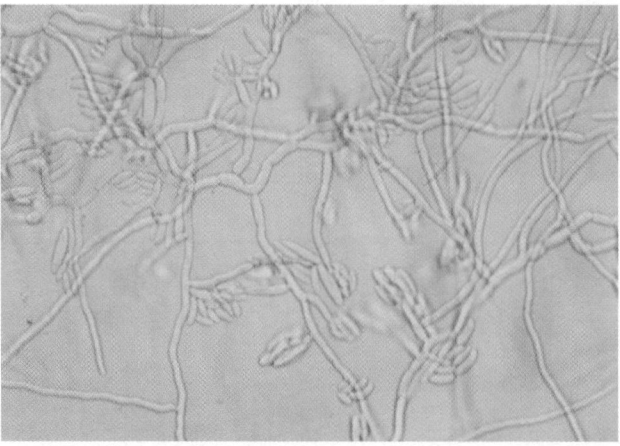

Figure 21 Branching septate hyphae of *Fusarium oxysporum*. *Source*: Courtesy of C. Ferrer.

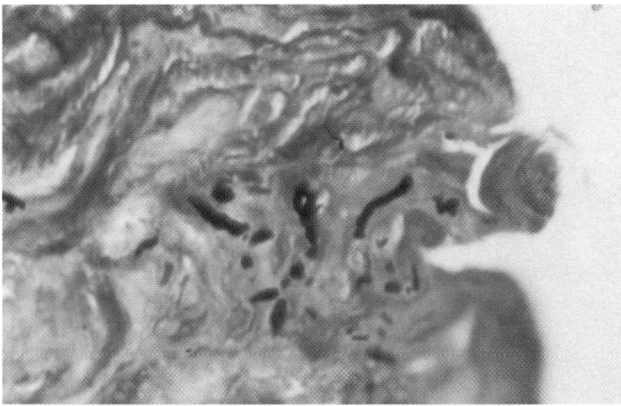

Figure 22 Gomori silver stain identifies hyphae of *Fusarium* sp. in tissue.

used as the test medium for antifungal susceptibility testing by the agar dilution method, since it is suitable for the testing of all antifungal drugs except 5FC.

Preparation of drugs: Ketoconazole is dissolved in 0.2 N HCl; itraconazole is dissolved in 0.2 N HCl in ethanol (0.2 mL of 1 N HCl in 0.8 mL of absolute ethanol), while fluconazole and amphotericin B can be reconstituted in sterile distilled water. From the stock solutions of 2560 µg of each compound per mL of solvent, serial twofold dilutions, from 1280 to 0.63 µg/mL for the azoles and from 320 to 0.2 µg/mL for amphotericin B, are prepared. The stock solutions are mixed with the melted and cooled agar in a proportion of 1:9 and poured into round petri dishes of 100 mm diameter or in square Petri dishes specially designed for this purpose. Thus, the final dilutions of the drugs in the medium will be from 128 µg/mL to 0.06 µg/mL for the azoles and 32 to 0.02 µg/mL for amphotericin B.

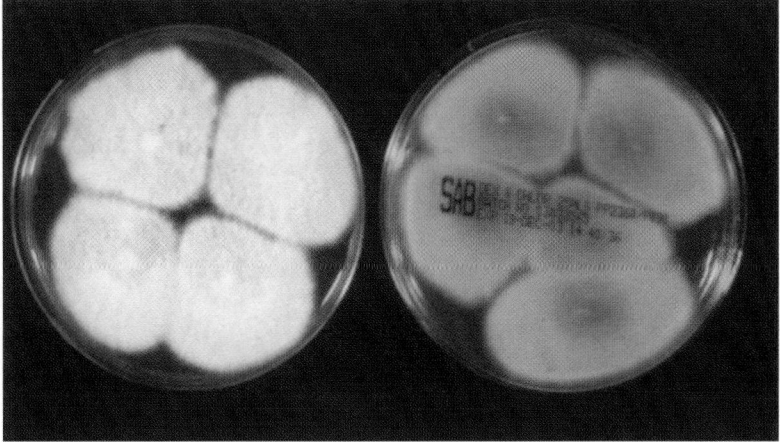

Figure 23 *Fusarium solani* plate culture. Colonies are flat, grayish-white, cream, buff or pinkish-purple in color (*reverse is pale cream*), floccose in texture, and attain a diameter of 30 mm in 1 week.

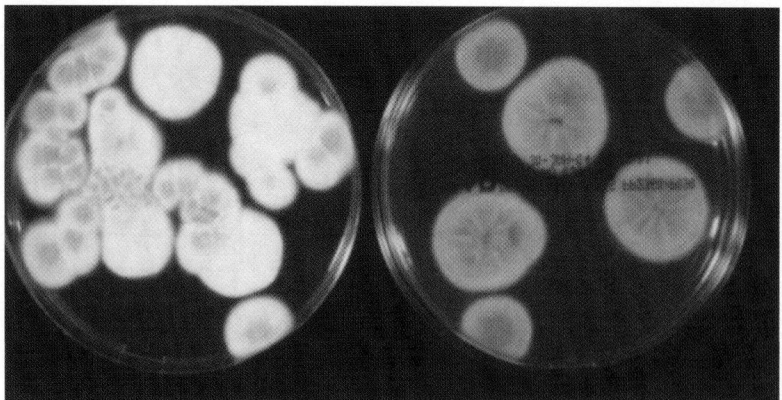

Figure 24   *Paecilomyces lilacinus* plate culture. Colonies are flat to domed, vinaceous to violet-white in color (*reverse is pale or deep purple*), densely floccose in texture, and attain a diameter of 30 mm in 1 week.

Test Procedure: Suspensions of the conidia are prepared in sterile isotonic saline and adjusted to a count of 0.4 to 0.5×10^4 colony forming units (CFU) per mL and the number of conidia determined by counting in a counting chamber. The reliability of this method for preparing appropriate concentrations of the conidia in the suspensions can be checked by inoculating known volumes of the diluted conidial suspensions onto plates of NSDA; after 48 hours, the colonies are counted. For the test, 5 µL of each conidial suspension is inoculated by a micropipette onto the agar; each drug dilution is tested in triplicate. Two drug-free control plates of Kimmig agar are also inoculated to serve as controls of the ability of the medium to support fungus growth. A solvent control plate containing a high concentration of the solvent without any drug (to rule out antifungal activity of the solvent) is also

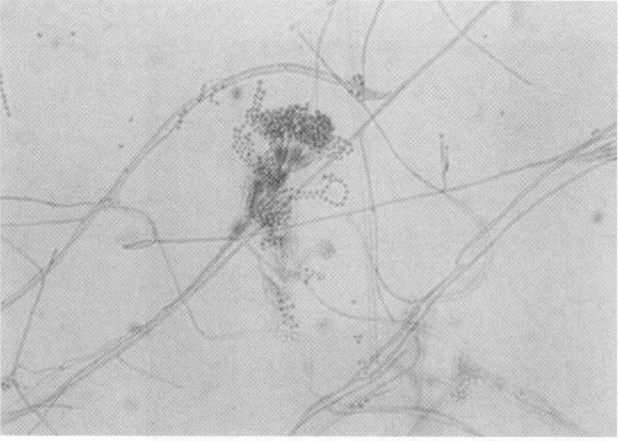

Figure 25   *Paecilomyces lilacinus* microscopy. Conidia (2.5–3 × 2–2.2 µm) are ellipsoidal to fusiform in shape, smooth-walled to slightly roughened in texture, hyaline in color (*purple in a mass*), arranged in divergent chains; they are borne on irregularly branched heads terminating in long phialides which consist of a swollen basal part, tapering into a thin neck.

inoculated. The Carshalton strain of *Candida kefyr* is also inoculated (concentration of 1×10^5 to 1×10^6 cells/mL) to serve as a control of intertest reproducibility. The plates are kept at 30°C and read at 24 and 48 hours following inoculation. The MIC is defined as the lowest concentration of the drug preventing growth of macroscopically visible colonies and is determined at the time when growth becomes apparent on the drug-free control plates. Hazy responses and pinpoint colonies are regarded as negative.

Broth Dilution Method (NCCLS M38-A)

Test medium: RPMI 1640 (R-6504, Sigma (Cinciwatti), U.S.A.) with L-glutamine and without sodium bicarbonate and buffered to pH 7.0 at 25°C with morpholine propane sulfonic acid (MOPS), is used as the test medium for antifungal susceptibility testing by broth dilution method (NCCLS, M38-A standard). RPMI 1640 is used as the test medium since it is a fully defined, commercially available synthetic medium, ensuring that there are only minimal variations of the medium from lot to lot and that quality control can be maintained. Moreover, RPMI 1640 is often used as a mammalian cell culture medium and thus it provides an in vitro milieu that is close to what is found in human body fluid(s) where the fungal pathogen produces infection.

Two vials of the pre-weighed medium (10.4 g each) are dissolved in 1600 mL distilled water in a 4 L conical flask. Then 69.06 g of MOPS buffer are added along with 40 g of glucose to the medium and stirred to dissolvecompletely. The pH of the medium is adjusted to 7.0 at 25°C with 1N NaOH, and then water is added to give a final volume of 2 L. This medium is filter sterilized and dispensed in sterile containers and stored at 4°C.

Preparation of drug solutions: Amphotericin B and ketoconazole are dissolved in dimethyl sulfoxide while itraconazole is dissolved in polyethylene glycol. From the stock solutions, serial twofold dilutions of each drug, namely 32 to 0.06 mg/L for amphotericin B and 128 to 0.25 mg/L for itraconazole and ketoconazole, are prepared in RPMI 1640 medium.

Test procedure: The broth dilution test is performed in sterile "U" bottom 96 well microtiter plates. Concentrationsof conidial suspensions of *Fusarium* spp., *A. flavus*, *A. fumigatus*, and *Curvularia* spp. are standardized spectrophotometrically (*Aspergillus* spp. 80–82% transmission; *Fusarium* spp. and *Curvularia* spp. 68–78% transmission). Appropriate positive (the Carshalton strain of *Candida kefyr*) and negative controls (drug free and solvent controls) are also included. The test plates are prepared in triplicate and are incubated at 35%C without agitation; readings are taken after 48 hours. The endpoint is determined as the lowest concentration of the drug that causes complete inhibition of fungal growth.

REFERENCES

1. Lundstrom M, Wejde G, Stenevi U, et al. Endophthalmitis after cataract surgery. A Nationwide Prospective study evaluating incidence in relation to incision type and location. Ophthalmol 2007; 114:866–70.
2. Seal DV, Bron AJ, Hay J. Ocular Infection: Investigation and Treatment in Practice. Martin Dunitz: London: 1998.
3. Yorston D, Foster A. Cutaneous anthrax leading to corneal scarring from cicatricial ectropion. Br J Ophthalmol 1989; 73:809–11.

4. Soysal HG, Kiratli H, Recep OF. Anthrax as the cause of preseptal cellulitis and cicatricial ectropion. Acta Ophthalmol Scand 2001; 79:208–9.

5. Romsaitong DP, Grasso CM. *Clostridium perfringens* endophthalmitis following cataract surgery. Arch Ophthalmol 1999; 117:970–1.

6. Iyer MN, Kranias G, Daun ME. Post-traumatic endophthalmitis involving *Clostridium tetani* and Bacillus sp. Am J Ophthalmol 2001; 132:116–7.

7. Nangia V, Hutchinson C. Metastatic endophthalmitis caused by *Clostridium perfringens*. Br J Ophthalmol 1992; 76:252–3.

8. Insler MS, Karcioglu ZA, Naugle T Jr. *Clostridium septicum* panophthalmitis with systemic complications. Br J Ophthalmol 1985; 69:774–7.

9. Eliott D, O'Brien TP, Green WR, et al. Elevated intraocular pressure, pigment dispersion and dark hypopyon in endogenous endophthalmitis from *Listeria monocytogenes*. Surv Ophthalmol 1992; 37:117–24.

10. Mistlberger A, Ruckhofer J, Raithel E, et al. Anterior chamber contamination during cataract surgery with intraocular lens implantation. J Cataract Refract Surg 1997; 23: 1064–9.

11. Leong JK, Shah R, McCluskey PJ, Benn RA, Taylor RF. Bacterial contamination of the anterior chamber during phacoemulsification cataract surgery. J Cataract Refract Surg 2002; 28:826–33.

12. Mames RN, Friedman SM, Stinson WG, Margo CE. Positive vitreous cultures from eyes without signs of infectious endophthalmitis. Ophthalmic Surg Lasers 1997; 28: 365–9.

13. Dickens A, Greven CM. Posttraumatic endophthalmitis caused by *Lactobacillus* (letter). Arch Ophthalmol 1993; 111:1169.

14. Hull DS, Patipa M, Cox F. Metastatic endophthalmitis:A complication of meningococcal meningitis. Ann Ophthalmol 1982; 14:29–30.

15. Saperstein DA, Bennett MD, Steinberg JP, et al. Exogenous *Neisseria meningitidis* endophthalmitis. Am J Ophthalmol 1997; 123:35–6.

16. Barnard T, Das A, Hickey S. Bilateral endophthalmitisas an initial presentation of meningococcal meningitis. Arch Ophthalmol 1997; 115:1472–3.

17. Abousaesha F, Dogar GF, Young BJ, OHare J. Endophthalmitis as a presentation of meningococcal septicaemia. Ir J Med Sci 1993; 162:495–6.

18. Margo CE, Mames RN, Guy JR. Endogenous *Klebsiella* endophthalmitis. Report of two cases and review of the literature. Ophthalmology 1994; 101:1298–1301.

19. Joondeph HC, Nothnagel AF. *Serratia rubidae* endophthalmitis following penetrating ocular injury. Ann Ophthalmol 1983; 15:1138–40.

20. Lam DS, Kwok AK, Chew S. Post-keratoplasty endophthalmitis caused by *Proteus mirabilis*. Eye 1998; 12 (Pt I):139–40.

21. Cunningham ET Jr, Whitcher JP, Kim RY. *Morganella morganii* postoperative endophthalmitis. Br J Ophthalmol 1997; 81:170–1.

22. Recchia FM, Baumal CR, Sivalingam A, et al. Endophthalmitis after pediatric strabismus surgery. Arch Ophthalmol 2000; 118:939–44.

23. Pach JM. Traumatic *Haemophilus influenzae* endophthalmitis. Am J Ophthalmol 1988; 106:497–8.

24. Al Faran MF. *Brucella melitensis* endogenous endophthalmitis. Ophthalmologica 1990; 201:19–22.

25. Penland RL, Boniuk M, Wilhelmus KR. Vibrio ocular infections on the U.S. Gulf Coast. Cornea 2000; 19:26–9.

26. Mindel JS, Rosenberg EW. Is *Helicobacter pylori* of interest to ophthalmologists? Ophthalmology 1997; 104:1729–30.

27. Kountouras J, Mylopoulos N, Boura P, et al. Relationship between *Helicobacter pylori* infection and glaucoma. Ophthalmology 2001; 108:599–604.

28. Gopal L, Ramaswamy AA, Madhavan HN, et al. Postoperative endophthalmitis caused by sequestered *Acinetobacter calcoaceticus*. Am J Ophthalmol 2000; 129:388–90.

29. Roussel TJ, Olson ER, Rice T, et al. Chronic postoperative endophthalmitis associated with *Actinomyces* species. Arch Ophthalmol 1991; 109:60–2.
30. Thomas PA. Current perspectives on ophthalmic mycoses. Clin Microb Rev 2003; 16: 730–97.
31. Weinberg,R.S. 1999. Uveitis. Update on therapy. Ophthalmology Clinics of North America 12 (1), 71–81.
32. Diekema DJ, Messer SA, Hollis RJ, et al. Evaluation of E test and Disk Diffusion methods compared with Broth Microdilution Antifungal Susceptibility testing of Posaconazole against clinical isolates of *Candida sp*. J Clin Microbiol 2007; 45:1974–7.
33. Serrano MC, Morilla D, Valverde A, et al. Comparison of E test with modified broth microdilution method for testing susceptibility of *Aspergillus sp*. to Voriconazole. J Clin Microbiol 2003; 41:5270–2.
34. Pfaller MA, Messer SA, Boyken L et al. In vitro susceptibility testing of filamentous fungi: comparison of E test and reference M38-A microdilution methods for determining Posaconazole MICs. Diag Microbiol Infect Dis 2003; 45:241–4.

THREE

Molecular Biology

MOLECULAR DIAGNOSIS OF OCULAR INFECTIONS

Consuelo Ferrer
Molecular Biology Laboratory, Vissum Instituto Oftalmológico de Alicante,
Alicante, Spain

Clinical diagnosis of ocular infections can be confirmed by numerous techniques based on the microbiological analysis of ocular samples (aqueous, vitreous, and corneal scrapings). Diagnostic techniques include standard microbiological tests (culture and stains) and culture-independent diagnosis tests such as those based on molecular biology methods. The capacity for detection and identification of genomic material in any type of sample has constituted an enormous step in the field of medicine, allowing diagnosis of many genetic or infectious diseases based on the DNA sequence. These molecular techniques have their advantages and disadvantages, offering speed and sensitivity in the detection and identification of pathogens, but at present they are still not standardized and their cost is high.

The two main determining factors in the microbiological diagnosis of ocular infections are: (*i*) the scarce quantity of the samples obtained, and (*ii*) the need for a quick diagnosis to avoid the loss of an eye or of vision. These two factors are resolved by molecular biology, for with a minimal quantity (microliter) and within a few hours we are able to determine if the infection is caused by a bacterium, fungus, virus, or parasite and to identify the species.

Polymerase Chain Reaction

The polymerase chain reaction (PCR) has outweighed other techniques used for diagnosis based on the study of nucleic acids such as Southern blot due to its simplicity, sensitivity, speed, and specificity. PCR allows amplification of the region of the genome selected for analysis in a minimal period of time, approximately one hour and half. PCR is based on the ability of DNA polymerase to copy a strand of DNA (1,2,3). The enzyme initiates elongation at the 3′ end of a short (primer) sequence joined to a longer strand of DNA. This is known as the target sequence. When two primers bind to a complementary strand of target DNA, the sequence between the two primers' binding sites is amplified exponentially with each cycle of PCR.

There are three steps to each cycle: (*i*) a DNA heat denaturation step, separation of the double strands of the target DNA by heat; (*ii*) a primer-annealing step, primers anneal to their complementary amplification target sequences at a lower temperature; and (*iii*) an extension reaction step, extension of the target sequences between the primers by DNA polymerase. At the end of each cycle, the PCR products are theoretically doubled (Fig. 1). A thermocycler is used for the procedure. Generally, 30 to 40 thermal cycles result in detectable amounts of a target sequence originally present in less than 100 copies, and the PCR is theoretically capable of detecting a single-copy of DNA (4,5).

The primers constitute the factor determining what genome and which part to amplify. These are short sequences of DNA, which can vary in size (usually between 10–25 nucleotides) and will hybridize specifically with their complementary sequence, permitting elongation of the chain by polymerase. For example, if we wish to know whether there is *P. acnes* in a vitreous sample, we select a fragment of the *P. acnes* genome (DNA target) flanked by short specific sequences. The primers

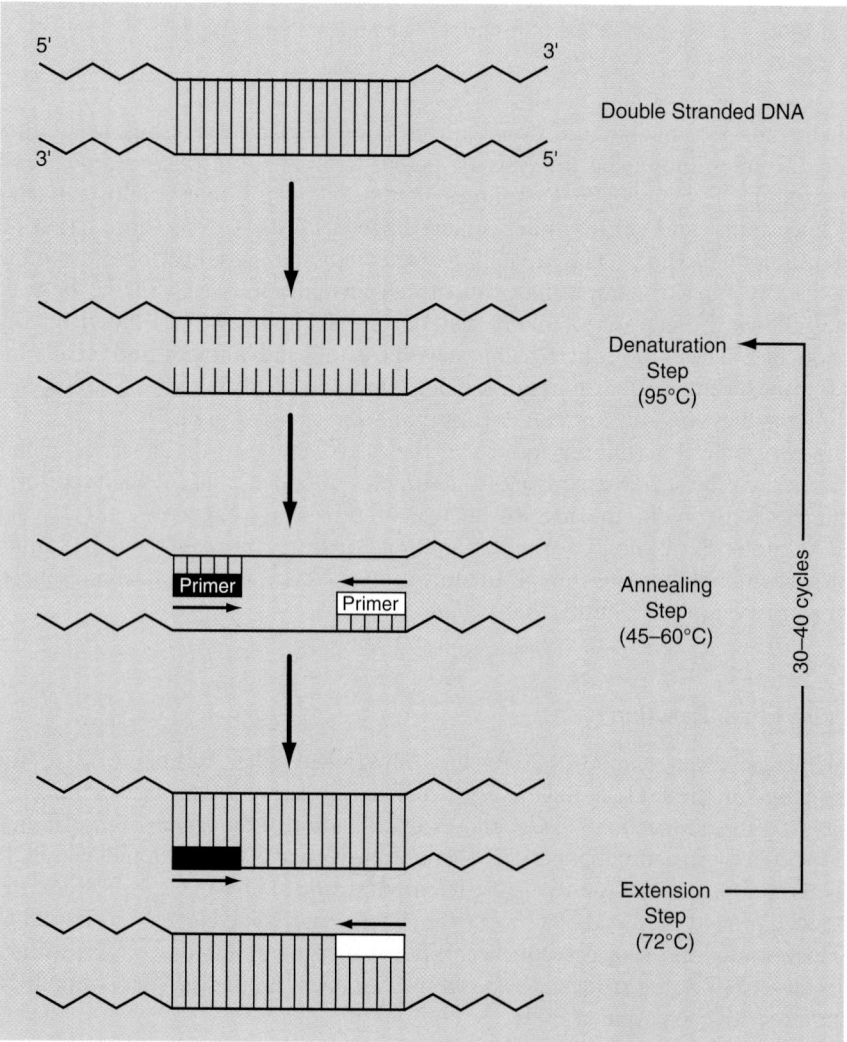

Figure 1   At the end of each cycle, the PCR products are theoretically doubled.

are bound to these two zones and the amplification of the flanked region is produced. If in the last cycle we finally obtain a product the size of the selected fragment, we can determine whether the sample has *P. acnes* DNA.

Depending on the genome region that we are amplifying, we can consider DNA ribosomal targets and the non-ribosomal targets. Among the non-ribosomal targets we can find those that amplify DNA regions codifying certain enzymes [e.g., *Candida* actin gene (6), and cytochrome P450(7)], those that amplify mitochondrial telomere (8), and methods using arbitrary primers PCR (AP-PCR)(9).

However, the area of the genome most commonly used as a target to amplify and detect bacterial or fungal DNA is the region within the ribosomal RNA (rRNA) complex. This region is used most frequently for two fundamental reasons: first, the ribosomal RNA genes have a relatively conserved nucleotide sequence, and second, they are present in a high number of copies, which achieves the sensitivity of the detection method. The number of copies of 16S rRNA gene in bacterial pathogens vary between 1 to 10, being *Staphylococcus epidermidis* 5, *S. aureus* 5–6, *Streptococcus pneumoniae* 4 (10). In yeast cells tandemly repeated rRNA genes are around 150 (11).

This bacterial rRNA complex is organized as shown in Figure 2: the small subunit 16S rRNA gene, the longer subunit 23S rRNA gene, and 5S rRNA gene. Among the three rRNA molecules (16S, 23S, and 5S rRNA), 16S rRNA gene has been used most often for bacterial characterization (12). This sequence contains alternating regions of sequence conservation and heterogeneity (13). The conserved regions can be used to create PCR amplification primers that anneal to ribosomal targets for all or most bacterial species. The 16S and 23S rRNA molecules consist of variable sequence motifs that reflect their phylogenetic origins. This sequence variability permits the design of probes at different taxonomic levels.

The organization of fungal rRNA complex includes a sequence coding for the 18S rRNA gene, an internal transcribed spacer region 1 (ITS1), the 5.8S rRNA gene coding region, another ITS (ITS2), and the sequence coding for the 28S rRNA gene (Fig. 2). The coding regions of 18S, 5.8S, and 28S nuclear rRNA genes evolved slowly, and are relatively conserved among fungi, providing a molecular basis of establishing phylogenetic relationships (14). Between these rRNA gene-coding regions are the ITS regions (ITS1 and ITS2), which evolved more rapidly, leading to

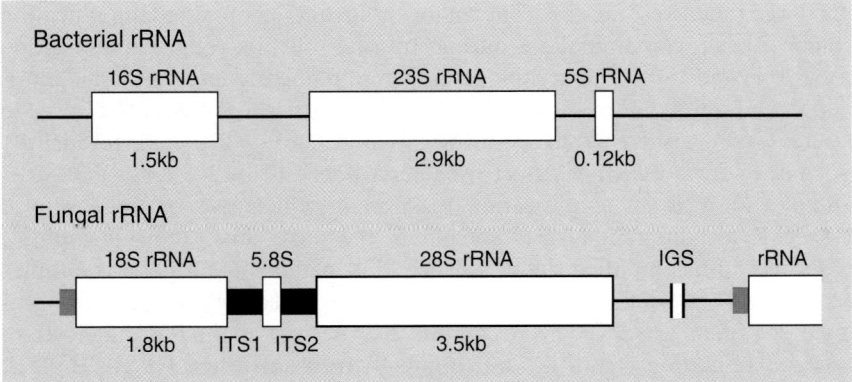

Figure 2 The bacterial rRNA complex is organized as shown: the small subunit 16S rRNA gene, the longer subunit 23S rRNA gene, and 5S rRNA gene. The organization of fungal rRNA complex includes the 18S rRNA gene, the ITS1, the 5.8S rRNA gene, another ITS (ITS2) and the 28S rRNA gene.

sequence variability among genera and species of fungi. Early studies in molecular testing used the rRNA genes complex as a target, and concentrated on the region of the 18S rRNA gene. The comparison of nucleotide sequences within this gene region has been successful for the separation of fungal genera and species. However, limited sequence variability within these rRNA genes, together with a need to compare large sequence regions, has led to a shift to the evaluation of the region between 18S rRNA gene and 28S rRNA gene as a target for molecular testing for diagnosis. This region includes the shorter spacer regions (ITS1 and ITS2) and the 5.8S rRNA gene.

The organization of the *Acanthamoeba* ribosomal complex is very similar to the fungal ribosomal complex. The region to amplify within a virus is specific for each one due to the fact that they do not have ribosomal genes or DNA-conserved regions that we can use to amplify with a broad-range virus PCR.

Conventional PCR with specific primers and other versions of PCR are used in the diagnosis of ocular infections. The most frequently used forms of PCR in molecular diagnosis are as follows:

Broad-Range PCR. This is the method used most frequently for the diagnosis of endophthalmitis and keratitis, due to the fact that we do not know which species of microorganism is the cause of the infectious process, and carrying out a specific PCR for every microorganism susceptible to produce infection would be a very long and costly process. This application uses conserved sequences within phylogenetically informative genetic targets to diagnose infection. Broad-range rRNA genes PCR techniques allow rapid and highly specific bacterial or fungal identification with a single pair of primers targeting the rRNA genes. Table 1 gives the primers that have been used to date for the detection of pathogens in bacterial endophthalmitis or keratitis by broad-range PCR, their sequences, and genome regions of *Escherichia coli* which hybridize. Table 2 shows the pan-fungal primers used in fungal endophthalmitis or keratitis, their sequences and the region of *Candida albicans* which hybridize. Figure 3 shows the position of the bacterial and fungal broad-range primers in the ribosomal DNA region.

Nested PCR. Designed mainly to increase sensitivity, nested PCR uses two sets of amplification primers (15). One set is used for the first round amplification, consisting of 25 to 30 cycles. Products of the first reaction are then subjected to a second round of amplification with another set of primers specific for a sequence within the product of the first pairs of primers (15,16). Nested PCR is highly sensitive due to the large total cycle number. The major disadvantage of nested amplification is the high risk of contamination during transfer of first-round amplification products to a second tube. This technique has been frequently used in the diagnosis of bacterial endophthalmitis, not only to increase the sensitivity but also to differentiate Gram-positive from Gram-negative bacteria (17), and to identify *P. acnes*, once the first PCR has detected bacterial DNA (Tables 3 and 4) (18–20).

Multiplex PCR is an amplification reaction in which two or more sets of primers specific for different targets are put in the same tube, allowing multiple target sequences to be amplified simultaneously. For diagnostic purposes, multiplex PCR can be used for detecting internal controls and for detecting multiple pathogens in a single specimen (21,22). This technique has been used in the diagnosis of intraocular inflammation, keratitis, and conjunctivitis for detected HVS-I, HVS-II, and VZV (23–25).

Visualization of the amplified product. In these three methods, after PCR the millions of target DNA copies that accumulate need to be visualized. The most frequently used technique is by electrophoresis with ethidium bromide staining. The

Table 1 Commonly Used Primers for Broad-Range PCR in Bacterial Endophthalmitis or Keratitis

Name[a]	Original reference	Sequence (beginnings or ends may have been modified from the indicated references by subsequent authors)	rRNA gene amplified	Position on E. coli rRNA gene	Fragment (bp)	Primers	Authors (reference)
1194f 1525r	Chen (72) Edwards (73)	5'-AACTGGAGGAAGGTGGGGAT-3' 5'-AGGAGGTGATCCAACCGCA-3'	16S	1168-1188 1520-1539	370	RW01 DG74	Anand (29)
27f 1175r	Edwards (73) Chen (72)	5'-TTGGAGAGTTTGATCCTGGCTC-3' 5'-ACGTCATCCCACCTTCCTC-3'	16S	4-25 1175-1194	1190	16SF 16SR	Carrol (17)
66f 1044r	Hykin (19) Okhravi (30)	5'-GGCGGCAKGCCTAAYACATGCAAGT-3' 5'-GACGACAGCCATGCASCACCTGT-3'	16S	42-66 1044-1067	1000	NF NR	Hykin (19)
27f 787r	Edwards (73) Wilson (74)	5'-TTGGAGAGTTTGATCCTGGCTC-3' 5'-GGACTACCAGGGTATCTAA-3	16S	4-25 789-806 506-528	800	U1 rU4	
64f 515r	Hykin (19) Lane (14) Relman (75)	5'-GGGCGTGCTTAAGACACATGCAAGTCG-3' 5'-GCGGCTGGCACGTAGTTAG-3'	16S	41-64 506-528	500	U2 rU3	
27f 556r	Edwards (73) Bergman (76)	5'-AGAGTTTGATCMTGGTCCAG-3' 5'-CGCTTTAGCCCARTNASTCCG-3'	16S	8-25 556-578	570	16S8FEVAR1 16S556RBVAR2	Kerkhoff (77)
27f 787r	Edwards (73) Wilson (74)	5'-AGTTTGATCCTGGCTCAG-3' 5'-GGACTACCAGGGTATCTAAT-3'	16S	8-25 787-806	800	8FLP 806R	Knox, (37,78), Ferrer (61,79)
27f 515r	Edwards (73) Lane (14) Relman (75)	5'-TTGAACGCTGGCGGCAGGCCT-3' 5'-TGCGTGCGCTTTACGCCCAGT-3'	16S	16-36 516-533	500	MF16 B515	
533f 1371r	Lane (14) Relman (75) Chen (72)	5'-GTGCCAGCAGCCGCGGTAA-3' 5'-AGGCCCGGGAACGTATTCAC-3'	16S	515-533 1371-1390	870	515FPL 13B	

(Continued)

Table 1 Commonly Used Primers for Broad-Range PCR in Bacterial Endophthalmitis or Keratitis *(Continued)*

Name[a]	Original reference	Sequence (beginnings or ends may have been modified from the indicated references by subsequent authors)	rRNA gene amplified	Position on *E. coli* rRNA gene	Fragment (bp)	Primers	Authors (reference)
926f 1371r	Lane (14) Chen (72)	5'-ACTCAAATGAATTGACGGGGGC-3' 5'-AGGCCCGGGAACGTATTCAC-3'	16S	911-930 1371-1390	480	MF91 13B	
806f 1371r	Wilson (74) Chen (72)	5'-CAAACAGGATTAGATACCC-3' 5'-CCCGGGAACGTATTCACCG-3'	16S	778-796 1368-1386	600	Bakt F2 Bakt rev3	Lohmann (36,62,81)
1194f 1525r	Chen (72) Edwards (73)	5'-AACTGGAGGAAGGTGGGGAT-3' 5'-AGGAGGTGATCCAACCGCA-3'	16S	1168-1188 1520-1539	370	RW01 DG74	Lohmann (35), Hollander (82)
27f 1175r	Edwards (73) Chen (72)	5'-TTGGAGAGTTTGATCCTGGCTC-3' 5'-ACGTCATCCCCACCTTCCTC-3'	16S	4-25 1175-1194	1190	16SF 16SR	Okhravi (34)
27f 1175r	Edwards (73) Chen (72)	5'-TTGGAGAGTTTGATCCTGGCTC-3' 5'-ACGTCATCCCCACCTTCCTC-3'	16S	4-25 1174-1194	1190	16SF 16SR	Okhravi (30)
66f 1044r	Hykin (19) Okhravi (30)	5'-GGCGGCAKGCCTAAYACATGCAAGT-3' 5'-GACGAGCAGCCATGCASCACCTGT-3'	16S	42-66 1044-1067	1000	NF NR	
27f 787r	Edwards (73) Wilson (74)	5'-TTGGAGAGTTTGATCCTGGCTC-3' 5'-GGACTACCAGGGTATCTAA-3	16S	4-25 789-806	800	U1 rU4	Therese (20)
64f 515r	Hykin (19) Lane (14) Relman (75)	5'-GGCGTGCTTAACACATGCAAGTCG-3' 5'-GCGGGCTGGCACGTAGTTAG-3'	16S	41-64 506-528	500	U2 rU3	

[a]Names are based on the 3' nucleotide position and the orientation (f, forward; r, reverse) according to Lane (14).

Table 2 Commonly Used Primers for Broad-Range PCR in Fungal Endophthalmitis or Keratitis

Original reference	Name[a]	Sequence	rRNA gene amplified	Position on *Candida albicans* rRNA gene	Fragment (pb)	Primers	Authors (reference)
Sandhu (83)	423f / 646r	5'-GTGAAATTGTTGAAAGGGAA-3' / 5'-GACTCCTTGGTCCGTGTT-3'	28S	404–423 (28S) / 646–663 (28S)	260	U1 / U2	Anand (59), Bagyalakshimi (83)
White (84)	1777f / 40r	5'-TCCGTAGGTGAACCTGCGG-3' / 5'-TCCTCCGGCTTATTGATATGC-3'	ITSs-5.8S	1759–1777 (18S) / 40–59 (28S)	550	ITS1 / ITS4	Ferrer (18,38,39,86), Tarai (87), Kumar (31,32)
Jaeger (28)	224f / 900r	5'-AGGGATGTATTTATTTAGATAAAAAATCAA-3' / 5'-CGCAGTAGTTAGTCTTCAGTAAATC-3'	18S	196–224 (18S) / 900–924 (18S)	740	Pffor / Pfrev2	Jaeger (28)
Makimura (87)	332f / 987r	5'-ACTTTCGATGGTAGGATAG-3' / 5'-TGATCGTCTTCGATCCCCTA-3'	18S	314–332 (18S) / 987–968 (18S)	650	B2 F / B4 R	Lohmann (36)
Sandhu (83)	64f / 825r	5'-ATCAATAAGCGGAGGAAAAG-3' / 5'-CTCTGGCTTCACCCTATTC-3'	28S	45–64 (28S) / 825–843 (28S)	800	P1 / P2	Lohmann (35)
Gaudio (88)	–	5'-CAGGGGAGGTAGTGACAATAAATA-3' / 5'-ACAAATCACTCCACCAACTAAGAA-3'	18S	452–476 (18S) / 1322–1346 (18S)	870	– / –	Gaudio (89)
White (84,84)	f / 40r	5'-GCATCGATGAAGAACGCAGC-3' / 5'-TCCTCCGCTTATTGATATGC-3'	5.8S-ITS2	(5.8S) / 40–59 (28S)	330	ITS3 / ITS4	Rishi (90)

[a] Names are based on the 3'nucleotide position and the orientation (f, forward; r, reverse) according to Lane 14.

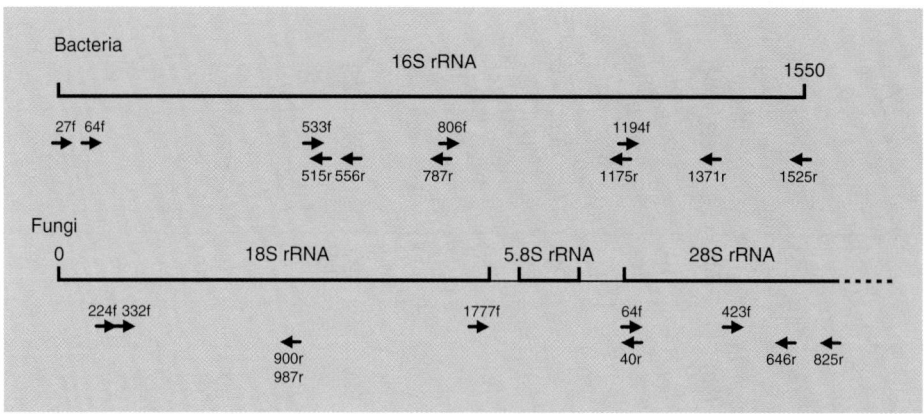

Figure 3   Position of the bacterial and fungal broad range primers in the ribosomal RNA region.

DNA fragments are separated according to their size in an agarose gel; the smaller DNA molecules move faster towards the anode than large DNA molecules (Tables 3 and 4).

Real-time PCR. Recently, a modification of the traditional PCR, real-time PCR, has been developed; its use in the diagnosis of ocular infection has still to be determined. In real-time PCR the amplification and detection take place in the same reaction chamber or tube, often simultaneously (9,26,27). This vastly reduces handling of the amplicon (and therefore the risk of carryover contamination). A thermocycler is combined with a fluorimeter, allowing the detection and quantification of PCR products the moment they are produced. These can be used as highly sensitive qualitative detection methods or as quantitative tests.

Currently, fluorescence is the best indicator for real-time PCR. The quantity of the initial target DNA determines the increase in fluorescence obtained during this cycle. Within real-time PCR we can find variants, depending on what is used as a marker.

1. SYBR® Green I (Qiagen, Hilden, Germany) is a double strand specific dye. SYBR Green I dye becomes fluorescent upon binding to double-stranded DNA, providing a direct method for quantifying PCR products in real time (Fig. 4A).
2. Probe hydrolysis. If a probe labeled with a fluorophore and a quencher is hydrolyzed during PCR and the labels are separated, fluorescence will increase. The most common implementation uses the 5′-exonuclease activity of the DNA polymerase to hydrolyze the probe and dissociate the labels (Fig. 4B). PCR with Tacman system (Perkin-Elmer Corp., Applied Biosystems, Foster City, California, U.S.A.) works with this method. The Taqman® probe is an oligonucleotide probe labeled with fluorescent dye on the 5′ end and quencher on the 3′ end. The close proximity of the dye and quencher results in negligible fluorescence when the probe is intact. However, during polymerization, the 5′-nuclease activity of Taq cleaves the probe, separating the dye and quencher, which allows detection of fluorescence (Fig. 4B). This technique has been used in the detection of adenovirus (101), HVS-I (102,103), VZV (102), and *Chlamydia trachomatis* (102) in ocular samples.
3. Molecular beacon. Primers labeled with a fluorophore and a quencher in a hairpin conformation can also be used to monitor PCR. The hairpins straighten

(Text continues on page 73.)

Table 3 Methods Used for Detection and Identification of Bacterial DNA in Bacterial Endophthalmitis or Keratitis

Authors (reference)	Infection	Yr. of publication	Target region	Product size (bp)	Type of assay	Method of detection	Level of identification	Method of identification
Hykin et al. (19)	E	1994	16S	760	Single step PCR	Agarose	Propionibacterium acnes	Nested PCR
Lohmann et al. (35)	E	1998	16S	370	Single step PCR	Agarose	Species level	DNA sequencing
Therese et al. (20)	E	1998	16S	470	Nested PCR	Agarose	Propionibacterium acnes	Nested PCR
Knox et al. (78)	K	1998	16S	800	Two PCR	Agarose	Species level	DNA sequencing
Knox et al. (37)	E	1999	16S	800	Two PCR	Agarose	Species level	DNA sequencing
Anand et al. (29)	E	2000	16S	370	Single step PCR	Agarose	Gram reaction	Hybridization
Carrol et al. (17)	E	2000	16S	1200	Nested PCR	Agarose	Gram reaction	nested PCR
Lohmann et al. (36)	E	2000	16S	600	Single step PCR	Agarose	Species level	DNA sequencing
Lohmann et al. (62,81)	K	2000	16S	600	Single step PCR	Agarose	Species level	DNA sequencing
Okhravi et al. (34)	E	2000	16S	1200	Single step PCR	Agarose	Species level	DNA sequencing
Okhravi et al. (30)	E	2000	16S	1000	Nested PCR	Agarose	Species level	RFLP DNA sequencing
Ferrer et al. (18)	E	2001	16S	760	Single step PCR	Agarose	Propionibacterium acnes	Nested PCR
Seo et al. (60)	K	2002	16S		Real- time PCR	Real- time PCR	Mycobacteium chelonae	Real- time PCR

(Continued)

Table 3 Methods Used for Detection and Identification of Bacterial DNA in Bacterial Endophthalmitis or Keratitis (*Continued*)

Authors (reference)	Infection	Yr. of publication	Target region	Product size (bp)	Type of assay	Method of detection	Level of identification	Method of identification
Kerkhoff et al. (77)	E	2003	16S	570	Single step PCR	Biotinylated primer	*Neisseria meningitidis*	Microtiter assay
Ferrer et al. (61)	K	2004	16S	800	Nested PCR	Agarose	*Propionibacterium granulosum*	DNA sequencing
Ferrer et al. (79)	E	2004	16S	800	Single PCR	Agarose	*Corynebacterium macginleyi*	DNA sequencing
Hollander et al. (82)	E	2004	16S	370	PCR	Agarose	*Propinibacterium acnes*	Southern blot
Uy et al. (91)	E	2005	16S	214	PCR	Agarose	*Achromobacter xylosoxidans*	DNA sequencing
Lai et al. (92)	E	2006	16S	587	PCR	Agarose	*Propinibacterium acnes*	Specific *P. acnes* PCR
Montero et al. (80)	E	2006	16S	800	Single PCR	Agarose	*Propinibacterium acnes*	DNA sequencing
Bagyalakshmi et al. (83)	E	2006	16S	500	Nested PCR	Agarose	Not done	Not done

Abbreviations: E, Endophthalmitis; K, Keratitis.

Table 4 Methods Used for Detection and Identification of Fungal DNA in Fungal Endophthalmitis and Keratitis

Authors (reference)	Infection	Year of publication	Target region	Product size (bp)	Type of assay	Method of detection	Level of identification	Method of identification
Alexandrakis (57)	E	1996	Cutinase gene of *Fusarium*	196	Single step PCR	Agarose	*Fusarium solani*	Specific PCR Southern blot hybridization
Alexandrakis (93)	K	1998	Cutinase gene of *Fusarium*	196	Single step PCR	Agarose	*Fusarium solani*	Specific PCR
Lohmann (35)	E	1998	28S rRNA gene	800	Single step PCR	Agarose	Not done	Not done
Okhravi (7)	E	1998	Cytochrome P450 L$_1$ A$_1$ gene of *Candida*	1000	Single step PCR	Agarose	Species of *Candida.*	Nested PCR RFLP
Hidalgo (9)	E	2000	Random	Variable (500-3200)	Random PCR	Agarose	*Candida albicans*	Length polymorphism of amplified products
Jaeger (28)	E	2000	18S rRNA gene	740	Single step PCR	Agarose	*Candida albicans Aspergillus fumigatus Fusarium solani*	Nested PCR
Anand (59)	E	2001	28S rRNA gene	260	Single step PCR	Agarose	Not done	Not done
Ferrer (18)	E/K	2001	ITS2	250	Nested PCR	Agarose	Species level	DNA sequencing
Ferrer (38)	K	2002	ITSs-5.8S rRNA gene	550	PCR	Agarose	*Alternaria alternata*	DNA sequencing
Gaudio (89)	K	2002	18S rRNA gene	870	Nested-PCR	Agarose	*Candida albicans Aspergillus fumigatus Fusarium oxysporum*	Nested-PCR

(Continued)

Table 4 Methods Used for Detection and Identification of Fungal DNA in Fungal Endophthalmitis and Keratitis *(Continued)*

Authors (reference)	Infection	Year of publication	Target region	Product size (bp)	Type of assay	Method of detection	Level of identification	Method of identification
Rishi (90)	K	2003	5.8S-ITS2 rRNA gene	330	PCR	Agarose	Not done	Not done
Ferrer (38)	E	2003	ITSs-5.8S rRNA gene	550	Single step PCR	Agarose	*Alternaria infectoria*	DNA sequencing
Ferrer (39)	E	2005	ITSs-5.8S rRNA gene	550	Single step PCR	Agarose	*Fusarium proliferatum*	DNA sequencing
Kumar (31,32)	K	2005	ITSs-5.8S rRNA gene	550	Single step PCR	Agarose	Genus level	SSCP
Bagyalakshmi et al. (83)	E	2006	28S rRNA gene	260	Single step PCR	Agarose	Not done	Not done
Tarai et al. (87)	E	2006	ITSs-5.8S rRNA gene	550	Single step PCR	Agarose	Not done	Not done
Kumar (33)	K	2006	28S rRNA gene	550	Single step PCR	Agarose	Genus	SSCP

Abbreviations: E, Endophthalmitis; K, Keratitis.

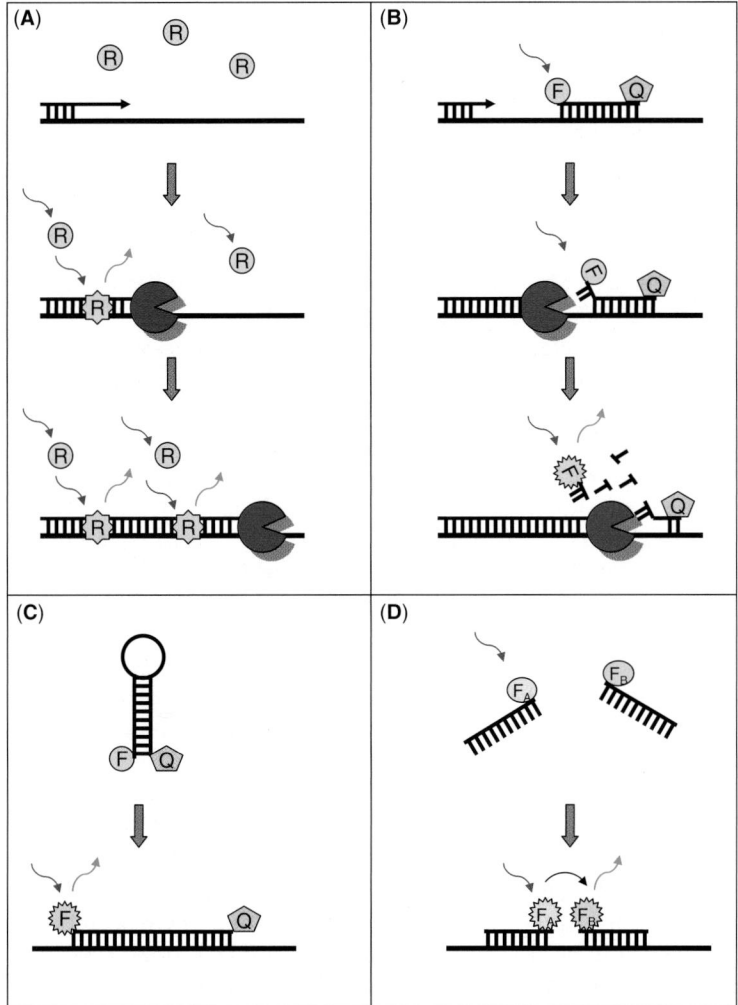

Figure 4 (**A**) SYBR® Green I dye becomes fluorescent upon binding to double-stranded DNA. (**B**) Probe hydrolysis. A probe is created by conjugating carboxyfluorescein fluorescent dye (FAM) and a fluorescent quencher, to a single-stranded DNA (ssDNA) oligonucleotide that is complementary to a conserved region of the target genome. No fluorescence is emitted so long as the probe is intact; when the probe is hydrolyzed during PCR and the labels are separated, fluorescence will increase. (**C**) Molecular Beacon. The hairpins (primers labeled with a fluorophore and a quencher) straighten out during amplification, resulting in an increase in fluorescence. (**D**) Hybridization probe. Hybridization probes change fluorescence on hybridization, usually by fluorescence resonance energy transfer. Two interacting fluorophores may be placed on adjacent probes or one may be placed on a primer and the other may be placed on a probe.

out during amplification, resulting in an increase in fluorescence. Any double-stranded product formed from the primers will be detected (Fig. 4C).

4. Hybridization probe. Hybridization probes change fluorescence on hybridization, usually by fluorescence resonance energy transfer. Two interacting fluorophores may be placed on adjacent probes. However, one may be placed on a primer and the other may be placed on a probe (Fig. 4D).

Real-time PCR has been extremely useful for studying viral agents of infectious disease and helping to clarify disputed infectious disease processes. Most of the assays presented in the literature allow an increased frequency of virus detection compared with conventional techniques, which makes the implementation of real-time PCR attractive to many areas of virology (102,103).

Identification of the Pathogen

Once the DNA region has been amplified, we can use various molecular techniques for the identification of the pathogen. These methods include utilization of genus or species-specific primers and oligonucleotide probes, length polymorphism of amplified DNA, restriction fragment length polymorphism (RFLP) analysis, and direct sequence analysis of amplified DNA. In real-time PCR, the detection, identification, and visualization of the results happens at the same time.

Genus or Species-Specific PCR. In this method, detection and identification is at the same time. PCR is carried out with genus or specific primers. Usually, first a PCR with universal or broad-range primers is performed, and when bacterial or fungal DNA is detected, a second PCR (nested PCR) is carried out with specific primers. This technique is frequently used for detection of *P. acnes* DNA (18–20) in delayed endophthalmitis (Table 3). PCR with universal fungal primers and subsequent nested PCR using species-specific primers has been used to differentiate *Candida albicans, Aspergillus fumigatus,* and *Fusarium solanii* (28). Also, then the first PCR with genus specific primers and nested PCR with species specific primers for different *Candida* species (7), have been used in the diagnosis of delayed fungal endophthalmitis (Table 4).

Hybridization Analysis of PCR Products with Genus or Species-Specific Oligonucleotide Probes. The most common way to investigate the sequence obtained by PCR is to hybridize it to one or more oligonucleotide probes. The best known method is Southern blotting. This technique allows the identification of the pathogen at genus or species level according to probes used by hybridization. PCR with broad-range primers followed by hybridization with oligonucleotide probes have been reported for the differentiation between Gram-positive and Gram-negative bacteria by Anand et al. (Table 3) (29).

Restriction Fragment Length Polymorphism Analysis. After amplification, the sequence composition of a PCR product can be investigated by restriction enzyme-mediated digestion of amplified DNA. Digestion of PCR products by restriction endonucleases may generate multiple fragments, which can be resolved by gel electrophoresis. The endonucleases recognize a specific sequence of nucleotides (4–8 nucleotides) and, when found, cleave the DNA. Variations of the compositions of the fragment alter the restriction site; the endonuclease may not recognize the site and will fail to digest the DNA, resulting in different sized restriction fragments and an alteration of the restriction fragment length polymorphism (RFLP) profile. Figure 5 shows a schematic representation of the ITS-5.8S rRNA gene fragment cut with three restriction endonucleases. RFLP analysis has been used by Okhravi et al. for identification at species level of bacteria (30) and fungi (7) causing endophthalmitis (Table 3 and 4).

Single-Strand Conformation Polymorphism Analysis

SSCP analysis of PCR-generated products can detect sequence variation in specific genetic loci. The amplification products of a target sequence are denatured into two

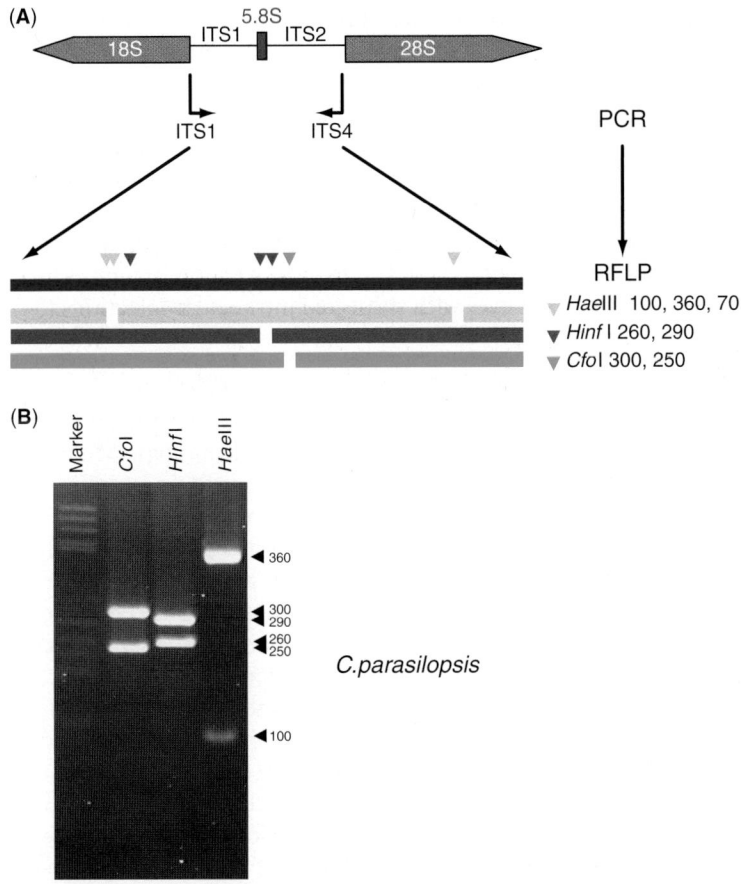

Figure 5 Schematic representation of the ITS-5.8S rRNA gene fragment cut with three restriction endonucleases.

single-stranded DNA fragments and subjected to non-denaturing polyacrylamide gel electrophoresis. PCR-SSCP is capable of detecting over 90% of all single base substitutions in a 200 bp fragment, based on changes in the mobility of the fragments. Mobility changes are the results of changes in the secondary structure of the DNA under non-denaturing conditions. SSCP-PCR has been applied to the identification of fungi causing keratitis (31–33).

For all of these methods, clinical suspicion of the causative agent is mandatory, because of the inherent limitation caused by the selection of specific primers or probes.

Nucleic Acid Sequence Analysis. This technique provides the most information, given that in this case the size, nucleotide composition, and the order of this nucleotide are taken into account. The technique is based on the synthesis of new DNA strands synthesized from the purification PCR amplicon beginning with one sequencing primer, and each strand having one more nucleotide. This is due to limited concentrations of dideoxynucleotide triphosphates (ddNTPs) into which four different fluorescent dyes have been incorporated mixed with unlabeled deoxynucleotides (dNTPs). Synthesis terminates whenever a ddNTP is incorporated into the strand. Accumulated fragments are separated by size using electrophoresis. During electrophoresis, labeled products are visualized by fluorescence, with each of the four fluorescent dyes indicating which of the terminal ddNTPs have been incorporated. Combining the terminal ddNTP information with the fragment size allows the determination of sequence information (Fig. 6).

This is the best method because suspicion of which microorganism is causing the infection is unnecessary. This technique has been used by numerous authors to identify the microorganism causing the infection in both bacterial endophthalmitis (Table 3) (30,34–37) and fungal endophthalmitis (Table 4) (18,38,39).

New, fastidious, or non-cultured pathogens have been identified directly from infected human tissue or blood by this method, and in the case of endophthalmitis there is evidence of bacterial involvement in eyes with suspected intraocular infections (34).

Strain Characterization

The ability to distinguish between strains of a bacterial or fungal pathogen is crucial for addressing many questions in clinical microbiology and epidemiology. Molecular methods are now almost universally used to characterize strains and to determine the relatedness between isolates causing disease in different contexts, whether it be different patients, hospitals wards, or countries. The molecular method most frequently used is Multilocus Sequence Typing (MLST). This method is based on the differences of the DNA sequence of internal fragments of seven housekeeping genes (genes that encode essential metabolic functions) between strains of the same species. For each locus, every different sequence is assigned a different allele number, to produce a strain identifier or allelic profile that corresponds to the allele numbers at the seven loci. Relationships between isolates may be visualized in the form of a dendrogram constructed from the matrix of pairwise differences between their allelic profiles. Two isolates with identical allelic profiles are considered to be sufficiently closely related to be described as members of the same clone, and isolates with similar allelic profiles that form a cluster of isolates on the dendrogram are considered to have resulted by minor diversification from a recent common ancestor and to be members of the same lineage or clonal complex. Epidemiological investigation of *Fusarium* keratitis occurring in the United States associated with ReNu Multiplus with MoistureLoc (Bausch & Lomb, Rochester, New York, U.S.A.) was carried out by this method (Appendix D) (40).

DISCUSSION

There are significant limitations to culture methods (4.) In vitro growth of an organism may be a lengthy process, possibly taking days, weeks, or even months,

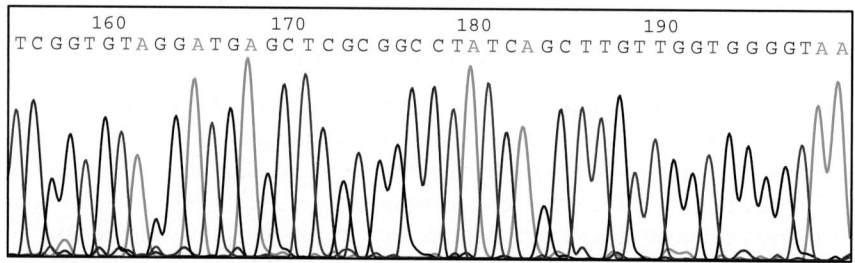

Figure 6 Combining the terminal ddNTP information with the fragment size allows the determination of sequence information.

and may even need subculture for final characterization, resulting in further delays. Due to the fastidious nature of many organisms, they may require specialized media or culture conditions, thus making routine screening of specimens impractical and severely limiting the identification of disease associations. The sensitivity and specificity that are desirable in a clinical assay are often lacking in these methods. In fact, it is now thought that only a small proportion of all bacterial species can be grown on artificial media (4,41), leading us to question what might be missing, both diagnostically and in the study of disease pathogenesis.

Molecular methods aid our understanding about which microorganisms are involved in endophthalmitis and their pathogenic mechanisms. More and more laboratories are using molecular biology techniques to support microbiological diagnosis and/or solve some of its limitations. This circumstance requires special attention to avoid false-positive or -negative results. The greatest disadvantages of molecular diagnosis are the false-positive results (42) the high cost and the lack of standardization of the techniques.

False results can be divided into three groups according to the origin:

- From the sample. Any disturbance while taking the sample may produce both false positives and negatives. The false positives that appear in bacterial PCR may be produced by contamination occurring during ocular surface puncture with the needle to take the sample or with contaminated povidone iodine used to disinfect the skin (42).
- From the laboratory. Contamination has been reported not only from specimens that have been processed adjacent to one another but also from sources such as contaminated instruments, clothing, room air, and even skin particles from laboratory workers. PCR assays have a reputation for producing false-positive reactions if there is specimen contamination or amplicon carryover. It is essential that negative controls are processed at every step to exclude false positives and that samples are extracted, amplified, and visualized in separate rooms using equipment designated for each area (including pipettes) to minimize the possibility of specimen contamination. UV irradiation has also been used to help destroy nucleic acids on laboratory surfaces (43–45). In real-time PCR the amplification and detection take place in the same reaction chamber or tube (27); this vastly reduces the handling of amplicon and therefore the risk of carryover contamination.
- From reagents. Some authors have described contamination problems in some commercially available laboratory reagents. It is widely known that many commercial polymerases have remains of bacterial DNA (46,47). When nested PCR is carried out with bacterial universal primers, false-positive results can appear as a result of these DNA remains or amplicon carryover. Decontamination measures have included the use of topical agents, such as sodium hypochlorite, and of decontaminants within mixes of PCR, such as isopsoralens and uracil-n-glycosylase (48–53).

The cost of PCR tests is due to the high degree of expertise needed to develop, run, and interpret the results of tests, expensive physical space requirements to set up and run the assays, and high capital and operating costs associated with many of the reagents. With real-time PCR, these costs are even higher.

The final disadvantage of molecular diagnosis is the lack of standardization between laboratories because these are recently developed techniques that are still not optimized.

Endophthalmitis

In the field of ophthalmology we may see how many chronic endophthalmitis cases have an unknown cause (negative culture) and are diagnosed as sterile endophthalmitis, even though the infection improves with antimicrobial therapy. It is quite likely that there was a microorganism causing this clinical picture and the culture was negative due to the fact that this species needs specific growth requirements, bacteria is sequestrated on solid surfaces (54), or located inside macrophages (55,56). PCR has detected new microorganisms in endophthalmitis, demonstrating evidence of various species of bacteria involved in the infection, whereas only one or none of the species detected by PCR grew in culture (34). In addition, there are studies in which the identification of the pathogen has been achieved at species level using molecular techniques, improving our knowledge of the microbial spectrum causing endophthalmitis (34,38).

Conventional PCR is the molecular method used for the laboratory diagnosis of endophthalmitis, but the target DNA has changed over time. If we analyze the early studies into molecular diagnosis of ocular infections, we find that these were based on DNA species-specific regions that tend to be unicopy genes, and only detect one species usually *Candida* (7), *Fusarium* (57), or *Aspergillus* (58). Later studies establish the ribosomal genes as the base of phylogenetics and taxonomy and as the target for the detection and diagnosis of the infection process. Ribosomal genes are beginning to be used as a target, thus increasing the sensitivity of the method (multicopy genes) and the spectrum detectable. This is extremely important, as ocular samples are very limited in size; furthermore, it is imperative to know if the origin of the infection is bacterial or fungal. For the application of species-specific primers, unless it is clear which species is the cause of infection, it is preferable to extend the target to be able to detect any species of bacteria or fungi and then identify it by molecular techniques.

There are various groups working with ribosomal genes as a target for the diagnosis of ocular infections. In bacterial endophthalmitis, all the groups work with 16S rRNA gene (Table 3), although they differ in the amplified zone. There is a greater variety in fungal endophthalmitis, as can be seen from Table 4. Anand and Lohmann use the 28S rRNA gene as target DNA (35,59); Okhravi used the 18S rRNA gene (28); and Ferrer works with the ITS/5.8S rRNA gene region (18,38,39). In all cases the sensitivity is higher than by culture, whatever the amplified region. However, sensitivity of detection by PCR does not depend only on the amplified ribosomal gene but also on the primers and the specific conditions for each primer. Figure 7 shows how sensitivity varies according to the primers used, where in both cases the amplified region is 16S rRNA gene of *Pseudomonas aeruginosa* but the zone and the primers are different.

There is a wide range of different techniques used for the identification of the microorganism causing the infection as shown in Tables 3 and 4. Depending on the type of infection produced, either technique of molecular identification can be applied. In late-presenting or delayed bacterial endophthalmitis, specific primers are often used to detect DNA of *P. acnes* (18–20); for endogenous endophthalmitis mostly caused by *Candida* and *Aspergillus,* we can apply techniques (nested, RFLP) that detect and directly differentiate their species (7,9,28,58). In these cases, we can base the identification on techniques that are *a priori* limiting us to a series of species, given that the spectrum produced by this infection is quite restricted.

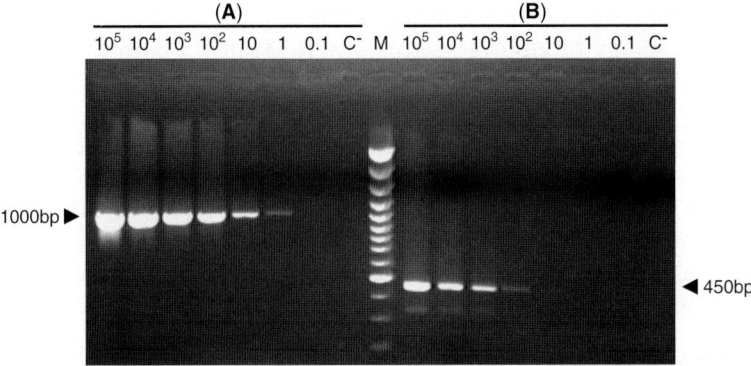

(A) 10^5 10^4 10^3 10^2 10 1 0.1 C⁻ M **(B)** 10^5 10^4 10^3 10^2 10 1 0.1 C⁻

1000bp ►

◄ 450bp

Figure 7 Sensitivity varies according to the primers used, where in both cases the amplified region is 16S rRNA gene of *Pseudomonas aeruginosa* but the zone and the primers are different.

However, in other endophthalmitis cases and particularly for posttraumatic endophthalmitis, the spectrum broadens so much that unless we have used broad-range primers for their detection, with identification performed by sequencing, it is highly likely that we will not be able to identify the pathogen. Even though the broadest spectrum is found in keratitis and posttraumatic endophthalmitis, there are also cases in which the causal agent in post-surgery endophthalmitis remains a mystery (34,59).

Keratitis

Bacterial Keratitis. When the clinical suspicion is bacterial, the diagnostic method that gives the best result is culture, with some exceptions. The cornea is continuously wet from tears due to blinking with innocuous bacteria that are part of the conjunctival biota on the surface. As we have seen in the previous section, PCR is able to detect DNA starting from one copy, so this biota can cause a false-positive result. In addition, culture provides a series of advantages such as the microbial count, so if there is only one colony, contamination is suspected. It also allows us to obtain antibiotic sensitivities to guide therapy and detect resistance. In what cases should the PCR be applied? It is best reserved for those cases where (*i*) the culture fails or (*ii*) there is very low probability of contamination during the sample being taken such as by corneal biopsy. Culture can fail mainly for two reasons: the first is when bacteria are difficult to isolate, either because special media is required (mycobacteria) (60), or because special conditions are needed [anaerobes like *Propionibacterium* spp. (61) or fastidious streptococci or nocardia (Chapter 2)]; the second reason, which is more common, is when the patient has been treated with antibiotics before the sample is collected (62). In both situations, PCR and DNA sequencing can help to establish the cause of the infection.

In addition, there are deep-seated intrastromal corneal infections such as those occurring after LASIK, the implantation of intracorneal rings, or in crystalline keratopathies. In these cases the culture is often negative because the nature of the microorganism makes its culture difficult (viz. *Mycobacterium* or encapsulated slow growing bacteria). Another problem is sampling; corneal biopsy is best used. Such deep stromal tissue, a priori, should be sterile with deep scraping. PCR used to amplify bacterial or fungal DNA can be useful in this situation, as the probability of contamination during the collection of the deep scraping sample is very low. Another useful alternative is histology or electron microscopy (EM) of the corneal biopsy, which can identify intrastromal microbes but not their species. Their combination with PCR, performed separately, is powerful.

It is observed, from the literature, that work carried out for the detection of bacterial DNA in the cornea is scarce due to the fact that PCR can only be applied in selected cases (60,61) due to contamination of the sample. In any case, the PCR broad primers are the same to detect bacterial DNA (Table 1). However, it is possible to use molecular biology to identify the microbe once bacteria have grown in the culture, with identification of the species or strains (63–65).

Fungal Keratitis. When a fungus is suspected as the cause of keratitis, direct microscopy of scrapings can help us to identify yeast cells or mycelia but does not determine the species. Furthermore, about 20% of cases are neither diagnosed by direct visualization nor culture (66). Therefore, molecular diagnosis is of great help, both to confirm the fungal etiology and to identify the species; this assists with correct antifungal therapy. Given that many fungi cause keratitis, the use of broad-spectrum primers is recommended for their detection and identification by sequencing. On the other hand, it is possible to use molecular biology to characterize strains and to determine the relatedness between isolates such as occurred in the *Fusarium* keratitis outbreak associated with ReNu MoistureLoc contact lens multipurpose solution in the United States (40).

Molecular biology can be a rapid method to detect and identify ocular pathogens at species level. Bacterial and fungal speciation constitutes an important aid for effective treatment, facilitating the application of species-specific therapy, thereby avoiding problems of drug resistance; furthermore, it establishes a more precise epidemiology. The lack of species identification in corneal infections in the literature prevents precise knowledge of antimicrobial therapy efficiency and species epidemiology. Therefore, whenever possible, we recommend the use of universal or broad range primers to detect DNA of the pathogen and its identification by sequencing (18,38,86). Use of PCR combined with automated DNA sequencing provides great advantages compared to other forms of molecular identification.

Acanthamoeba **Keratitis**. In this case the diagnosis by PCR is highly suitable because cultures take several days to grow and may yield a false negative result. However, direct microscopy of a wet field scraping should also be used as it is quick (minutes), cheap (Chapter 6, cornea), and often positive. The most frequently used region to detect *Acanthamoeba* DNA is a specific region of the 18S rRNA gene (67,68) (Table 5). A genus-specific primer to amplify *Acanthamoeba* DNA can be used both for conventional and real-time PCR. Thus, in the diagnosis of *Acanthamoeba* keratitis PCR can increase the sensitivity as well as the speed of diagnosis.

Viral Keratitis. Viral keratitis can be a secondary process to infection in other parts of the eye, such as the conjunctiva, so both corneal and conjunctival samples may be sent to the laboratory. In addition, in these cases, it is highly probable that determination of several infectious agents is requested (adenovirus, *Chlamydia*, HSV-I, etc.), so a 'multiple' PCR is usually performed, which is a PCR including specific primers for different pathogens, in order to establish the causal agent in a single sample (or by nested PCR). There are reports on the optimization of multiple PCR for the determination in a single step of adenovirus, *Chlamydia*, and herpes simple virus (HSV) (24); or herpes simple virus, varicella-zoster virus (VZV), and cytomegalovirus

Table 5 Commonly Used Primers for PCR in Diagnosis of Acanthamoeba, Chlamydial, and Viral Keratitis

Microorganisms	Primers	Sequence	DNA Region amplify	Fragment (bp)	Authors (reference)
Acanthamoeba	P1 GP f	5'-GTTTGAGGCAATAACAGGT-3'	18S rRNA	253	Lehmann (67)
	P1 GP r	5'-CCTAGTAAGCGCGAGTC-3'			
	P2 GP f	5'-TCCCCTAGCAGCTTGTG-3'	18S rRNA	290	
	P2 GP r	5'-GTTAAGGTCTCGTTCGTTA-3'			
	(forward)	5'-GGCCCAGATCGTTTACCGTGAA-3'	18S rRNA	468	Pasricha (68)
	(reverse)	5'-TCTCACAAGCTGCTAGGGGAGTCA-3'			
Chlamydia trachomatis	(forward)	5'-GGATAGAGTAGTGGTCATCT-3'	KDO transferase intergenic region	160	Girjes (94)
	(reverse)	5'-GATAGCGTAGTCGGATT-3'			
	KL1	5'-TCCGGAGCGAGTTACGAAGA-3'	Plasmid	240	Elnifro (24)
	KL2	5'-AATCAATGCCCGGGATTGGT-3'			
HVS	Forward	5'-ATCCGAACGCAGCCCGCTG-3'	Glycoprotein D	400	Read (95)
	Reverse	5'-TCCGG(G/C)GGGAGCAGGGTGCT-3'			
	Forward	5'-CCGGCGTCAGCGAGGATAAC-3'	Glycoprotein D	265	Read (95)
	Reverse	5'-AGCTGTATA(G/C)GGGCGACGGTG-3'			
HVS-1	YSI	5'-GACTACCTGGAGATCGAG-3'	Polymerase de DNA	211	Elnifro (24)
	YS2	5'-TGAACCCGTACACCGAGT-3'			
HVS-2	R1	5'-ACGTACTACCGGCCTCAG-3'	US-4	258	Druce (96)
	R1	5'-CCACCTCTACCCACAAC-3'			
	R2	5'-CCGCGCCTGCCGTCAGCCCATCCTC-3'	US-4	225	
	R2	5'-AGACCCACGTGCAGCTCGCCG-3'			
VZV	R1	5'-ATGTCCGTACAACATCAACT-3'	Matrix	450	Druce (96)
	R1	5'-GGAGACTGGATATGCACGCATC-3'			
	R2	5'-CGATTTTCCAAGAGAGACGC-3'	Matrix	100	
	R2	5'-GGTGGAGACGACTTCAATAGC-3'			
Adenovirus	AdTU7	5'-GCCACCTTCTTCCCCATGGC-3'	Hexon conserved region	1000	Uchio (97), Saitoh-Inagawa (98), Takeuchi (99,100)
	AdTU4	5'-GTAGGGTTGCCGGCCGAGAA-3'			
	AdnU-S	5'-TTCCCCATGGCNCACAACAC-3'	Hexon region conserved	950	
	AdnU-A	5'-GCCTCGATGACGCCGCGGTG-3'			
	ADRJC1	5'-GACATGACTTTCGAGGTCGATCCCATGGA-3'	Hexon region	160	Elnifro (24), Cooper (101)
	ADRJC2	5'-CCGGCTGAGAAGGGTGTGCGCAGGTA-3'			

(CMV) (25); or HSV-1, -2, VZV, CMV, Epstein-Barr virus (EBV), and HHV-6 (69); HVS-I, HVS-2, and VZV (70,71). In all these cases the size of the amplified band determines the causal agent (101,102).

ROUTINE USE OF PCR TO DIAGNOSE OCULAR VIRAL AND CHLAMYDIAL INFECTIONS

Regis P. Kowalski
Charles T. Campbell Ophthalmic Microbiology Laboratoy,
University of Pittsburgh, Pittsburgh, Pennsylvania, U.S.A.

Introduction

In this era of managed care and high technology, diagnostic tests must be highly sensitive for detecting infection, highly specific for distinguishing from other infections, and cost-effective for the processing laboratory. Many diagnostic laboratories are under budgetary restraints and concentrate on daily high-volume testing, batch low-volume testing, and refer infrequent testing to other laboratories. The challenge for the ophthalmologist is to locate diagnostic laboratories that will deliver testing that will assist in managing ocular infections. Once located, the practitioner must clarify the microbiologic requirements of their practice and the results necessary to conclude a diagnosis. The ophthalmologist will fail if there is no communication with the microbiology laboratory.

Specimens need to be submitted in specific media; the incorrect medium will deem a specimen as "unacceptable." A frequent mistake by many physicians is to submit a soft-tipped applicator or commercial culturette to a laboratory and request the detection of every imaginable pathogen. This generally results in a complete waste of time for the patient, diagnostic laboratory, and ophthalmologist. The laboratory should provide all media and transport material.

Fortunately, the detection of virus using DNA PCR technology is available to most ophthalmologists. Many medical centers now have molecular facilities that process ocular specimens for PCR testing. It still requires communication between the ophthalmologist and molecular diagnostic facility to produce a beneficial relationship. The molecular facility will need to validate their testing for ocular samples. It is possible that certain molecular facilities will resist this extra work because it may not be worthwhile for validation, if there is only a small increase in workload volume.

Most molecular diagnostic facilities routinely test for HSV (1 and 2), VZV, CMV, EBV, and *Chlamydia trachomatis* DNA using PCR. These are also frequent ocular pathogens. The University of Pittsburgh Medical Center Molecular Facility (Pittsburgh, Pennsylvania, U.S.A.) also tests for adenovirus DNA, but this may not be offered by other molecular facilities. A description follows to educate the ophthalmologist in effectively using PCR to diagnose ocular viral and chlamydial infections.

Collection of Patient Samples for PCR Testing

Samples for PCR testing are easily collected from the conjunctiva and cornea using soft-tipped applicators made of dacron or using a kimura spatula to collect epithelium or tissue. Intraocular specimens from the aqueous or vitreous can be collected with a needle and syringe (see Chapters 6 and 8). The collected specimens can be transferred to 2.0 mL or less of viral transport medium. Bartels Chlamydia Transport Medium (Jamestown, New York, U.S.A.) has proven to be suitable for PCR testing of virus and chlamydia (102). Some molecular laboratories may have other medium for transport; thus, it is important that this information be obtained by the submitting ophthalmologist. Specimens should be transported to the processing facility as soon as possible, but delays under refrigerated conditions generally do not adversely affect the sample for PCR testing. Delays under variable temperature conditions could affect other supplement testing, such as cell culture isolation.

Selected clinical samples (HSV, adenovirus) for PCR should also be tested with cell culture isolation to confirm a definitive diagnosis. PCR results should not always be considered definitive and are a function of collection after onset of infection. For example, a specimen collected one day after adenovirus infection onset is more definitive for indicating infection compared to a sample collected seven days after onset.

A problem in testing ophthalmic samples with PCR is the use of diagnostic stains such as fluoroscein and rose bengal. Real-time PCR uses TacMan® probes to detect amplified product, and these stains will interfere with the probes (102) The use of stains cannot be avoided and this limitation must be brought to the attention of the molecular facility. Most laboratories use extraction techniques to remove protein and concentrate DNA for detection; these techniques generally will remove the stains from the sample (102).

All molecular diagnostic facilities should be expected to produce PCR results within 24 to 72 hours. Numerous delays beyond 72 hours should be questioned and deemed unacceptable for patient care by the ophthalmologist.

PCR Testing for Adenovirus

The Charles T. Campbell Ophthalmic Microbiology Laboratory at the University of Pittsburgh Medical Center has routinely processed ocular specimens for adenovirus DNA using PCR since 2001. PCR was able to detect adenovirus DNA from 91 submitted clinical samples, and these highly correlated with cell culture and shell vial testing. In a separate study to validate real-time PCR, it was found that in comparison with cell culture, PCR was 85% sensitive, 98% specific, 98% positivity predictive, 85% negatively predictive, and 92% efficient (102). These descriptive statistics indicate that adenovirus PCR testing is not definitive and should be confirmed with cell culture isolation. It must be noted that this study included all culture-positive samples collected at different times of onset of infection (one day to two weeks). PCR results from samples collected within three days of onset of infection could reasonably be definitive without supplementary confirmation. The laboratory supplements testing with A549 cell culture isolation and a three-day shell vial test for viable virus. Rapid enzyme immunoassays (Adenoclone, Adenotest, or RPS) have not produced high sensitivity to supplant cell culture or PCR testing.

PCR Testing for Herpes Simplex Virus (HSV)

We have routinely processed ocular specimens for HSV (1 and 2) DNA using PCR since 2004. The laboratory supplements testing with A549 cell culture isolation and a one-day shell vial test (ELVIS—Enzyme Linked Virus Inducible System) for viable virus. PCR was able to detect HSV (1 and 2) DNA from 51 submitted clinical samples and these did not correlate with cell culture or shell vial testing. Of these 51, 27 (53%) were PCR-positive and cell culture-positive; 24 (47%) were PCR-positive and cell culture-negative. Three PCR negative samples were positive for HSV in cell culture. Only three samples were determined to be HSV-2. These results suggest that cell culture isolation is not reliable for indicating the presence of HSV in ocular samples and it also poses the question of the importance of positive PCR results when cell culture is negative. Should patients with positive PCR and culture-negative results be treated for HSV infection? This question is further complicated when samples are submitted from patients who have a low suspicion for HSV infection but the keratitis includes a remote differential for HSV infection. Kaufman reported the presence of HSV DNA (without cell culture confirmation) in the tears of healthy volunteers (103). This study supports the idea that HSV may be present in the eye and not causing disease. In a separate study to validate real-time PCR, it was found that in comparison with cell culture, PCR was 98% sensitive, 100% specific, 100% positivity predictive, 91% negatively predictive, and 95% efficient (102). All of this information suggests that the diagnosis of non-classical HSV ocular infection based on positive PCR results alone should be viewed with caution. Positive cell culture and other clinical support should be considered prior to accepting a HSV diagnosis based on positive PCR results alone.

PCR Testing for Varicella-Zoster Virus (VZV)

We have routinely processed ocular specimens for VZV DNA using PCR since 2001. The laboratory supplements PCR testing with A549 cell culture isolation. Although A549 cells are acceptable for cell culture isolation of VZV, fibroblast cell lines such as MRC5 may be better. PCR was able to detect VZV DNA from 28 submitted clinical samples and these did not correlate with cell culture isolation. Of the 28 clinical specimens, only 4 (14%) were positive for cell culture isolation. In a separate study to validate real-time PCR, it was found that in comparison with PCR positive results from another molecular diagnostic laboratory, PCR was 100% sensitive, 100% specific, 100% positivity predictive, 100% negatively predictive, and 100% efficient (102). These results indicate that PCR is the best test for detecting VZV in ocular samples and cell culture is not reliable for making a laboratory diagnosis.

PCR Testing for *Chlamydia trachomatis*

PCR is the best test for detecting *Chlamydia trachomatis* in comparison with other laboratory diagnostic tests (102,104–106). PCR is highly sensitive and specific, while other tests such as giemsa staining for inclusion bodies within cells, indirect immunofluorescence, enzyme immunoassays, and serum titers are less sensitive and specific and only support or are indicative of a chlamydial infection. PCR testing for chlamydial DNA is available widely because *Chlamydia trachomatis* is a sexually transmitted disease that is routinely tested in many clinics. The Roche Amplicor® CT/NG kit is commercially available to detect *Chlamydia trachomatis* and *Neisseria gonorrhoeae* DNA. This system demonstrated 88% sensitivity and 100% specificity with ocular specimens (104). The addition of PCR to the system for *Neisseria gonorrhoeae* is an advantage to detect an infrequent cause of ocular infection that may be difficult to isolate on routine bacterial media. A *Neisseria gonorrhoeae*

PCR-positive, culture-negative specimen was reported for one patient. We have routinely processed ocular specimens for *Chlamydia trachomatis* DNA using PCR since the inception of the Roche Amplicor® kit and the ophthalmic laboratory no longer supplements PCR testing with any other diagnostic test. In a separate study to validate real-time PCR, it was found that in comparison with cell culture, PCR was 94% sensitive, 100% specific, 100% positivity predictive, 98% negatively predictive, and 99% efficient. Sensitivity based on PCR from another molecular diagnostic laboratory was 86% (102).

PCR Testing for Cytomegalovirus (CMV), Epstein-Barr Virus (EBV), and Other Entities

There is no validation data for using PCR to detect the DNA of CMV, EBV, or other viral entities from ocular samples. This should not prevent a molecular diagnostic laboratory from testing for these infectious agents as long as validation for the test itself with specimens from other body sites has been established. Vitreous and aqueous samples have been successfully tested for CMV and this data does support that DNA detection is feasible (107). Many laboratories will report the PCR result with the limitation that the testing was not validated for ocular specimens.

Caution should be used for interpreting EBV ocular infection by the presence of DNA. Plugfelder reported that EBV can be detected in the lacrimal glands and tears of normal patients (108). We do not expect EBV DNA to be normally present in the aqueous or vitreous; thus, a positive test in these samples should be considered significant.

With the recognition of new primers and probes for many pathogens, PCR has unlimited potential for diagnosing unusual and infrequent infections. Many molecular diagnostic laboratories have the capability to refer unusual requests to laboratories that test for infrequent pathogens. In conclusion, it must be reemphasized that communication is the key for fully implementing PCR in ophthalmology.

MOLECULAR ANALYSES OF *ACANTHAMOEBA* AND CORRELATION OF MOLECULAR GENOTYPES WITH MORPHOLOGICALLY BASED CLASSIFICATION SYSTEMS AND PATHOGENICITY

Gregory C. Booton
Department of Molecular Genetics, Ohio State University, Columbus, Ohio, U.S.A.

Challenges of Morphological Taxonomy in *Acanthamoeba*

Classification schemes of *Acanthamoeba* isolates based on morphological characteristics have led to considerable confusion because of the limited number of stable characters that can be used and because of the morphological plasticity of the organisms (109). Molecular approaches using immunological methods and isozyme electrophoresis has only been partially helpful. Identification of acanthamoebae within ocular and other tissues can be difficult and time consuming even for trained microscopists. In histological preparations, the amoebae look very similar to keratoplasts, neutrophils, or monocytes, and this can often lead to false-negative and/or false-positive results. It has been estimated, for example, that 70% of *Acanthamoeba* keratitis (AK) cases are misdiagnosed clinically as viral keratitis (110–113). Thus, many patients are initially treated with inappropriate drug therapies. This lag time may hamper disease resolution as several studies have found that earlier diagnosis and treatment results in lower morbidity and better visual outcome (111,113–114).

Continuing problems with classification of species has led to confusion over which species are the agents of disease. At one extreme, it is speculated that all *Acanthamoeba* strains are potentially pathogenic to humans, causing a variety of conditions including AK, skin and lung infections, and fatal encephalitis cases (115–118). Recent evidence suggests that this does not seem to be the case in AK infections that seem to involve tear IgA deficiencies, although it might be for patients with more serious immunological deficiencies. The role of host immunological deficiencies in the pathogenicity of *Acanthamoeba* spp. is also being investigated (119–122). Another avenue of investigation is based primarily on the use of ribosomal RNA gene sequences for the identification of genetic lineages within this genus (123). Using this approach, as discussed below, it appears that AK may be caused by a more limited number of rDNA genotypes than is the case with GAE. If this conclusion is correct, it then becomes important to know whether the selectivity is due to the relative abundance of particular genotypes in the environment, or because some lineages are virulent and others are not.

Nuclear rDNA Molecular Analysis Leads to Reliable Identification of *Acanthamoeba* Genotypes

Fifteen genotypic classes (originally called sequence types) of *Acanthamoeba* have been identified, or proposed, by phylogenetic analyses of more than 200 nuclear 18S rDNA sequences from 24 of the 25 named species and many unnamed strains (124–128). A genotype in *Acanthamoeba* was arbitrarily defined as including all strains whose 18S rRNA gene sequences are less than 5% divergent from one another in a standard alignment of the sequence. Thus, different strains within a "genotype" may have different 18S rDNA sequences. The overall topology of genotype relationships is shown in Figure 8. The strains used have included isolates from AK, other human diseases, and the environment. These isolates have included morphological species-type strains, whenever available. Unfortunately, type strains are not always available. Therefore, numerous other isolates were also investigated. The 18S rRNA genes of this genus are relatively large (compared to other eukaryotic ribosomal small subunit rRNA genes) due to a number of highly variable expansion regions and, in a few cases, the presence of introns (124,129–131). Although most strains have a gene of about 2300 bp, some are longer than 3,000 bp. The highly variable regions are not useful for phylogenetic analysis of the genus as a whole, primarily because of the problems associated with alignment, but are extremely useful for looking at relationships among the more closely related strains. The problematic identification of some proposed new genotypes is still under evaluation based on genotype definitions, on the alignment of sequences, and on the availability of samples. That said, the data have successfully illuminated relationships between some genotypes and many of the named morphological species of *Acanthamoeba*.

The 18S rDNA sequence data confirm that *A*. sp. [to be named] (T1), *A. palestinensis* (T2), *A. griffini* (T3), *A. lenticulata* (T5), *A. astronyxis* (T7), *A. tubiashi* (T8), *A. comandoni* (T9), *A. culbertsoni* (T10), *A. hatchetti* (T11), and *A. healyi* (T12), are species identifiable on the basis of both morphological and molecular characteristics. The data indicate, however, that *A. castellanii, A. lugdunensis, A. polyphaga,* A. *rhysodes,* A. *divionensis,* A. *mauritaniensis,* A. *quina,* A. *terricola,* and

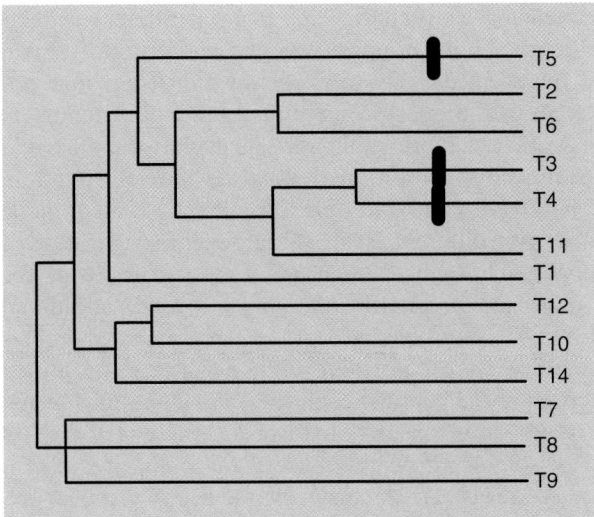

Figure 8 18S rDNA gene tree for *Acanthamoeba* genotypes.

A. triangularis are so closely related that they cannot be distinguished reliably on the basis either of morphology or 18S rDNA sequences. The last two species also were identical based on enzyme electrophoresis (132). All are members of the genotypic class T4. This T4 group is temporarily referred to as the "*A. castellanii* species complex" (G. Booton, unpublished data, 2002). This conclusion also is supported by RFLP studies of 18S rDNA (riboprints) (133). However, as discussed below, recent work suggests that subdivision of the T4 complex now may be partially achieved based on statistically significant clusters of strains sharing nearly identical DNA sequences of 16S mitochondrial rRNA genes, thus extending the correlation between DNA sequences and morphological-type strains (134).

The results also indicate that *A. pustulosa* (T2), *A. pearcei* (T3), and *A. stevensoni* (T11) are invalid synonyms for other species names. More recent studies suggest three additional *Acanthamoeba* genotypes. Two of these genotypes, T13 and T14, are associated with isolates of unnamed species (126–127). The third new genotype (T15) is associated with the species *Acanthamoeba jacobsi*, which is genetically very distinct from other *Acanthamoeba* genotypes (135).

Measures of relationships among strains identified using "complete" nuclear 18S rDNA sequences are the most reliable means available at present to discriminate between new isolates. These sequences should be used as the standard for evaluating relationships based on other approaches. Sequence alignments can reveal and correct for the presence of inserts such as introns, which have been found in some T3, T4, and T5 isolates (124,129–131,136, 137). Introns can distort apparent relationships between strains that are determined based on 18S rDNA restriction fragment length polymorphisms (RFLPs).

Mitochondrial 16S rDNA Sequences Supports and Further Resolves the Nuclear 18S rDNA Phylogeny

Several authorities have questioned the validity of phylogenetic trees based only on 18S rDNA gene sequences (138). Thus, the validity of the nuclear 18S phylogenetic gene tree was evaluated by sequencing a different gene, from a different cellular compartment. The 16S rRNA gene of mitochondria was chosen because it has a similar function to 18S rDNA, but seemed likely to evolve at a different rate and would be under somewhat different evolutionary constraints in the mitochondrion. In addition, mitochondrial genes have a more rapid rate of evolution and could potentially help to resolve relationships within *Acanthamoeba* species found in genotype T4. Also, mitochondrial DNA may be to able to determine if identifiable sub-lineages of T4 are associated with different levels of pathogenicity, or type of infection, caused by *Acanthamoeba*. Thus, the 16S mtRNA gene from 26 of the strains of *Acanthamoeba* previously used for 18S rDNA analysis, plus 42 additional strains, were analyzed (134, 139). All genotypes except T6 were included in this study. The genetic lineages identified were consistent with the lineages identified using the nuclear gene, thus increasing the confidence in the 18S rDNA phylogeny. In contrast to the nuclear gene results, however, the phylogenetic analysis of the 16S rDNA T4 sequences resolved the isolates into eight clusters, or sublineages of T4. Individual strains of T4 *Acanthamoeba* within each cluster shared nearly identical 16S rDNA sequence. In two cases, the clusters included morphological-type species, providing a correlation between genotype and morphology. Thus, the mitochondrial rDNA has proven useful in subdividing the T4 lineage, which is important for two reasons. First, this result may aid in determining how individual subsets correlate with morphological species and, second, whether these "subclades" within the T4 genotype have epidemiological significance.

Large-Scale Nuclear 18S rDNA Analysis Supports mt 16S Results

An analysis of a large data set of partial and complete nuclear 18S rRNA gene sequences is presented in Figure 9. This analysis resulted in seven subclades within T4, most of which correspond to subclades established by the mitochondrial 16S rRNA study. Thus, nuclear gene sequences support the substructure within T4 suggested by the mitochondrial 16S rDNA analyses. One subclade of particular interest within T4 (Sc2 in Fig. 9) occurs in both the nuclear and the mitochondrial analysis. This subclade contains the widely studied strain *Acanthamoeba castellanii* Neff. *A. castellanii* Neff and the other isolates in this T4 subclade are environmental isolates, and are not associated with AK, or other non-*Acanthamoeba* keratitis (NAK), infections. Consequently, *A. castellanii* Neff's general use as the model isolate for *Acanthamoeba* studies, including AK, may not be especially valid, as it relates to the potentially pathogenic isolates of *Acanthamoeba*. The pattern of subclades within T4 may reflect the potential pathogenicity of *Acanthamoeba* isolates. That is, when the T4 genotype subclades are compared, there are differences in those subclades associated with AK and with NAK infections. Whereas six of the seven T4 subclades (the exception being the Sc2 subclade containing *A. castellanii* Neff) contain numerous isolates from AK infections, only three of the sublclades are associated with more than one isolate from NAK infections. A fourth subclade contains only a single isolate of a NAK infection within the subclade. Therefore, the determination of gene sequence differences that separate the *A. castellanii* Neff subclade from the other subclades within T4, and a determination of how they may be correlated to pathogenicity, is of a great deal of interest and the focus of ongoing studies (AKpath vs. NONpath, AKpath vs NAKpath, etc.).

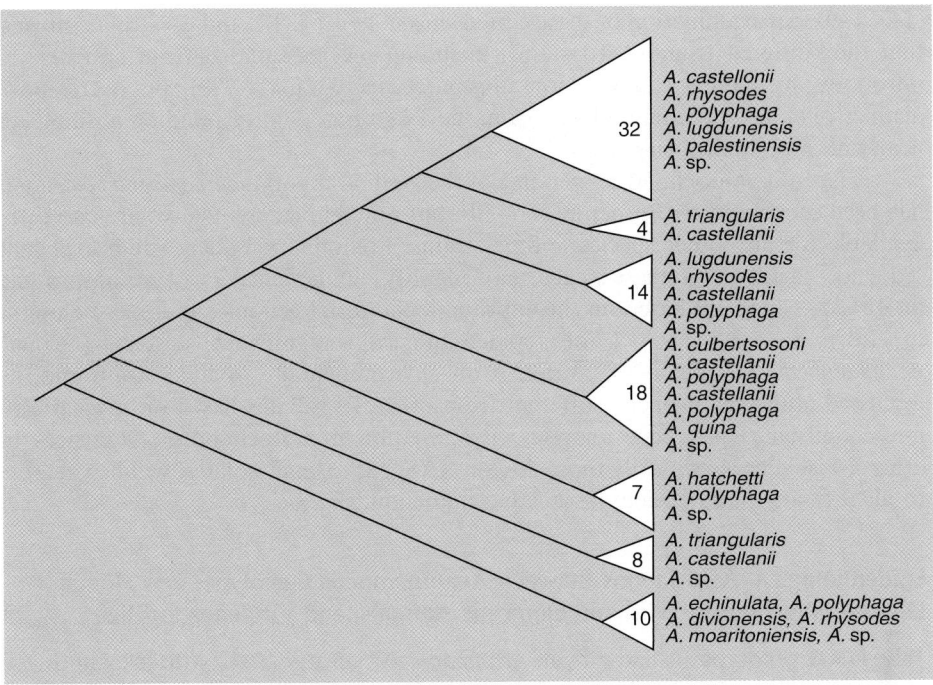

Figure 9 Genotype T4 subclades.

Whole Mitochondrial Genome RFLPs (mtRFLPs) Reveal Sequence Variation Among all Strains Tested

RFLPs are electrophoretic patterns obtained following digestion of DNA with restriction enzymes that cut at specific sites and produce fragments of varying length depending on the distribution of target sites in the DNA. A number of studies examining *Acanthamoeba* mtRFLPs have indicated that there is extensive interstrain sequence variation in the total mitochondrial genome among most or all strains tested (140–146). Thus, future comparative sequencing of the mitochondrial genomes from selected strains will undoubtedly reveal sequence diversity and, most likely, phylogenetic diversity of gene content that can be correlated with the phylogeny of the rDNA genes that have been studied to date. In addition, epidemiological significance with regard to AK and NAK pathogenic and non-pathogenic genotypes can also be investigated.

Mitochondrial transfer RNA (tRNA) gene sequence variation, although only available for genotypes T1–T4, is consistent with the 16S and 18S rDNA phylogenies. In the mitochondrial genome of *Acanthamoeba*, 5–6 tRNA gene sequences are in a spacer region adjacent to the 16S rRNA gene. Thus, it was relatively easy to sequence these genes when studying the rDNA and 80 tRNA genes sequences, including 5 different genes from each of 16 strains representing genotypes T1-T4 (139). Because tRNA genes are small and relatively conserved in sequence, these sequences were not especially useful for phylogenetic analyses. However, prior work had discovered corrective editing of tRNA amino acid acceptor stem base pairs that were mismatched due to the nucleotide sequence of the corresponding tRNA gene. The phenomenon had only been studied in the non-pathogenic T4 Neff strain

(147–149). An examination of these edited sites in five T1, T2, and T3 strains showed that they differed from 12 T4 stains, including several isolated from patient AK infections, by having slightly more mismatches (10–11 vs. 7–9), but were more distinctive in having different bases from the T4 strains at corresponding positions in nearly all mismatched sites (139).

The entire mitochondrial genome of the Neff strain of *Acanthamoeba castellanii* has been sequenced by Burger et al. (150), but no complete sequences are currently available for any other species and/or strains. Preliminary evidence indicates that sequence variation of these genes may identify similar clusters of strains as the mtRFLPs, which are based on the entire mitochondrial genome, and may be useful in efforts to subdivide T4 into epidemiologically significant clusters. Further, *Acanthamoeba castellanii* Neff strain is an environmentally isolated strain and, as discussed above, is distinctly different from other T4 isolates based on nuclear and mitochondrial rRNA gene analyses. Examination of mitochondrial sequences in other T4 isolates, specifically those AK or NAK pathogenic isolates, will be valuable to identify any pathogenicity-associated mitochondrial loci.

Epidemiological Associations Between *Acanthamoeba* Genotypes and Human Disease, Infections of Wild and Domestic Animals, and Virulence to Mice

18S rDNA genotypes isolated from *Acanthamoeba* keratitis (AKpath) are almost all identified as genotype T4. The species comprising T4 genotypes are of special interest because nearly all keratitis isolates that have been examined belong to this single lineage (125,151). There are few exceptions of keratitis isolates from other genotypes that have been observed and most all of those rare exceptions are identified as genotypes T3 (*A. griffini*) and T11 (*A.* sp.) (152). Genotypes T3, T4, and T11 are very closely related and are not resolved in some phylogenetic analyses, which may explain why these three genotypes have all been observed in AK infections. A single isolate from genotype T6 from an AK infection also has been reported (153). In addition, a single isolate of T5 associated with AK has recently been reported (154). Two interesting questions arise. The first question is why other *Acanthamoeba* genotypes have not, or very rarely, been found in AK. Two AK isolates have been classified as morphological species *A. culbertsoni* and *A. hatchetti,* but these may be species misidentifications, because the morphological-type strains are classified as T10 and T11, respectively (155–156). Because the rDNA genotypes of these particular strains have not been tested, their classification is uncertain. However, previous molecular analyses have identified other strains that have valid T4 genotypes but were identified morphologically as *A. culbertsoni* or *A. hatchetti.* Thus, these were apparently misidentified morphologically (125). The second question arises because initially the differentiation of species based on 18S rDNA within genotype T4 was unreliable. Is there some way to reliably subdivide the T4 group of strains and, if so, how does such subdivision relate to pathogenesis, ecological niche, or other relevant biological factors?

Human CNS infections (NAKpath) involve genotypes other than T3 and T4. *A. palestinensis*, *A. astronyxis*, and *A. culbertsoni* have been reported from CNS infections based on criteria other than DNA sequences (156–157). If the identifications are correct, and they may not be, then T2, T7, and T10 are able to cause GAE. However, DNA sequencing has definitely confirmed the genotypes of *A.* species strain V006 (T1), *A.* sp. (T10), and *A. healyi* (T12), which were isolated from GAE infections (124). *A. castellanii* and *A. rhysodes* (T4) have both previously been reported in CNS infections (157). However, until recently, isolates were not available to test whether T4 strains also are able to cause GAE. Recent examination of seven *Acanthamoeba* samples obtained from HIV-infected patients, including skin, lung, and GAE samples, have shown that these isolates are genotype T4 (158). Thus, these results demonstrate that in addition to its prevalent involvement in AK, genotype T4 is capable of causing other types of amebic infections. Because these infections occur in severely immunocompromised patients, whereas keratitis occurs in apparently healthy patients, it would not be surprising to discover that a broader spectrum of genotypes would infect the less healthy patients. Many more GAE isolates are available now, and are being examined to further address this question.

Animal Experiments Suggest Strains of Genotypes T5, T10, and T11 Are Potentially Pathogenic

The intracerebral or intranasal injection of mice has been used to test the virulence of *Acanthamoeba* strains. In one study, 36 strains representing 19 species were used in experimental infection of mice (159). In this study, *A. culbertsoni* A-1 (T10), *A. hatchetti* BH-2 (T11), and *A. lenticulata* strains PD2S and 7327 (T5) were virulent in at least one of the tests. The rest of the strains, which included genotypes T2, T3, T4, T5, and T7, were not virulent in these tests. The results of T5 indicate with this type of testing that a single genotype can include both virulent and nonvirulent strains. Nearly all T5 isolates that have been examined have been from environmental sources. Recently, a single genotype T5 isolate associated with an AK infection has been described (154). Thus, a survey of the ribosomal rDNA sequences from a larger set of strains that have been tested for virulence in mice is needed to determine whether there is any consistent relationship between genotype and virulence to the host. In combination with studies on genotypes that have been found in human infections, a survey of lineages that are virulent to mice would help determine how well this virulence predicts pathogenicity in humans (158).

Environmental Isolates Contain Genotypes Not Observed in Infections

A large number (>80) of *Acanthamoeba* environmental isolates have now been examined at the molecular level. A substantial portion of these isolates are T4, the genotype associated with the vast majority of AK infections. However, in addition to disease-associated genotypes, environmental isolates contain other distinct genotypes thus far not found in disease isolates. Most prominently (i.e., associated with a large number of environmental isolates) is genotype T5 (*A. lenticulata*), which has been found in numerous environmental samples, often associated with polluted or sewage dump locations. However, except for the recent identification of a single T5 AK case (154), genotype T5 has not been previously observed in an AK or NAK infection isolate. An argument can be made that many of the T5 isolates are found in environmental locations that are unlikely to permit exposure to humans; however, T5 isolates have been found in close proximity to T4 strains in beach sampling, and tap water, surveys (160). Therefore, at least in some cases, T5 strains are in locations where interactions with individuals are possible, if not likely. In addition to genotype T5 we have also found T2 (*A. palestinensis*), T7 (*A. astronyxis*), T8 (*A. tubiashi*), and T9 (*A. comandoni*) from environmental samples, none of which have to date been isolated from disease cases. Three of these (T7, T8, and T9) represent the *Acanthamoeba* morphological group I taxa, which are larger in size than the other

Acanthamoeba species, and whose distinct morphological differences has been strongly supported by molecular analyses. While thus far not found in disease samples, *A. lenticulata* (T5) strains have been shown to be virulent in animal model testing. In contrast, when T2 and T7 were tested on animal models, they were not virulent. However, isolates of some genotypes that are clearly pathogenic in some cases (T3 and T4) have strains that were not virulent in animal models (158). In addition to human infections, *Acanthamoeba* infections of tissues from many other animals including fish, birds, horse, reptiles, cattle, anteaters, dogs, rabbits, and water buffalo have been reported (156,161–164).

Acanthamoeba Isolates Can Be Summarized into Genotypic Groups

As discussed above, the large number of *Acanthamoeba* isolates that have been examined now permits a better understanding of the pathogenic and non-pathogenic species/strains of this ameba genus. AK infections are dominated by genotype T4, and less often by genotypes T3 and T11. Single cases caused by T5 and T6 have been reported. The genotype group associated with the vast majority of AK cases (T3, T4, and T11) we refer to as the AK pathogenic group (AKpath). NAK infections caused by *Acanthamoeba* also contain a majority of T4 samples, but also contain genotypes T1, T10, and T12, genotypes thus far not found in AK infections. We refer to this group as the NAK pathogenic group (NAKpath). Lastly, whereas many of the above listed genotypes are found in environmental isolates (e.g., T3, T4, and T6), environmental isolates also contain genotypes not yet found from disease samples. These include T2, T5, T7, T8, and T9. This group we refer to as the Non-pathogenic group (NONpath).

Acanthamoeba Proteins Potentially Associated with Variable Pathogenicity

In recent years a number of studies have sought to determine those proteins that may be correlated with pathogenicity in different strains of *Acanthamoeba*. These proteins and their corresponding genes provide a novel target for future studies to investigate the observed variable pathogenicity of *Acanthamoeba* isolates. In the following section we briefly review a number of these recently identified proteins and summarize their potential in future analyses.

Mannose-Binding Protein (MBP)

A number of recent studies have examined a protein found in *Acanthamoeba* with the ability to bind to mannose (120). This mannose-binding protein is hypothesized to be an important protein involved in the pathogenicity of *Acanthamoeba* because of its potential to facilitate initial binding to infected host cell membranes (165–167). The gene sequence of the mannose binding protein has now been determined in an *Acanthamoeba* isolate (168). The sequence determination of this gene will provide researchers with a means to study a variety of pathogenic and non-pathogenic strains of *Acanthamoeba* using this gene, whose protein product may play an important role in the variable pathogenicity of this genus. Variation in primary gene sequence, which will result in variable MBP following mRNA translation, may be correlated with the variable pathogenicity of *Acanthamoeba* species/strains/isolates.

Profilin

The protein-coding gene for profilin has been identified as a prospective target for study because it has the potential to be associated with increased pathogenicity in

Acanthamoeba. Profilin is involved in actin polymerizaton and therefore is associated with the movement rate of *Acanthamoeba* spp. It is possible that different profilin alleles (and their protein products) may be better adapted for faster movement, which may lead to increased pathogenicity. Profilin has been isolated and sequenced from *Acanthamoeba castellanii* Neff and has been studied extensively in this strain (169–171). *A. castellanii* Neff is an environmental sample and alleles found in *A. castellanii* Neff may differ in primary gene sequence, and subsequently in structure and function of the protein following translation, from those in the more pathogenic isolates. Profilin has been shown to be concentrated at the leading edge of movement of *Acanthamoeba* (172). This result is in agreement with profilin's role in actin polymerization and movement. Three closely related genes have been found in *Acanthamoeba*: profilin I, profilin Ib, and profilin II, and are potential targets of study in *Acanthamoeba* isolates (172).

Serine Protease

Higher levels of protease activity have been observed in pathogenic *Acanthamoeba* species versus non-pathogenic taxa (165,173). A subtilisin-like serine protease gene has been found in the pathogenic *Acanthamoeba healyi,* and the primary gene sequence was determined by Hong et al. (174). A much smaller partial sequence also was obtained from an *Acanthamoeba castellanii* isolate. These represent samples from the T12 and T4 genotypes, respectively.

Actophorin

In addition to mannose-binding protein, profilin, and the putative serine protease-like gene that have been identified in *Acanthamoeba*, another gene with a potential pathogenic role in *Acanthamoeba* has also has been identified. Quirk et al. (175) determined the primary sequence of actophorin from *Acanthamoeba castellanii*. Actophorin is an actin-depolymerizing factor (ADF). It has high similarity with proteins from vertebrates and echinoderms that also have actin-severing properties. Given its biochemical properties, it has the potential to act synergistically with profilin. However, actophorin has only been sequenced from a single *Acanthamoeba* genotype (T4), so it is impossible to know the level of variation of this gene in other *Acanthamoeba* genotypes.

Several other proteins with potential for association with pathogenicity have been identified in *Acanthamoeba*. These include a secreted cysteine proteinase locus (176) that has been identified in several strains and a cyst specific protein (177). There is also nucleotide sequence information available on over 350 Expressed Sequence Tag (EST) sequences (178). Analyses of these EST sequences indicate that about 40% show highly significant sequence similarity to known proteins, including several that would play a part in cell motility and membrane attachment, factors that are involved in pathogenicity.

Summary

Recent advances using molecular biological techniques and analytical methods have shown that particular genotypes of *Acanthamoeba* may be more pathogenic than others. This methodology is robust, repeatable, and much less subjective than previous classification methods that have heavily relied upon relatively plastic morphological characters. These methods allow for the accurate genotypic determination of new isolates that are obtained from AK and other infections. This acquired genotypic data can be correlated with infection type and provides further insight into the relative pathogenicity of various *Acanthamoeba* genotypes, and the types of disease they are capable of producing. Further, recent investigations into proteins that may interact with mammalian cell–surface binding proteins has led to another avenue of investigation into the particular genetic variation that may be found in more pathogenic genotypes of *Acanthaomeba*. Investigation of this variation at the gene level will allow for further elucidation of the specific characteristics of those *Acanthamoeba* that are more pathogenic, and should facilitate faster and more efficient treatment of infections caused by *Acanthamoeba*, including keratitis.

REFERENCES

1. Saiki RK, Scharf S, Faloona F, et al. Enzymatic amplification of beta-globin genomic sequences and restriction site analysis for diagnosis of sickle cell anemia Science 1985; 230:1350–1354.
2. Mullis KB, Faloona FA. Specific synthesis of DNA in vitro via a polymerase-catalyzed chain reaction Methods Enzymol 1987; 155:335–350.
3. Saiki RK, Gelfand DH, Stoffel S, et al. Primer-directed enzymatic amplification of DNA with a thermostable DNA polymerase Science 1988; 239:487–491.
4. Fredricks DN, Relman DA. Application of polymerase chain reaction to the diagnosis of infectious diseases Clin Infect Dis 1999; 29:475–486.
5. White TJ, Madej R, Persing DH. The polymerase chain reaction: clinical applications Adv Clin Chem 1992; 29:161–196.
6. Kan VL. Polymerase chain reaction for the diagnosis of candidemia J Infect Dis 1993; 168:779–783.
7. Okhravi N, Adamson P, Mant R, et al. Polymerase chain reaction and restriction fragment length polymorphism mediated detection and speciation of Candida spp causing intraocular infection Invest Ophthalmol Vis Sci 1998; 39:859–866.
8. Nosek J, Tomaska L, Rycovska A, Fukuhara H. Mitochondrial telomeres as molecular markers for identification of the opportunistic yeast pathogen Candida parapsilosis J Clin Microbiol 2002; 40:1283–1289.
9. Hidalgo JA, Alangaden GJ, Eliott D, et al. Fungal endophthalmitis diagnosis by detection of Candida albicans DNA in intraocular fluid by use of a species-specific polymerase chain reaction assay J Infect Dis 2000; 181:1198–1201.
10. Klappenbach JA, Saxman PR, Cole JR, Schmidt TM. rrndb: the Ribosomal RNA Operon Copy Number Database Nucleic Acids Res 2001; 29:181–184.
11. Maleszka R, Clark-Walker GD. Yeasts have a four-fold variation in ribosomal DNA copy number Yeast 1993; 9:53–58.
12. Kolbert CP, Persing DH. Ribosomal DNA sequencing as a tool for identification of bacterial pathogens Curr Opin Microbiol 1999; 2:299–305.
13. Woese CR. Bacterial evolution Microbiol Rev 1987; 51:221–271.
14. Lane DJ, Pace B, Olsen GJ, Stahl DA, Sogin ML, Pace NR. Rapid determination of 16S ribosomal RNA sequences for phylogenetic analyses Proc Natl Acad Sci U S A 1985; 82:6955–6959.
15. Haqqi TM, Sarkar G, David CS, Sommer SS. Specific amplification with PCR of a refractory segment of genomic DNA Nucleic Acids Res 1988; 16:11844.
16. Schmidt B, Muellegger RR, Stockenhuber C, et al. Detection of Borrelia burgdorferi-specific DNA in urine specimens from patients with erythema migrans before and after antibiotic therapy J Clin Microbiol 1996; 34:1359–1363.

17. Carroll NM, Jaeger EE, Choudhury S, et al. Detection of and discrimination between Gram-positive and Gram-negative bacteria in intraocular samples by using nested PCR J Clin Microbiol 2000; 38:1753–1757.

18. Ferrer C, Colom F, Frases S, Mulet E, Abad JL, Alio JL. Detection and identification of fungal pathogens by PCR and by ITS2 and 5.8S ribosomal DNA typing in ocular infections J Clin Microbiol 2001; 39:2873–2879.

19. Hykin PG, Tobal K, McIntyre G, Matheson MM, Towler HM, Lightman SL. The diagnosis of delayed post-operative endophthalmitis by polymerase chain reaction of bacterial DNA in vitreous samples J Med Microbiol 1994; 40:408–415.

20. Therese KL, Anand AR, Madhavan HN. Polymerase chain reaction in the diagnosis of bacterial endophthalmitis Br J Ophthalmol 1998; 82:1078–1082.

21. Bej AK, Mahbubani MH, Miller R, DiCesare JL, Haff L, Atlas RM. Multiplex PCR amplification and immobilized capture probes for detection of bacterial pathogens and indicators in water Mol Cell Probes 1990; 4:353–365.

22. Geha DJ, Uhl JR, Gustaferro CA, Persing DH. Multiplex PCR for identification of methicillin-resistant staphylococci in the clinical laboratory J Clin Microbiol 1994; 32: 1768–1772.

23. Chichili GR, Athmanathan S, Farhatullah S, et al. Multiplex polymerase chain reaction for the detection of herpes simplex virus, varicella-zoster virus and cytomegalovirus in ocular specimens Curr Eye Res 2003; 27:85–90.

24. Elnifro EM, Cooper RJ, Klapper PE, Yeo AC, Tullo AB. Multiplex polymerase chain reaction for diagnosis of viral and chlamydial keratoconjunctivitis Invest Ophthalmol Vis Sci 2000; 41:1818–1822.

25. Jackson R, Morris DJ, Cooper RJ, et al. Multiplex polymerase chain reaction for adenovirus and herpes simplex virus in eye swabs J Virol Methods 1996; 56:41–48.

26. Whitcombe D, Newton CR, Little S. Advances in approaches to DNA-based diagnostics Curr Opin Biotechnol 1998; 9:602–608.

27. Whitcombe D, Brownie J, Gillard HL, et al. A homogeneous fluorescence assay for PCR amplicons: its application to real-time, single-tube genotyping Clin Chem 1998; 44: 918–923.

28. Jaeger EE, Carroll NM, Choudhury S, et al. Rapid detection and identification of Candida, Aspergillus, and *Fusarium* species in ocular samples using nested PCR J Clin Microbiol 2000; 38:2902–2908.

29. Anand AR, Madhavan HN, Therese KL. Use of polymerase chain reaction (PCR) and DNA probe hybridization to determine the Gram reaction of the infecting bacterium in the intraocular fluids of patients with endophthalmitis J Infect 2000; 41:221–226.

30. Okhravi N, Adamson P, Matheson MM, Towler HM, Lightman S. PCR-RFLP-mediated detection and speciation of bacterial species causing endophthalmitis Invest Ophthalmol Vis Sci 2000; 41:1438–1447.

31. Kumar M, Mishra NK, Shukla PK. Sensitive and rapid polymerase chain reaction based diagnosis of mycotic keratitis through single stranded conformation polymorphism Am J Ophthalmol 2005; 140:851–857.

32. Kumar M, Shukla PK. Use of PCR targeting of internal transcribed spacer regions and single-stranded conformation polymorphism analysis of sequence variation in different regions of rRNA genes in fungi for rapid diagnosis of mycotic keratitis J Clin Microbiol 2005; 43:662–668.

33. Kumar M, Shukla PK. Single-stranded conformation polymorphism of large subunit of ribosomal RNA is best suited to diagnosing fungal infections and differentiating fungi at species level Diagn Microbiol Infect Dis 2006; 56:45–51.

34. Okhravi N, Adamson P, Carroll N, et al. PCR-based evidence of bacterial involvement in eyes with suspected intraocular infection Invest Ophthalmol Vis Sci 2000; 41:3474–3479.

35. Lohmann CP, Heeb M, Linde HJ, Gabel VP, Reischl U. Diagnosis of infectious endophthalmitis after cataract surgery by polymerase chain reaction J Cataract Refract Surg 1998; 24:821–826.

36. Lohmann CP, Linde HJ, Reischl U. Improved detection of microorganisms by polymerase chain reaction in delayed endophthalmitis after cataract surgery Ophthalmology 2000; 107:1047–1051.

37. Knox CM, Cevallos V, Margolis TP, Dean D. Identification of bacterial pathogens in patients with endophthalmitis by 16S ribosomal DNA typing Am J Ophthalmol 1999; 128:511–512.

38. Ferrer C, Montero J, Alio JL, Abad JL, Ruiz-Moreno JM, Colom F. Rapid molecular diagnosis of posttraumatic keratitis and endophthalmitis caused by Alternaria infectoria J Clin Microbiol 2003; 41:3358–3360.

39. Ferrer C, Alio J, Rodriguez A, Andreu M, Colom F. Endophthalmitis caused by *Fusarium* proliferatum J Clin Microbiol 2005; 43:5372–5375.

40. Chang DC, Grant GB, O'Donnell K, et al. Multistate outbreak of *Fusarium* keratitis associated with use of a contact lens solution JAMA 2006; 296:953–963.

41. Amann RI, Krumholz L, Stahl DA. Fluorescent-oligonucleotide probing of whole cells for determinative, phylogenetic, and environmental studies in microbiology J Bacteriol 1990; 172:762–770.

42. Ugahary L, van de SW, van Meurs JC, van BA. An unexpected experimental pitfall in the molecular diagnosis of bacterial endophthalmitis J Clin Microbiol 2004; 42:5403–5405.

43. Fairfax MR, Metcalf MA, Cone RW. Slow inactivation of dry PCR templates by UV light PCR Methods Appl 1991; 1:142–143.

44. Cone RW, Fairfax MR. Protocol for ultraviolet irradiation of surfaces to reduce PCR contamination PCR Methods Appl 1993; 3:S15–S17.

45. Ou CY, Moore JL, Schochetman G. Use of UV irradiation to reduce false positivity in polymerase chain reaction Biotechniques 1991; 10:442, 444, 446.

46. Hughes MS, Beck LA, Skuce RA. Identification and elimination of DNA sequences in Taq DNA polymerase J Clin Microbiol 1994; 32:2007–2008.

47. Rand KH, Houck H. Taq polymerase contains bacterial DNA of unknown origin Mol Cell Probes 1990; 4:445–450.

48. Cimino GD, Metchette KC, Tessman JW, Hearst JE, Isaacs ST. Post-PCR sterilization: a method to control carryover contamination for the polymerase chain reaction Nucleic Acids Res 1991; 19:99–107.

49. Isaacs ST, Tessman JW, Metchette KC, Hearst JE, Cimino GD. Post-PCR sterilization: development and application to an HIV-1 diagnostic assay Nucleic Acids Res 1991; 19: 109–116.

50. Espy MJ, Smith TF, Persing DH. Dependence of polymerase chain reaction product inactivation protocols on amplicon length and sequence composition J Clin Microbiol 1993; 31:2361–2365.

51. Rys PN, Persing DH. Preventing false positives: quantitative evaluation of three protocols for inactivation of polymerase chain reaction amplification products J Clin Microbiol 1993; 31:2356–2360.

52. Meier A, Persing DH, Finken M, Bottger EC. Elimination of contaminating DNA within polymerase chain reaction reagents: implications for a general approach to detection of uncultured pathogens J Clin Microbiol 1993; 31:646–652.

53. Sharma S, Das D, Anand R, Das T, Kannabiran C. Reliability of nested polymerase chain reaction in the diagnosis of bacterial endophthalmitis Am J Ophthalmol 2002; 133:142–144.

54. Pinna A, Sechi LA, Zanetti S, Delogu D, Carta F. Adherence of ocular isolates of staphylococcus epidermidis to ACRYSOF intraocular lenses A scanning electron microscopy and molecular biology study. Ophthalmology 2000; 107:2162–2166.

55. Abreu JA, Cordoves L, Mesa CG, Mendez R, Dorta A, De la Rosa MG. Chronic pseudophakic endophthalmitis versus saccular endophthalmitis J Cataract Refract Surg 1997; 23:1122–1125.

56. Margo CE, Pavan PR, Groden LR. Chronic vitritis with macrophagic inclusions A sequela of treated endophthalmitis due to a coryneform bacterium. Ophthalmology 1988; 95:156–161.

57. Alexandrakis G, Sears M, Gloor P. Postmortem diagnosis of Fusarium panophthalmitis by the polymerase chain reaction Am J Ophthalmol 1996; 121:221–223.

58. Anand AR, Madhavan HN, Sudha NV, Therese KL. Polymerase chain reaction in the diagnosis of Aspergillus endophthalmitis Indian J Med Res 2001; 114:133–140.

59. Anand A, Madhavan H, Neelam V, Lily T. Use of polymerase chain reaction in the diagnosis of fungal endophthalmitis Ophthalmology 2001; 108:326–330.

60. Seo KY, Lee JB, Lee K, Kim MJ, Choi KR, Kim EK. Non-tuberculous mycobacterial keratitis at the interface after laser in situ keratomileusis J Refract Surg 2002; 18:81–85.

61. Ferrer C, Rodriguez-Prats JL, Abad JL, Alio JL. Unusual anaerobic bacteria in keratitis after laser in situ keratomileusis: diagnosis using molecular biology methods J Cataract Refract Surg 2004; 30:1790–1794.

62. Lohmann CP, Winkler von MC, Gabler B, Reischl U, Kochanowski B. [Polymerase chain reaction (PCR) for microbiological diagnosis in refractory infectious keratitis: a clinical study in 16 patients] Klin Monatsbl Augenheilkd 2000; 217:37–42.

63. Pinna A, Sechi LA, Zanetti S, et al. Bacillus cereus keratitis associated with contact lens wear Ophthalmology 2001; 108:1830–1834.

64. Frueh BE, Dubuis O, Imesch P, Bohnke M, Bodmer T. Mycobacterium szulgai keratitis Arch Ophthalmol 2000; 118:1123–1124.

65. Weber E, Carlotti A, Furtado RM. Microbial identification using a sequencing-based system: bacterial and fungal case studies Eur J Parent Pharmaceutical Science 2006; 11: 45–52.

66. Dunlop AA, Wright ED, Howlader SA, et al. Suppurative corneal ulceration in Bangladesh. A study of 142 cases examining the microbiological diagnosis, clinical and epidemiological features of bacterial and fungal keratitis Aust N Z J Ophthalmol. 1994; 22:105–10.

67. Lehmann OJ, Green SM, Morlet N, et al. Polymerase chain reaction analysis of corneal epithelial and tear samples in the diagnosis of Acanthamoeba keratitis Invest Ophthalmol Vis Sci 1998; 39:1261–1265.

68. Pasricha G, Sharma S, Garg P, Aggarwal RK. Use of 18S rRNA gene-based PCR assay for diagnosis of acanthamoeba keratitis in non-contact lens wearers in India J Clin Microbiol 2003; 41:3206–3211.

69. Robert PY, Traccard I, Adenis JP, Denis F, Ranger-Rogez S. Multiplex detection of herpesviruses in tear fluid using the "stair primers" PCR method: prospective study of 93 patients J Med Virol 2002; 66:506–511.

70. van Gelderen BE, Van der LA, Treffers WF, van der GR. Detection of herpes simplex virus type 1, 2 and varicella zoster virus DNA in recipient corneal buttons Br J Ophthalmol 2000; 84:1238–1243.

71. Tenorio A, Echevarria JE, Casas I, Echevarria JM, Tabares E. Detection and typing of human herpesviruses by multiplex polymerase chain reaction J Virol Methods 1993; 44: 261–269.

72. Chen K, Neimark H, Rumore P, Steinman CR. Broad range DNA probes for detecting and amplifying eubacterial nucleic acids FEMS Microbiol Lett 1989; 48:19–24.

73. Edwards U, Rogall T, Blocker H, Emde M, Bottger EC. Isolation and direct complete nucleotide determination of entire genes Characterization of a gene coding for 16S ribosomal RNA. Nucleic Acids Res 1989; 17:7843–7853.

74. Wilson KH, Blitchington RB, Greene RC. Amplification of bacterial 16S ribosomal DNA with polymerase chain reaction J Clin Microbiol 1990; 28:1942–1946.

75. Relman DA, Falkow S. Identification of uncultured microorganisms: expanding the spectrum of characterized microbial pathogens Infect Agents Dis 1992; 1:245–253.

76. Bergmans AM, Groothedde JW, Schellekens JF, van Embden JD, Ossewaarde JM, Schouls LM. Etiology of cat scratch disease: comparison of polymerase chain reaction detection of Bartonella (formerly Rochalimaea) and Afipia felis DNA with serology and skin tests J Infect Dis 1995; 171:916–923.

77. Kerkhoff FT, van der ZA, Bergmans AM, Rothova A. Polymerase chain reaction detection of Neisseria meningitidis in the intraocular fluid of a patient with endogenous endophthalmitis but without associated meningitis Ophthalmology 2003; 110:2134–2136.

78. Knox CM, Cevellos V, Dean D. 16S ribosomal DNA typing for identification of pathogens in patients with bacterial keratitis J Clin Microbiol 1998; 36:3492–3496.

79. Ferrer C, Ruiz-Moreno JM, Rodriguez A, Montero J, Alio JL. Postoperative Corynebacterium macginleyi endophthalmitis J Cataract Refract Surg 2004; 30: 2441–2444.

80. Montero JA, Ruiz-Moreno JM, Rodriguez AE, Ferrer C, Sanchis E, Alio JL. Endogenous endophthalmitis by Propionibacterium acnes associated with leflunomide and adalimumab therapy Eur J Ophthalmol 2006; 16:343–345.

81. Lohmann CP, Gabler B, Kroher G, Spiegel D, Linde HJ, Reischl U. Disciforme keratitis caused by Bartonella henselae: an unusual ocular complication in cat scratch disease Eur J Ophthalmol 2000; 10:257–258.

82. Hollander DA, Dodds EM, Rossetti SB, Wood IS, Alvarado JA. Propionibacterium acnes endophthalmitis with bacterial sequestration in a Molteno's implant after cataract extraction Am J Ophthalmol 2004; 138:878–879.

83. Bagyalakshmi R, Madhavan HN, Therese KL. Development and application of multiplex polymerase chain reaction for the etiological diagnosis of infectious endophthalmitis J Postgrad Med 2006; 52:179–182.

84. Sandhu GS, Kline BC, Stockman L, Roberts GD. Molecular probes for diagnosis of fungal infections J Clin Microbiol 1995; 33:2913–2919.

85. White TJTBSLaST. In 1.M.A. Innins DHGJJS&TJW, editor. PCR protocols A guide to methods and applications. 1990: 315–322.

86. Ferrer C, Munoz G, Alio JL, Abad JL, Colomm F. Polymerase chain reaction diagnosis in fungal keratitis caused by Alternaria alternata Am J Ophthalmol 2002; 133:398–399.

87. Tarai B, Gupta A, Ray P, Shivaprakash MR, Chakrabarti A. Polymerase chain reaction for early diagnosis of post-operative fungal endophthalmitis Indian J Med Res 2006; 123:671–678.

88. Makimura K, Murayama SY, Yamaguchi H. Detection of a wide range of medically important fungi by the polymerase chain reaction J Med Microbiol 1994; 40:358–364.

89. Gaudio PA, Gopinathan U, Sangwan V, Hughes TE. Polymerase chain reaction based detection of fungi in infected corneas Br J Ophthalmol 2002; 86:755–760.

90. Rishi K, Font RL. Keratitis caused by an unusual fungus, Phoma species Cornea 2003; 22:166–168.

91. Uy HS, Matias R, de la CF, Natividad F. Achromobacter xylosoxidans endophthalmitis diagnosed by polymerase chain reaction and gene sequencing Ocul Immunol Inflamm 2005; 13:463–467.

92. Lai JY, Chen KH, Lin YC, Hsu WM, Lee SM. Propionibacterium acnes DNA from an explanted intraocular lens detected by polymerase chain reaction in a case of chronic pseudophakic endophthalmitis J Cataract Refract Surg 2006; 32:522–525.

93. Alexandrakis G, Jalali S, Gloor P. Diagnosis of *Fusarium* keratitis in an animal model using the polymerase chain reaction Br J Ophthalmol 1998; 82:306–311.

94. Girjes AA, Carrick FN, Lavin MF. Single DNA sequence common to all chlamydial species employed for PCR detection of these organisms Res Microbiol 1999; 150: 483–489.

95. Read SJ, Kurtz JB. Laboratory diagnosis of common viral infections of the central nervous system by using a single multiplex PCR screening assay J Clin Microbiol 1999; 37:1352–1355.

96. Druce J, Catton M, Chibo D, et al. Utility of a multiplex PCR assay for detecting herpesvirus DNA in clinical samples J Clin Microbiol 2002; 40:1728–1732.

97. Uchio E, Aoki K, Saitoh W, Itoh N, Ohno S. Rapid diagnosis of adenoviral conjunctivitis on conjunctival swabs by 10-minute immunochromatography Ophthalmology 1997; 104:1294–1299.

98. Saitoh-Inagawa W, Oshima A, Aoki K, et al. Rapid diagnosis of adenoviral conjunctivitis by PCR and restriction fragment length polymorphism analysis J Clin Microbiol 1996; 34:2113–2116.

99. Takeuchi S, Itoh N, Uchio E, et al. Adenovirus strains of subgenus D associated with nosocomial infection as new etiological agents of epidemic keratoconjunctivitis in Japan J Clin Microbiol 1999; 37:3392–3394.

100. Takeuchi S, Itoh N, Uchio E, Aoki K, Ohno S. Serotyping of adenoviruses on conjunctival scrapings by PCR and sequence analysis J Clin Microbiol 1999; 37:1839–1845.

101. Cooper RJ, Yeo AC, Bailey AS, Tullo AB. Adenovirus polymerase chain reaction assay for rapid diagnosis of conjunctivitis Invest Ophthalmol Vis Sci 1999; 40:90–95.

102. Kowalski RP, Thompson PP, Kinchington PR, Gordon YJ. The Evaluation of the SmartCycler® II System for the Real-Time Detection of Viruses and Chlamydia from Ocular Specimens, Arch Ophthalmol 2006; 124:1135–1139.

103. Kaufman HE, Azcuy AM, Varnell ED, Sloop GD, Thompson HW, Hill JM. HSV-1 DNA in tears and saliva of normal adults Invest Ophthalmol Vis Sci 2005; 46:241–247.

104. Kowalski RP, Uhrin M, Karenchak LM, Sweet RL, and Gordon YJ. The evaluation of the Amplicor™ test for the detection of chlamydial DNA in adult chlamydial conjunctivitis Ophthalmology 1995; 102:1016–19.

105. Sheppard JD, Kowalski RP, Meyer MP, Amortegui AJ, and Slifken M. Immunodiagnosis of adult chlamydial conjunctivitis Ophthalmology 1988; 95:434–443.

106. Tantsira JG, Kowalski RP, and Gordon YJ. The evaluation of the Kodak Surecell Chlamydia Test for the laboratory diagnosis of adult inclusion conjunctivitis Ophthalmology 1995; 102:1035–37.

107. Pathanapitoon K, Ausayakhun S, Kunavisarut P, Pungrasame A, Sirirungsi W. Detection of cytomegalovirus in vitreous, aqueous, and conjunctiva by polymerase chain reaction (PCR) J Med Assoc Thailand 2005; 88:228–232.

108. Pfugfelder SC, Crouse CA, Pereira I, Atherton SS. Amplification of Epstein-Barr virus genomic sequences in blood cells, lacrimal glands, and tears from primary Sjogren's syndrome patients Ophthalmology 1990; 97:976–984.

109. Visvesvara GS. Classification of *Acanthamoeba* Rev Infect Dis 1991; 13:S369–S372.

110. Johns KJ, O'Day DM, Head WS, Neff RJ, Elliott JH. *Herpes simplex* masquerade syndrome: *Acanthamoeba* keratitis Curr Eye Res 1987; 6:207–212.

111. Bacon AS, Dart JS, Ficker LA, Matheson MM, Wright P. *Acanthamoeba* keratitis. The value of early diagnosis Ophthalomol 1993; 100:1238–1243.

112. Goodall KA, Brahma, Ridgeway A. *Acanthamoeba* keratitis: Masquerading as adenoviral keratitis Eye 1996; 10:643–644.

113. Seal DV, Bron AJ, Hay J. 1998. Ocular Infection Investigation and Treatment in Practice. Martin Dunitz, Ltd., London. 269 pp.

114. Bacon AS, Frazer DG, Dart JK, Matheson M, Ficker LA, Wright P. A review of 72 consecutive cases of *Acanthamoeba* keratitis Eye 1993; 7:719–725.

115. Tan B, Weldonlinne M, Rhone DP, Penning CL, Visvesvara GS. *Acanthamoeba* infection presenting as skin lesions in patients with the acquired-immunodeficiency-syndrome. Arch Path Lab Med 1993; 117:1043–1046.

116. Niederkorn JY, Alizadeh H, Leher H, McCulley JP. The pathogenesis of *Acanthamoeba* keratitis. Microbes and Infection 1999; 1:437–443.

117. Marciano-Cabral, Cabral GA. *Acanthamoeba* spp. As Agents of Disease in Humans Clin Microbiol Rev 2003; 16:273–207.

118. Marciano-Cabral F, Puffenbarger R, Cabral GA. The increasing importance of *Acanthamoeba* infections. J Eukarot Microbiol 2000; 47:29–36.

119. He YG, Mellon J, Pidherney M, Alizadeh H, McCulley JP, Neiderkorn J. A Pig Model of *Acanthamoeba* keratitis Invest. Ophthalmol Vis Sci 1991; 32:1182–1182.

120. Leher H, Silvany R, Alizadeh H, Huang J, Niederkorn JY. 1998. Mannose Induces the Release of Cytopathic Factors from *Acanthamoeaba castellanii*. Infection and Immunity. 66:5–10.

121. Hurt M, Apte S, Leher H, Howard K, Niederkorn J, Alizadeh H. Exacerbation of *Acanthamoeba* keratitis in animals treated with anti-macrophage inflammatory protein 2 or antineutrophil antibodies. Infection and Immunity 2001; 69:2988:2995.

122. Hurt M, Proy V, Niederkorn JY, Alizadeh H. The Interaction of *Acanthamoeba castellanii* Cysts with Macrophages and Neutrophils The Journ Parasitol 2003; 89: 565–572.

123. Sogin ML, Ellwood HJ, Gunderson JH. Evolutionary diversity of eukaryotic small-subunit rRNA genes. PNAS USA 1986; 83:1383–1387.

124. Gast RJ, Fuerst PA , Byers TJ. Discovery of group I introns in the nuclear small subunit ribosomal RNA genes of *Acanthamoeba*. Nuc Acids Res 1994; 22:592–596.

125. Stothard DR, Schroeder-Diedrich JM, Awwad MH, Gast RJ, Ledee DR, Rodriguez-Zaragoza S, Dean CL, Fuerst PA, Byers TJ. The evolutionary history of the genus *Acanthamoeba* and the identification of eight new 18S rRNA gene sequence types. J Euk Micrbiol 1998; 45:45–54.

126. Horn M, Fritsche TR, Gautom RK, Schleifer K-H, Wagner M. Novel bacterial endosymbionts of *Acanthamoeba* spp. related to the *Paramecium caudatum* symbiont *Caedibacter caryophilus*. Environ Micro 1999; 1:357–367.

127. Gast RJ. Development of an *Acanthamoeba*-specific reverse dot-blot and the discovery of a new ribotype. J Euk Microbiol 2001; 48:609–615.

128. Horn M, Fritsche TR, Linner T, Gauton RK, Harzenetter MD, Wagner M. Obligate bacterial endosymbionts of *Acanthamoeba* spp. related to the β-*Proteobacteria*: proposal of '*Candidatus* Protobacter acanthamoebae' gen. nov., sp. nov. Intl J Syst Evol. Micro 2002; 52:599–605.

129. Schroeder-Diedrich JM, Fuerst PA, Byers TJ. Group-I introns with unusual sequences occur at three sites in nuclear 18S rRNA genes of *Acanthamoeba lenticulata*. Curr Genet 1998; 34:71–78.

130. Chung DI, Kong HH, Yu HS, Hong YC, Yun HC, Kim TW, Choi HJ. *Acanthamoeba* sp. KA/E4 18S ribosomal RNA gene Unpublished. GenBank reference number AF349045, 2001.

131. Kim SY, Yu HS, Moon EK, Kong HH, Hahn TW, Chung DI. 18s rDNA sequence analyses of Acanthamoeba isolates from infected corneas of amoebic keratitis patients in Korea Unpublished) GenBank accession number AY148954, 2002.

132. De Jonckheere JF. Isoenzyme and total protein analysis by agarose isoelectric focusing, and taxonomy of the genus *Acanthamoeba*. J Protozol 1983; 30:701–706.

133. Chung DI, Yu HS, Hwang MY, Kim TH, Kim TO, Yun HC, Kong HH. Subgenus classification of *Acanthamoeba* by riboprinting Korean. J Parasitol 1998b; 36:69–80.

134. Ledee DR, Booton GC, Awwad MH, Sharma S, RK, Niszl IA, Markus MB, Fuerst PA, Byers TJ. Advantages of Using Mitochondrial 16S rDNA Sequences to Classify Clinical Isolates of *Acanthamoeba*. Invest Ophth Vis Sci 2003; 44:1142–1149.

135. Hewitt MK, Robinson BS, Monis PT, and Saint CP. Identification of a New *Acanthamoeba* 18S rRNA Gene Sequence Type, Corresponding to the Species *Acanthamoeba jacobsi* Sawyer, Nerad, and Visvesvara, 1992 (Lobosea; Acanthamoebidae) Acta Protozool 2003; 42:325–329.

136. Chung DI, Kong H-H, Yu H-S, Hahn T-W. Molecular genetic analyses of *Acanthamoeba* isolates from infected corneas and contact lens storage cases in Korea. Proceed 4th Japan-Korea Parasitologists Seminar 1998a;62–70.

137. Chung D, Kong H, Liu H. Genetic Diversity of *Acanthamoeba* Isolated from Marine Sediment of Sunchen and Gangjin in Korea Unpublished. GenBank accession numbers AY173013, AY173014, AY173015, AY176047, 2002.

138. Maley LE, Marshall CR. The Coming of Age of Molecular Systematics Science. 1998; 279:505–506.

139. Ledee DR. Interspecific mitochondrial rRNA and tRNA gene variation in *Acanthamoeba*: New insights into phylogeny, taxonomy, RNA editing and epidemiology Ph.D. Dissertation, The Ohio State University, Columbus, Ohio 1995; 153pp.

140. Kong HH, Shin JY, Yu HS, Kim J, Hahn TW, Hahn YH, Chung DI. Mitochondrial DNA restriction fragment length polymorphism (RFLP) and 18S small-subunit ribosomal DNA PCR-RFLP analyses of *Acanthamoeba*. J Clin Microbiol 2002; 40: 1199–1206.

141. Yu H-S, Hwang M-Y, Kim T-O, Yun H-C, Kim T-H, Kong H-H, Chung DI. Phylogenetic relationships among *Acanthamoeba* spp. based on PCR-RFLP analyses of mitochondrial small subunit rRNA gene. Kor J Parasitiol 1999; 37:181–88.

142. Kanno M, Yagita K, Endo T, Miyata K, Araie M, Tsura T. Restriction enzyme analysis of mitochondrial DNA of *Acanthamoeba* strains isolated from corneal lesions. Jpn J Ophthalmol 1998; 42:22–26.

143. Hahn TW, Chung DI, Yu HS, Kong HH, Kim JH. Riboprint and mitochondrial DNA RFLP findings of four ocular isolates from *Acanthamoeba* keratitis *Invest Ophth*. Vis Sci 1997; 38:5015, Part 2.

144. Gautom RK, Lory S, Seyedirashti S, Bergeron DL, Fritsche TR. Mitochondrial DNA fingerprinting of *Acanthamoeba* spp. isolated from clinical and environmental sources J Clin Microbiol 1994; 32:1070–1073.

145. Yagita K, Endo T. Restriction enzyme analysis of mitochondrial DNA of *Acanthamoeba* strains in Japan. J Protozool 1990; 37:570–575.

146. Bogler SA, Zarley CD, Burianck LL, Fuerst PA, Byers TJ. Interstrain mitochondrial polymorphism detected in *Acanthamoeba* by restriction endonuclease analysis *Molec Biochem*. Parasitol 1983; 8:145–163

147. Lonergran KM, Gray MW. Predicting Editing of Additional Transfer-RNAs in *Acanthamoeba castellanii* Mitochondria. Nuc Acids Res 1993; 21:4402–4402.

148. Price DH, Gray MW. A novel nucleotide incorporation activity implicated in the editing of mitochondrial transfer RNAs in *Acanthamoeba castellanii*. RNA 1999a; 5: 302–317.

149. Price DH, Gray MW. Confirmation of predicted edits and demonstration of unpredicted edits in Acanthamoeba castellanii mitochondrial tRNAs. Curr Genet 1999b; 35:23–29.

150. Burger GK, Plante I, Lonergan KM, Gray M. The mitochondrial DNA of the amoeboid protozoan, *Acanthamoeba castellanii*: Complete sequence, gene content and genome organization. J Mol Biol 1995; 245:522–537.

151. Schroeder J, Booton GC, Hay J, Niszl IA, Seal DV, Markus MB, Fuerst PA, Byers TJ. Use of Subgenic 18s Ribosomal DNA PCR and Sequencing For Genus and Genotype Identification Of Acanthamoebae From Humans With Keratitis And From Sewage Sludge. J Clin Microbiol 2001; 39:1903–1911.

152. Ledee DR, Hay J, Byers TJ, Seal DV, Kirkness CM. *Acanthamoeba griffini*: Molecular characterization of a new corneal pathogen. Invest Ophth Vis Sci 1996; 37:544–550.

153. Walochnik J Haller-Schober E-M, Kõalli H, Picher O, Obwaller A, Aspõck H. Discrimination between Clinically Relevant and Nonrelevant Acanthamoeba Strains Isolated from Contact Lens-Wearing Keratitis Patients in Austria. J Clin Microbiol 2000; 38:3932–3936.

154. Spanakos G, Tzanetou K, Miltsakakis D, Patsoula E, Mamamou-Lado E, Vakalis NC. Genotyping of pathogenic Acanthamoebae isolated from clinical samples in Greece-Report of a clinical isolate presenting T5 genotype Parasitiol. Int 2006; 55:147–149.

155. Mannis MJ, Tamaru R, Roth AM, Burns M, Thirkill C. *Acanthamoeba* sclerokeratitis Arch Ophthalmol 1986; 104:1313–1317.

156. John DT. Opportunistically pathogenic free-living amoebae. In: Kreier, J (ed) Parasitic Protozoa, 2nd ed., Academic Press, San Diego 1993; 1:143–246.

157. Martinez MS, Gonzalez-Mediero G, Santiago P, Rodriguez DE Lope A, Diz J, Conde C, Visvesvara GS. Granulomatous Amebic Encephalitis in a Patient with AIDS: Isolation of *Acanthamoeba* sp. Group II from Brain Tissue and Successful Treatment with Sulfadiazine and Fluconazole. J Clin Microbiol 2000; 38:3892–3895.

158. Booton GC, Visvesvara GC, Byers TJ, Kelly DJ, Fuerst PA. Identification and Distribution of *Acanthamoeba* Species Genotypes Associated with Nonkeratitis Infections Journal of Clinical Microbiology 2005; 43:1689–1693.

159. De Jonckheere JF. Growth characteristics, cytopathic effect in cell culture, and virulence in mice of 36 type strains belonging to 19 different *Acanthamoeba* spp Appl. Environ. Micro 1980; 39:681–685.

160. Booton GC, Rogerson A, Bonilla TD, Seal DV, Kelly DJ, Beattie TK, Tomlinson A, Lares-Villa F, Fuerst PA, Byers TJ. Molecular and Physiological Evaluation of Subtropical Environmental Isolates of *Acanthamoeba* spp., Causal Agent of *Acanthamoeba* Keratitis. Journal of Eukaryotic Microbiology 2004; 51:192–200.

161. Visvesvara GS, Booton G, Kelly DJ, Fuerst P, Sriram R, Finkelstein A, Garner MM. In vitro culture, serologic and molecular analysis of *Acanthamoeba* isolated from the liver of a keel-billed toucan (*Ramhastos sulfuratus*) Vet Parisitol, in press.

162. Dubey JP, Benson JE, Blakeley KT, Visvesvara GS, Booton GC, Dubey JP. Disseminated *Acanthamoeba* sp. infection in a dog. Veterinary Parasitiology 2005; 128:183–187.

163. Coke RL, Carpenter JW, Aboellail T, Armbrust L, Isaza R. Dilated cardiomyopathy and amebic gastritis in a giant anteater (Myrmecophaga tridactyla). J Zoo Wildl Med 2002; 33:272–279.

164. Dyková I, Lom J, Schroeder-Diedrich JM, Booton GC, Byers TJ. Aconthamoeba stains isolated from organs of freshwater fishes. J Pavasitol 1999; 85:1106–13.

165. Khan NA, Jarroll EL, Panjwani N, Cao Z, Paget TA. Proteases as Markers for Differentiation of Pathogenic and Nonpathogenic Species of *Acanthamoeba*. J Clin Micro 2000; 38:2858–2861.

166. Yang Cao ZZ, Panjwani N. Pathogenesis of *Acanthamoeba* Keratitis: Carbohydrate-Mediated Host-Parasite Interactions. Infect Immun 1997; 65:439–445.

167. Cao Z, Jefferson DM, Panjwani N. Role of Carbohydrate-mediated Adherence in Cyotpathogenic Mechanisms of *Acanthamoeba*. J Biol Chem 273:15838–15845. *Acanthamoeba* strains isolated from organs of freshwater fishes. J Parasitiol. 1998; 85: 1106–1113.

168. Garate M, Cao Z, Bateman E, Panjwani N. Cloning and Characterization of a Novel Mannose-binding Protein of *Acanthamoeba*. J Biol Chem 2004; 279:29849–29856.

169. Ampe C, Vandekerckhove J, Brenner SL, Tobacman L, Korn ED. The amino acid sequence of *Acanthamoeba* profilin. J Biol Chem 1985; 260:834–840.

170. Ampe C, Sato M, Pollard TD, Vandekerckhove J. The primary structure of the basic isoform of *Acanthamoeba* profilin. Eur J Biochem 1988; 170:597:601.

171. Archer SJ, Vinson VK, Pollard TD, Torchia DA. Secondary structure and topology of *Acanthamoeba* profilin I as determined by heteronuclear magnetic resonance spectroscopy. Biochemistry 1993; 32:6680–6687.

172. Bubb MR, Baines IC, Korn ED. Localization of actobindin, profilin I, profilin II, and phosphatidylinositol-4,5-bisphosphate (PIP2) in *Acanthamoeba castellanii*. Cell Motil Cytoskeleton 1998; 39:134–146.

173. Khan NA, Jarroll EL, Paget TA. Molecular and Physiological Differentiation Between Pathogenic and Nonpathogenic *Acanthamoeba*. Curr Microbiol 2002; 45:197–202.

174. Hong YC, Hwang MY, Yun HC, Yong TS, Chung DI. Isolation and characterization of a cDNA encoding a mammalian cathepsin L-like cysteine proteinase from *Acanthamoeba healyi* Unpublished GenBank accession number 2002. AF462309.

175. Quirk S, Maciver SK, Ampe C, Doberstein SK, Kaiser DA, VanDamme J, Vandekerckhove JS, Pollard TD. Primary structure of and studies on *Acanthamoeba* actophorin. Biochemistry 1993; 32:8525–8533.

176. Na BK, Kim JC, Song CY. Characterization and pathogenetic role of proteinase from *Acanthamoeba castellanii*. Microbial Pathogenesis 2001; 30:39–48.

177. Hirukawa Y, Nakato H, Izumi S, Tsuruhara T, Tomino S. Structure and expression of a cyst specific protein of *Acanthamoeba castellanii* Biochemica et Biophysica. ACTA-Gene Structure Expression 1998; 1398:47–56.
178. Kong HH, Hwang MY, Yu HS, Hong YC, Kim TO, Chung DI. New insight into biology of *Acanthamoeba healyi* by expressed sequence tag analysis. Unpublished 1999.

FOUR

Pharmacology and Pharmacokinetics

This chapter was written in association with Suzanne Gardner, Atlanta, Georgia, U.S.A.

INTRODUCTION

The science of pharmacokinetics mathematically describes drug behavior in living tissue. It provides a basis for understanding how best to achieve a desired pharmacologic response. Drug penetration into the eye is influenced by many factors, including characteristics of the antibiotic administered (e.g. lipophilicity versus hydrophilicity), route of drug administration, frequency of dosing, degree of ocular inflammation, and surgical status of the eye.

LIPOPHILICITY/HYDROPHILICITY

Antibiotics, like other drugs, are chemical compounds that can be characterized as to their solubility in water. A coefficient named the lipid/water solubility coefficient (or ether/water solubility coefficient) describes to what extent the compound is freely soluble in either kind of solvent. It is this solubility ratio that determines how well the drug will penetrate cell membranes and be retained within compartments, how easily it can be dissolved in nontoxic solvents for human use, how safely it can be administered and what the pharmacokinetic characteristics will be in regard to a specific targeted tissue.

The greater the lipophilicity of the compound, the greater the penetration through the lipid containing membranes of cell walls. However, to administer a drug with a very high lipid/water solubility ratio requires dissolution in a solvent that is not aqueous, and these solvents themselves are usually toxic to cell membranes and would also be noxious when administered via the bloodstream. Therefore, dissolution of highly lipophilic compounds in specific solvents is feasible in the laboratory, but not feasible for administration in clinical medicine without specially developed delivery systems.

Familiar examples of a relatively lipophilic antibiotic is chloramphenicol, levofloxacin, and clarithromycin. Highly lipophilic antibiotics are fewer in number than the relatively hydrophilic antibiotics used extensively in clinical medicine. The penicillins, cephalosporins, and aminoglycosides are relatively water-soluble and these comprise the majority of antibiotics used. The new group of quinolone antibiotics (ciprofloxacin, ofloxacin, levofloxacin, gatifloxacin and moxifloxacin) is relatively more lipophilic than cephalosporins, but data specific to each antibiotic should guide the clinical use and expectations.

DIFFUSION

Simple diffusion is the most common method of drug transfer from one space or tissue of the body to another or from one compartment to another. It involves the movement of solute particles from an area of higher concentration to one of lower concentration by transfer through tissues and interfaces. Once diffusion has occurred, drug concentrations within adjacent compartments become stabilized according to the drug's partition coefficient (Fig. 1).

Diffusion is affected by factors such as the concentration of drug presented at a tissue interface (concentration gradient), molecular weight of the drug, membrane permeability (inflamed or non-inflamed), tissue binding, active transport mechanisms, surgical status of the eye, and others including formulation of the drug itself.

EPITHELIAL BARRIER

The epithelia of the bulbar conjunctiva and cornea are relatively impermeable to even water-soluble agents of small molecular size applied topically. Proprietary ophthalmic preparations such as gentamicin sulphate are available as drops (0.3%) or ointments at high concentration (e.g., 0.3–1.0%) relative to their effective antimicrobial concentration and are thus active in the treatment of surface ocular infection (e.g., conjunctivitis). Antibiotics with a relatively high lipid–water solubility coefficient (such as chloramphenicol or the poorly dissociated sulphonamides,

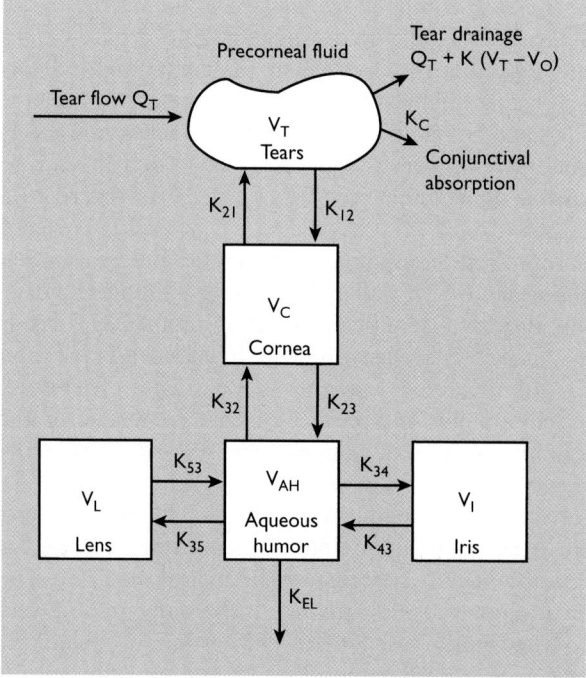

Figure 1   Partition coefficients for drug transfer from topical drops to aqueous humor. *Source*: Ref. 1.

levofloxacin and moxifloxacin) can penetrate the conjunctiva and cornea much better to enter deeper tissues at more effective concentrations. For a topically applied antibiotic to penetrate to the aqueous humor, it has to pass through the various layers of the cornea, as illustrated in Figure 2.

If the surface epithelium is breached, as with corneal ulceration, water-soluble drugs can diffuse into the anterior segment of the eye in high concentration. This can be enhanced by the use of fortified eye drops whose concentrations exceed those of commercial preparations. However, this route is ineffective in providing useful concentrations in the posterior segment of the eye (e.g. vitreous) because there is a diffusion barrier across the lens-zonule compartment and across the anterior vitreous.

The barrier offered by the epithelium of the ocular surface can be circumvented therapeutically by delivering a bolus of drug under the conjunctiva (subconjunctival injection; see Table 1) or more deeply into the orbit (sub-Tenon's injection). These periocular routes will deliver effective antimicrobial concentrations into the anterior segment of the eye (cornea and anterior chamber) and to some extent into the posterior segment (the vitreous) but should *not* be relied upon to deliver a sufficiently high concentration of antibiotic to satisfactorily treat bacterial endophthalmitis (2).

PENETRATION OF THE DIFFERENT QUINOLONE ANTIBIOTICS INTO THE ANTERIOR CHAMBER

Penetration of the different quinolone antibiotics—ciprofloxacin, ofloxacin, levofloxacin, moxifloxacin, and gatifloxacin—into the anterior chamber from the topical drop route or by oral therapy is given in Table 2. The rates vary widely between the different compounds, with levofloxacin and moxifloxacin achieving the highest levels in the aqueous humor due to their relative lipophilicity. They are considered later in this chapter.

BLOOD–OCULAR BARRIERS

Barriers Influencing the Entry of Drugs into the Eye

There are two main barrier systems in the eye (17). The first regulates exchange between blood and aqueous humor in which inward movement predominates—the

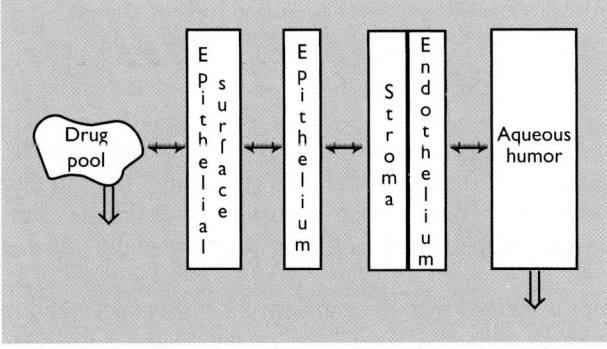

Figure 2 Barriers to drug penetration. *Source:* Ref. 1.

Table 1   Aqueous Humor (AH) Levels of Antibiotics in µg/mL Achieved by the Subconjunctival and Topical Routes

Antibiotic	Subconjunctival injection/ AH level achieved	Topical administration/AH level achieved
Gentamicin	10–40 mg gave 4 µg/mL in aqueous for 4 hr Suggested dose is 40 mg	3 mg/mL (0.3%), TL in cornea & AC for 1 hr
Amikacin	25 mg gave 19 µg/mL in aqueous for 4 hr	Level far below TL of *Pseudomonas* infections (10 µg/mL)
Penicillin G	300–600 mgs (1 × 10[6] units), gave persistent TL in AC for 12–24 hr	NP
Methicillin	20–40 mg gave TL for 4 hr Can give 100mg	10% solution, no TL achieved
Oxacillin	100 mg gave 70 µg/mL in aqueous for 4 hr	NP
Dicloxacillin	50 mg gave levels almost 100 x MIC for *S. aureus* at 4 hr Can give 100mg	NP
Ampicillin	2.5–250 mg gave TL up to 4–6 hr Suggested dose is 100 mg	0.5% solution, aqueous levels better than systemic administration, compares favorably to subconjunctival administration
Cephaloridine	50 mg gave extremely high TL for 4 hr Can give 100mg	NP
Cefuroxime	125 mg gave 20µg/mL	NP
Ceftazidime	100 mg	NP
Erythromycin	10–20 mg gave excellent TL but caused persistent chemosis	2.5% solution–TL 4 hr without significant local reaction
Lincomycin	75 mg gave excellent TL for 12 hr	NP
Clindamycin	34 mg	NP
Vancomycin	25 mg gave TL >4 hrs with mild chemosis (can be irritating)	50 mg/mL (5% solution), AC levels similar to subconjunctival injection results
Chloramphenicol	1.2 mg gave good TL for at least 30 min	2.5–50 mg/mL (0.25 to 5%) gave various TL depending on conditions and investigator
Ciprofloxacin	1 mg gave level of 0.9 µg/mL after 1 hr; level of 0.009 µg/ mL after 10 hr	NP

Abbreviations: AC, anterior chamber; NP, not performed; TL, therapeutic level; AH, aqueous humor.
Source: From Ref. 2.

blood-aqueous barrier (Fig. 3). Aqueous humor is secreted into the posterior chamber by the ciliary processes and flows through the pupil into the anterior chamber to leave via the trabecular meshwork. Important diffusional solute exchange takes place between the aqueous and not only the surrounding tissues but also with the vitreous.

   The blood aqueous barrier is formed by two layers of cells: the endothelium of the iris blood vessels and the non-pigmented inner layer of the ciliary epithelium. These cell layers impair transport of blood proteins to maintain an osmotic and chemical equilibrium. However, the ciliary vascular endothelium is permeable to

Table 2 Effect of Oral and Topical Doses of Quinolones on Aqueous Humor (AH) Levels in µg/mL

Quinolone drug	Oral dose pre-op	Topical dose pre-op	AH levels µg/mL	Number of replicates
GAT (3)	400 mg GAT 16 and 4 hr before sampling		1.34 GAT (1.36 vitreous) (5.14 serum)	n = 24
LVFX (4)		5 drops LVFX 0.5% over 2 h (30 min intervals)	0.68 LVFX	n = 32
LVFX (4)	200 mg LVFX 2 hr pre-op		0.42 LVFX	n = 35
LVFX (4)	3x 200 mg _VFX 18, 6, and 2 hr pre-op		1.3 LVFX	n = 34
LVFX (4)	3x 200 mg _VFX 18, 6, and 2 hr pre-op	5 drops LVFX 0.5% over 2 hr (30 min intervals)	1.86 LVFX	n = 35
LVFX (5)OFX		4 drops 0.5% LVFX over 1 hr (15 min intervals) 4 drops 0.3% OFX over 1 hr (15 min intervals)	1.14 LVFX 0.62 OFX	n = 69
LVFX (6)CIP		4 drops 0.5% LVFX over 1 hr (15 min intervals) 4 drops 0.3% CIP over 1 hr (15 min intervals)	0.92 LVFX 0.12 CIP	n = 59
CIP (7) LVFX MOX	2 × 500 mg CIP 24 and 12 hr pre-op 1 × 500 mg LVFX 6 hr pre-op 1 × 500 mg MOX 6 h pre-op		0.5 CIP 1.5 LVFX 2.33 MOX	
MOX (8) OFX		1 drop 30 min prior to collection in rabbits with 0.3% MOX or 0.3% OFX	1.8 MOX 0.8 OFX	n = 10
LVFX (9)	1 × 400 mg LVFX on day of surgery		0.78 LVFX	n = 35
LVFX (10) LOM		6 drops 0.5% LVFX or 0.3% LOM over 1.5 hr (15 min intervals)	0.6 LVFX 0.23 LOM	n = 59
LVFX (11) CIP		8 drops 0.5% LVFX or 0.3% CIP over 2 days (6hr intervals) 5 drops 0.5% LVFX or 0.3% CIP over 1 hr (12 min intervals) and the combination	0.28 LVFX 0.07 CIP 1.14 LVFX 0.19 CIP 1.62 LVFX 0.24 CIP	n = 93
LVFX (12)	2x 500 mg _VFX 12 hr pre-op	3 drops of LVFX 0.3% (15 min intervals)	1.9 LVFX	n = 17
LVFX (13)			1.49 LVFX	...
LVFX (14)	200 mg 4–6 hr pre-op		0.68 LVFX	...
LVFX (15)		1 drop of LVFX 0.5% ×2 before surgery (every 15 min)	Stromal Levels (µg/g) LVFX 18.2 OFX 10.8 CIP 9.9	n = 65
LVFX (16)		1 drop of LVFX 0.5% x2 before surgery (every 30 min) & 1 drop 0.5% every 5 min after surgery x3	4.4 LVFX	n = 30

Abbreviations: CIP, ciprofloxacin; GAT, gatifloxacin; LVFX, levofloxacin; LOM, lomifloxacin; MOX, moxifloxacin; NOR, norfloxacin; OFX, ofloxacin.

blood-borne solutes that give a concentration of 1% of proteins found in the plasma. The ciliary body has a pivotal role in regulating all inner ocular fluids via its large surface covered by the ciliary processes. There is active secretory transport of sodium, chloride, and bicarbonate providing the osmotic force for the production of the aqueous as well as maintaining ascorbic acid at a higher level than in plasma. The ciliary body is also responsible for a free continuous flow of ions, amino acids, and vitamins from the aqueous into the vitreous.

(**A**) Epithelial barrier. The corneal epithelium restricts the entry of water-soluble drugs into the cornea and aqueous humor. The barrier is breached by an epithelial defect and, if the epithelium is intact, is bypassed by a subconjunctival injection.

(**B**) Bulk flow of aqueous humor from the eye, and the presence of an intact lens and zonule retard the diffusion of drugs from the anterior chamber into the vitreous humor.

(**C**) Blood-aqueous barrier: limits the entry of drugs from the blood into the aqueous.

(**D**) The epithelia of the iris and ciliary body pump anionic drugs out of the aqueous.

(**E**) Blood-retinal barrier: Limits the entry of drugs into the vitreous from the systemic circulation. *Externally* is the pigment epithelial barrier.

(**F**) *Internally* is the retinal capillary endothelial barrier. There is an outward pumping of anions across the retina by the retinal pigment epithelium and the endothelial cells of the retinal vessels (17).

In contrast, the second barrier in the posterior segment affects outward movement from the retina into the blood—the blood-retinal barrier (Fig. 3). It is responsible for the homeostasis and microenvironment of the retina. It also serves to remove waste products of metabolic activity from within the eye. There are no diffusional barriers between the extracellular fluid of the retina and the adjacent vitreous nor with the posterior chamber of the anterior segment so that the functions

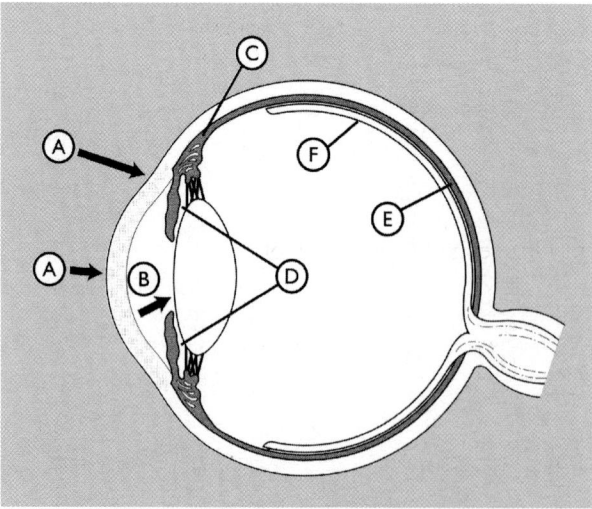

Figure 3 Barriers to penetration of antibiotics into the eye. *Source*: Derived from Ref. 17.

of both anterior and posterior segments work together. The blood-retinal barrier is formed by a combination of the endothelium of retinal blood vessels, the retinal pigment epithelium, and retinal glial cells. The endothelial cells of the retina have functional and morphological differences compared with those of other organs. They have narrow tight junctional structures composed of a complex called the zonulae occludentes. They impair cellular transport of hydrophilic compounds. The retinal pigment epithelium is in the outer blood-retinal barrier and is also sealed with extensive zonulae occludentes. High activities of barrier selective enzymes contribute to a protective function, but essential nutrients are transported into the vitreous by carrier-selective mechanisms.

Equally important are the active transport mechanisms that remove ions from the vitreous. The microenvironment of the retina resembles the extracellular fluid of the brain. It is regulated by active transport processes located within the barrier cells. There are three main sites of transport out of the eye: the endothelial cells of the retinal vessels, the retinal pigment epithelium, and the ciliary epithelium. Within the retinal barrier, there is active transport of potassium to blood with a net magnesium flux in the opposite direction. Organic anions such as prostaglandins and related compounds, that are produced but not metabolized within the eye, and anionic drugs such as penicillins and cephalosporins and zwitter ions such as ciprofloxacin, are removed from the extracellular fluid of the retina by active transport across the blood-retinal barrier cells similarly to the renal tubular clearance mechanism.

This active transport system is inhibited by probenecid, which can be used to maintain higher levels of anionic drugs in the eye; however, this may interfere with outward transport of prostaglandins. Use of oral probenecid (500 mg every 12 hr) retards outward transport of penicillins, cephalosporins, and ciprofloxacin across the retinal capillary endothelial cells, thus extending the half-life of the drug within the vitreous. This has been demonstrated in normal monkey eyes when the half-life of cefazolin was extended from 7 to 30 hours. Probenecid also raises the plasma level by blocking secretion of these drugs by the proximal renal tubule.

Cationic compounds and drugs such as the aminoglycosides, (gentamicin, tobramycin), vancomycin, erythromycin and rifampicin are not actively transported out of the eye by the blood-retinal barrier and have to leave the posterior compartment by diffusing forward into the anterior chamber and aqueous humor. Ciprofloxacin and the other quinolones are removed by both the active and passive transport systems (Fig. 4).

The barriers referred to the globe (Fig. 3) do not exist for other orbital structures. Thus, infections within the orbit or ocular adnexae (the eyelids, lacrimal gland, and nasolacrimal system) are readily treatable with systemic antibiotics.

The result of the blood-retinal barrier described above is an active transport system for the anionic drugs (penicillin, cephalosporins) out of the eye, giving a half-life of approximately 8 hours when given intravitreally, as opposed to the passive diffusion forward of the cationic drugs (gentamicin and other aminoglycosides, vancomycin, rifampicin, and erythromycin), giving a half-life of approximately 24 hours. This means that a combination injection given intravitreally of a cephalosporin, viz. ceftazidime, and an aminoglycoside or vancomycin, have different half-lives, which must be considered in the management of the clinical infection (Fig. 4). However, for reasons given below, in the presence of inflammation removal of the actively absorbed drug, e.g. cephazolin, is delayed, resulting in a longer half-life than without inflammation. In contrast, the passively transported drug eliminated by forward diffusion, e.g., aminoglycosides, (gentamicin, tobramycin), has enhanced removal. The combination

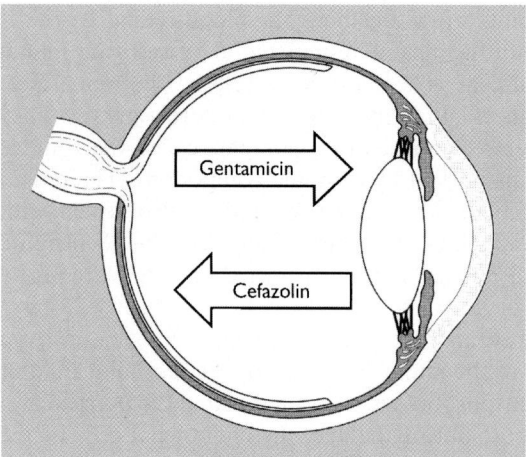

Figure 4   Active (*cefazolin*) and passive (*gentamicin*) transport of antibiotics out of the eye.

of a cephalosporin and an aminoglycoside given intravitreally thus becomes acceptable in practice for treatment of the acutely infected and inflamed eye (Refer to Chapter 8, Endophthalimitis).

EFFECTS OF INFLAMMATION

These barriers affect the ocular distribution of drugs, the potential of systemically administered drugs to enter the vitreous, as well as the retention of drugs after direct delivery by vitreous tap.

In the *uninflamed* eye, the highest aqueous concentration using the systemic route will be achieved by lipid-soluble drugs such as chloramphenicol or the quinolones (ciprofloxacin, ofloxacin, levofloxacin, gatifloxacin, and moxafloxacin). Negligible concentrations will be achieved using water-soluble drugs, particularly those in anionic form (such as the penicillins and cephalosporins), which are actively transported out of the eye.

In the *inflamed* eye, after systemic antibiotic administration, the concentrations achieved in the ocular compartments are increased, owing to partial breakdown of the blood-aqueous barrier, so that effective antimicrobial concentrations may be reached in the aqueous. However, concentrations reached in the vitreous with parenteral therapy are always much lower than those achieved in the aqueous, even in the inflamed eye with endophthalmitis. Vitreous concentrations after high-dose systemic therapy in such situations will be sub-therapeutic after initial doses and lower than those achievable by subconjunctival injection.

The highest antibiotic levels are gained by direct injection into the vitreous, which is always favored in the treatment of serious endophthalmitis (Table 3; see also Chapter 8). However, giving the *same* antibiotic by the intravenous route as that given by an intra-vitreal tap will prolong effective levels in the vitreous and should be practiced in the treatment of acute bacterial endophthalmitis.

Infectious endophthalmitis creates sufficient intraocular inflammation to compromise the function of the retinal pump mechanism that helps to eliminate antibiotics such as cephazolin, thereby prolonging their half-life in the vitreous of the

Table 3 Intravitreal Doses of Antibiotics[a]

Antibiotic	Intravitreal injection (μg)	Duration (h)	Intravenous dose (mg)
Amikacin	400	24–48	15m g/k g/24 hr
Ampicillin	500	24	1 g/4 hr
Amphotericin	5 or 10	24–48	0.1–1.0 m g/k g/24 hr
Carbenicillin	2000	16–24	3 g/4 hr
Cefazolin	2000	16	1 g/4 hr
Ceftazidime	2000	16–24	1.5 g/6 hr
Cefuroxime	2000	16–24	1.5 g/6 hr
Clindamycin	1000	16–24	0.75 g/8 hr
Ciprofloxacin[b]	200	—	0.2 g/12 hr
Erythromycin	500	24	0.5 g/6 hr
Gentamicin	200	48	5m g/k g/24 hr
Methicillin	2000	40	1 g/4 hr
Miconazole	10	24	—
Oxacillin	500	24	2 g/4 hr
Vancomycin	1000	72	1 g/12 hr

[a] Maximum intravitreal injection volume for each antibiotic is usually 0.1 mL.
[b] Very limited use only; not used by the authors.

normal eye from approximately 6.5 to 10.4 hours. At the same time, surgical alteration of the eye, such as lensectomy and/or vitrectomy, altogether change the inner anatomy of the eye, affecting overall elimination characteristics as described below.

The relative effects of inflammation, aphakia, and vitrectomy on the half-life of cephazolin are best studied in experimental animal models that simulate these conditions (Figs. 5–8). It can be seen that the half-life prolongation given by inflammation in the phakic eye does not occur after aphakia and vitrectomy.

Low levels of cefazolin were found within the vitreous cavity of uninflamed albino rabbit eyes after intravenous administration, but with inflammation, much higher levels were achieved (Figs. 5 and 7). This was true, irrespective of the surgical status of the eye. It should be noted that after repeated systemic doses with the inflamed eye (Fig. 7), intravitreal antibiotic levels rise, approaching therapeutic levels, and these may be considered useful as an adjunct to the declining drug levels

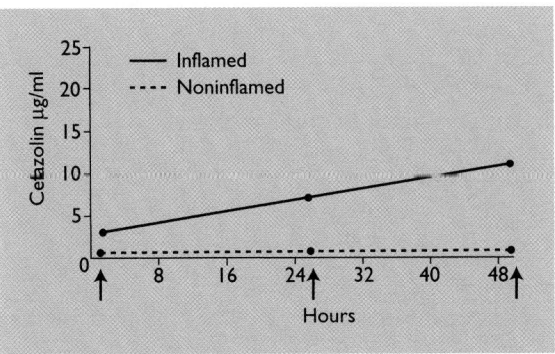

Figure 5 Antibiotic levels in vitreous of inflamed and non-inflamed phakic eyes of rabbits given *intravenous* cefazolin.

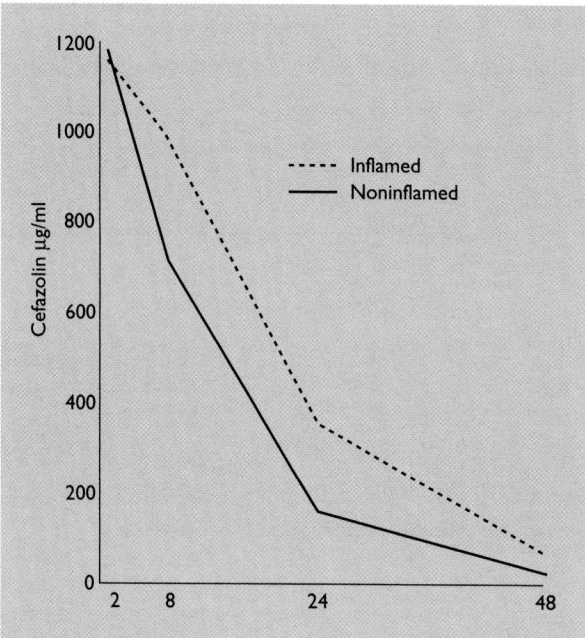

Figure 6 Antibiotic levels in vitreous of inflamed and non-inflamed phakic eyes of rabbits given *intravitreal* cefazolin.

seen within the vitreous cavity after an intravitreal injection. The use of repeated systemic doses, and an increase in dose, where safe and tolerable to the patient, are factors often overlooked in clinical treatment strategies for infectious endophthalmitis.

In contrast, gentamicin, which is not removed by the retinal pump mechanism, does not show any prolongation of half-life in the vitreous of the inflamed or infected eye but instead has a reduction of 50% from 20 to 10 hours because of its elimination by the anterior route. A similar reduction is given with aphakia but this is not further increased with inflammation or infection. This finding was also repeated with vancomycin, when aphakia and vitrectomy were found to reduce vitreous

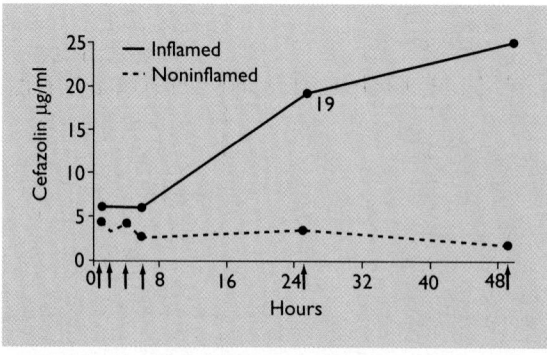

Figure 7 Antibiotic levels in vitreous of inflamed and non-inflamed aphakic vitrectomized eyes of rabbits given *intravenous* cefazolin.

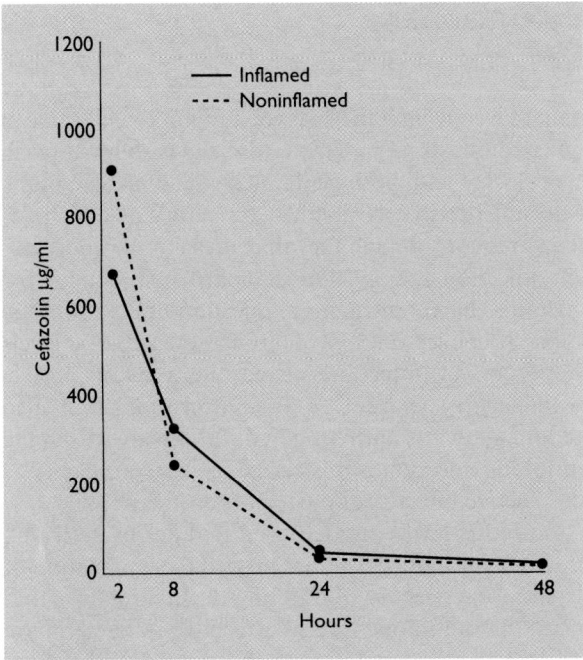

Figure 8 Antibiotic levels in vitreous of inflamed and non-inflamed aphakic vitrectomized eyes of rabbits given *intravitreal* cefazolin.

vancomycin levels, compared with phakic non-vitrectomized eyes, so that only 25 μg/mL remained out of a dose of 2 mg given 48 hours earlier (18). This finding of the shortest half-life occurring with aphakia and vitrectomy was also shown by Case et al. with clearance studies of 3H-fluorouracil (19).

A clinical parallel to the treatment of infectious endophthalmitis is seen in the treatment of meningitis, where higher CSF drug levels are also achieved through inflamed meninges and after repeated systemic doses. Here, also, drug penetration through the meninges is proportional to its degree of lipophilicity.

REPEATED SYSTEMIC DOSING

When serum levels of an antibiotic can be increased or sustained by repeated intravenous administration of high but safe doses, vitreous drug penetration is further increased. Repeated, or higher, systemically administered doses raise the *concentration gradient*, that is, the concentration of antibiotic present in the serum that interfaces with the tissues of the blood-eye barrier. When inflammation is also present, intravitreal penetration of an antibiotic can be increased.

The exact increments in drug penetration, or precise intraocular antibiotic levels, are not quantified because the degree of ocular inflammation due to infection is difficult, if not impossible, to standardize in experimental animal models of endophthalmitis. Therefore, drug levels commonly reported in the literature often reflect assays after only a single systemic antibiotic dose has been administered, a situation that is not analogous to the clinical setting, While useful, these reports do not adequately describe the penetration of antibiotics after multiple doses have been given!

INTRAVITREAL INJECTION OF ANTIBIOTICS, VITRECTOMY, AND TOXICITY

Many factors are described above that severely limit the levels of antibiotic achievable in the vitreous cavity after intravenous administration. This is also the case for topical, subconjunctival, and intra-muscular dosing with both gentamicin and amikacin in the normal and aphakic eye, which do not produce levels that are sufficiently high for effective bactericidal therapy in the vitreous (although they may do so in the aqueous).

Direct intravitreal antibiotic injection is now the "standard of practice" for treatment of infectious endophthalmitis. Direct intravitreal injection of an antibiotic in the highest non-toxic dose (Table 3) delivers high, instantaneous antibiotic levels where they are needed—at the focus of infection inside the vitreous space. "Debulking" of the vitreous with surgery (partial or full vitrectomy) will also remove part of the infectious load and allow the antibiotic to diffuse easily within the cavity. The goal of intravitreal antibiotic injection is to sterilize the vitreous cavity as early as possible to reduce the destructive effect of the bacteria on the retina.

With intravitreal injection, antibiotic levels are delivered that are bactericidal, at least for a period of time. These high antibiotic levels are needed for eradication of bacteria in this closed avascular space. The purpose of injecting the highest dose that is nontoxic to the retina is to ensure that antibiotic levels destructive to bacteria are delivered in a single injection, and that they are maintained above a sub-therapeutic level for as long as possible, while the intravitreal level of antibiotic is declining. Bacteria kill rates are a function not only of drug concentration but also of contact time with the bacterium as well as the numbers involved.

When diagnosed early, there will be a smaller number of actively replicating bacterial cells than with a late diagnosis when there is a large number of static bacterial cells present. Antibiotics such as cephalosporins that act against the bacterial cell wall require replicating cells for their bactericidal effect. The same situation applies to aminoglycosides that act on the cell ribosome but to a lesser extent. Treatment at a late stage involves a changed environment with pH change, hypoxia, nutrient shortage, and non-replicating bacteria, which additionally reduces the effectiveness of the antibiotic. This is when vitrectomy is needed to "evacuate the pus."

The concentration of antibiotic in the vitreous cavity after an intravitreal injection is generally calculated by dividing the dose injected by the average volume of the vitreous, in humans 4.5 mL (refer to end of chapter). However, the presence of infection/inflammation, aphakia, and vitrectomy effect the immediate and early antibiotic levels distributed throughout the vitreous space after intravitreal injection.

After instantaneous antibiotic levels are established by intravitreal injection, the drug half-life then determines the rate of decline or reduction of drug levels within the vitreous space (Figs. 6, 8). Drug elimination is effected by factors including the degree of inflammation, the elimination route for the antibiotic in question (anterior or posterior), tissue binding of drug, and the surgical status of the eye such as aphakia or vitrectomy.

Inflammation prolongs the half-life of cefazolin after intravitreal injection in the non-surgical eye. In this instance, as described above, damage to the retinal pump mechanism actually serves to prolong the retention of those drugs within the vitreous cavity that are eliminated via the posterior route. However, drugs eliminated via the anterior route are not subject to this effect, and inflammation will decrease their half-life or retention after an intravitreal injection, as will removal of the cataractous lens.

In aphakic, vitrectomized eyes the rate of antibiotic decline in the vitreous space can still be faster (Fig. 8). The vitreous gel that sequesters the injected bolus of drug, and promotes gradual diffusion throughout the vitreous cavity, has been removed. Figure 8 demonstrates the more rapid decline in antibiotic levels in the aphakic/vitrectomized eyes. In phakic eyes, the relative retention of cefazolin is seen in inflamed eyes.

Interpretation of these models suggests that the use of a cephalosporin and an aminoglycoside (gentamicin) or vancomycin in combination by the intravitreal route to treat endophthalmitis in the aphakic vitrectomized eye will yield approximately similar half-lives of 10 hours. This can be supplemented from the intravenous route, since there is better penetration in the inflamed eye (Figs. 5, 7). By 48 hours, however, most of the antibiotic will have been removed, so if there is a limited clinical response, repeat intravitreal injection should be therapeutic and safe at this time.

The increasing use of vitrectomy demands that the toxicity of the injected dose and the method of intravitreal injection be reconsidered. When injected into the vitreous gel, gradual diffusion of antibiotic occurs. However, where vitrectomy has been performed, the injection should be given slowly so that these relatively high concentrations of antibiotic, or their vehicles, are not "squirted" with force onto the retina itself. This is especially important with gentamicin. The dose of 0.4 mg in 0.1 mL has been reduced to 0.2 mg because of suspected macular infarction due to toxicity, as has the dose of amikacin (0.2 mg), although it is not equipotent and amikacin is less effective per mg as a bactericidal drug than gentamicin. Gentamicin and other antibiotics should be reduced in concentration by a further 50% (for gentamicin to 0.1 mg) if given intravitreally when a full vitrectomy is performed.

Most cases of gentamicin toxicity occurred after a 0.4 mg dose in eyes that had undergone a vitrectomy. The retinal toxicity of gentamicin may have been exacerbated due to the methods used to prepare the intraocular injection in some places. By irregular means of creating the intraocular injection (drawing up concentrated commercial products within one tuberculin syringe in "series"—in poor order) in the syringe, and not going through the more cumbersome serial dilutions first in separate vials that also dilute out the toxins in the vehicle, clinicians have deviated in the past from standard clinical procedures and recommendations.

However, a more recent survey of retinal specialists has suggested that amikacin or low-dose gentamicin can still cause macular toxicity (20). To further investigate this issue, critical details were investigated from 13 patients who received intravitreous injections of 0.2 to 0.4 mg of amikacin sulfate or 0.1 to 0.2 mg of gentamicin sulfate for prophylaxis or treatment of endophthalmitis and suffered toxicity (20). These cases suggest that amikacin and low-dose gentamicin can cause macular infarction. The causative dose cannot be ascertained in any of the cases, but the dilutions were prepared by hospital pharmacists using typewritten protocols, a practice that helps to prevent errors. Several cases differed from previously reported cases of aminoglycoside toxicity in that the involvement of the macula was quite discrete. Most of the patients suffered severe visual loss, but two patients, in whom most of the non-perfusion was adjacent to the macula and in whom some of the perifoveal capillaries were spared, recovered 20/50 visual acuity. These cases emphasize the potential hazards of the intravitreous use of aminoglycosides. A toxic reaction can occur even when injection of low doses is intended and precautions are made to avoid dilution errors. A localized increase in concentration in dependent areas of the retina may play a role in aminoglycoside toxicity. If some of the perifoveal capillaries are spared, retention of some central vision is possible. Therefore, consideration should be given to substituting ceftazidime for aminoglycosides (gentamicin) for the treatment and prophylaxis of endophthalmitis to avoid toxicity to the macula (Chapter 8).

LIPOSOMAL PREPARATION OF DRUGS

Drugs such as amphotericin B can be incorporated into liposomes to enhance drug transport into the retina (and brain) by the intravenous route, but the disadvantage of this technique is the rapid uptake of the liposomes by the liver, lungs, and spleen. It is debatable whether to treat fungal endophthalmitis with intravitreal amphotericin B and systemic use of liposomal amphotericin B (Ambizone) instead of deoxycholate amphotericin (Fungizone) (see Chapter 9). A recent study, using a rabbit model of inflammatory uveitis, has found that liposomal amphotericin B penetrated better to the cornea (mean concentration of amphotericin 2.4 +/−1.5 µg/g) compared to deoxycholate amphotericin B (the standard IV preparation), which yielded only 0.5 +/−0.2 µg/g (20); amphotericin could not be detected in the corneas of non-inflamed eyes with either preparation. Amphotericin can be repeated intravitreally every 48 hours according to the clinical response and can as well be injected into the anterior chamber (intracameral route) at 10 µgs in 0.1 mL. Clinicians should be aware of the possibility of retinal damage with more than two doses.

INTRACELLULAR AND EXTRACELLULAR LEVELS OF ANTIBIOTICS

Once an antibiotic or another drug reaches a particular tissue within the eye, its concentration will be partitioned between the cellular and extracellular space. There is a big difference in cellular penetration between the various antibiotics that, surprisingly, is not related to their ionic charge. Their values are listed in Table 5 for experimental concentrations of drugs (µg/mL) within polymorphonuclear (PMN) cells and extra-cellular (EC) tissue. PMN cells are used in laboratory experiments and are representative of levels expected within macrophages.

The antibiotics can be classified according to whether their PMN/EC ratio is less than or greater than 1 (Table 4). Those antibiotics with a ratio of <1 have a lower concentration within the cell than within the extracellular tissue. In contrast, those antibiotics with a ratio >1 become concentrated within the cell. The penicillins, cephalosporins, and all the aminoglycosides have a ratio of <1. The ratio is highest for azithromycin (also representative of clarithromycin) at 226, clindamycin at 43, erythromycin at 14, and the various quinolones at between 10.9 and 3.7. Chloramphenicol, which has been much used in the past due to its lipophilicity, and therefore cell membrane penetration, only has a ratio of 2.2.

Chronic low-grade infection, often presenting as a 'hypopyon uveitis' that develops 6 weeks or more after phacoemulsification cataract surgery and lasts for months, can be due to 'saccular' plaque or capsular bag endophthalmitis (see Chapter 8). The bacteria causing this type of infection are usually *P. acnes*, diphtheroids, or coagulase-negative staphylococci. They are found within macrophages lining the capsule fragment (22–25). To eradicate these intracellular bacteria, illustrated in Chapter 8, there is need to use antibiotics that penetrate well into macrophages such as the erythromycin derivatives azithromycin and clarithromycin.

Table 4 Experimental Concentrations of Drugs (µg/mL) within Polymorphonuclear (PMN) Cells and Extracellular Tissue

	PMN cells	Extracellular tissues	Ratio PMN/ extracellular
Ampicillin	8	100	0.08
Cephalothin	<1	10	<0.1
Cloxacillin	32.5	100	0.33
Fusidic acid	16	40	0.4
Penicillin	5	10	0.5
Gentamicin	15	18	0.8
Chloramphenicol	22	10	2.2
Rifampicin	47	20	2.4
Cipro[c]/Lomefloxacin[f]	10.0	2.7	3.7
Levo[a,b]/Ofloxacin[e]	30	5	6
Gemi[g]/Moxifloxacin[d]	55	5	8/10.9
Erythromycin	254	18	14
Clindamycin	434	10	43
Azithromycin (clarithromycin)	2260	10	226

[a] Mandell GL, Coleman E. Uptake, transport, and delivery of antimicrobial agents by human polymorphonuclear neutrophils. Antimicrob Ag Chemother 2001; 45: 1794–1798.
[b] Perea EJ, Garcia I, Pascual A. Comparative penetration of lomefloxacin and other quinolones into human phagocytes. Am J Med 1992; 92: 48S–51S.
[c] Smith RP, Baltch A, Franke MA, Michelsen PB, Bopp LH. Levofloxacin penetrates human monocytes and enhances intracellular killing of *Staphylococcus aureus* and *Pseudomonas aeruginosa*. J Antimicrob Chemother 2000; 45: 483–493.
[d] Vazifeh D, Bryskier A, Labro M-T. Mechanism underlying levofloxacin uptake by human polymorphonuclear neutrophils. Antimicrob Ag Chemother 1999; 43: 246–252.
[e] Garcia I, Pascual A, Ballesta S, Perea EJ. Uptake and intracellular activity of ofloxacin isomers in human phagocytic and non-phagocytic cells. Int J Antimicrob Agents 2000; 15: 201–205.
[f] Garcia I, Pascual A, Ballesta S, Joyanes P, Perea EJ. Intracellular penetration and activity of Gemifloxacin in human polymorphonuclear leukocytes. Antimicrob Ag Chemother 2000; 44: 3193–3198.
[g] Pascual A, Garcia I, Ballesta S, Perea EJ. Uptake and intracellular activity of moxifloxacin in human neutrophils and tissue-cultured epithelial cells. Antimicrob Ag Chemother 1999; 43: 12–20.
Source: From Ref. 1.

Clarithromycin was first used for this purpose in an individual case of 'saccular' (plaque) endophthalmitis, confirmed by PCR, which responded well to treatment and resulted in a quiet eye without the need for any surgical removal of the lens (26). Since then clarithromycin has been used similarly by others (27) and in a randomized trial of therapy for chronic subacute endophthalmitis (28), when patients responded much better to this therapy than those who did not receive clarithromycin.

Clarithromycin is given orally as 500 mg every 12 hours and is well absorbed. It penetrates well into both the anterior and posterior segments (29) and is then concentrated into macrophages at up to 200-fold (Table 4). To examine the penetration of clarithromycin in ocular tissues, 21 patients who underwent elective cataract surgery received a single 500-mg dose of clarithromycin orally preoperatively. The concentrations of clarithromycin in the aqueous fluid 4, 8, 10, 12, and 22 hours after administration were: (mean +/− SD) 0.13+/−0.05, 0.137+/−0.11, 0.074+/−0.03, 0.06+/−0.02, and 0.074+/−0.04 µg/mL. Another 21 patients who underwent elective retina/vitreous surgery received 500 mg every 12 hours orally for 3 days preoperatively. The concentrations in vitreous 3, 6, 8, 11, and 24 hours after

Table 5 Expected Antibiotic Sensitivities (MICs) of Pathogenic Bacteria for Causing Ocular Infection

	Penicillin	Amp/ amoxicillin	Flu/ cloxacillin	Cefuroxime	Ceftazidime	Fusidic acid	Ofloxacin/ levo/gati/ moxifloxacin[a]	Oxytetracycline	Chloramphenicol	Vancomycin	Gentamicin/ amikacin	Metronidazole
S. aureus[b/c]	R	R	S (0.1)	S (4)	(S) (6)	S (0.1)	S (0.4)	(S) (4)	S (4)	S (2)	S (0.25)	R
CNS[b/c]	R	R	S (0.1)	S (4)	(S) (6)	S (0.1)	S (0.2)	(S) (4)	S (4)	S (2)	S (0.25)	R
Streptococci[b/c]	S (0.03)	S (0.03)	S (0.1)	S (0.1)	S (0.25)	(S) (1–16)	(S) (1–4)	(S) (2)	S	S (0.2)	R	R
P. acnes and *corynebacteria*[b/c]	S	S	S	S	S	S	S	S		S (1)	R	R
Bacillus sp.	R	R	R	R	R	R	S (0.25)	(S)	(S)	S	S	R
N. meningitidis	(S) (1.0)	S	R	S	S	S (0.5)	S (<0.1)	S (1)	S (2)	R	(S) (4)	R
H. influenzae	R	S (0.25)	R	S (0.1)	S (0.1)	R	S (<0.1)	S (1)	S (0.5)	R	S (0.5)	R
Coliforms	R	(S) (2–R)	R	S (4)	S (0.1)	R	S (0.25)	(S) (8)	(S) (6)	R	S (0.5)	R
Ps. aeruginosa	S	R	(S)	S	S (4)	R	S (0.5)	R	R	R	S (2)	R
Anaerobic streptococci	S (0.3)	S (0.3)	S (1)	S (0.5)	S (0.5)	R	R	(S) (0.3/R)	S (4)	S (0.5)	R	S
Clostridia sp.[b/c]	(S) (0.1/ R)	R	R	S (0.5)	R	R	R	(S) (2/R)	(S) (8)	S (0.5)	R	S
Bacteroides sp.					R	R	R			R		S

[a] Only moxifloxacin is effective against *Enterococcus faecalis*, with enhanced activity for other streptococci.
[b] Sensitive to clindamycin.
[c] Sensitive to erythromycin and azithromycin.
Abbreviations: S, sensitive; (S), moderately sensitive (intermediate); R, resistant.

administration were: (mean $+/-$ SD) $0.11+/-0.02$, $0.257+/-0.13$, $0.27+/-0.21$, $0.307+/-0.26$, and $0.108+/-0.07\,\mu g/mL$. The mean concentration of clarithromycin in the iris was $6.2\,\mu g/g$, demonstrating its concentration into cells (28).

EFFECT OF DIFFERENT DOSES OF QUINOLONES ON AQUEOUS HUMOR (AH) LEVELS IN μg/ML

There is currently great interest in use of quinolones given by the topical route for treating bacterial conjunctivitis and prophylaxis of endophthalmitis following phacoemulsification cataract surgery. Levofloxacin is a new quinolone that is both lipophilic and hydrophilic (water soluble). This allows it to penetrate through the corneal epithelial barrier into the stroma and anterior chamber in good levels reaching $4.4\,\mu g/mL$ (mean value) in the aqueous humor at 60 minutes after the third postoperative dose (16); two doses of 0.5% drops were given half-hourly in the 1 hour before cataract surgery and three doses of 0.5% drops were given every 5 minutes immediately after surgery. Samples of aqueous humor were collected at different times after cataract surgery for assay of levofloxacin by HPLC. Different regimens have been investigated prior to cataract surgery with use of frequent drop therapy, oral therapy, or the combination thereof, which has penetrated well both into the stroma and into the anterior chamber (Table 2).

These studies have consistently shown that levofloxacin is better absorbed into the anterior chamber by the topical or oral route than ofloxacin by a twofold factor and by a four- to eightfold factor greater than ciprofloxacin. Levofloxacin is the levoracemer of ofloxacin and is the active component. Absorption for moxifloxacin may be similar to levofloxacin (Table 2).

A frequent drop regime of levofloxacin 0.5%, 4 times every 15 minutes in the one hour prior to cataract surgery, has given levels of $1.1\,\mu g/mL$ in the aqueous humor, which can be boosted by additional oral therapy to $1.8\,\mu g/mL$ (Table 2) or, as described above, with pulsed dosing to $4.4\,\mu g/mL$ (16). This latter peak level has appeared later than expected, at 60 minutes, so that all previous studies (Table 2) will have under-reported their peak values; in addition, these previous studies only included samples of AH collected at the time of surgery and not afterwards.

Kowalski et al. (30) have found that the MIC90 of most bacteria causing corneal infection, and hence a source of postoperative infection, to be $<1.0\,\mu g/mL$ (Fig. 9). Effective levels of levofloxacin are maintained in the tears and conjunctiva for up to 6 hours or more with a single dose of 0.5% (Fig. 9) (31). In addition, levofloxacin has a rapid action that can be expected to reduce the bacterial count on the conjunctiva prior to surgery, when applied two or four times every 15 minutes, by up to 2 $\log_{10}$, or 100-fold.

Moxifloxacin has given equally good penetration to levofloxacin. Moxifloxacin is a new quinolone effective against methicillin-resistant *Staphylococcus aureus* (MRSA) that is not susceptible to levofloxacin and has better activity against streptococci. Moxifloxacin prevented experimental *Staphylococcus aureus* endophthalmitis in the rabbit when given by the topical route as 0.5% drops, four times every 15 minutes, prior to inoculation of the anterior chamber with bacterial cells (32). It is our belief, however, that moxifloxacin should be reserved for treatment of keratitis or endophthalmitis, due to bacteria resistant to other antibiotics, and that levofloxacin, an earlier and more widely used quinolone that

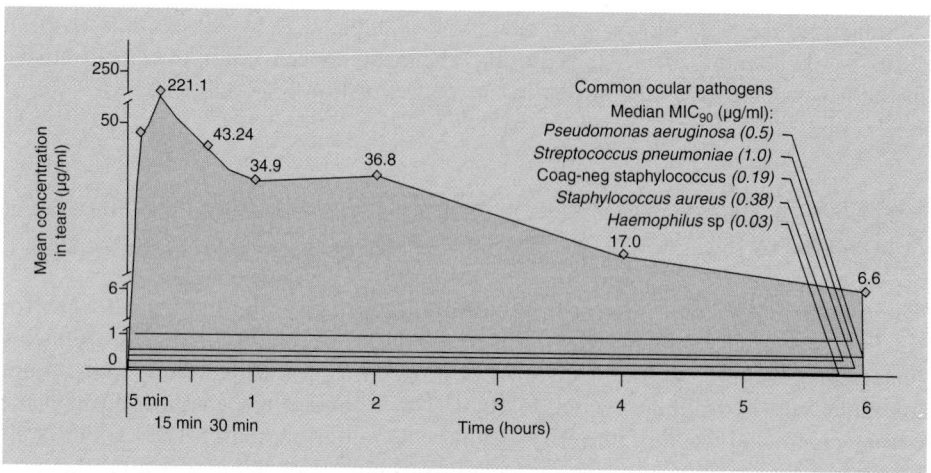

Figure 9   Graph of levofloxacin concentration in tears with time over six hours in relation to MICs of common ocular pathogens. *Source*: From Refs. 30 and 31.

penetrates well into the eye, should be used for surgical prophylaxis. There is good scientific reason to argue for such separate use of these two quinolones.

MINIMUM INHIBITORY CONCENTRATIONS (MICs) AND MINIMUM BACTERICIDAL CONCENTRATIONS (MBCs) OF ANTIBIOTICS FOR PATHOGENIC BACTERIA

Antibiotics have an inhibitory (bacteriostatic) and killing (bactericidal) effect on the bacterial cell. Bacteriostatic antibiotics such as chloramphenicol, tetracycline, low-dose erythromycin, and sulphonamides bind to ribosomes to inhibit messenger RNA translation and therefore protein production. The bacterial cell cannot divide but remains viable, the killing effect depending on polymorphonuclear (PMN) cells and macrophages. Other antibiotics such as penicillins and cephalosporins, which attach to penicillin-binding proteins, prevent cell wall formation resulting in cell death. The aminoglycosides (gentamicin, tobramycin, and amikacin) bind to ribosomes irreversibly with a bactericidal effect not requiring PMNs. Laboratory testing involves assessing the MIC and MBC and usually refers to the MIC_{90} and the MBC_{90} when the values quoted reflect those likely to be found for 90% of the isolates. Tables 5 and 6 quote the MIC_{90} values expected for bacteria and fungi pathogenic for the eye. It is meant as a Quick Guide for the clinician, but a microbiology laboratory should always be consulted for an individual case because of hospital, local, and regional differences in bacterial resistance. For bactericidal antibiotics, the MIC value is similar to the MBC in most cases except for Lancefield group G betahemolytic streptococci, when the MBC can be much higher than the MIC for penicillins, and clindamycin should be used adjunctively. For bacteriostatic antibiotics, there is no bactericidal effect, only an inhibitory effect, so that killing of the sessile cell depends instead on the immune response. *It is important to achieve the MBC level within the vitreous cavity, rather than the MIC, for gaining control of the infection.*

Table 6 Expected Antibiotic Sensitivities (MICs) of Pathogenic Fungi for Causing Ocular Infection

	Polyene antibiotics Amphotericin/ Natamycin	Imidazole antibiotics Miconozole/Fluconazole/ Voriconazole et al.	5-fluorocytosine
Canidida albicans and yeasts	S	S	S
Aspergillus sp.	S	S	R
Fusarium sp. and filamentous fungi	(S)	(S)	R
Scedosporium apiospermum	R	(S)	R

CHALLENGE FOR THE FUTURE

A challenge in the medical management of infectious endophthalmitis, aside from early surgical intervention, remains the ability to administer, as soon as possible, the most effective antibiotic against the microorganism (a drug of choice), in a maximum nontoxic dose by direct intravitreal injection, that may be supplemented with systemically administered doses, preferably of the *same* antibiotic. An understanding of how a drug is likely to behave, in light of all the factors described above, will contribute to the evolution of safe and effective antibiotic regimens for the treatment of infectious endophthalmitis. In addition, the regimen will also include corticosteroids (Chapter 8), or newer anti-inflammatory drugs (Chapter 8), to suppress inflammation within the retina.

GLOBE SIZE AND INTRAVITREAL INJECTIONS

Teichmann (33) and Fechner and Teichmann (34) have considered the variation in the volume of vitreous, from 1.7 to 16.5 mL, for globe sizes of axial lengths from 16 mm to 34 mm, respectively. These are given as whole globe volumes (on average 80% is vitreous) in Table 7 together with the expected frequency and degree of myopia or hypermetropia. For the 90% of eyes that have an axial length between 21 mm and 26.0 mm, the expected volume of the whole globe varies from 4.2 to

Table 7 Frequency of Volume of Globe and Vitreous in Relation to the Axial Length and Lens Power

Frequency	◁——— 95% ———▷ ◁——— 90% ———▷								
Axial length (mm)	16	18	20	22	24	26	28	30	32	34
External volume of eye (ml)	2.1	3.0	4.2	5.6	7.2	9.2	11.4	14.1	17.2	20.6
Volume of vitreous (ml)	1.7	2.4	3.4	4.5	5.8	7.4	9.1	11.3	13.8	16.5
Dioptic power			+10.8	+5.4	0	−5.4	−10.8	−16.2	−20	

Source: Adapted from Ref. 33.

9.2 mL, respectively, of which the vitreous constitutes 75% to 85%, giving an average value for the vitreous from 4.0 mL to 7.5 mL. For the average eye with an axial length of 23.5 mm, the globe volume would be 6.5 mL (vitreous volume of 5.2 mL). The effect of injecting an antibiotic would be to dilute the expected concentration in the myopic, long eye [26 mm, -7 diopters (D)] by approximately 38%. This range of error should be considered by the clinician but is probably acceptable in practice, because the antibiotic toxicity studies were done in rabbit eyes with a volume of 1.4 mL, which is lower than the vitreous volume of a human hyperopic eye. Provided the correct intravitreal dose of antibiotic has been administered with proper technique, no toxicity should be expected.

REFERENCES

1. Patton TF. Ocular drug disposition characteristics. Ophthalmic Drug Delivery Systems. Washington, D.C. American Pharmaceutical Association, 1980: 28–55.
2. Peyman G, Lee P, Seal DV. Endophthalmitis: Diagnosis and Management. London & New York: Taylor & Francis, 2004: 1–278.
3. Hariprasad SM, Mieler WF, Holz ER. Vitreous and aqueous penetration of orally administered gatifoxacin in humans. Arch Ophthalmol 2003; 121:345–350.
4. Kobayakawa S, Tochikubo T, Katayama Y. Penetration of levofloxacin into human aqueous humor. Ophthalmic Res 2003; 35:97–101.
5. Koch HR, Kulus SC, Ropo A, Geldsetzer K. Corneal penetration of fluoroquinolones: aqueous humour conentrations after topical application of levofloxacin 0.5% and ofloxacin 0.3% eye drops. J Cat Refract Surg 2005; 31:1377–85.
6. Colin J, Simonpoli S, Geldsetzer K, Ropo A. Corneal penetration of levofloxacin into the human aqueous humor: a comparison with ciprofloxacin. Acta Ophthalmol Scan 2003; 81:611–613.
7. Garcia-Saenz MC, Arias-Puente A, Fresnadillo-Martinez MJ, Carrasco-Font C. Human aqueous humor levels of oral ciprofloxacin, levofloxacin and moxifloxacin. J Cataract Refract Surg 2001:27; 1969–74.
8. Robertson SM, Sanders M, Jasheway D et al. Penetration and distribution of moxifloxacin and ofloxacin into ocular tissues and plasma following topical ocular administration to pigmented rabbits. Presented at ARVO 2003 (abstract 1454).
9. Dong B, Muhtaseb M, Mearza AA, et al. Aqueous level of ofloxacin following oral administration using capillary zone electrophoresis. Presented at ARVO 2003 (abstract 1458).
10. Yamada K, Yamada M, Mochizuki H, et al. Aqueous humor levels of topically applied levofloxacin, norfloxacin, and lomefloxacin in the same human eyes. J Cataract Refract Surg. 2003 Sep; 29(9):1771–5.
11. Bucci FA. An in vivo study comparing the ocular absorption of levofloxacin and ciprofloxacin prior to phacoemulsification. Am J Ophthalmol 2004; 137:308-12.
12. Fiscella RG, Nguyen TKP, Cwik MJ, et al. Aqueous and vitreous penetration of levofloxacin after oral administration. Ophthalmology 1999; 106:2286–2290.
13. Sasaki K, Mitsui Y, Fukuda M, Ooishi M, Ohashi Y. Intraocular penetration mode of five fluoroquinolone ophthalmic solutions evaluated by the newly proposed parameter of AQCmax. Atarashii Ganka (J Eye) 1995; 12:787–790.
14. Inoue S, Misaki M, Matsumura K. Intraocular penetration of DR-3355. Atarashii Ganka (J Eye) 1992; 9:487–490.
15. Healy D, Holland EJ, Nordlund ML, et al. Concentrations of Levofloxacin, Ofloxacin and Ciprofloxacin in Human Corneal Stromal Tissue and Aqueous Humor after topical administration. Cornea 2004; 23:255–63.

16. Sundelin K, Stenevi U, Seal DV, Gardner S, Geldsetzer K, Ropo A. Pharmacokinetic profile of intensive drop application of levofloxacin 0.5% for cataract surgery. Paper in preparation, 2007.

17. Cunha-Vaz JG. The blood-ocular barriers: past, present, and future. Documenta Ophthalmologica 1997:93; 149–157.

18. Pflugfelder AC, Hernandez E, Fleisler St, et al. Intravitreal vancomycin: retinal toxicity, clearance and interaction with gentamicin. Arch Ophthalmol 1987; 105:831–37.

19. Case JL, Peyman GA, Barrada A, Hendricks R, Fiscella R, Hindi M. Clearance of intravitreal 3H-Fluorouracil. Ophthalmic Surg 1985; 16:378–381.

20. Campochiaro PA, Lim JI. Aminoglycoside toxicity in the treatment of endophthalmitis. The Aminoglycoside Toxicity Study Group. Arch Ophthalmol 1994; 112:48–53.

21. Goldblum D, Rohrer K, Fruch BE, Theurillat R, Thormann W, Zimmerli S. Corneal concentrations following systemic administration of amphotericin B and its lipid preparations in a rabbit model. Ophthalmic Res 2004; 36:172–26.

22. Abreu JA, Cordoves L. Endoftalmitis sacular. Arch Soc Esp Oftalmol 2001; 76:5–6.

23. Abreu JA, Cordovés L. Chronic or saccular endophthalmitis: Diagnosis and management. J Cataract Refract Surg 2001; 27:650–1.

24. Warheker PT, Gupta SR, Mansfield DC, Seal DV, Lee WR. Post-operative saccular endophthalmitis caused by macrophage-associated staphylococci. Eye 1998; 12 1019–1021.

25. Kresloff MS, Castellarin AA, Zarbin MA. Endophthalmitis. Surv Ophthalmol 1998:43; 193–224.

26. Warheker PT, Gupta SR, Mansfield DC, Seal DV. Successful treatment of saccular endophthalmitis with clarithromycin. Eye 1998; 12:1017–1019.

27. Karia N, Aylward GW. Letter on use of clarithromycin. Ophthalmol 2001; 108:634.

28. Okhravi N, Guest S, Matheson MM, et al. Assessment of the effect of oral clarithromycin on visual outcome following presumed bacterial endophthalmitis. Curr Eye Res 2000: 21(3); 691–702.

29. Al-Sibai MB, Al-Kaff AS, Raines D, El-Yazigi A. Ocular penetration of oral clarithromycin in humans. J Ocul Pharmacol Ther 1998; 14:575–583.

30. Kowalski RP, Dhaliwal DK, Karenchak Lm et al. Gatifloxacin and moxifloxacin: an in vitro susceptibility comparison to levofloxacin, ciprofloxacin and ofloxacin using bacterial keratitis isolates. Am J Ophthalmol 2003; 136:500–05.

31. Raizman MB, Rubin JM, Graves AL et al. Tear concentrations of levofloxacin following topical administration of a single dose of 0.5% levofloxacin ophthalmic solution in healthy volunteers. Clin Ther 2002; 24:1439–50.

32. Kowalski RP, Romanowski EG, Mah FS, Yates KA, Gordon YJ. Intracameral Vigamox (moxifloxacin 0.5%) is non-toxic and effective in preventing endophthalimitis in a rabbit model. Am J opthalmol 2005; 140:497-504.

33. Teichmann KD. Intravitreal injections: Does the globe size matter? J Cat Refract Surg 2002; 28:1886–89.

34. Fechner PU, Teichmann KD. Ocular Therapeutics—Pharmacology and Clinical Application. Slack Inc.: Thorofare, New Jersey, U.S.A., 1998: 11, 451–70.

FIVE

Cellulitis, Blepharitis, and Dacrocystitis

ORBITAL (POSTSEPTAL) CELLULITIS

Orbital cellulitis is a retroseptal infection of the extraocular orbital contents presenting with pain, lid edema, proptosis, and diplopia due to impaired extraocular muscle function. A few cases follow penetrating injury or are secondary to panophthalmitis, but the majority occur in association with sinusitis. The condition commonly affects children, where spread to the orbit occurs across the thin orbital plate of the ethmoid bone. Retroseptal infection requires multidisciplinary management because of the risk of extension to the eye or cranial cavity. Subperiosteal abscess with displacement of the globe, or cavernous sinus thrombosis with headache and neck stiffness, can develop. Loculated pus must be drained. Delayed or inadequate treatment may lead to blindness or death.

Pathogens from the sinuses include *Streptococcus pyogenes*, *Staphylococcus aureus*, *Haemophilus influenzae* (Fig. 1) and anaerobic bacteria if there is a chronic sinusitis. Phycomycetes may be involved in diabetic patients (see Chapter 9).

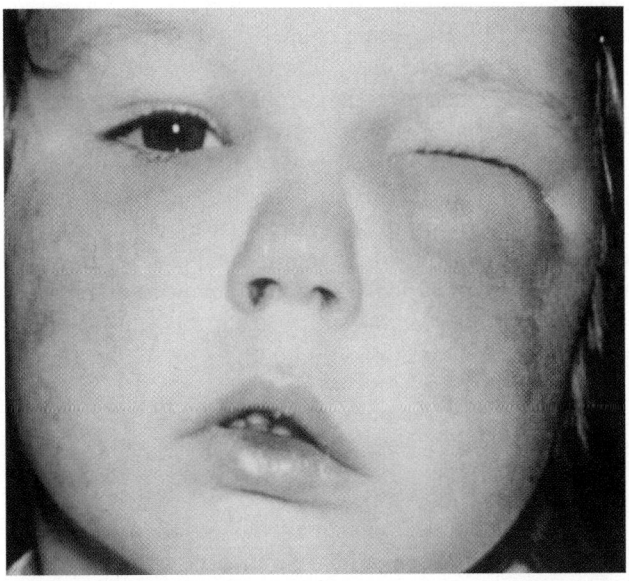

Figure 1 Cellulitis of the orbit in a child due to *Haemophilus influenzae*. *Source*: Courtesy of the late Dr. K. Rodgers.

Posttraumatic orbital cellulitis is often polymicrobial when *Clostridia* sp. can cause gas gangrene (1); differentiation from necrotizing fasciitis and other types of microbial gangrene needs to be made (2). Systemic antibiotic therapy is required in large doses for which purpose Augmentin (amoxicillin and clavulanic acid) combined with a quinolone (levofloxacin, ofloxacin, or ciprofloxacin) or vancomycin combined with a cephalosporin and metronidazole are suitable. If extension occurs into the meninges then antibiotics that penetrate into the central nervous system are required, such as cefotaxime, levofloxacin, chloramphenicol, co-trimoxazole (Septrin®), or rifampicin in high dosage.

Proptosis with cellulitis may be due to an enlarging hydatid cyst, from infection with the dog tapeworm *Echinococcus granulosus*, which requires surgical removal and treatment with albendazole (Eskazole®) (800 mg/day in divided doses for 56 days, both pre- and postsurgery) to reduce the risk of recurrence. Investigate with the intradermal (Casoni) skin test or measure complement-fixing antibodies.

PRESEPTAL CELLULITIS

Preseptal cellulitis may resemble orbital cellulitis. Intense lid swelling is present, but the presence of normal ocular movements and the absence of globe inflammation in preseptal disease help to distinguish the two. Eye movement can usually be detected despite a tense swelling of the lids. The infection lies anteriorly to the orbital septum but delay in therapy can allow extension to the orbit (periorbital cellulitis). The differential diagnosis can be resolved by magnetic resonance imaging. In a study of 35 children, preseptal cellulitis was associated with sinusitis in 17, ocular infection in 11 (seven conjunctivitis and four dacryocystitis), and an infected wound in six. Of those with ocular infection, 73% were less than two years old (3).

Parenteral therapy is designed to cover the common causative organisms: *H. influenzae*, *S. aureus*, *Streptococcus pneumoniae*, and *Strep. pyogenes*. *H. influenzae* is the commonest cause of orbital cellulitis in young children and in this age group Augmentin or cefuroxime are the drugs of choice. In view of the emergence of multiple-resistant strains of *H. influenzae*, consideration should be given to the use of a "third generation" cephalosporin such as cefotaxime, particularly when the clinical response is poor or resistant organisms are isolated from nasal swabs. In adults, therapy is directed against streptococci and *S. aureus* with high-dose intravenous benzylpenicillin alternating with flucloxacillin or, in advanced cases, clindamycin or vancomycin. It may also be due to *Bacillus anthracir* (anthrax) (Fig. 2). Anaerobic cellulitis occurs following trauma (Fig. 2), particularly with human or animal bites, and the latter can include infection by *Pasteurella* sp. or group G beta-hemolytic streptococci for which Augmentin® is the antibiotic of choice.

NECROTIZING FASCIITIS

This is a potentially lethal infection of the deep subcutaneous tissue of the eyelid, above the fascial layer initially appearing as a preseptal cellulitis. It is caused by beta-hemolytic streptococci of Lancefield's group A (*Strep. pyogenes*) or groups C and G, together with *S. aureus* in 25% of cases. They penetrate the skin in small numbers to cause a necrotizing cellulitic lesion, with septic thrombophlebitis of dermal vessels,

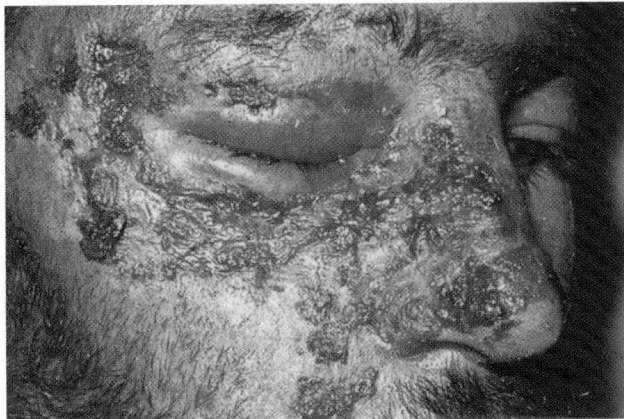

Figure 2 Swelling of the right upper and lower lids, with characteristic ring of vesicles, due to infection by the anthrax bacillus in Iran.

which spreads upwards over the forehead or downwards over the cheek (Fig. 3). Initially, there is ulcerative necrosis at the site of invasion that later extends over and undermines other subcutaneous tissues erratically, depending on the vessels thrombosed. The first 48 hours are the most vital for the patient, particularly if the pathogenesis of the infection is missed. Patients usually suffer from profound toxemia, may develop the "toxic shock syndrome," and may die from septicemia. This is a medical emergency and patients must be treated with large doses of benzylpenicillin [adult: 3 MU, (or 1.8G) 4 hourly] in order to obtain adequate tissue perfusion in the presence of the thrombosing infection, together with flucloxacillin or another anti-staphylococcal antibiotic such as clindamycin. Gram-negative bacteria and anaerobes do not normally initiate this infection at this site.

The spreading edge of the necrotizing infection can advance by as much as 1 inch per hour and should be marked with a pen. If it has not stopped advancing

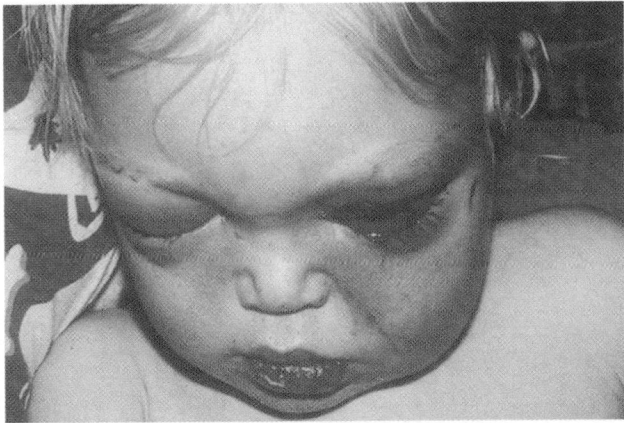

Figure 3 Acute necrotizing fasciitis due to *Streptococcus pyogenes* of the upper eyelid, with rapid spreading inflammation throughout the left cheek marked with a pen, and sympathetic edema of the right eyelids in a two-year-old child. *Source*: Courtesy of Dr. F. Ghanchi.

after 24 hours, then surgical debridement is indicated to remove the infected subcutaneous tissue and to prevent further spread due to failure of the antibiotics to penetrate the thrombosed tissue. If the infection responds within 48 hr, then the edge regresses considerably, and ulcerative necrosis is confined to the lid (Fig. 4). After a further 5 days, the characteristic eschar forms and the subcutaneous swelling resolves (Fig. 5). The eschar will slough off eventually, often leaving a damaged lid requiring reconstructive surgery. Early debridement of the lids has been advocated by some, but others, including ourselves, recommend conservative therapy as described above. The patient should be investigated for latent diabetes or immunodeficiency during convalescence.

A rare necrotizing lid lesion encountered in North and South America that may resemble necrotizing fasciitis is loxoscleromatis. This may be induced during sleep by the bite of the black African widow spider *Lactodectus geometricus* (Fig. 6), also known as the "reclusive" spider, which survives inside bedroom closets!

BLEPHARITIS

Marginal blepharitis is an inflammation of the lid margin, which may be anterior or posterior. Anterior blepharitis involves the lash line, while posterior blepharitis involves dysfunction of the meibomian glands. Both are strongly associated with skin disease, chiefly seborrhoeic, atopic denmatitis and rosacea.

The term 'staphylococcal' blepharitis was coined by Thygeson 60 years ago to describe an anterior blepharitis with lash collarettes and crusting, lid ulceration, and folliculitis, associated with a positive culture for *S. aureus*. This appears to be a distinct clinical entity. McCulley has extended the term to include those patients with lid signs and any positive culture for staphylococci, but since coagulase-negative staphylococci are an almost universal commensal on the normal lid, this approach loses its utility, and we prefer to confine the diagnosis to those patients showing the above clinical features, with positive *S. aureus* culture. It should be remembered that cultures fluctuate from day to day, so that where the clinical morphology is strongly

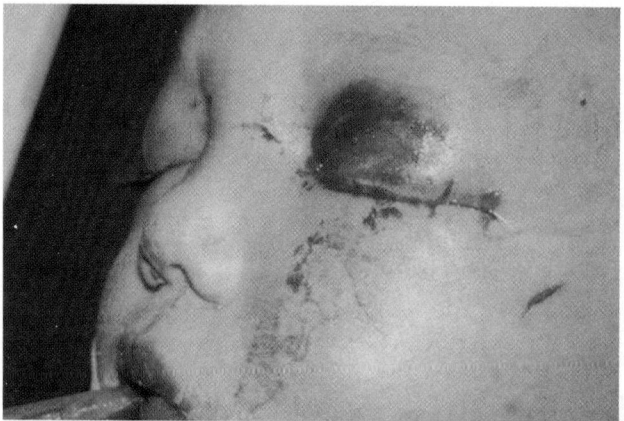

Figure 4 Close up of left lids and cheek two days after starting antibiotics, showing characteristic blistering; discharging necrotic lesion limited to the left upper lid with control of surrounding edema. *Source*: Courtesy of Dr. F. Ghanchi.

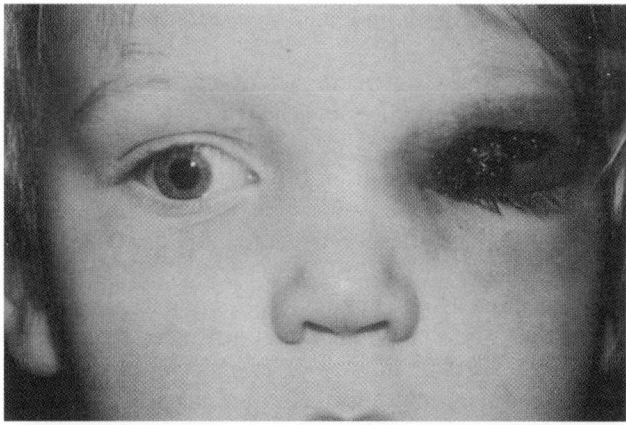

Figure 5 Typical eschar on lid five days after starting antibiotics, requiring later surgery. *Source*: Courtesy of Dr. F. Ghanchi.

suggestive it is worth repeating the culture if the initial culture was negative for *S. aureus* with a lid scrub. Similarly, since *S. aureus* may colonize the normal lid margin (6–15%) without giving rise to blepharitis, it follows that positive culture alone is not sufficient for the diagnosis.

Seventy-six percent of patients with blepharitis associated with atopic dermatitis have a positive lid culture for *S. aureus* (4), and ulcerative blepharitis is almost entirely confined to patients in this group. Positive cultures for *Candida* sp. are also found, but only in the ulcerative subgroup, a finding possibly related to corticosteroid use. Most patients with blepharitis due to retinoid toxicity are also positive for *S. aureus*. The blepharitis of atopy presents in relative youth (19–31 years) and may wax and wane with the activity of the dermatitis. Blepharitis accompanying seborrhoeic dermatitis preferentially affects young men in their second and third

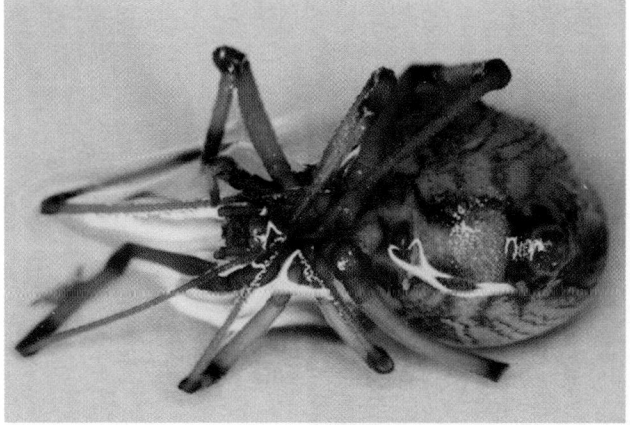

Figure 6 African widow spider (*Lactodectus geometricus*), the cause of loxosclerma, or necrotic eyelid, from the bite of this "reclusive" spider, found in N. and S. America and Africa, which travels by night out of its resting place inside cupboards.

decades. Rosacea presents later, in the seventh and eighth decades, with a greater frequency in women.

Several groups have suggested ways by which lid bacteria might contribute to both forms of blepharitis, either directly by infection and the release of exotoxins, or indirectly through their action on lipids. Groden et al. (5) found an increased colonization of the lids with a wide range of organisms in patients with blepharitis. Others have shown that several bacterial species additional to *S. aureus* produce esterases and lipases capable of splitting cholesterol esters into cholesterol and fatty acids. Such fatty acids are conceived as potentially irritant to the eye, perhaps by the formation of soaps. They have shown in vitro that cholesterol stimulates the growth of some *S. aureus* strains and that, in the normal population, those individuals whose meibomian secretions are rich in cholesterol show twice as many staphylococcal strains on their lid margins. This mechanism could therefore increase the risk of colonization, infection, and immunization.

Mondino et al. (6) developed a model of blepharitis by immunizing rabbits with staphylococcal cell wall antigen, while Ficker et al. (7), found enhanced cell-mediated immunity (CMI) to protein-A of *S. aureus*, that exceeded far exceed the percentage who were culture positive. This suggests that CMI may provide the mechanism for inflammatory lid disease during episodes of recolonization. In this view, eradication of *S. aureus* from the lids by antimicrobial therapy will reduce the opportunity of stimulating CMI-induced reactions to this organism (8), and the propensity for causing marginal ulceration.

Anterior and posterior blepharitis often occur together, in part because of their mutual association with skin disease. Posterior blepharitis takes the form chiefly of obstructive meibomian gland disease (MGD), which may be cicatricial or non-cicatricial, and meibomian seborrhea. MGD is not an infective condition and any role that bacteria may have on symptoms is indirectly through the action of their products of lipolysis (9).

Blepharitis is diagnosed by clinical examination. The lid margin is examined with the slit-lamp for evidence of folliculitis and collarettes (cuffing of infected secretion around the lashes). In the acute condition, there will be beads of pus and an ulcerated margin. When chronicity ensues, there is a loss or misdirection of lashes,

Figure 7   Acute folliculitis of upper and lower right lid margins due to *Staphylococcus aureus*.

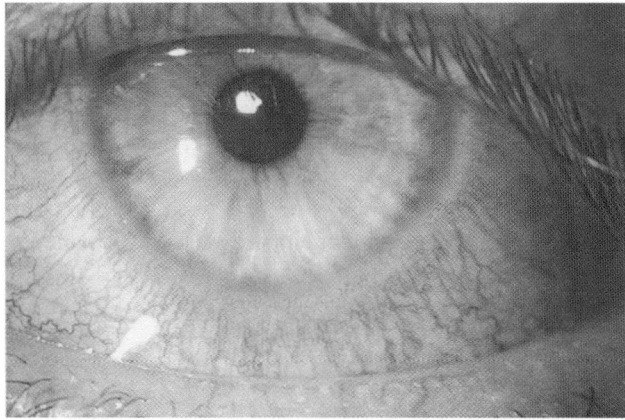

Figure 8 Acute toxic limbitis and conjunctivitis from blepharitis due to *Staphylococcus aureus.*

telangiectasia, and a swollen lid margin. There may be a history of recurrent marginal ulceration.

Culture of the lid margin requires "scrubbing" with a swab soaked in sterile broth, and plating out directly on blood agar and a selective medium for *S. aureus*. If *S. aureus* is present, treatment should commence with Fucithalmic® (topical 1% fusidic acid in gel). If laboratory cultures demonstrate fucidin resistance, tetracycline or Polytrim ointment can be used instead if the isolate is sensitive. In acute blepharitis (Figs. 7–9), flucloxacillin or erythromycin can also be given by mouth (500 mg qds for four days) and whole body bathing commenced with chlorhexidine (Hibiscrub, AstraZeneca, Macclesfield, UK).

Coagulase-negative staphylococci (CNS) may be cultured but should not be considered as pathogenic. CNS colonize over 80% of normal and blepharitic lid margins and are part of the normal lid flora; the presence of CNS alone is not an indication for treatment.

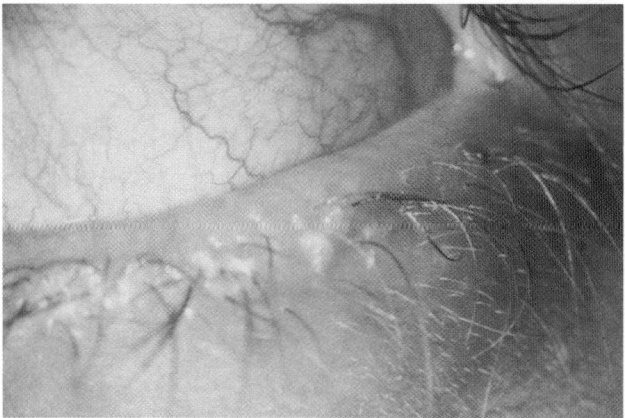

Figure 9 Close-up of patient in Figure 8 showing acute folliculitis of the left lower lid margin.

In chronic anterior blepharitis, lid hygiene should be attended to regularly, using lid scrubs with dilute baby lotion, and misdirected lashes must be removed. Intermittent therapy should be given with topical Fucithalmic (or less efficiently by Polytrim®) to suppress the presence of *S. aureus* on the lids. Patients should be encouraged to wash with antiseptic soaps to suppress carriage of *S. aureus* at other skin sites, especially axillary and perineal; chlorhexidine (Hibiscrub) is the most efficacious product and gives a persistent anti-staphylococcal effect on skin.

If no apparent response occurs, the blepharitic patient should be investigated in more detail. An algorithm has been established to assist with this (Fig. 10) (10). Patients are divided into those with and without enhanced CMI to *S. aureus*. This is

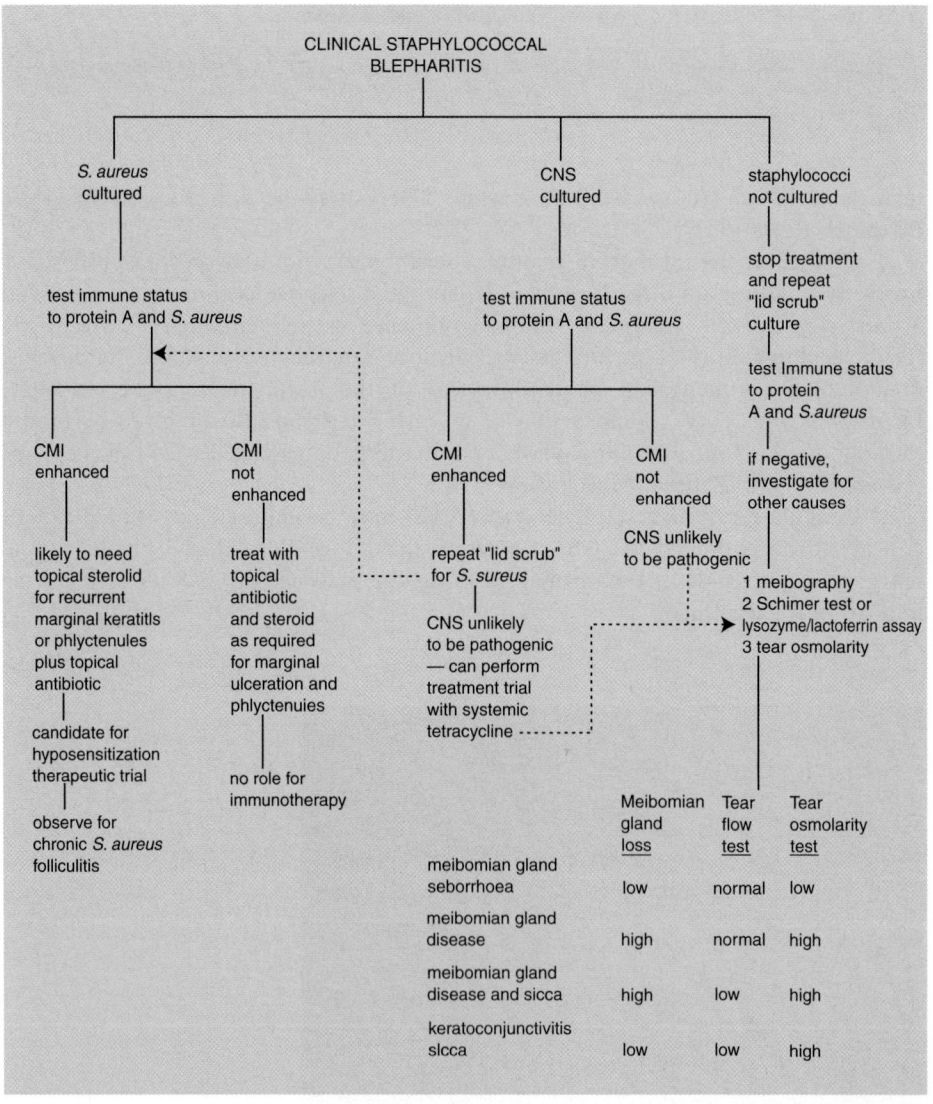

Figure 10 Therapeutic decision making in the management of chronic blepharitis. *Source*: Adapted from Ref. 10.

determined by injecting 0.1 mL of killed *S. aureus* cells intradermally (10^8/mL preserved with 0.5% phenol) and 2.5 nanograms of protein A (Sigma Chemicals, London, U.K.) into the forearm. An induration reaction is read at 48 hours; enhancement is indicated by a positive result of > 5 mm for killed *S. aureus* cells and > 20 mm for protein A (7). The presence of *S. aureus* on the lid margin should be suppressed in enhanced patients, especially if associated with a marginal infiltrate or ulcer on the cornea (Fig. 11); this situation occurs more commonly in association with rosacea.

ATOPY

Atopic blepharitis should not be confused with staphylococcal blepharitis, particularly in children where, in staphylococcal blepharitis, the lid signs can be mild but the associated corneal signs much more severe with marginal infiltrates, vascularization, corneal scarring, and visual impairment. In one recent study, these corneal signs were particularly prevalent in Asian and Middle Eastern children (11). The two may coexist, but atopic keratoconjunctivitis (AKC) is a separate disorder (4).

Seventy percent of normal, non-inflamed lids in patients with atopic dermatitis are colonized with *S. aureus* (4), with 50% being heavily colonized. Additionally, 60% of these atopic subjects had colonization of their conjunctiva and nasal mucosa with *S. aureus*. The skin of patients with atopic dermatitis is also heavily colonized by *S. aureus* but the reason for this association is not understood. Huber-Spitzy et al. (12) found the highest rate (76%) of *S. aureus* isolation from lids of atopic patients aged 19 to 31 years with blepharitis and atopic dermatitis, whether or not the blepharitis was accompanied by ulcerative or squamous signs. These results are similar to our own for the atope without blepharitis, which therefore questions whether the clinical signs were primarily due to allergic disease or *S. aureus*. Huber-Spitzy et al. (12) found that *Candida* sp. was isolated from atopic lids with ulcerative blepharitis, and suggested the presence of "superinfection" with frequent use of corticosteroids requiring antimycotic treatment. Management of concomitant

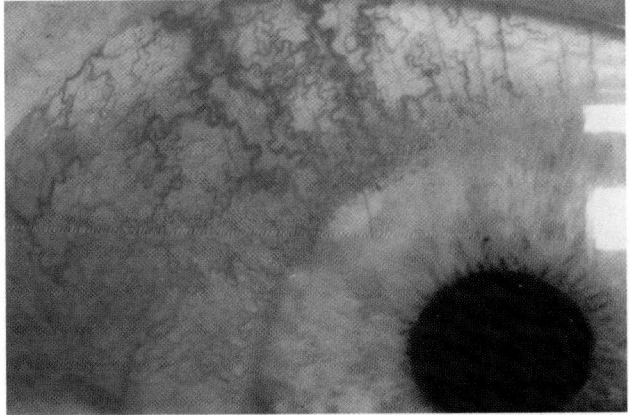

Figure 11   Marginal infiltrate and conjunctival reaction characteristically associated with blepharitis due to *Staphylococcus aureus*.

S. aureus blepharitis in the atope follows the same procedures outlined above. While tear IgE levels are high, they are not directed against *S. aureus* antigens even in AKC patients (4).

ROSACEA

Rosacea is associated with blepharitis (Fig. 12). The lids are often colonized by *S. aureus*, although it is not clear why the rosacea patient, like the atope, is more susceptible to this colonization than others. In addition these patients often appear to have enhanced CMI to *S. aureus* (10,13) as well, but it is not clear whether this is cause or effect. In a randomized, placebo-controlled cross-over trial (13), symptomatic improvement of blepharitis in mild rosacea occurred in 90% of patients receiving topical fusidic acid (Fucithalmic), possibly due to its cyclosporin-type effect, compared to 50% in those on oral oxytetracycline. A similar 50% response to tetracycline in rosacea had been found by Bartholomew et al. (14), but in their series 25% responded to placebo. Such symptomatic improvement did not occur in non-rosacea blepharitis (12). There was no response to fusidic acid and a 25% response to tetracycline. Rosacea-associated blepharitis should be distinguished as a separate group and treated accordingly. Rosacea patients also develop an inflammatory keratitis, which requires separate management with topical corticosteroid.

CANALICULITIS

Recurrent unilateral conjunctivitis due to an antibiotic-sensitive bacterium such as *H. influenzae* can be the presenting symptom of a canaliculitis due to *Actinomyces* sp. (formally Streptothrix or Leptothrix), or *Arachnia propionica* (Fig. 13). The organism does not usually invade the canaliculus wall but forms a 'fungal' ball that obstructs the lumen. The canaliculus provides a microaerophilic environment that supports the growth of non-fastidious anaerobic bacteria, and becomes infected with endogenous flora.

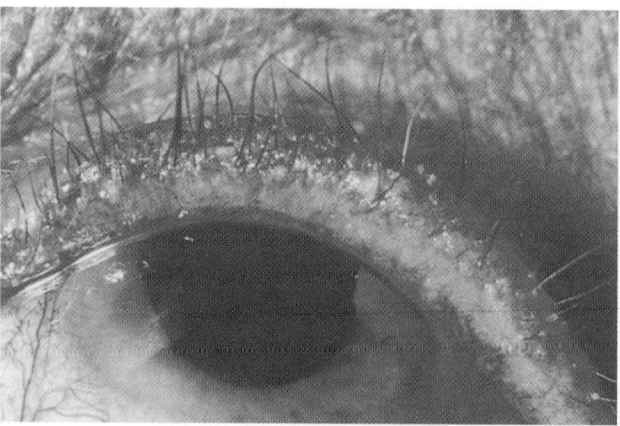

Figure 12 Telangiectiasia and folliculitis of lid margin, with chronic *Staphylococcus aureus* infection, of a patient with rosacea.

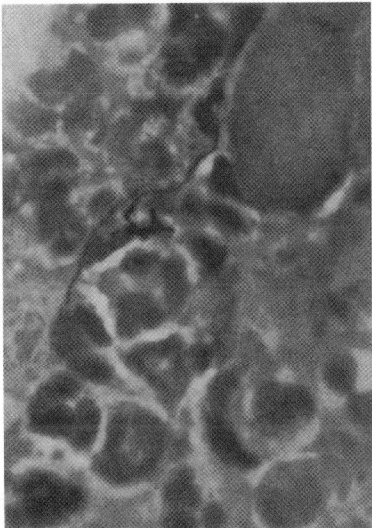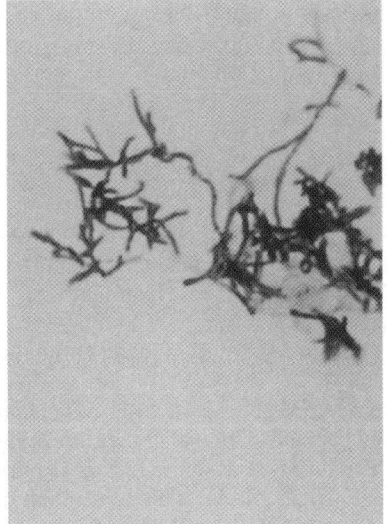

Figure 13　Gram stain of pus from the lacrimal canaliculus and from the laboratory culture showing branching Gram-positive bacilli identified as *Arachnia propionica*.

Pus, massaged along the canaliculus to the punctum, can be Gram stained to show typical, branching Gram-positive bacilli. Prolonged anaerobic culture on blood agar plates is necessary to demonstrate actinomycetes. Thioglycolate broth should also be inoculated, since oxygen tension decreases with depth, so that the actinomycete ("breadcrumb") colonies grow and float at the appropriate level. Sensitivity tests should be performed.

Actinomycetes and *Arachnia* spp. are usually sensitive to penicillin, tetracycline, and erythromycin but resistant to gentamicin and other aminoglycosides often used ineffectively to attempt cure. Initial treatment involves irrigating the canaliculus with penicillin. If this fails, then surgery is needed. The canaliculus is opened and debrided. All material removed should be Gram stained and cultured. The canaliculus should be treated with 5% povidone iodine for 5 min as an effective antiseptic. The canaliculus should be syringed daily for seven days with penicillin and the patient reviewed at three months. Occasionally, repeat surgery and further povidone iodine and penicillin are required to effect a cure.

DACRYOCYSTITIS

Acute dacryocystitis is caused by stasis, due to distal obstruction of the nasolacrimal system. *S. aureus* or streptococci are the usual causes and infection may subside on systemic chemotherapy alone. However, drainage of a lacrimal sac abscess is occasionally needed and ultimately the obstruction must be relieved by dacryocystorhinostomy (DCR).

Chronic dacryocystitis commonly involves Gram-negative bacteria as "secondary" pathogens. It can only be effectively treated by relieving the nasolacrimal duct obstruction.

The postoperative infection rate in patients undergoing DCR is greatly reduced by intraoperative IV cefuroxime (750 mg), or a five-day course of oral cefalexin 250 mg four times daily.

REFERENCES

1. Rehany U, Dorenboim Y, Lefler E, et al. *Clostridium bifermentans* panophthalmitis after penetrating eye injury. Ophthalmol 1994; 101:839–42.
2. Kingston D, Seal DV. Current hypotheses on synergistic microbial gangrene. Br J Surg 1990; 77:260–264.
3. Aidan P, Francois M, Prunel M, et al. Or bital cellulitis inchildren. Arch Pediatr 1994; 1: 879–85.
4. Tuft S, Kemeney M, Buckley R, Ramakrishnan M, Seal D. Role of *S.aureus* in chronic allergic conjunctivitis. Ophthalmology 1992; 99:180–184.
5. Groden LR, Murphy B, Rodnite J, et al. Lid flora in blepharitis. Cornea 1991; 10:50–53.
6. Mondino BJ, Caster AL, Dethlefs B. A rabbit model of staphylococcal blepharitis. Arch Ophthalmol 1987; 105:409–12.
7. Ficker L, Ramakrishnan M, Seal D, Wright P. Role of cell-mediated immunity to staphylococci in blepharitis. Am J Ophth 1991; 111:473–479.
8. Seal DV. Ficker L. Immunology and therapy of marginal ulceration as a complication of chronic blepharitis due to *S. aureus*. In: J Lass (ed.). Advances in Corneal Research. Selected Abstracts of the World Congress on the Cornea IV. New York: Plenum Press, 1997, pp. 19–25.
9. Bron AJ, Tiffany JM. The evolution of lid margin changes in blepharitis. In: J Lass (ed.). Advances in Corneal Research. Selected Abstracts of the World Congress on the Cornea IV. New York: Plenum Press, 1997, pp 26–31.
10. Seal DV, Ficker L, Wright P. Chapter 61. Staphylococcal blepharitis. In: JS Pepose, GN Holland, KR Wilhelmus (eds). Ocular Infection and Immunity. Chicago: Mosby Year Book, 1996: 788–798.
11. Viswalingam ND, Rauz S, Morlet N, Dart JKG. Blepharoconjunctivitis in children: diagnosis and treatment. Brit J Ophthalmol 2005; 89: 400–3.
12. Huber-Spitzy V, Buhler-Sommeregger K, Arocker-Mettinger E. Ulcerative blepharitis in atopic patients—is *Candida* the causative agent. Brit J Ophthalmol 1992; 76:272–7.
13. Seal D, Wright P, Ficker L, et al. Placebo-controlled trial of Fusidic acid gel and Oxytetracycline for recurrent blepharitis and rosacea. Brit J Ophthalmol 1995; 79:42–45.
14. Bartholomew RS, Reid BJ, Chessborough MJ, et al. Oxytetracycline in the treatment of ocular rosacea: a double-blind trial. Brit J Ophthalmol 1982; 66:386–88.

SIX

Conjunctiva, Cornea, and Anterior Chamber

CONJUNCTIVA

Conjunctivitis can be due to bacteria, viruses, fungi, helminths, and protozoa. Non-infective forms are frequently caused by allergy, dry eye, or toxicity. Toxicity may be due to preservatives, (thiomersal, benzalkonium chloride), associated with eye drops or contact lens (CL) wear. Rarely it may be due to systemic medication (Stevens-Johnson syndrome) or circulating bacterial toxins, as in toxic shock syndrome due to *Staphylococcus aureus* (TSST-1) and *Streptococcus pyogenes* (exotoxin A).

Clinical Presentation

Bacterial Conjunctivitis

Bacterial conjunctivitis (Table 1) commonly has an acute or subacute manifestation with redness, discharge, swelling, tearing, and irritation. The vision is not normally affected. Also, pain is an uncommon finding and may direct to differential diagnoses such as (epi-)scleritis. The discharge can be mucopurulent or just purulent, consisting of cellular (leukocytes, bacteria, epithelial cells) and non-cellular (fibrin, protein, mucus) material. There is no strict association of the type of discharge and the etiology of conjunctivitis; a mucopurulent exudate is most commonly seen in bacterial conjunctivitis (Fig. 1).

Formation of pseudo-membranes can occur in severe bacterial conjunctivitis especially due to *H. influenzae*, *Streptococcus pneumoniae*, and *Corynebacterium diphtheriae*. Since *C. diphtheriae* infection is rarely seen in the western world (due to diphtheria immunization), *Streptococcus* spp. are more likely to cause membranes, including also other mucosal surfaces, e.g., nasopharynx and mouth.

Neisseria gonorrhoeae or *meningitidis* may present hyperacutely, with massive lid swelling and a characteristic, profuse yellow-green discharge (Fig. 2). In the neonate this infection can progress rapidly to keratitis and perforation, leading to blindness.

Enquiry should always be made of an associated urethritis or proctitis, which may be the clue for chlamydial infection but need not be present.

Bacterial conjunctivitis is essentially a clinical diagnosis but may be supported by identification of the causative organism with smear and/or culture. Laboratory investigations are necessary when the conjunctivitis is chronic (>2 weeks), does not improve with antimicrobial medication and in very young children. Also,

Table 1 Bacteria Causing Conjunctivitis

Common isolates (>5% of patients)
Usually considered non-pathogenic
   Staphylococcus epidermidis
   Corynebacterium spp.
   Propionibacterium acnes
   Peptococcus spp.

	Frequently associated findings/conditions
Usually pathogenic	
Staphylococcus aureus	Blepharitis
Streptococcus pneumoniae	Sinus disease
Beta-hemolytic streptococci	
(Lancefield groups A, C & G)	
Peptostreptococcus spp.	
Haemophilus influenzae	Intrinsic throat flora, often in children

Less-frequent isolates (<5 % of patients)

Bacteroides spp.	
Escherichia coli	Contact lens wear
Klebsiella spp.	Contact lens wear
Listeria monocytogenes	"Farmer's eye" (rural, farmyard dust)
Moraxella spp.	Damaged ocular surface
Neisseria gonorrhoeae	Genital infection
Neisseria meningitidis	Throat infection
Proteus spp.	Old age (men > women)
Pseudomonas spp.	Contact lens wear
Streptococcus pyogenes	Throat infection

perioperatively occurring conjunctivitis requires laboratory tests to identify the pathogen with a view to preventing later endophthalmitis. Conjunctival swabs should be cultured on blood and chocolate agars (and a selected gonococcal agar if thought relevant). Culture should take place in carbon dioxide at 37°C for 48 hours. Pathogens should be identified and sensitivity tests performed. Treatment should be selected from preparations listed in Appendix A.

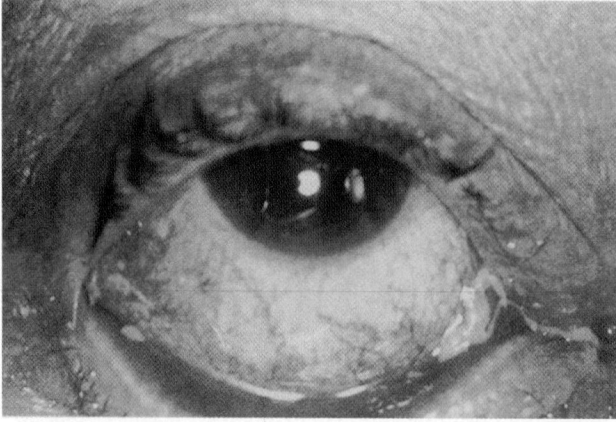

Figure 1 Purulent conjunctivitis in an adult eye typical of bacterial infection.

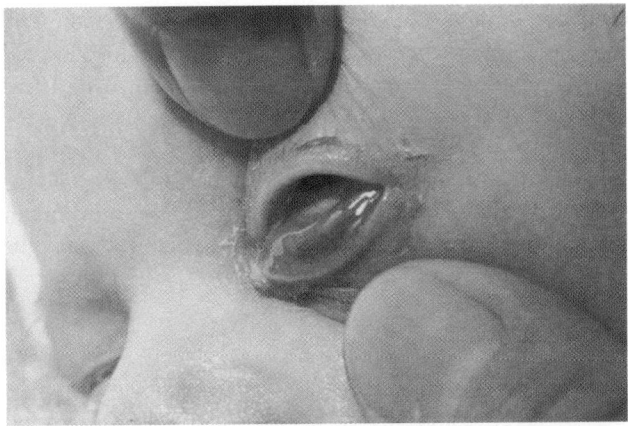

Figure 2 Severe meningococcal conjunctivitis in the left eye of a baby, showing the characteristic abundant green pus of an infection by neisseria. *Source*: Courtesy of the late Dr. K. Rodgers.

Viral Conjunctivitis

Adenovirus infection, like chlamydial conjunctivitis to a lesser degree, may present with a follicular conjunctivitis, usually absent from bacterial infections. This may be seen as acute follicular conjunctivitis or pharyngoconjunctival fever (PCF) (both more common in children), or as epidemic keratoconjunctivitis (EKC) or acute hemorrhagic conjunctivitis. Adenovirus is associated with epidemics from shipyards, close living quarters, and eye clinics (via tonometers and staff handling of patients). Early diagnosis is required to bring outbreaks to a quick halt.

Since a rapid confirmation of suspected EKC is of epidemiologic and clinical importance, an enzyme immunoassay (ELISA) detecting viral antigen from conjunctival swabs has been made available as a diagnostic test (Adenoclone, Cambridge BioScience). This test system demonstrated 81% sensitivity and 100% specificity. Refer also to Ch. 3.

Acute follicular conjunctivitis (Fig. 3) is a self-limiting condition without sequelae, usually caused by adenovirus 1, 2, 4–6, and 19. Conjunctivitis begins unilaterally, but commonly becomes bilateral. In its acute form it lasts up to 21 days,

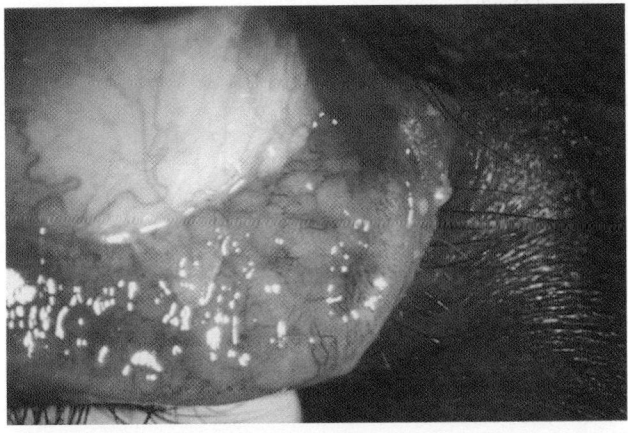

Figure 3 Acute follicular conjunctivitis due to viral infection.

although full recovery may take 28 days or longer. PCF is caused chiefly by adenovirus 3 and 7 with features including fever, conjunctivitis, pharyngitis, coryza, headache, diarrhea, rash, and lymphadenopathy occurring over a course of 10 to 14 days, usually without sequelae. A mild keratitis may occur. PCF is responsible for community-based, summer epidemics of "swimming pool conjunctivitis" in young adults. Management is palliative but polymerase chain reaction (PCR) should be undertaken if possible to confirm the diagnosis (Chapter 3). This is important in relation to potential epidemics. Adenoviruses are highly contagious and may be transmitted between patients by doctors and their equipment.

Acute Hemorrhagic Conjunctivitis (AHC)

Enterovirus 70, associated with occasional paralysis, and coxsackie A24 variant virus are associated with epidemics of AHC especially in South East Asia and India. Adenovirus 11 may also give rise to a form of AHC. AHC caused by enterovirus 70 was first recognized in West Africa in 1969 and rapidly spread to the Eastern hemisphere, with a second pandemic from 1980 to 1982. In the United Kingdom in 1971, 76 cases occurred with enterovirus 70 but this was exceptional. In Singapore, AHC was due to coxsackie A24 vanout virus in 1970, but enterovirus 70 in 1971 and 1990. A large outbreak occurred in Hong Kong due to coxsackie A24v virus in 1988, which continued for two months; recent outbreaks with Coxsachie A24v virus have occured in Singapore (2005) and India (2007).

The infection involves an incubation period of 18 to 36 hours, with the sudden onset in one eye, followed by the other the same day. There is lid swelling, photophobia, irritation, and a seromucous discharge, becoming watery on the second day. Preauricular lymphadenopathy develops in 65% of patients. The tarsal conjunctiva is hyperemic with small petechiae and is edematous in the lower fornix. Small follicles develop on the second day in the lower, temporal conjunctiva and last up to 10 days. The bulbar conjunctiva is edematous and shows subconjunctival hemorrhage, which often starts in the upper temporal portion one day after onset, and is the characteristic sign (Fig. 4). Bleeding can vary in size from a pinpoint to the whole of the bulbar conjunctiva, and is exacerbated by clinical manipulation on examination. Bleeding decreases after the second day and absorbs gradually over one

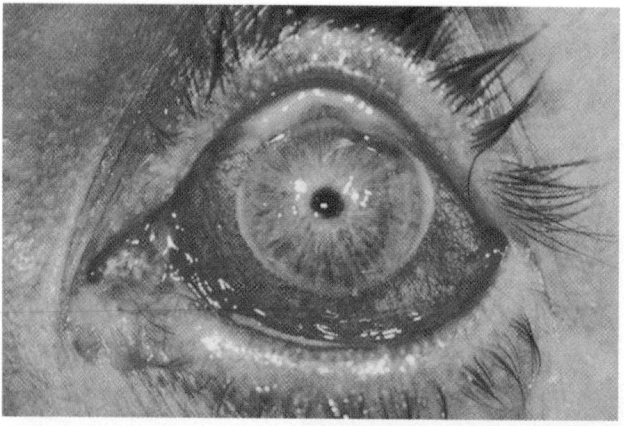

Figure 4   Acute hemorrhagic conjunctivitis due to viral infection (enterovirus 70 or coxsackie A24 variant).

week. The cornea shows fine punctate epithelial keratitis but nummular opacities do not occur. The infection is highly contagious, with rapid spread among people in homes and workplaces including hospitals. Spread is mainly person-to-person but during the acute infection the virus can be detected occasionally in throat washings and stool specimens. Diagnosis is usually clinical. Virus culture should be performed if possible. Treatment is symptomatic.

Conjunctivitis in measles, mumps, dengue, glandular fever, and Hepatitis A are associated with systemic virus infection, Herpes simplex with primary blepharoconjunctivitis. Herpes zoster is associated with shingles of the Vth nerve.

Presumed viral conjunctivitis. If indicated, a suspected viral etiology can be confirmed by laboratory tests. Swabs can be used for immunofluorescent tests directed against viral antigen or for PCR-based diagnosis (see Chapter 3); a combination of techniques can provide a rapid and sensitive result. This is much cheaper and quicker than attempting culture in cell lines.

The differential diagnosis between bacterial and viral conjunctivitis is summarized in Table 2.

Presumed Chlamydial Conjunctivitis

The diagnosis of trachoma is made clinically (refer below). For epidemiological surveys and research purposes, various laboratory assays can be used for the diagnosis of ocular *C. trachomatis* infection, including staining of conjunctival scrapings for intra-cytoplasmic inclusions (Bedson bodies), ELISA, and PCR. Also, for trachoma inclusion conjunctivitis (TRIC), the diagnosis is based on clinical suspicion but should be confirmed by identifying the antigen using a monoclonal antibody test, or PCR (Table 3). For both test systems it is important to collect conjunctival cells by scraping.

Table 2 Differential Diagnosis: Bacterial/Viral Conjunctivitis

	Bacterial conjunctivitis	Viral conjunctivitis
Systemic symptoms	Usually none (Chlamydia: genital infection)	May have symptoms of respiratory or systemic viral illness
Pre-auricular adenopathy	Infrequent	Common (adenovirus), but can be absent
Character of discharge	Usually mucopurulent, but may be watery	Usually watery, but may be mucopurulent
Conjunctival reaction	Papillary or mixed papillary/follicular	Follicular or mixed papillary/follicular
Superficial corneal infiltrates	Uncommon	Common in adenoviral keratoconjunctivitis

Table 3 Diagnostic Tests for Ocular Chlamydia Infection

	Sensitivity	Specificity
Direct immunofluorescence	80–99%	81–99%
Immunoenzymatic technique (ELISA)	85–97%	97–98%
Polymerase chain reaction (PCR)	88–100%	95–100%
Cell culture	65–85%	100%

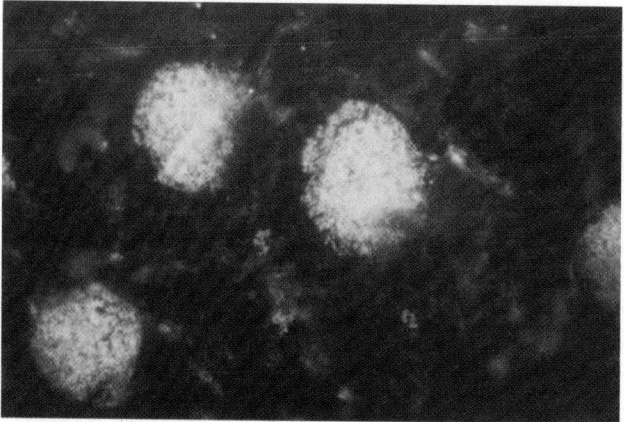

Figure 5 Conjunctival scrape stained with Giemsa showing intracytoplasmic chlamydia, within conjunctival epithelial cells, fluorescing non-specifically under ultraviolet light.

Bedson bodies are complete chlamydial cells lacking a conventional bacterial cell wall. This method becomes more sensitive if the smear is observed under ultraviolet light, when the stained chlamydia fluoresce yellow, non-specifically (Fig. 5). They can be cultured in tissue culture cells but not on solid agar.

Chlamydia-specific immunoglobulin A (IgA) is produced in tears when the conjunctival cells are infected with chlamydia. As chlamydia multiply intracellularly, this response is to a soluble antigen. It is not protective, and recurrent infection occurs in its presence. It merely demonstrates that chlamydial infection has taken place at some time. Serologic tests are not indicated in chlamydial conjunctivitis since many individuals have pre-existing antibodies.

Other Causes of Conjunctivitis

Fungi
Fungal keratitis may be associated with an adjunctive conjunctivitis, but fungal conjunctivitis per se does not usually occur.

Protozoa
Acanthamoeba sp. associated with CL wear or rural corneal trauma and keratitis, and other protozoa such as microsporidia. *Acanthamoeba* conjunctivitis and (epi-)scleritis is secondary to the keratitis infection and should be managed with it (see below).

Helminths
Thellazia capillaris and *Th. californensis* larvae, associated with birds in the Middle East and United States.
Loa Loa, a subconjunctival worm occurring in tropical countries.

Fly Larvae
Oestris ovis inhabits the nostrils of sheep and invades ocular tissues of humans in unhygienic living conditions.

If the conjunctiva contains a Thellazia worm, it should be removed with forceps. Thellazia infection is a zoonosis from birds that occurs when fly-borne larvae are deposited on the conjunctival surface; it has been reported from California, the Middle East, and India.

Therapy of Conjunctivitis

For bacterial conjunctivitis, therapy should include antibacterial drops such as quinolones (ofloxacin, levofloxacin) or Polytrim (polymyxin/trimethoprim) (Appendix A). If treatment fails, conjunctival specimens should be collected, together with specimens for chlamydia if thought appropriate. Empirical treatment with topical tetracycline ointment for chlamydial infection can be given while awaiting laboratory results, and systemic tetracycline or erythromycin 250 mg four times daily should be given for two weeks to eradicate chlamydial carriage at other sites. This regime can be shortened by use of clarithromycin or azithromycin. The use of corticocorticosteroids combined with antibiotics is controversial, but their use is contraindicated for fungal or viral keratitis when they contribute to progression of the infection by inhibiting the immune response. Herpes conjunctivitis should be treated with topical acyclovir.

Special Forms of Conjunctivitis

Ophthalmia Neonatorum

Ophthalmia neonatorum is defined as any purulent discharge from the eyes during the first 28 days of life. Ophthalmia neonatorum in the developed countries occurs in 8.2% to 12% of live births, with gonococcal infection now rare. Elsewhere, the incidence of gonococcal conjunctivitis varies from 0.04% of live births in the West to 1.0% in parts of Africa. The incidence of neonatal chlamydial ophthalmia in London has been estimated to be less than 1% (Fig. 6). Typical bacteria isolated include: *Haemophilus influenzae* (17%), *Staphylococcus aureus* (17%), *Streptococcus pneumoniae* (11%), and *Enterococcus* (8%). The microbial causes are summarized in Table 4.

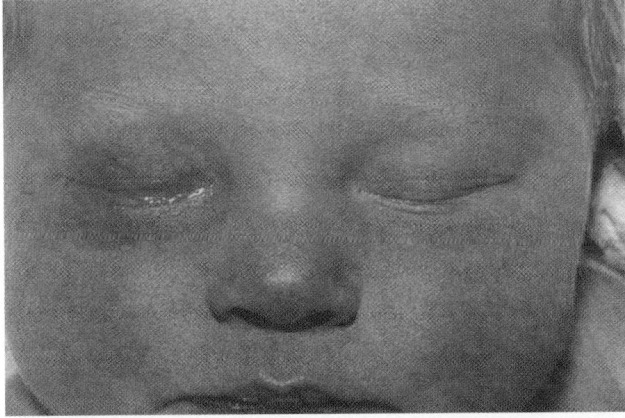

Figure 6   Characteristic swollen upper and lower lids of the right eye of a baby due to infection by *Chlamydia trachomatis*.

Table 4 Causes of Ophthalmia Neonatorum

Chemical
Credé prophylaxis (silver nitrate)
Erythromycin toxicity

Chlamydial
Chlamydia trachomatis

Bacterial
Staphylococcus aureus
Staphylococcus epidermidis
Viridans group streptococci
Group D streptococci
Haemophilus influenzae, H. parainfluenzae, and *H. haemolyticus*
Neisseria gonorrhoeae and *N. meningitidis*
Branhamella catarrhalis
Escherichia coli
Klebsiella pneumoniae

Viral
Herpes simplex

Nosocomial
Staphylococcus aureus
Pseudomonas aeruginosa

Irritative
Birth trauma

Neonatal prophylaxis against gonococcal infection is provided currently in the United States (and, in the past, in the United Kingdom), with one drop of silver nitrate (1% Crede's solution) in each eye. Silver nitrate causes more ocular symptoms in the first week of treatment than an agent such as oxytetracycline or erythromycin.

Systemic treatment is essential for the treatment of gonococcal keratoconjunctivitis in neonates. It is recommended to use benzylpenicillin in two daily doses for seven days. In recent years, however, the increasing frequency of isolation of penicillin-resistant strains has led to the use of beta-lactamase stable cephalosporins such as ceftriaxone 25–40 mg/kg intravenously every 12 hours for three days combined with topical saline lavage and antibiotic ointment (e.g., gentamicin, ofloxacin, or levofloxacin). Single-dose intramuscular therapy may be appropriate when there is no corneal involvement. As in the case with neonatal chlamydia ophthalmia, the infection must be treated systemically; topical therapy alone is not adequate.

Trachoma and Other Ocular Chlamydial Diseases

Trachoma. This year (2007) marks a number of milestones for trachoma. It is 100 years since chlamydia, the bacteria that causes trachoma, was first identified in cytology smears and 50 years since it was first cultured. It also marks 10 years since the World Health Organization (WHO) launched the Global Initiative to Eliminate Blindness from Trachoma as a public health problem (GET 2020). At present, good progress is being made in reducing the amount of trachoma worldwide, but 100 years ago trachoma was a major problem that faced all countries. In Europe it was an important cause of infection and disability in troops, and also involved the urban poor in the slums created by the industrial revolution. Now trachoma has been eliminated from most developed regions, but it still affects some 150 million people living in 56 developing countries and accounts for blindness in 5 to 10 million. There is still much work to do to eliminate trachoma as a major blinding disease.

Unfortunately, those areas in which trachoma persists are the most difficult to treat, such as the mountainous regions in Morocco's desert. There is need to tackle extreme poverty associated with lack of running water, failure to separate sewerage, and living with a high fly count. Until public health hygiene is tackled, trachoma cannot be eliminated. Where public health hygiene has been tackled, single oral dose treatment with azithromycin can eliminate trachoma from all the village inhabitants. The donation of azithromycin by Pfizer, and its distribution by the International Trachoma Initiative (ITI), is a model of public–private enterprise partnership.

Following research performed in the 1990s showing the increased effectiveness of single dose oral azithromycin versus 6 weeks of topical tetracycline ointment (1–4), the International Trachoma Initiative (ITI) was set up, to which Pfizer donated 10 million treatments (capsules) of azithromycin from 1998 to 2003. In November 2003, Pfizer donated a further 135 million treatments (capsules) of azithromycin for the period from 2003 to 2008. Whole communities have been effectively studied to establish which members need antibiotics to control trachoma (5–7).

Ocular infection by *Chlamydia trachomatis* takes three forms: trachoma, adult chlamydial ophthalmia or TRIC, and neonatal chlamydial ophthalmia (NCO). Classic blinding endemic trachoma in developing countries is usually associated with infections by serotypes A, B, Ba, or C, while trachoma/inclusion conjunctivitis (TRIC) is caused by serotypes D to K. In hyperendemic areas, 30% to 50% of the population has active disease and 10% of the population exhibits blinding sequelae. Although infection may be encountered as early as the second month of life, active inflammatory disease is most common in the preschool age group. It is at this stage that the infection leads to the typical conjunctival scarring, which results in entropion (Fig. 7) and trichiasis. In turn this causes recurrent microbial keratitis from repeated corneal trauma. Blinding sequelae due to repeated corneal scarring from infection and trauma occurs after the age of 40 years. The disease may be less

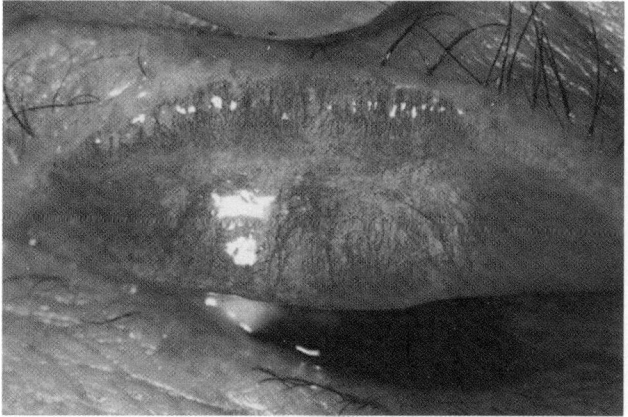

Figure 7   "Scaphoid" lid of trachoma, showing typical conjunctival scarring resulting in entropion. *Source*: Courtesy of Mr. I. A. Mackie.

severe and present as Herbert's pits, the sequelae of sloughed limbal follicles that cicatrize (Fig. 8).

Several findings distinguish trachoma from other forms of chronic conjunctivitis.

- The superior tarsal conjunctiva shows more severe follicular reaction than the inferior palpebral conjunctiva.
- It is the only conjunctivitis that manifests with Herbert's pits.

The upper tarsus has been selected to grade the degree of conjunctival inflammation by the WHO. Non-specialists are able to grade inflammation respecting five hallmarks (Fig. 9) that allow information gathering for population-based surveys.

- Trachomatous inflammation follicular (TF): defined as the presence of five or more follicles in the upper tarsal conjunctiva.
- Trachomatous inflammation intense (TI): presenting with intense inflammatory thickening of the tarsal conjunctiva.
- Trachomatous scarring (TS): the presence of scarring in the tarsal conjunctiva.
- Trachomatous trichiasis (TT): eyelashes rub on the ocular surface.
- Corneal opacity (CO): visible corneal opacity over the pupil.

Public health intervention is required to prevent spread of the infection at the childhood stage. Its frequency is reduced by improved hygiene with running water and a reduction in the fly population.

TRIC. This is also known as inclusion conjunctivitis and is by far the most common sexually transmitted infectious agent and most prevalent between the ages of 15 to 30 years. It usually occurs concomitantly with sexually transmitted chlamydial urethritis, cervicitis, or proctitis 2 to 19 days after exposure to the organism.

The disease affects both eyes, presenting with follicular conjunctivitis, scanty mucopurulent discharge, and palpable pre-auricular lymph nodes. In contrast to trachoma, the follicular reaction is predominant in the lower palpable conjunctiva and lower fornix. Also, in TRIC cornea involvement may occur with marginal and

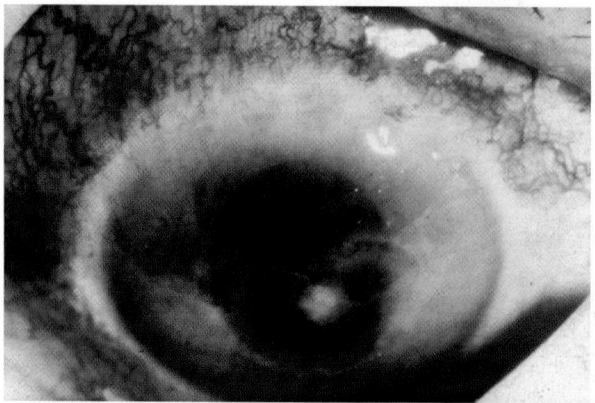

Figure 8 Herbert's pits of limbal chlamydial infection, in a patient with adjunctive *Acanthamoeba* keratitis.

catarrhal infiltrates (Fig. 10) and subepithelial nummuli similar to adenoviral keratoconjunctivitis. Keratitis is more likely to be found in the upper limbal area of the cornea, which helps to differentiate it from disseminated infiltrates in epidemic keratoconjunctivitis. In addition, a limbal micropannus is frequently seen, which distinguishes TRIC from adenoviral infection.

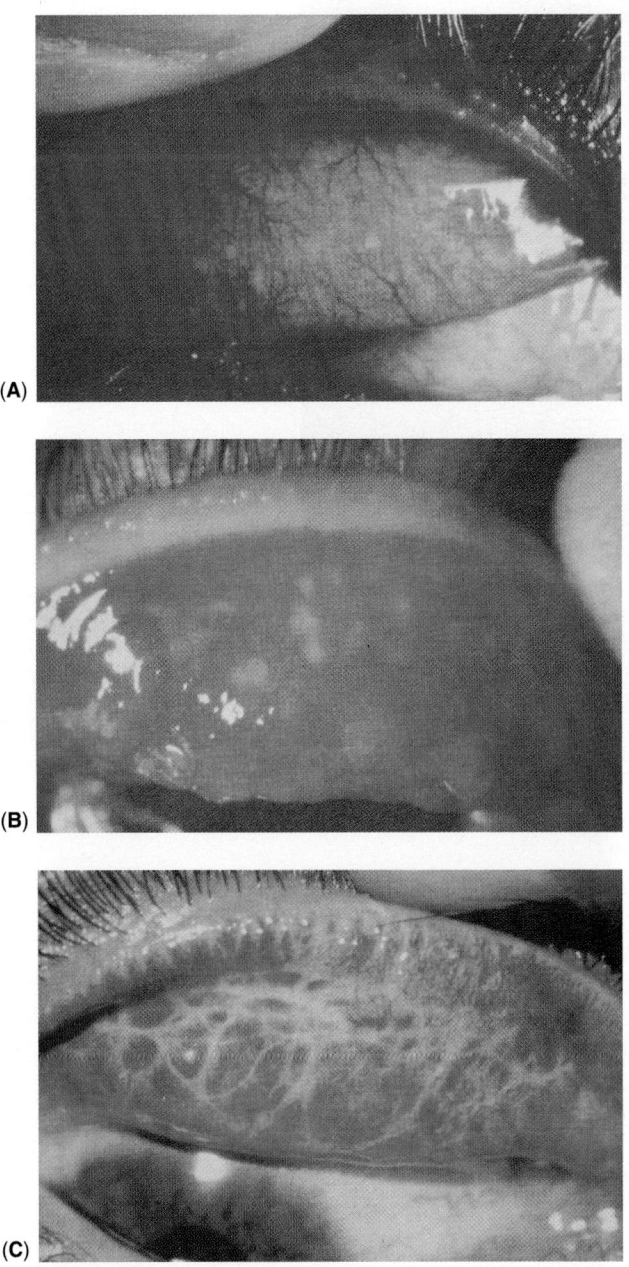

(A)

(B)

(C)

Figure 9A–E (*Continues on next page*)

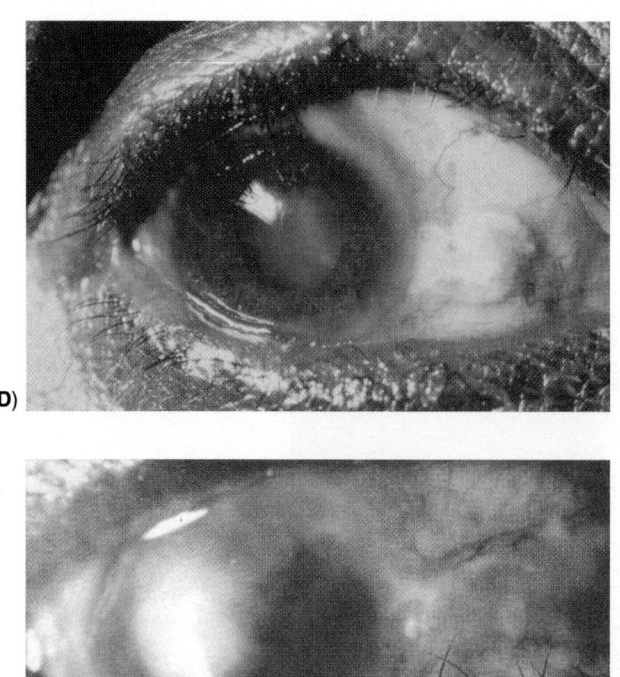

(D)

(E)

Figure 9A–E Composite WHO figure of different stages of trachoma. (**A**) Trachomatous inflammation follicular (TF) defined as the presence of five or more follicles in the upper tarsal conjuctiva. (**B**) Trachomatous inflammation intense (TI) presenting with intense inflammatory thickening of the tarsal conjunctiva. (**C**) Trachomatous scarring (TS)—the presence of scarring in the tarsal conjunctiva. (**D**) Trachomatous trichiasis (TT)—eyelashes rub on the ocular surface. (**E**) Corneal opacity (CO)—visible corneal opacity over the pupil.

Neonatal Inclusion Conjunctivitis (NIC). Neonatal chlamydial infection arises from passage in the birth canal. It rarely occurs immediately after birth, but is seen in most patients between 5 to 14 days postpartum.

Neonatal inclusion conjunctivitis differs clinically from adult chlamydial conjunctivitis in several respects. Pseudo-membranes and a hyper-purulent discharge is often observed in newborns. Because of not-yet developed lymphoid tissue in newborns, there is no follicular response. In up to 50% of infected children, systemic infection, especially of the respiratory tract, occurs and requires systemic antibiotic treatment.

Treatment

Trachoma

Chlamydia trachomatis is unresponsive to the aminoglycosides neomycin and gentamicin. It is partially sensitive to both chloramphenicol and penicillin, which

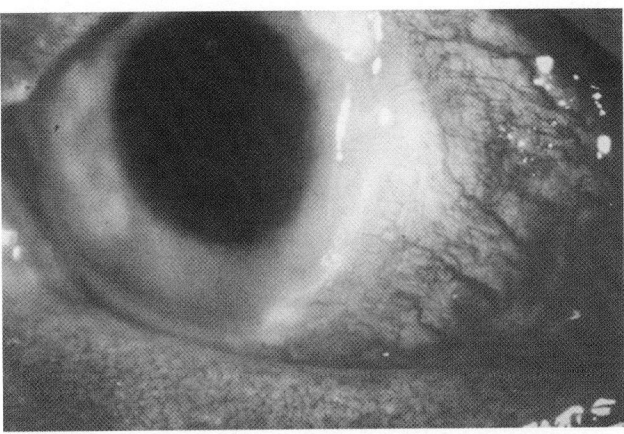

Figure 10 Marginal ulceration associated with severe inclusion conjunctivitis due to chlamydial infection, usually derived in caucasians from genital infection.

adversely affect growth in subsequent cultures. *Chlamydia trachomatis* is fully sensitive to the tetracyclines, erythromycin, and other macrolides including azithromycin, rifampicin, and the quinolones, especially ofloxacin. It is also sensitive to chlorhexidine (MIC 5 mg/mL).

Treatment of the early active inflammatory stages of trachoma is highly effective with a single oral dose of azithromycin (1,7), probably due to the high intracellular concentration that azithromycin achieves (Chapter 4), and is now the treatment of choice. Topical treatment with tetracycline or erythromycin ointment or quinolone drops can be effective locally but must be given three times daily for five weeks and is now considered second-choice. In trachoma, treatment of other family members, or whole villages, is necessary to prevent re-infection.

In an effort to combat trachoma, the "SAFE" strategy has been introduced by the WHO based on: *S*urgery for in-turned lashes, *A*ntibiotics for active disease, *F*acial cleanliness, and *E*nvironmental improvement (8). Surgical intervention on trichiasis and/or entropion often has an immediate impact on prevention of blindness. The demonstration that a single oral dose of azithromycin is as effective as six weeks of topical tetracycline is an important advance in trachoma control. In endemic areas, public health efforts, including provision of running water, and tackling entropion and trichiasis surgically, are important measures to prevent blindness caused by trachoma.

TRIC

Topical antibiotics are effective but they do not eliminate the genital reservoir of the disease. Systemic therapy is necessary for the treatment or prophylaxis of the systemic manifestations of chlamydial disease, e.g., cervicitis, uveitis, proctitis, upper respiratory or ear disorders, and Reiter's disease.

For TRIC, treatment with oral tetracycline, erythromycin, or rifampicin for two weeks has been recommended, but long-acting once-daily tetracyclines, such as doxycycline and minocycline, offer convenience and improved compliance, as fewer doses are required and dietary constraints are not necessary. Oral azithromycin therapy is effective with a single dose. The patient should be checked for genital carriage of chlamydia. In addition, in TRIC the patient's partner(s) should be checked for genital carriage and treated with erythromycin or azithromycin if positive.

Neonatal Conjunctivitis (NIC)

Topical erythromycin is used for the prevention of neonatal conjunctivitis. However, povidone-iodine has a broad spectrum of antimicrobial effectiveness and has given excellent results. In cases of chlamydial pneumonitis, systemic treatment is recommended. A suitable treatment for NIC is oral erythromycin at 50 mg/kg per 24 hours in four divided doses for two to three weeks. Topical therapy is used adjunctively but is inadequate on its own.

CORNEA

Microbial Keratitis

A standardized approach is essential to the documentation of corneal disease, and a simple scheme is illustrated here (Figure 11).

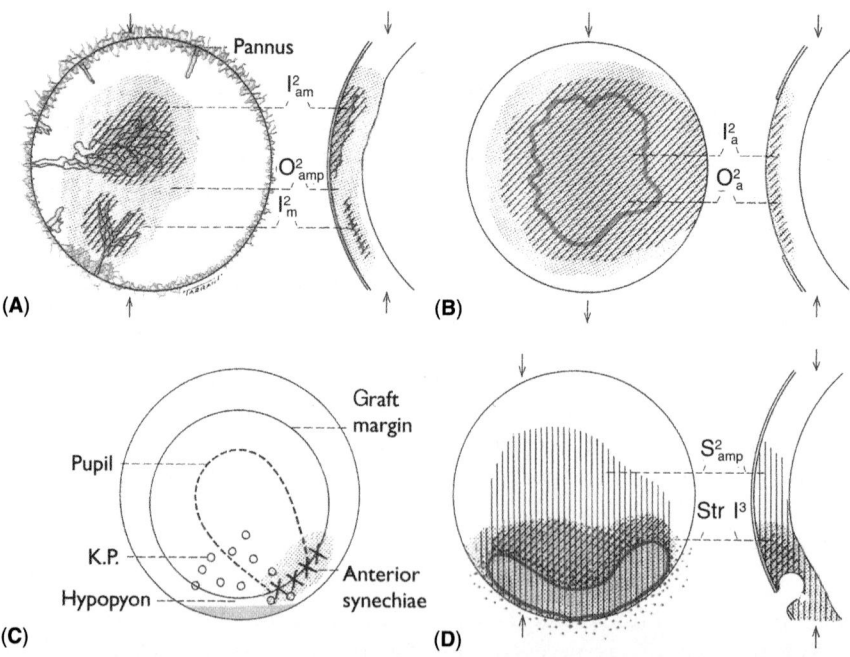

Figure 11 **(A)** Corneal infiltration and edema with deep and superficial vascularization. **(B)** Anterior stromal infiltration and edema with an overlying erosion as stained with fluorescein. **(C)** Corneal graft with corneal edema related to the site of anterior synechia at the graft/host junction. The pupil is deformed and elongated, there are keratic precipitates over the lower portion of the graft, and a hypopyon is present. **(D)** Cornea showing extensive stromal scarring and an elongated marginal ulcer below. The ulcer is undermined on one side and shelved on the other. It has been stained with bengal rose. There is an infiltrate at the axial margin of the ulcer.

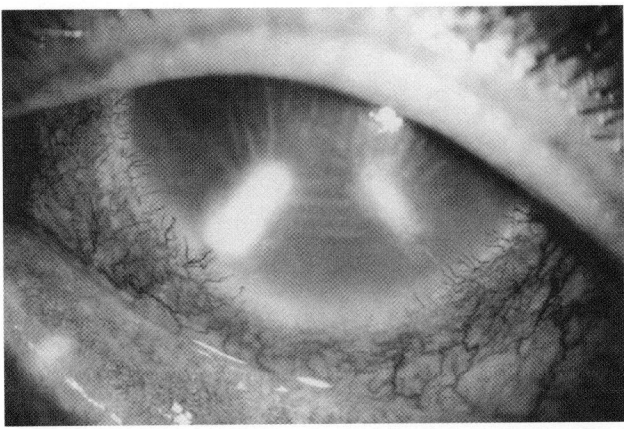

Figure 12 *Staphylococcus aureus* keratitis derived from infection of radial keratotomy incisions.

Clinical Presentation

Suppurative bacterial keratitis presents clinically as a corneal stromal infiltrate or abscess with an overlying epithelial defect (Figs. 12, 13). Patients complain of pain, lid swelling, redness, discharge, photophobia, and blurred vision. Bacterial invasion of the cornea causes inflammation and corneal ulceration, which may be followed by neovascularization. When the activity of proteinases, derived from bacteria or host neutrophils, leads to stromal melting, formation of a descmetocele may occur (Fig. 14) with the risk of corneal perforation. Some bacteria, particularly *S. pneumoniae* and other alpha-hemolytic streptococci, may induce focal infection only with minimal inflammation under an intact epithelium. These bacteria proliferate and multiply between the stromal lamellae, resulting in a needle-like or fern-like network opacity. This type of bacterial infection is called crystalline keratopathy. It can also occur in large ulcers due to other causes such as *Acanthamoeba* (Fig. 15).

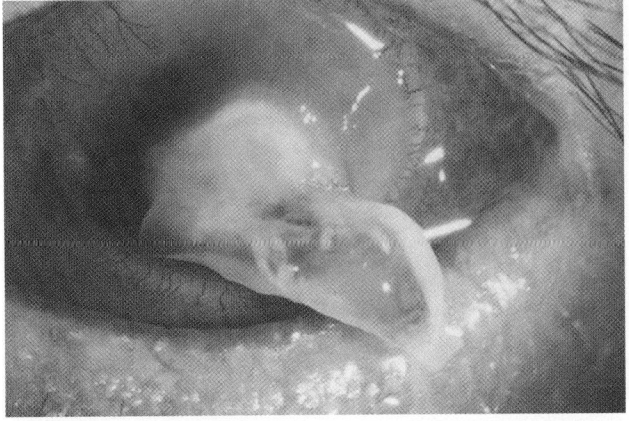

Figure 13 Moraxella infection of the cornea, causing a slough, in a patient with exposure-induced drying.

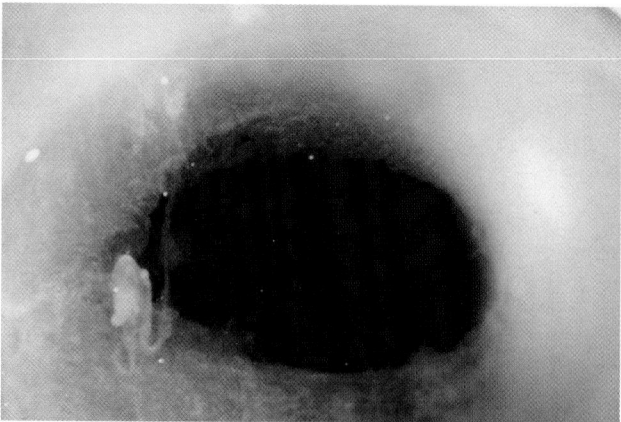

Figure 14 Suppurative keratitis descmetocele due to *P. aeruginosa* in a contact lens wearer.

Until recently, trauma was the most common underlying cause for bacterial keratitis. However, due to the wide spread use of soft CL wear, this is becoming a major and most important risk factor for the development of microbial keratitis.

Common organisms causing infection are given below.

- Gram-positive cocci
  Staphylococcus aureus
  Coagulase-negative staphylococci
  Streptococcus pneumoniae
  Streptococcus pyogenes
  Streptococcus viridans group
  Anaerobic streptococci (rare)
- Gram-negative cocci/diplobacilli
  Neisseria gonorrhoeae
  Neisseria meningitidis
  Moraxella sp.

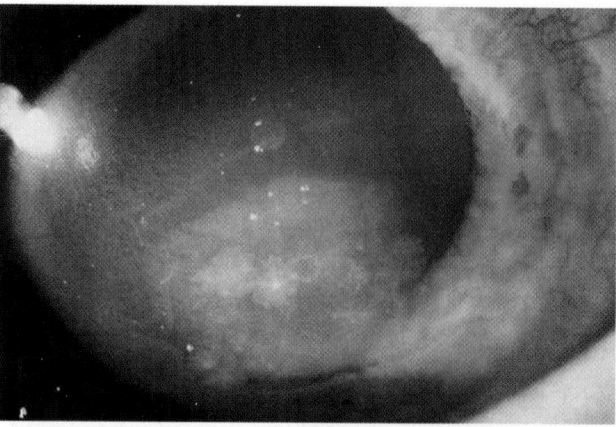

Figure 15 Crystalline keratopathy due to streptococci in a patient with a large corneal ulcer due to *Acanthamoeba* sp. at 12 weeks.

- Gram-positive rods
  Corynebacterium diphtheriae (rare)
  Diphtheroids
- Gram-negative rods
  Acinetobacter sp.
  Escherichia coli
  Klebsiella pneumoniae
  Morganella morganii
  Proteus sp.
  Pseudomonas aeruginosa
  Serratia marcescens
  Anaerobes: *Capnocytophaga* sp.
  Other enteric bacteria
- Acid-fast bacteria
  Mycobacterium chelonae
  Nocardia asteroides

Suppurative bacterial keratitis is usually central, but may be peripheral. Because corneal thickness is only about 0.5 mm, such an ulcer may rapidly progress to perforation within 24 hours of onset. In the aphakic or pseudophakic eye with a capsulotomy, access to the vitreous space is facilitated and a secondary endophthalmitis may supervene. There is an urgent need to treat bacterial keratitis with high doses of effective antibiotic. Supplementary corneal transplantation, or other reconstructive surgery, may be required at a later stage in the management of corneal scarring or perforation. The presentation in children is similar to that in adults (9).

Bacteria account for over 80% of ulcerative keratitis occurring in northern climates and 60% in southern climates where fungal keratitis is more common. Mixed bacterial and fungal infection frequently occur in the tropics and occasionally as well in northern climates especially when associated with rural injuries. In temperate climates, Gram-positive bacteria predominate, *Candida* sp. is an occasional pathogen, and *Acanthamoeba* is often associated with CL wear. This compares with *Pseudomonas aeruginosa* and filamentous mycelial fungi, which predominate in tropical and semi-tropical areas (Table 5) (10).

In tropical areas, *Acanthamoeba* is usually a non-CL-associated infection or can be detected as a chronic microbial keratitis. Within a hot country such as India, there can be variation in isolates with greater detection of *Aspergillus* sp. in Northern India, where summers are hot and dry, compared with filamentous fungi such as *Lasiodiplodia theobromae* in Southern India where there is a humid tropical climate (11). The same situation applies between Northern and Southern states in the United States (12)

Interstitial keratitis occurs with congenital syphilis as a consequence of systemic bacterial infection (Fig. 16). Lyme disease presents as bilateral keratitis with indistinct infiltrates scattered throughout the corneal stroma (Fig. 17). It resolves with corticocorticosteroid treatment without sequelae, suggesting an immunopathological basis.

Cogan-1 syndrome is an important differential diagnosis presenting with interstitial keratitis, hearing loss, and other neurological symptoms. If not treated with immunomodulatory agents, permanent hearing loss will result (13).

Table 5 Relationship of Climate to the Expected Ratio of Fungal to Bacterial Isolates from Keratitis Studies

Place of study	Annual rainfall (inches)	Mean temperature (°F) Jan./July	Fungal/bacterial ratio
Scotland	44	38/58	<1/34
Western Australia	6–34	85/60	<1/34
New York	44	38/84	1/99
South Africa	28	72/52	1/37[a]
Hong Kong	87	60/83	1/17
South California	12	55/69	1/12
Singapore	90	79/81	1/5
Bangladesh	80	65/83	1/2.5
South Florida	64	68/82	1/2
South India	50	76/87	1/2[b]
Accra, Ghana	30	80/78	1/1
Paraguay	54	82/64	1/1[b]

[a] Mixed fungus/bacteria culture in 14% of cases.
[b] Mixed polymicrobial culture of bacteria and fungi in 25% of cases.
Source: From Ref. 10.

In the past, suppurative keratitis was due to trauma, dry eyes, or existing corneal disease such as stromal Herpes simplex (Fig. 18, 19). This still happens now (14). However, with the growth of CL use in recent years, there has been a rapid increase in CL-associated microbial keratitis to 50% of all cases in the United States (15) and up to 35% in Europe (16,17).

Most of this microbial keratitis had been bacterial in origin until the early 1990s, when *Acanthamoeba* predominated in the United Kingdom, but has since become rare again. A similar epidemic of *Acanthamoeba* cases is currently occurring in the Chicago area. While the U.K. experience was due to storing CLs in tap water with a low-dose chlorine tablet, in Chicago the causation is thought due to use of weakly-chlorinated tap water in CL hygine.

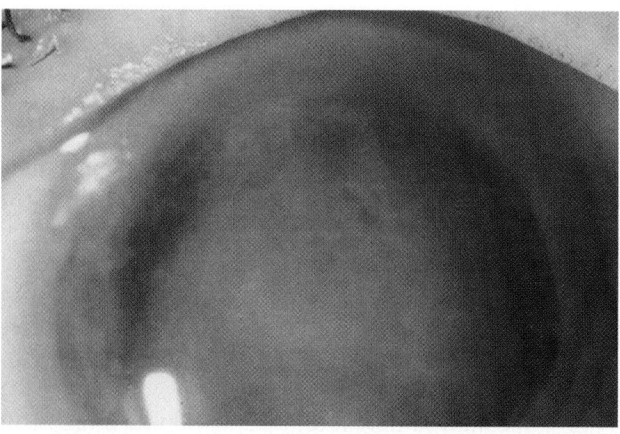

Figure 16 Interstitial keratitis in an adult due to congenital syphilitic infection.

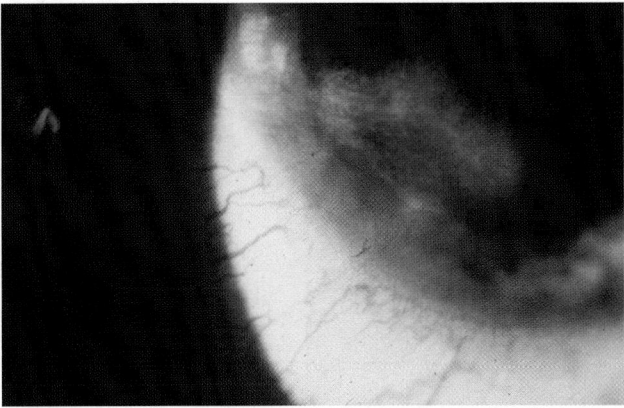

Figure 17 Lyme disease interstitial keratitis.

In general, the risk of microbial keratitis is much less for hard versus soft lens wearers and is greater with extended wear than daily wear. In a large recent cohort study in Hong Kong, and in multi-center case-controlled studies elsewhere, the overall risk for ulcerative keratitis with extended wear (overnight) hydrogel lenses was four to five times greater than that for daily wear lenses (18). The incidence of CL-associated microbial keratitis (presumed infection) was found to be approximately 1 in 500 with extended wear (overnight) of hydrogel lenses and 1 in 2500 with daily wear of hydrogels (18), showing that extended wear is a major risk factor for infection.

The bacteria responsible for CL-associated keratitis include most of those usually associated with suppurative keratitis, but Gram-negative bacteria are more commonly encountered than Gram-positive and *P. aeruginosa* is more frequent than other Gram-negative bacteria.

Silicone hydrogel (SH) contact lenses [Focus Night and Day (lotrafilcon A, FDA group 1), CibaVision; PureVision (balafilcon A, FDA group 1), Bausch & Lomb;

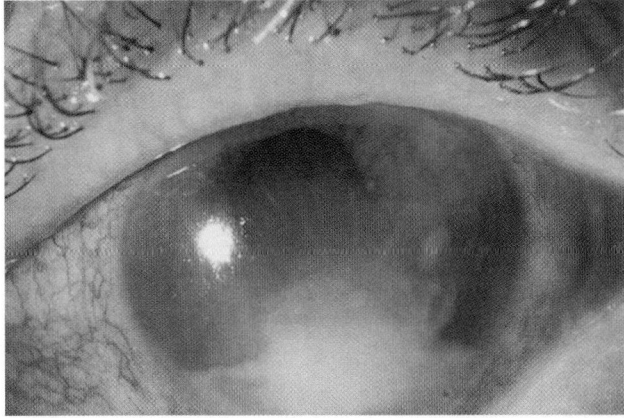

Figure 18 *Staphylococcus aureus* keratitis and hypopyon in a patient with pre-existing stromal keratitis due to Herpes simplex virus.

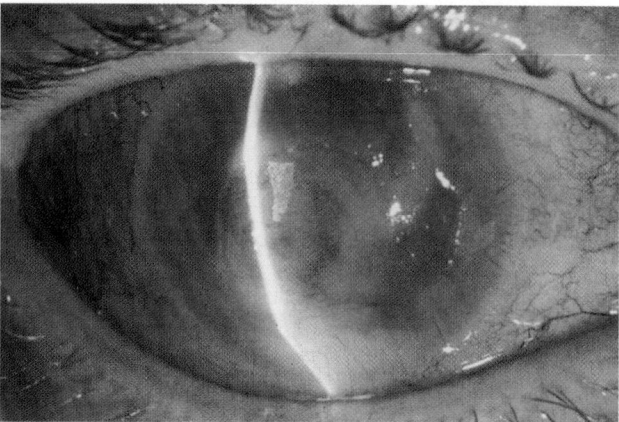

Figure 19 Coagulase-negative staphylococcus (CNS) keratitis in an eye with persistent epithelial defects and an anesthetic cornea due to bullous keratopathy.

Acuvue Advance (galyfilcon A, FDA group 1, Johnson & Johnson)] have been introduced recently after many years of research by industry to increase the oxygen diffusion to the cornea but yet to maintain the comfort factor of the existing hydrogel lens (19). These CLs have proven able to reduce corneal edema that occurs with overnight wear of existing hydrogels and which was originally thought to be the main predisposing factor for microbial keratitis associated with lens wear, in particular with overnight wear. They have also proven comfortable to wear. However, at a recent conference at ARVO 2006, the following "per annum" figures were provided by authors:

CibaVision Post-Marketing SH Lens Study (Oliver Schein, U.S.A.):
▪ Microbial keratitis with severe vision loss: < 1/2500
▪ Microbial keratitis overall: 1/500

Moorfields Eye Hospital (MEH) London Study December 2003 to December 2005 (John Dart, U.K.):
▪ Cases: MEH Accident and Emergency
▪ Controls: Hospital-based contact lens wearer (CLW) and population-based CLW
▪ Presumed microbial keratitis 367 cases, one third had a corneal scrape with 60 culture-positive results (50%), three (approx. 1% overall) *Acanthamoeba* and three (approx. 1% overall) filamentary fungus
▪ Types of contact lenses worn: Rigid 5, Daily Wear soft CL 160, Weekly replacement soft CL 127, SH 64, others 11
▪ Risk factor for overnight wear: x5.4 with no difference between SH and hydrogel disposables such as Acuvue (FDA group 4)
▪ Risk factor for one to four weekly soft CL wear: 0.6
▪ Risk factor for daily wear of SH lenses: 0.6
▪ Daily disposable lens wear had a greater risk for microbial keratitis than all other soft CL types! This is presumably because they were not being disposed of daily!
▪ 10% of patients with microbial keratitis lost vision in the eye.

Australian Study (Fiona Stapleton, CCLRU, Australia):
- 287 eligible cases for analysis
- 14% of patients with microbial keratitis lost vision in the eye.
  One third had mild, peripheral lesions.
- Half had a corneal scrape, of which 50% were culture-positive. 46% of these cultures were due to *Ps. aeruginosa* and three were due to *Acanthamoeba*.

Risk Factors for Microbial Keratitis:
- Rigid gas permeable: 1/10,000
- Daily wear (soft) hydrogel CL: 1.7/10,000
- Daily disposable (soft) hydrogel CL: 2.5/10,000
- Extended wear (overnight) (soft) hydrogel CL: 20/10,000
- Daily wear SH: 9.8/10,000
- Extended wear (overnight) SH: 25.7/10,000

Overnight wear of the contact lens was thus proven again to be the major risk factor for microbial keratitis but very surprisingly with *no* benefit of the new SH lens materials. These figures above are very disappointing for extended (overnight) wear of silicone hydrogel lenses, which have given similar increased rates of microbial keratitis as for existing hydrogel lenses. The hypothesis for the causation of CL-associated keratitis being due to lack of oxygen alone needs to reassessed, as other causation factors now need to be recognized.

First-generation but not second generation SH CLs have been found more sticky for adhesion of *Acanthamoeba* than hydrogel CLs (20). This is discussed in more detail below. Interestingly, silicone intraocular lenses (IOLs) used in phacoemulsification cataract surgery have been found in a recent large multi-center study (Chapter 8) to give a greater risk of postoperative endophthalmitis (p <0.002) than use of acrylic IOLs. The same silicone adhesion factor could be responsible between CL and IOL.

Contamination of contact lens care solutions is an important potential source of keratitis, and home-made solutions are a major risk factor. Recently, an epidemic of fungal keratitis caused by *Fusarium* sp. was reported in soft contact lens wearers, which could be linked to a new multi-purpose cleaning and storage solution—ReNu with MoistureLoc™ (Appendix D).

Soft contact lens wear also increases the risk of microbial keratitis in corneal graft patients. This can be due to "crystalline" keratopathy, an infection associated with corneal grafts, or a compromised cornea including *Acanthamoeba* infection, in which there are sheets of streptococci deposited between the stromal lamellae, giving the macroscopic appearance of ice crystals (see above); the condition responds slowly to antibiotics, perhaps because there is a unique lack of polymorphonuclear cell infiltration within the cornea or the streptococci are surrounded by a biofilm.

Orthokeratology

In addition, in overnight orthokeratology (OK) used for the temporary reduction of myopic refractive error, an increasing number of patients have been reported with microbial keratitis. Most cases of keratitis in OK were reported from East Asia (80%) and most affected patients were Asian (88%), with a peak age range from 9 to 15 years (21). Although *Pseudomonas aeruginosa* was the predominant organism implicated in almost every second infectious keratitis case, an alarmingly high frequency of *Acanthamoeba* infection (approx. 30%) was found and linked to tap water rinsing.

Infections Following Laser In Situ Keratomileusis

A steadily increasing number of infections following laser in situ keratomileusis (LASIK) has been reported in recent years, leading to moderate or severe visual reductions in visual acuity in every second eye (22). Gram-positive bacteria and mycobacterium are the most common causative organisms. Gram-positive infections were more likely to present less than seven days after LASIK, and they were associated with pain, discharge, epithelial defects, and anterior chamber reactions. Fungal infections were associated with redness and tearing on presentation. Mycobacterial infections were more likely to present 10 or more days after LASIK surgery. Severe reductions in visual acuity were significantly more associated with fungal infections. An early flap lift and repositioning within three days of symptoms was recommended for better visual outcome.

Diagnosis

Cultures and smears are not essential in every case of suspected infectious keratitis. In patients with only a small (< 2 mm), peripheral and non–sight-threatening lesion, laboratory workup may not be required. However, when the infiltrate is central, large, progressive, and sight–threatening, cultures and smears are essential. An ulcer that is > 4 mm^2 in area is much more likely to be infected (23). When performing the scrape, a sample for a Gram stain should be obtained first and then another collected for culture. In contact lens wearers, the lens, storage case, and care solutions should also be cultured.

A blade is best for a corneal scrape, but calcium alginate swabs can be used instead except for deep stromal infection. Even then a corneal biopsy may be needed. *Acanthamoeba* and filamentous fungi such as *Fusarium* sp. have been detected in the cornea in vivo by confocal microscopy. In particular for *Acanthamoeba*, confocal microscopy allows the differentiation of cysts (rounded bodies, 10 and 20 μm of diameter, high reflectivity) from trophozoites (more irregular shape, high reflectivity) (24). The coinfection in several patients of *Acanthamoeba* and Herpes simplex virus has lead to a postulation that the presence of the parasite in the cornea can represent in some cases an opportunistic infection. Figure 20 is a flow chart for the investigation of microbial keratitis.

Scrapes from the affected cornea are best obtained following one drop of unpreserved proparacaine, which has the least inhibitory effect on bacterial growth. Preservatives will inhibit bacterial replication, while some anesthetics have intrinsic antimicrobial properties. In children under five years, sedation (e.g., oral chloral hydrate) is useful. The surface material from the ulcer should be debrided using a swab. This material may be plated onto blood or chocolate agar, but it is less helpful than material from later scrapes since it usually contains only cellular debris and mucus. The second scrape should be taken for microscopy and the third scrape for culture. Using a platinum Kimura spatula, a large-gauge sterile needle, or a disposable surgical blade, the base and edge of the ulcer is firmly scraped. A freshly sterile instrument is used for each sample. In the presence of deep ulceration a small trephine may be useful to obtain an adequate specimen. One method to obtain

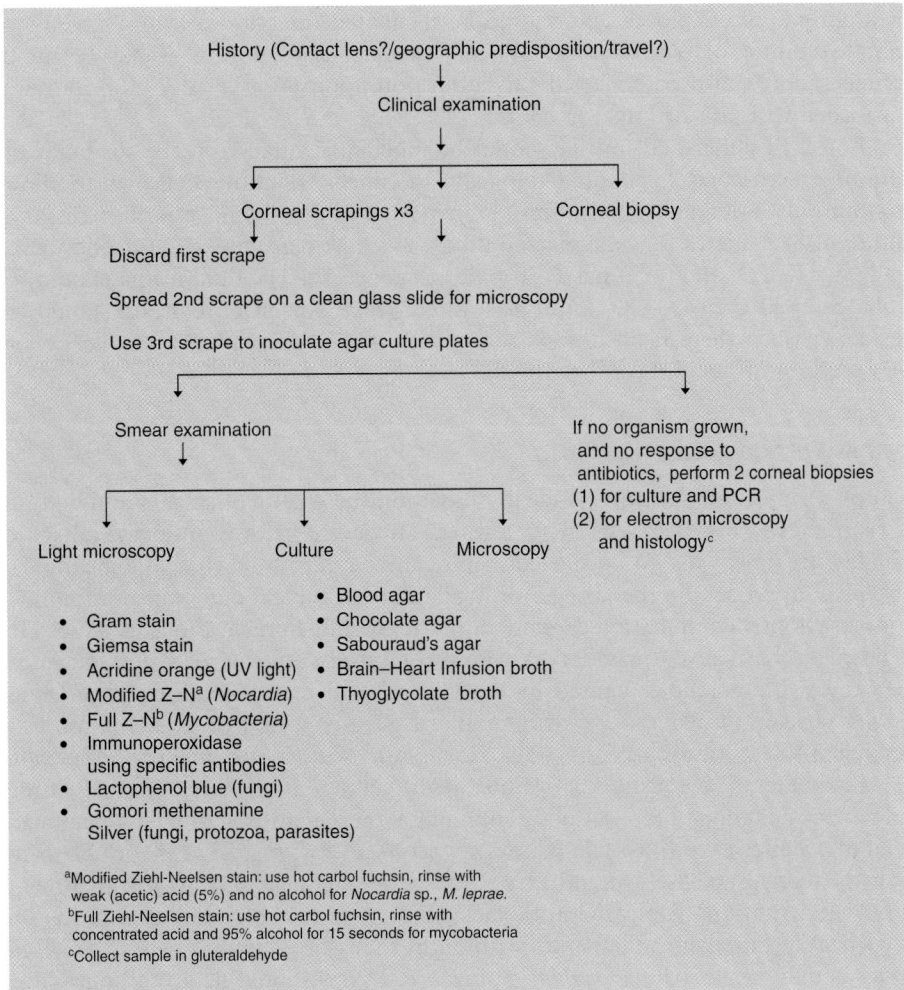

History (Contact lens?/geographic predisposition/travel?)

↓

Clinical examination

↓

Corneal scrapings x3   Corneal biopsy

Discard first scrape

Spread 2nd scrape on a clean glass slide for microscopy

Use 3rd scrape to inoculate agar culture plates

Smear examination   If no organism grown, and no response to antibiotics, perform 2 corneal biopsies
(1) for culture and PCR
(2) for electron microscopy and histology[c]

Light microscopy   Culture   Microscopy

- Blood agar
- Gram stain
- Giemsa stain
- Acridine orange (UV light)
- Modified Z–N[a] (*Nocardia*)
- Full Z–N[b] (*Mycobacteria*)
- Immunoperoxidase using specific antibodies
- Lactophenol blue (fungi)
- Gomori methenamine Silver (fungi, protozoa, parasites)
- Chocolate agar
- Sabouraud's agar
- Brain–Heart Infusion broth
- Thyoglycolate broth

[a]Modified Ziehl-Neelsen stain: use hot carbol fuchsin, rinse with weak (acetic) acid (5%) and no alcohol for *Nocardia* sp., *M. leprae*.
[b]Full Ziehl-Neelsen stain: use hot carbol fuchsin, rinse with concentrated acid and 95% alcohol for 15 seconds for mycobacteria
[c]Collect sample in gluteraldehyde

Figure 20 Flow chart for investigation of microbial keratitis.

organisms from a deeper corneal stroma infection is to pass a 6–0 silk suture through the affected area.

The base and edge of the ulcer are most likely to yield organisms. Some bacteria are mainly found at the edge of an active process (e.g., *Streptococcus pneumoniae*, *Pseudomonas aeruginosa*), whereas others are more often present within a corneal ulcer (e.g., *Moraxella* sp.).

The material gathered should be firmly spread on to a clean glass slide, to create a thin film, which is air dried. The second scrape should be similarly used for a second slide film. The third scrape should be plated on to blood, chocolate, and Sabouraud's agars. Fluid media (preferably brain-heart infusion) should be inoculated with the same blade. In addition, a Lowenstein-Jensen slope and Kirschner broth should be inoculated if the keratitis is chronic, although the atypical *Mycobacterium chelonae* will grow on blood agar incubated for 1 week at 37°C.

For a chronic ulcer, blood agar should be incubated for one week in 4% CO_2 to facilitate culture of *Nocardia* sp. When *Acanthamoeba* keratitis is suspected, a specimen should also be collected for culture on appropriate media (non-nutrient agar seeded with live or killed *E. coli*).

For each patient, all media should be inoculated directly at the slit lamp or operating microscope. If possible, duplicate specimens should be collected to allow for culture at different temperatures. Transport medium should not be necessary. Culture of agar plates for bacteria should always take place at 37°C for one week, ideally in 5% CO_2 and for fungi at 30°C for three weeks. The fluid media should be incubated at 30°C in 4% CO_2 for at least three weeks. Anaerobic cultures should be considered when there is an unsatisfactory response to therapy; *Propionibacterium acnes* requires up to 14 days for growth.

Treatment of Suppurative Keratitis

Therapy of infectious keratitis is ideally based on the results of smears and culture and subsequent in vitro sensitivity testing. However, before culture results are available, treatment has to be started empirically based on the clinical appearance. A frequent approach, in the absence of availability of topical quinolone antibiotics, was to use hourly combination drop therapy with fortified preparations, produced in the hospital pharmacy (Appendix A), whose combined antibacterial spectra covered most infective possibilities caused by the Gram-positive and Gram-negative (non-acid fast) bacteria listed above. A common and effective empirical combination was gentamicin (or tobramycin) forte 1.5% (15 mg/mL) with cefuroxime (or cephazolin or ceftazidime) 5% (50 mg/mL). Equivalent success is now recognized using commercial quinolone preparations containing either levofloxacin 0.5% (5 mg/mL) (Oftaquix, Santen), ciprofloxacin (Ciloxan, Alcon) 0.3% (3 mg/mL) (25) or ofloxacin (Exocin, Allergan) 0.3% (3 mg/mL) (26) at the same frequency. Ofloxacin treatment causes less irritation. Topical ciprofloxacin may leave microcrystalline deposits on the corneal surface, which take up to six months to dissolve, but has two- to fourfold greater activity against *Pseudomonas aeruginosa*. Drops should be given hourly day and night for the first three days, then two-hourly by day. Successful eradication of bacterial infection is reported in about 90% of patients treated in this way. Topical administration of levofloxacin (0.5% solution), a third generation quinolone, resulted in a two- to threefold greater penetration into human corneal stroma and aqueous humor tissues than ofloxacin (0.3%) or ciprofloxacin (0.3%) (27). However, the mean intracorneal concentrations of all three agents following two drops exceeds the MIC90 for the majority of pathogens causing bacterial keratitis.

The so-called 'fourth' generation of DNA gyrase inhibitors (quinolones) that have been launched in the United States (moxifloxacin, Alcon; gatifloxacin, Allergan), but are not available in Europe, may offer an expanded anti-infective spectrum. In particular, moxifloxacin is effective against methicillin-resistant *Staphylococcus aureus* (MRSA) bacteria and it, and gatifloxacin may be more effective against experimental gram-negative keratitis (28). However, comparative clinical trials are needed to confirm any advantage of these new drugs, which should be kept in reserve for treatment of known bacteria and not used for prophylaxis (28,29). Moxifloxacin is more effective than other quinolones against streptococci.

Subconjunctival injections are not necessary, provided an intensive fortified drop regimen is used, as the latter will produce therapeutic levels in the cornea that are sustained and without large fluctuation. If frequent topical applications are not possible, as in a child or a disturbed individual, then subconjunctival injections of gentamicin 40 mg and cephazolin 100 mg can be used as an alternative, delivered under a general anesthetic. Inclusion of adrenaline 0.3 mL (of 1:1000) in 1 mL of solution prolongs the effective concentration of antibiotic in the cornea and aqueous from about 6 hours to over 24 hours. Other potential regimens are given in Appendix A.

Treatment may be possible on an outpatient basis, especially in the United States, but this approach is not feasible for all patients, and in the United Kingdom it is usual to admit patients to hospital. Antibiotic ointment may be given at night in the later stages of therapy once infection is under control, but in the acute stages it may interfere with absorption from drop therapy. Systemic antibiotics have no place in the management of bacterial keratitis in the absence of limbal involvement or perforation.

Antibiotics are modified according to the results of cultures and the evaluation of the clinical response to initial therapy. If there is a clear clinical response, the same regime should be continued. Microbiological sensitivities may be misleading because they are performed on lower tissue antibiotic levels than can be achieved in the cornea during topical therapy. Therapy should be reduced by increasing the interval between drops every three to four days, and not by reducing their concentration. The decision to terminate therapy is based on clinical response and the virulence of the causative organism.

If there is no response, all topical therapy should be stopped to allow the various drugs and preservatives to leach from the tissues. After 24 or 48 hours the clinical condition is reappraised, and the cornea is scraped again. On this occasion a full search must be made for more exotic or fastidious organisms that may be unusual or have special cultural requirements, such as Nocardia (Figs. 21, 22), Mycobacteria (Figs. 23, 24) and microaerophilic or anaerobic bacteria (Fig. 25). A corneal biopsy may be required to identify the organism in the deeper stroma and has been successfully used for microaerophilic streptococci, *Fusarium* sp. and *Acanthamoeba*. If no organism is identified, a second-line broad-spectrum empirical antibiotic regime should be started to include antimicrobial action against resistant streptococci, Nocardia, and Mycobacteria. This may include topical vancomycin 50 mg/mL (5%) plus amikacin 50 mg/mL (5%) and trimethoprim 0.5% (given as Polytrim ointment) [or levofloxacin (0.5%), ciprofloxacin or ofloxacin at 3 mg/mL

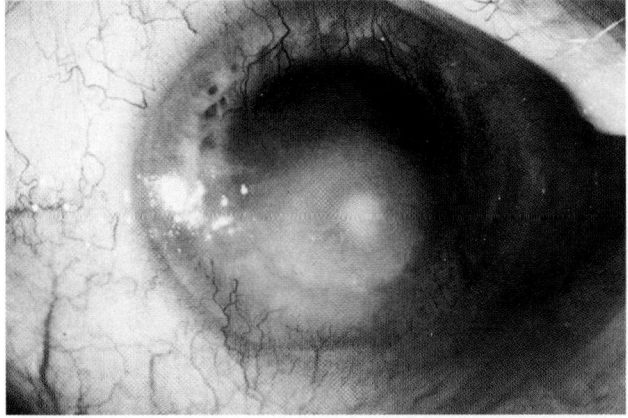

Figure 21 Chronic progressive keratitis due to *Nocardia* sp.

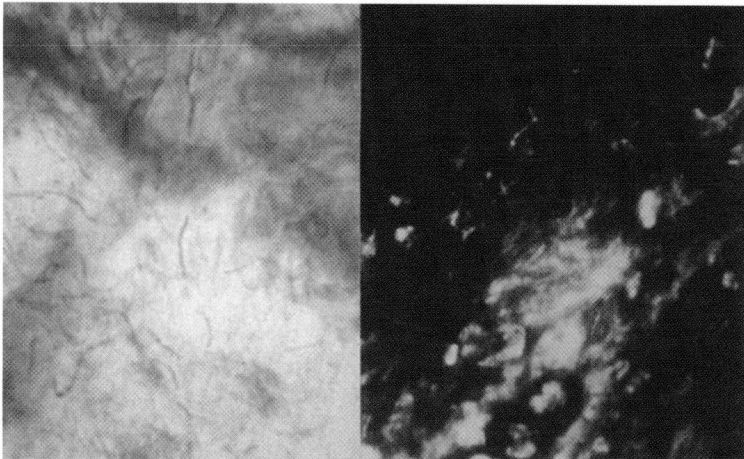

Figure 22   Composite picture of a biopsy corneal scraping showing long slender red bacilli of *Nocardia* sp. by the modified Ziehl-Neelsen stain (*left*) and fluorescing bacilli in ultraviolet light by the acridine orange stain (*right*).

(0.3%) or erythromycin 0.5% ointment or rifampicin 2.5% ointment instead] or a "fourth" generation fluoroquinolone (gatifloxacin 0.3%; moxifloxacin 0.5%) if available (United States only).

If perforation of the cornea is imminent, then a graft is required, but it is better to complete medical treatment first of all and then to graft a "quiet" eye, as the chances of success are improved. A lamellar graft may be suitable as a temporary expedient, and provides a bandage effect, but it is liable to become repeatedly infected at the host–graft interface with pathogenic bacteria including *P. aeruginosa* and streptococci.

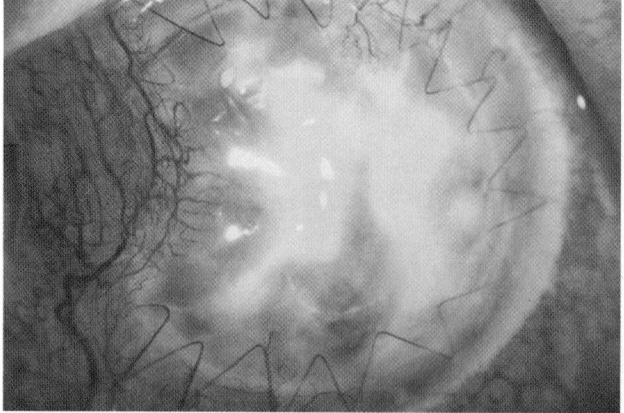

Figure 23   Chronic progressive keratitis in a graft due to *Mycobacterium chelonae* infection, treated unsuccessfully by a conjunctival flap but which then responded to amikacin. *Source*: Reproduced from Br J Ophthalmol 71, 690—3.

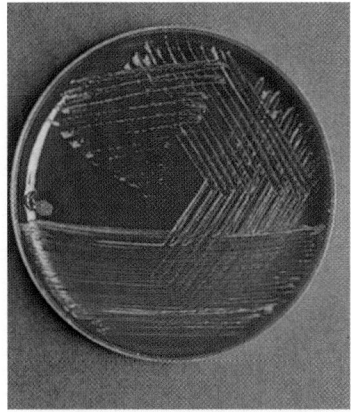
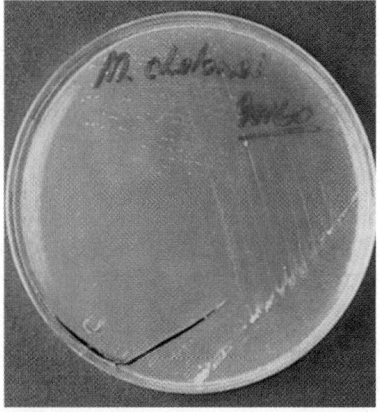

Figure 24 Composite picture of agar plate culture of *Mycobacterium chelonae* showing satisfactory growth after 1 week on chocolate agar (*left*) and Sabouraud's agar (*right*); note large and small colonies within a pure culture.

Suppurative Bacterial Keratitis: Specific Antibiotic Regimens for Topical or Periocular Therapy

Initial therapy is given below.

To treat unknown organism(s) in a new case:

Levofloxacin (0.5%) or ofloxacin 0.3% as monotherapy (or in the U.S., moxifloxacin 0.5%, gatifloxacin 0.3%) will treat the following organisms:

- *Staphylococcus aureus*
- Streptococci
- *Klebsiella* sp.
- *Haemophilus influenzae*
- Coagulase-negative staphylococci
- *Proteus* sp.
- *Pseudomonas aeruginosa*
- Other Enterobacteriaceae

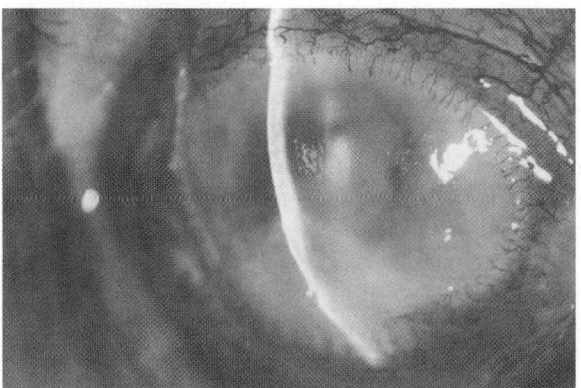

Figure 25 *Peptococcus* (anaerobic streptococcus) keratitis in a patient with Sjögren's syndrome.

To Treat Pseudomonas aeruginosa Infection

Levofloxacin (0.5%), ofloxacin (0.3%) or in the United States gatifloxacin 0.3% or moxifloxacin 0.5%. If quinolones are not available, use gentamicin (1.5%) plus ticarcillin or piperacillin (5%) and/or ceftazidime (5%). Treat with antibiotics for at least four weeks and use corticosteroids with great caution because of recognized relapse of this infection.

Note: ticarcillin and piperacillin should not be used when penicillin allergy is suspected. Cephalosporins should also be avoided when there is a history of an anaphylactic reaction to penicillin.

To Treat Mycobacterial and Nocardial Keratitis

Treatment of *Mycobacterium chelonae* infection requires topical amikacin or a fluoroquinolone (levofloxacin 0.5%, ofloxacin 0.3%, gatifloxacin 0.3%, moxifloxacin 0.5%). This mycobacterium, which causes a chronic keratitis and may follow photorefractive corneal surgery, is resistant to the common anti-tuberculous drugs.

Treatment of Nocardia infection, presenting as a chronic refractory keratitis of several months duration, is fraught with problems common with the disease at other sites. It usually requires a combination of surgery to debulk the infectious load (lamellar or penetrating keratoplasty), plus antibiotics. Antibiotics often fail on their own, despite apparent full sensitivity in vitro. A combination of topical amikacin (always) plus erythromycin and/or vancomycin and/or trimethoprim has been used successfully. Isolates are resistant to penicillin but may be sensitive to sulphonamides. The new generation of fluoroquinolones, or macrolides—azithromycin and clarithromycin—may prove useful.

Fungal Keratitis

Treatment is described in detail in Chapter 9. In addition, keratitis due to *Fusarium* sp., associated with contact lens multipurpose disinfecting solutions, is described in full in Appendix D. A rapid and accurate diagnosis of mycotic keratitis improves the chances of complete visual recovery, as otherwise the prognosis can be poor (11).

In keratitis due to *Candida albicans*, an ocular condition predisposes to the infection (e.g., dry eye, defective eyelid closure, neuroparalytic keratitis) or systemic immunosuppression or diabetes mellitus (Figs. 26 and 27). Candida invades diseased epithelium, sometimes secondary to herpes keratitis or injury.

Filamentary fungal keratitis, such as that due to *Fusarium* sp., may involve any area of the cornea and typically exhibits the following features: firm (sometimes dry) elevated slough, hyphate lines extending beyond the ulcer edge into the normal cornea, multifocal granular (or feathery) grey-white "satellite" stromal infiltrates, "immune ring," minimal cellular infiltration in the adjacent stroma, and mild iritis. An endothelial plaque and hypopyon generally does not occur within the first week, but the presence of a hypopyon in an indolent ulcer may suggest a fungal etiology or progressive infection towards an endophthalmitis (refer to Appendix D).

Topical 5-fluorocytosine (5-FC) can be used successfully as a 1% suspension in the treatment of *Candida albicans* keratomycosis. For yeast infection (*Candida* sp.), clotrimazole (or another imidazole) at 1% in arachis oil eye drops (Appendix A) is used. For hyphal infection (*Aspergillus* sp., *Fusarium* sp.), natamycin (5%) or amphotericin (0.15% to 0.3%) are used.

Imidazoles have been used effectively in the topical treatment of keratomycosis but are fungistatic only. Clotrimazole, miconazole, and econazole are effective

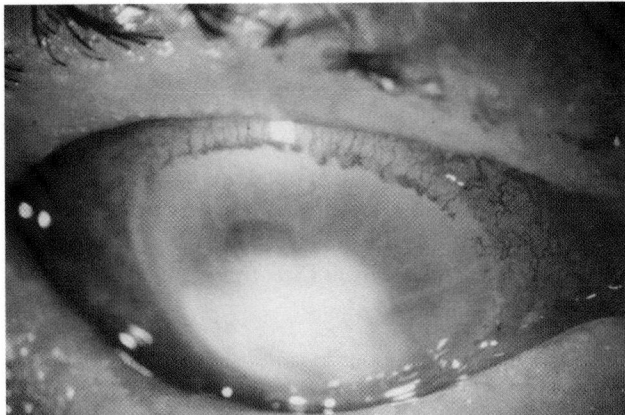

Figure 26 Corneal perforation due to *Candida albicans* in a neuroparalytic eye.

against *Candida* sp. and *Aspergillus* sp. but not against most *Fusarium* sp. They can be locally toxic. Ketoconazole is well absorbed after oral administration and is generally well tolerated, although hepatotoxicity is problematic. It has been used effectively in oculomycosis caused by *Fusarium* combined with another antifungal drug to prevent the emergence of resistance.

Recently, the second generation of triazoles such as voriconazole and posaconazole appear to be promising (30,31) including the treatment of *Fusarium* keratitis (30). Bunya et al. (32) describe the treatment of 9 cases of fungal keratitis treated with topical and oral voriconazole. Two patients wore contact lenses that used multipurpose cleaning and disinfecting solutions. Isolates included *Fusarium* (3), *Candida albicans* (3), *Alternaria* (1), *Scopulariopsis* (1), and *Pseudallescheria boydii* (*Scedosporium apiospermum*) (1). Patients were treated topically and orally for 4 to 17 weeks (mean 10 weeks). Two patients had to discontinue topical voriconazole

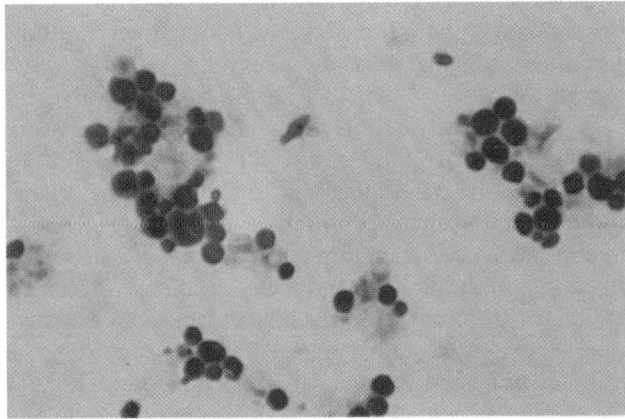

Figure 27 Gram stain of corneal scrape showing budding yeast cells of *Candida albicans*.

because of ocular toxicity (burning sensation). Five out of seven who maintained therapy were treated satisfactorily and the keratitis resolved.

Natamycin (Pimaricin) is a tetraene (polyene) antifungal drug that has been used in the topical treatment of a wide range of fungi causing keratitis, including *Fusarium* sp. A 5% suspension (Natacyn) is available commercially in the United States and India; elsewhere amphotericin, another polyene, is used instead. Amphotericin is toxic to the ocular surface and is not well absorbed into the stroma in the absence of an ulcer. Polyene drugs are fungicidal but *Fusarium* sp. are resistant to it in >50% of isolates, while *Pseudallescheria boydii* (*Scedosporium apiospermum*) and *Paecilomyces lilacinus* are always resistant.

Acanthamoeba Keratitis

In the United Kingdom ten years ago, *Acanthamoeba* infection occurred in 1 in 6720 soft contact lens wearers (1 per 96,000 population) per year (33) but this epidemic proportion, largely due to tap water rinsing and storage of lenses in weak chlorine solutions, has now receded to only one case presenting approximately every eight months at Moorfields Eye Hospital casualty department in London, where it represents only 1% of overall contact lens–associated cases of microbial keratitis (refer to figures above). While it is also a relatively rare infection in the United States, there is currently an epidemic of cases in the Chicago area for reasons yet to be identified, with 40 cases presenting over the last two years (34).

For both the United Kingdom and United States, it is predominantly an infection of contact lens wearers. In India it is more commonly associated with corneal trauma and mud splashing; presentation is late (35); the role of co-existing trachoma remains to be elucidated (36). In China the infection occurs with either contact lens wear or rural trauma (37).

The source of the ameba in most cases for the contact lens wearer is the domestic water supply (38–40). Domestic tap water should not be used to wash the contact lenses or their storage cases; it has been found to contain *Acanthamoeba* in between 15% and 30% of samples in the United Kingdom (39) but at a much lower rate of 2.8% in the south Florida water supply in the United States (40). This difference in rates by a factor of 10 is thought due to the U.K. plumbing system of maintaining a storage tank of cold water in the roof for household supplies, instead of tapping off the main pipe with a pressure reducing valve as happens in the United States. Similarly, contact lens wearers who have been exposed frequently to hot tubs or natural springs are at risk of developing *Acanthamoeba* keratitis.

Hartmannella vermiformis has also been isolated from the cornea of two soft contact lens wearers with "amebic-type" keratitis, in Dublin and Glasgow, and hence should now be considered a potential ocular pathogen (41,42).

A high index of suspicion must be maintained for all contact lens–related keratopathies presenting with epithelial disturbances or infiltrations with a "snow-storm" appearance on slit-lamp microscopy, multiple superficial abscesses, or dendritiform ulcers. A keratoneuritis (infiltration around the corneal nerves) seen on slit-lamp microscopy is diagnostic for the condition, which gives rise to an excessive amount of pain and photophobia (Fig. 28). The initial diagnosis may be confused with adenovirus punctate keratopathy and be missed. In adenovirus infection, the nummular stromal infiltrates appear at least nine days after the punctate keratopathy while in *Acanthamoeba* infection they present within the first 8 days. Dendritiform ulceration due to Herpes simplex is rare in the young contact lens wearer and *Acanthamoeba* should be

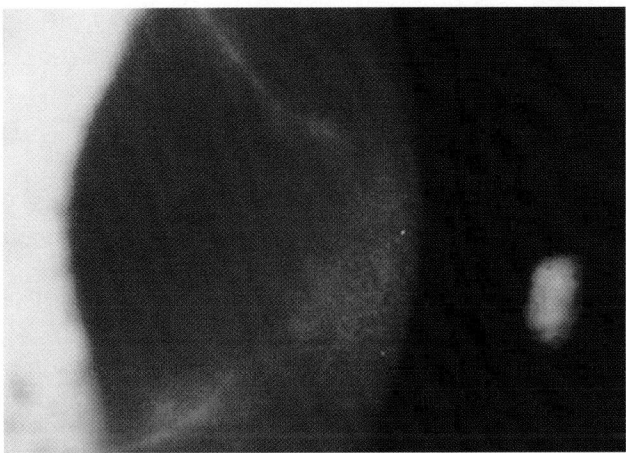

Figure 28 Close-up of epithelial lesion and perineuritis in early *Acanthamoeba* infection (the first 4 weeks) in a soft contact lens wearer.

considered first (Fig. 29). Early diagnosis greatly improves the outcome. The differential diagnosis of an inflammatory non-infective keratitis associated with wearing a contact lens, such as tight fit lens syndrome, should also be considered.

If unrecognized, the infection progresses, and at four to eight weeks there is anterior stromal infiltration, which may remain in the central cornea or give rise to a classic ring abscess (Fig. 30), accompanied by limbitis, episcleritis, and occasionally scleritis (Fig. 31). Epithelial scrapings will reveal ameba, but if missed, the infection proceeds to a large, deep infiltrated ulcer. The latter may be secondarily infected with streptococci and produce a "crystalline keratopathy" (Fig. 32), or a deep ring abscess. At this late stage, in vivo confocal microscopy (Fig. 33) may identify the ameba within the stroma in either their trophozoite or cyst stages.

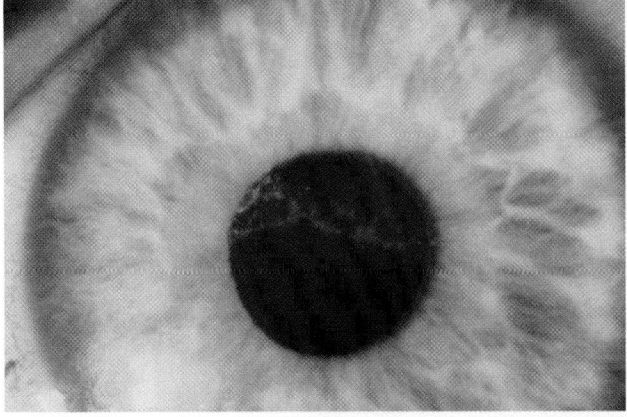

Figure 29 Dendritiform lesion in the epithelium, and infiltration around a nerve (*perineuritis*), of the central cornea demonstrating early *Acanthamoeba* infection (the first 4 weeks) in a soft contact lens wearer.

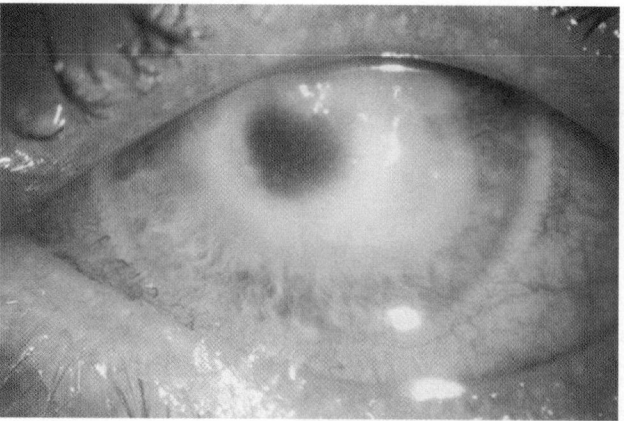

Figure 30 Classic ring abscess of anterior stromal infiltration of later *Acanthamoeba* infection (8 weeks) in a soft contact lens wearer.

Corneal biopsy is another effective way to confirm the diagnosis; two biopsies should be collected, one for culture and one for histology (using gluteraldehyde as fixative instead of formalin to process for electron microscopy), when scrapings should also be made before and after the biopsy. Tissue sections can be examined by electron microscopy when trophozoites or cysts are easily seen (Fig. 34). A severe scleritis develops in a few patients, which can be particularly difficult to treat.

It is important to make an early diagnosis, when the infection is limited to the epithelium or anterior stroma. Scrapes should be collected with a disposable sterile scalpel blade placed in a conical tube containing 2 mL saline (without potassium hydroxide). The infiltrated epithelium often lifts off easily as a sheet of cells. The tube is agitated on a vibrator, then centrifuged, and the deposit inspected by wet-field microscopy at x40. Wet mounts of epithelial scrapings have been found useful for identifying early cysts, and establishing the diagnosis by light microscopy (Figs. 35 and 36). Diagnosis of the ameba in corneal scrapings by PCR is described in Chapter 3.

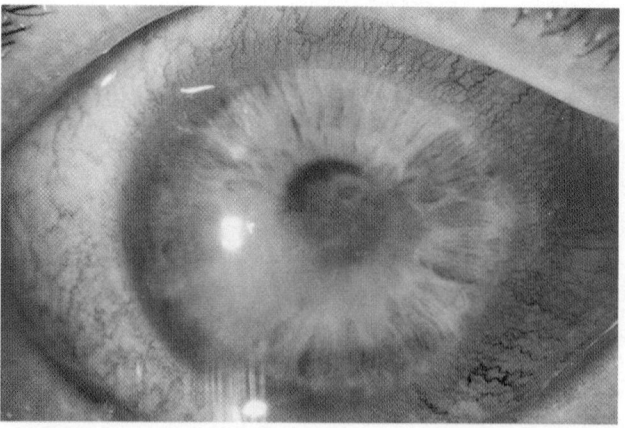

Figure 31 Anterior stromal infiltration, limbitis, and episcleritis of later *Acanthamoeba* infection (4–8 weeks) in a soft contact lens wearer.

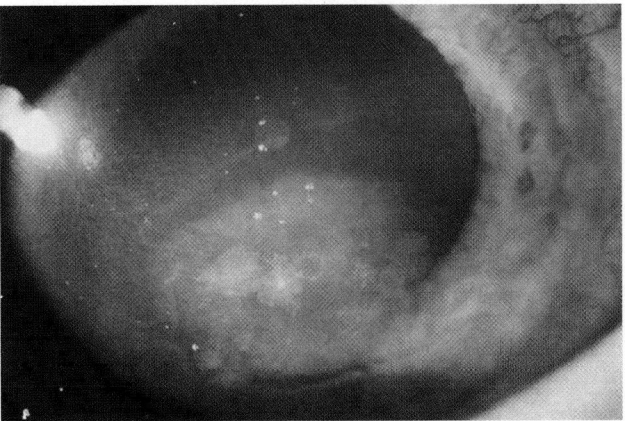

Figure 32 Crystalline keratopathy due to streptococci in a patient with a large corneal ulcer due to *Acanthamoeba* sp. at 12 weeks.

Acanthamoeba does not invade the epithelial cells themselves but is found between them (43), where the host defense depends on phagocytosis by macrophages (44). Drug targeting must therefore aim for penetration of the drug into the stroma and to the internalized ameba within the cysts. Penetration into the epithelial cell is not required (43). This is well demonstrated by the effectiveness of the cationic antiseptics, chlorhexidine, and polyhexamethylene biguanide (45), which do not penetrate within the epithelial cell.

Culture and Drug Sensitivity Testing

For culture, scrapes should be inoculated directly onto non-nutrient agar, made up in Page's amebal saline (AS). This involves use of Modified Neff's AS. A separate stock solution of each component is made up by dissolving in 100 mL of glass-distilled water: NaCl 1.20 g, $MgSO_4.7H_2O$ 0.04 g, $CaCl_2.2H_2O$ 0.04 g, Na_2HPO_4 1.42 g, KH_2PO_4 1.36 g. The final dilution is prepared by adding 10 mL of each stock solution to 950 mL glass-distilled water to make 1 L in total of AS. To prepare the

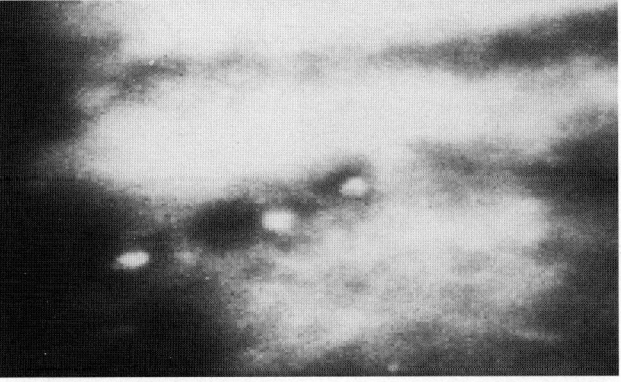

Figure 33 Confocal microscopy in vivo, of the corneal stroma showing cysts of *Acanthamoeba*.

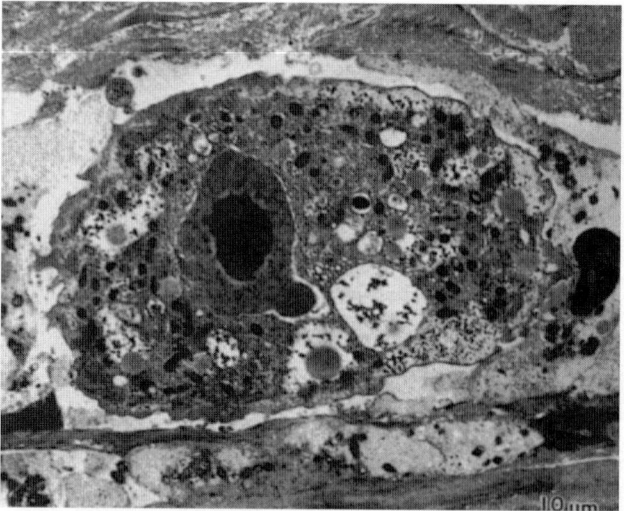

Figure 34 Transmission electron micrograph of a trophozoite of *Acanthamoeba* sp. in corneal stroma. *Source*: Reproduced from Ophthalmol Clin N Am 7, 605–16.

non-nutrient amebal agar, 1 L of AS is added to 15 g of plain agar, which is autoclaved and dispensed into 100 mL bottles or smaller. This can be stored at room temperature. It should be boiled before use and fresh plates poured of hot agar. In this way fungal contamination can be avoided. Plates are inoculated directly from corneal scrapings. Seeding of the agar plate with heat-killed *Klebsiella* sp. or *E. coli*, as a food source, is not usually required with use of AS agar.

The AS agar plate should be inoculated with the corneal scrape sample directly from the eye. The epithelium should be inspected with an inverted microscope for the presence of typical trophozoites or cysts, as well as intermediate forms, of *Acanthamoeba*. Plates are incubated at 25°C and 32°C for four weeks, wrapped up

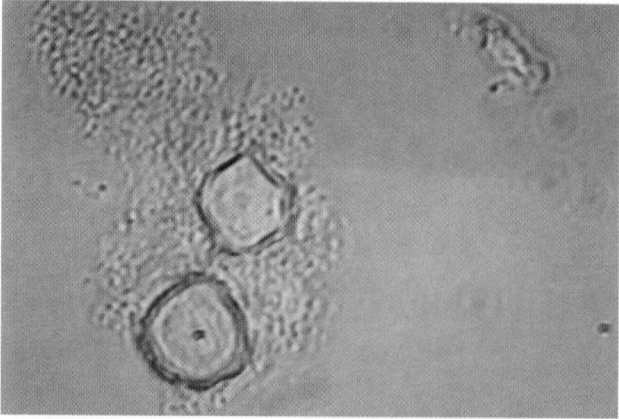

Figure 35 Wet preparation, viewed by phase contrast microscopy with a green filter, of a corneal scraping showing two early cysts of *Acanthamoeba*, collected from early keratitis with epithelial infiltration in a soft contact lens wearer.

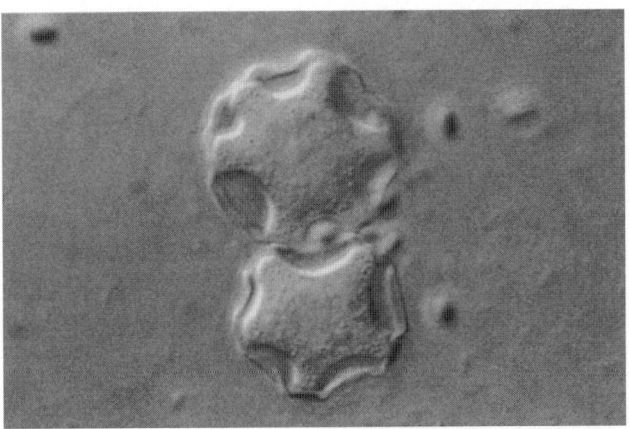

Figure 36 Cysts of *Acanthamoeba griffini* viewed by the Nomarski technique with light microscopy. *Source:* Courtesy of the late T. Byers.

in clean disposable plastic bags to avoid fungal growth. Fresh agar should not be kept for more than 24 hours before inoculation. If facilities for culture are not available, corneal scrape specimens may be mailed in saline to a suitable laboratory.

If non-nutrient agar without Page's saline is used, then the plate should be inoculated again with a turbid suspension (on a swab) of heat-killed *Klebsiella aerogenes*, or other coliforms, as a nutrient source for the ameba. The plate should be incubated at 32°C for four weeks; it should not be incubated at higher temperatures. Ameba will usually, but not always, be visible by low-power light microscopy after one week; after two weeks the whole plate is covered by the typical double-walled, star-shaped cysts. Each point of the star is the ostiole through which the internalized ameba receives exteroceptive cues; it is normally plugged with mucopolysaccharide.

For drugs to be effective, they have to penetrate to the internalized ameba within the cyst either through the ostioles or pores within the cyst wall. For in vitro drug sensitivity testing, trophozoites and cysts of *Acanthamoeba* can either be cultured axenically in fluid medium without bacteria using a proteose peptone glucose (PPG) solution or on amebal saline agar plates. Dilutions of drugs are carried out in microtiter plates and *Acanthamoeba* cultures of trophozoites and cysts are added (45). The plates are incubated at 32°C and the concentrations are recorded of dilutions with surviving cells. This assesses the minimum trophozoitical amebicidal concentration (MTAC), and the minimum cysticidal concentration (MCC). If this is not possible, isolates should be sent to a reference laboratory.

Treatment

Treatment should start with 0.02% (200 μg/mL) chlorhexidine digluconate in physiological saline (45) and Brolene (propamidine isethionate) 0.1% (1000 μg/mL) in physiological saline. If chlorhexidine is unavailable, polyhexamethylene biguanide (PHMB) 0.02% can be used but it is not licensed as a drug for ocular use. Hexamidine (Desmodine) can be used as an alternative, commercially available diamidine drug instead of propamidine.

These drugs are given every hour, day and night for the first three days, reducing to two-hourly by day only. This requires the patient's admission to hospital. Adjunctive therapy includes oral flurbiprofen, for both non-corticosteroidal anti-inflammatory and analgesic effects, and topical mydriasis. Thereafter, combination therapy is given three-hourly by day for up to two months and then four-hourly by day for two months more. Control is rapidly gained but treatment is needed for two to six months in some patients, partly because drop therapy is not an ideal vehicle with which to treat this infection.

If diagnosed early, medical cure is possible with complete recovery of vision. One week after starting therapy, however, there may be a corneal reaction to antigens or toxins from lysis of dead ameba, with localized stromal edema and anterior chamber activity, which lasts up to three weeks. Although this may be suppressed with corticocorticosteroids, their use is not encouraged. Cases presenting early can be treated with chlorhexidine for several weeks only and then observed. Prolonged treatment with chlorhexidine can be toxic (45).

Cases presenting late, with considerable pain, ring abscess, inflammation, and episcleritis, may need the introduction of corticocorticosteroids, although inflammation and pain can often be satisfactorily controlled by a non-corticosteroidal anti-inflammatory drug such as flurbiprofen. This necessary measure will prolong the treatment period, but without it the patient may find the pain intolerable. The pain of keratoneuritis may demand analgesics for a time. Adjunctive immunosuppression has been advocated for the management of *Acanthamoeba* scleritis.

The immunopathogenesis of *Acanthamoeba* infection in the cornea has been recognized recently. This may present in a subacute form as Wessley rings (see Chapter 1) or as a late inflammatory phenomenon as an area of epithelial and anterior stromal infiltration around the healed scar, after successful chemotherapy (45). This late effect is settled quickly with corticosteroids. Similar features are known to occur in association with other infections of the cornea. With *Onchocerca volvulus* punctate keratitis, for example, an inflammatory infiltrate comprising lymphocytes and eosinophils with concomitant localized edema was a common feature of diethylcarbamazine-treated patients, the reaction being consequent to a cell-mediated immune response to the dead microfilaria, which were localized in the center of ill-defined opacities. The development of such cell-mediated immunity to the presence of *Acanthamoeba* antigens in the cornea may also be responsible for the severe scleritis suffered by some patients with chronic *Acanthamoeba* keratitis, which usually requires high-dose corticosteroid therapy to suppress it. Some workers consider that the role of corticosteroids in *Acanthamoeba* infection is ambiguous, while others advise against their use in acute infection, because of the risk of compromising the host inflammatory response against *Acanthamoeba*; these workers reserve them for later complications. The host immune response is thought to depend on macrophages (44) that phagocytose the ameba, unless inhibited by corticocorticosteroids. Our experience suggests that corticosteroids have a role in controlling late immuno-inflammatory responses, when the ameba have been killed and antigen remains bound to the corneal stroma, but this aspect requires further elucidation.

Prevention of *Acanthamoeba* Keratitis

Contact lens (CL) storage cases are contaminated with a polymicrobial biofilm, as well as *Acanthamoeba* from the domestic water supply. *Acanthamoeba* is also found as cysts in airborne dust and dirt. Prevention involves use of acanthamebicidal disinfectants in cleaning and storage case solutions, of which the best is overnight use of hydrogen peroxide 3%. Chlorine is ineffective against cysts. Multi-purpose solutions are considered in Appendix A (with details of the Chicago outbreak and withdrawal of Complete MoisturePlus). Storage cases for CLS should never be washed in tap water, but with boiled-cooled water only, and they should be stored dry when not in use. This is important because coliform bacteria die quickly in dry conditions when the ameba cannot multiply. *Acanthamoeba* cysts are killed at a temperature of 70°C.

Adherence of *Acanthamoeba* to Different Types of Contact Lens Materials

A number of studies have been conducted to investigate the effect of different CL materials on the adherence of *Acanthamoeba* (20,46–49). This showed that rigid lenses had the least adherence and that, for hydrogels, the FDA group 4 (ionic, high water content at > 50%) were by far the most adherent. FDA group 4 CLs were also found associated with *Acanthamoeba* infection more frequently in the U.K. epidemic in the 1990s than other FDA groups 1 to 3, albeit in association with use of chlorine tablets and tap water for disinfection (33). This adherence was further increased if the CLs were worn, due to protein coating (48), or exposed to *P. aeruginosa* to form a biofilm on them (49). Studies with salicylate found that it inhibited attachment of *Acanthamoeba* to CLs (47). Recent work with first-generation SHs and first- and second-generation SH CLs has found that first-generation SH materials (46) were much more adherent for *Acanthamoeba* than for FDA group 4 hydrogel CLs but that second-generation SH CLs did not give this effect (20). The reformed surface of second-generation SH CLs should be explored in future studies, as it is important to identify the adherence factor of first-generation SH CLs that is suppressed in second-generation SH materials. Is this due to altered surface chemistry or other surface properties? Interestingly, a large multi-center European study has just found that silicone IOLs were more frequently associated with postoperative endophthalmitis ($p < 0.003$) than were acrylic IOLs. This data is reviewed in Chapter 8.

Microsporidial Keratoconjunctivitis

Microsporidia are obligate intracellular protozoa that may cause keratitis in both immunocompetent and immunocompromised persons, even when major attention has been drawn to HIV-infected patients. Current evidence indicates that microsporidiosis is a common opportunistic infection in patients with AIDS. The gastrointestinal tract and cornea are most frequently involved. It presents as a unilateral corneal stromal keratitis without epitheliopathy or iritis occurring in an immunocompetent patient. A bilateral superficial punctate keratoconjunctivitis can present in AIDS patients. There is no established treatment of deep stromal microsporidial infection. It has been suggested that topical propamidine isethionate 0.1% and systemic itraconazole is effective in microsporidial keratoconjunctivitis. A flow chart for the laboratory investigation of parasitic keratitis is given in Fig. 37.

Phlyctenular Keratoconjunctivitis

Phlyctenular keratoconjunctivitis is a nodular inflammatory process caused by a delayed-type hypersensitivity reaction to bacterial antigens. It presents as uni- or bilateral small, round, grey or yellow, raised inflammatory, hyperemic nodule(s) close to the limbus, with engorged conjunctival vessels (Figs. 38, 39). It causes irritation and tearing. After approximately 10 days, the nodule may develop a fine

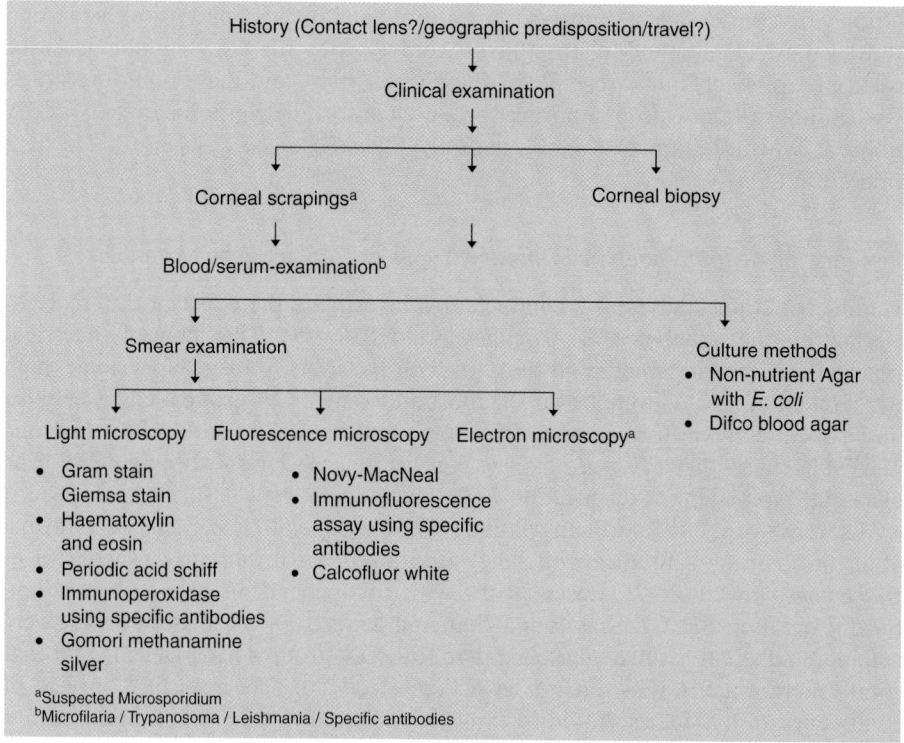

History (Contact lens?/geographic predisposition/travel?)

Clinical examination

Corneal scrapings[a]   Corneal biopsy

Blood/serum-examination[b]

Smear examination   Culture methods
- Non-nutrient Agar with *E. coli*
- Difco blood agar

Light microscopy   Fluorescence microscopy   Electron microscopy[a]

- Gram stain Giemsa stain
- Haematoxylin and eosin
- Periodic acid schiff
- Immunoperoxidase using specific antibodies
- Gomori methanamine silver

- Novy-MacNeal
- Immunofluorescence assay using specific antibodies
- Calcofluor white

[a]Suspected Microsporidium
[b]Microfilaria / Trypanosoma / Leishmania / Specific antibodies

Figure 37   Flow chart of laboratory investigations in suspected parasitic keratitis.

ulcer 1–3 mm diameter and then regress over a few days. There is no scarring when the lesion is on the conjunctiva but, if it occurs on the peripheral cornea, there can be the development of a fascicular ulcer that creeps over the cornea towards the center supplied by vessels that follow it; in the severest cases with multiple phlyctens, a ring ulcer can form when the whole cornea is endangered. Unlike trachomatous pannus, there is no special predilection for the upper part of the cornea.

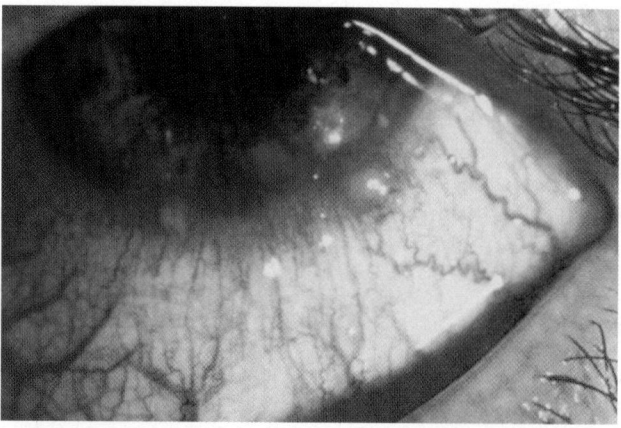

Figure 38   Multiple perilimbal phlyctens, with a conjunctival response, in an Indian patient presumed due to *Staphylococcus aureus*.

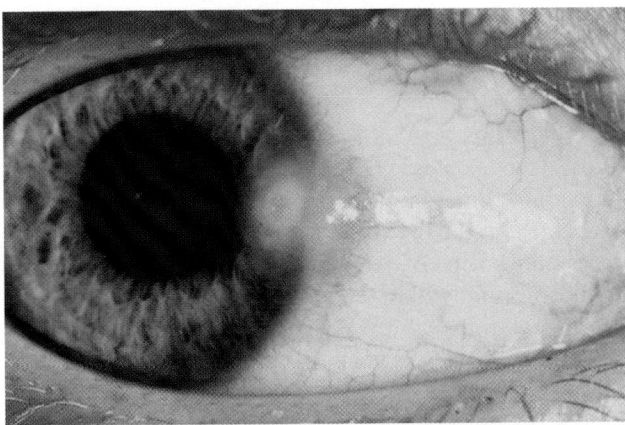

Figure 39 Single perilimbal phlycten with surrounding limbitis in a Caucasian patient presumed due to *Staphylococcus aureus*

Whereas in the past hypersensitivity to *Mycobacterium tuberculosis* was the most common agent producing phlyctenulosis, infection with *Staphylococcus aureus* in association with chronic blepharitis is now much more common. Other agents such as chlamydia, fungi, and parasites have also been related to phlyctenulosis. Since the possibility of an association with tuberculosis still exists, efforts have to be taken to exclude this infection including a detailed history, and, if suspected, a Mantoux (PPD) test and a chest X ray should be done. If an infectious etiology is suspected, corneal smears and culture should be taken.

When due to a non-tuberculous cause, it responds to topical corticocorticos-teroids but often recurs. Treatment of a concomitant chronic anterior blepharitis by lid hygiene is an important part of therapy and should be explained to the patient. Unfortunately, if often recurs. A tuberculous phlycten can also be associated with bilateral ulcerative keratitis, stromal infiltration, and anterior uveitis. Treatment is required with systemic anti-tuberculous chemotherapy and topical prednisolone acetate 1%, when a good recovery can be expected.

"Topical Anesthetic Abuse" Keratitis

Topical anesthetic abuse is a self-inflicted injury. Anesthetic drops may be started due to pain from epithelial erosions, which can be due to excessive exposure to sunlight (UV) irradiation. It can present as a differential diagnosis of *Acanthamoeba* keratitis and be further complicated by therapy for this infection. Further complication can arise from secondary infection of the inflamed cornea by *Candida albicans*. This type of keratitis can also be self-inflicted as a psychological disorder, (keratitis artefacta), mostly in young women, when perfume can be used as the noxious agent.

Virus Infection

Herpes simplex virus and adenovirus account for 1% of all acute conjunctivitis in an ophthalmic casualty department. Antiviral agents are available for the treatment of H. simplex and H. zoster infections, but not for adenovirus infection. A number of other ocular viral infections occur for which there is no specific antiviral therapy but topical antibiotics are often prescribed to reduce the risk of secondary bacterial infection.

Herpes Simplex Eye Disease

Humans are the only natural reservoir for Herpes simplex. Primary ocular infection occurs when a non-immune individual comes into close contact with someone who is shedding virus. Skin or mucosal entry on the face results in zosteriform spread along Vth cranial nerve axons with establishment of latency in the trigeminal ganglion. Latency can follow asymptomatic infection within the territory of the first division of the Vth nerve dermatome or inoculations of the neighboring second and third divisions. Virus may be grown from keratitis specimens in culture, and viral particles have been demonstrated within keratocytes, suggesting the existence of a non-neuronal latency.

Establishment of latency by a given strain of HSV 1 greatly reduces the likelihood of further HSV 1 infection, and recurrences are due to the same strain. Also, while patients who have been infected by HSV 2 usually show resistance to HSV 1 infection, the reverse is not the case; patients who show immunity to HSV 1 may still be susceptible to oculogenital infection by HSV 2. Rarely, both HSV 1 and 2 have been isolated from the same cornea in an AIDS patient.

For HSV 1, the onset of seropositivity is in the first 5 years of life, as maternal antibody subsides. For HSV 2 this is at puberty, reflecting sexual transmission of the virus. Eighty percent of genital herpes is caused by HSV 2, which causes a small proportion of herpes keratitis. Maternal transmission of type 2 occurs in about 1:7500 live births and spread is commonly via the genital tract, or, less often, transplacental. Neonatal ocular infection occurs in about 1:100,000 deliveries and can cause conjunctivitis, keratitis, microphthalmia, cataract, uveitis, chorioretinitis, optic neuritis, and cortical blindness following encephalitis. Late sequelae include motility disorders, chorioretinal scarring (28%), and optic atrophy (22%). Keratitis is relatively resistant to treatment, and chorioretinitis has a poor visual prognosis.

Most ocular HSV disease is caused by HSV 1. Primary infection manifests with gingivostomatitis, pharyngitis, rhinitis, or tonsillitis, with fevers, chills, myalgias, lymphadenopathy, and vesicular eruptions that clear in 7 to 10 days without scar (Fig. 40). Initial episodes involve lid or conjunctiva in 54%, superficial cornea in

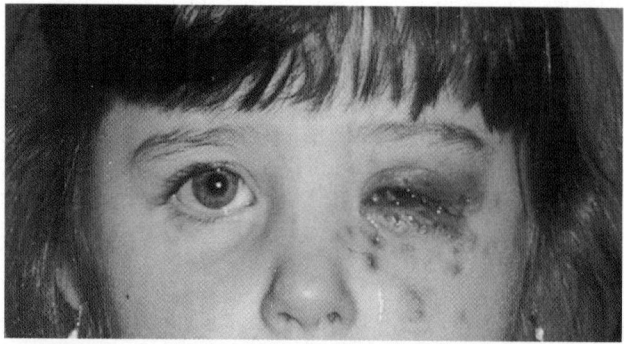

Figure 40 Primary Herpes simplex virus infection in a child showing lesions on the upper and lower lids and cheek.

63%, deeper cornea in 6%, and uveitis in 4%. There may be a vesicular blepharitis, follicular or pseudomembranous conjunctivitis, and a punctate or dendritic keratitis. HSV-specific antibody is found in tears at a greater concentration than in saliva, suggesting local production.

Reactivation of latent virus in immune subjects causes ocular disease associated with peripheral viral shedding and is termed "recurrent herpetic eye disease." It includes epithelial keratitis (Figs. 41 and 42), stromal keratitis (disciform and necrotizing), trabeculitis, keratouveitis, secondary glaucoma, and, rarely, acute retinal necrosis. Not all primary eye disease is followed by recurrent disease and not all recurrent disease is unilateral. Bilateral disease occurs in 2% of cases. It is favored by atopy, systemic diseases such as lumbar zoster, pulmonary tuberculosis, and malaria, or malnutrition with vitamin A deficiency. In Africa, it causes a severe herpes keratitis, which may be bilateral, in children with measles. Atopy is common in those with bilateral disease (25–40%) and disease is more frequently bilateral and severe in the immunosuppressed, with more frequent and prolonged recurrences in AIDS patients.

Pathogenesis of keratitis: dendritic ulcer is due to replication of virus within the corneal epithelium, giving rise to cell lysis and the characteristic branching lesions with terminal bulbs. Inadvertent corticosteroid use may broaden this into a geographic ulcer (Fig. 43). Disciform keratitis is due to a corneal endotheliitis, with an additional delayed-type hypersensitivity, which explains its inflammatory features, including iritis. Stromal keratitis, often with ulceration, is due to an immune response to viral antigens, and the severity of the host stromal response depends partly on the strain of virus and partly on host genetic factors.

Therapy of Herpetic Eye Disease

The first generation antivirals, available for the topical treatment of dendritic ulcer, were idoxuridine (IDU), adenine arabinoside (ARA-A), and trifluorothymidine (F3T). F3T is available commercially in the United States but not in the United Kingdom. However, a 1.0% solution in normal saline can be prepared from the dry powder in many hospital pharmacies. IDU is a thymidine analogue incorporated

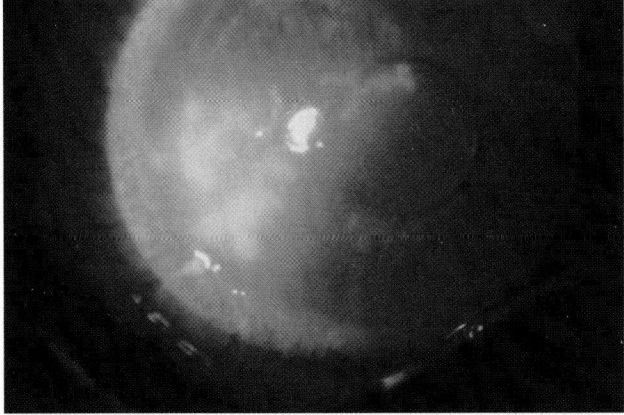

Figure 41   Epithelial dendritic ulcer due to Herpes simplex virus infection. *Source*: Courtesy of Prof. C. M. Kirkness.

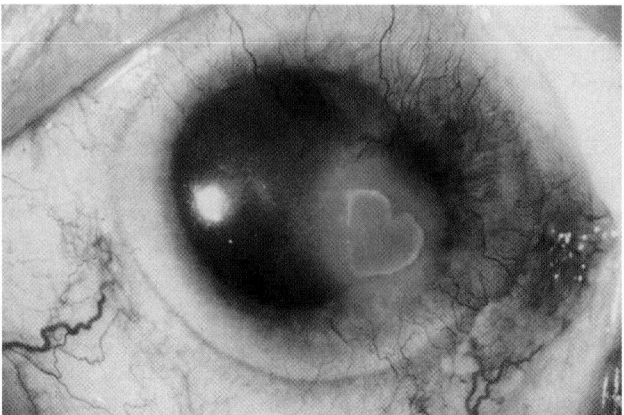

Figure 42   Herpes simplex keratitis with epithelial and stromal disease, complicated by associated uveitis.

into viral DNA, where it inhibits thymidine monophosphate kinase, virus encoded thymidine kinase, and DNA polymerase. ARA-A is a purine analogue that inhibits DNA polymerase. All three agents are extremely effective in blocking herpes virus replication; but, as they are incorporated into the DNA of both infected and uninfected cells, they show significant toxicity with prolonged use, least for F3T. Toxic signs include follicular conjunctivitis, with lid thickening, ptosis, punctal stenosis and epiphora, hurricane and filamentary keratopathy, and "ghost" dendritic changes. Corneal wound strength is reduced and epithelial healing impaired. Corneal epithelial changes appear preferentially on the graft after keratoplasty. Although there is cross-toxicity between first generation antivirals, there is not usually cross-allergenicity among either first- or second-generation drugs.

The second-generation antivirals are those activated by virus-induced enzymes and which therefore exert their actions chiefly in infected cells. Thus, acyclovir is phosphorylated by viral thymidine kinase and converted to the active triphosphate

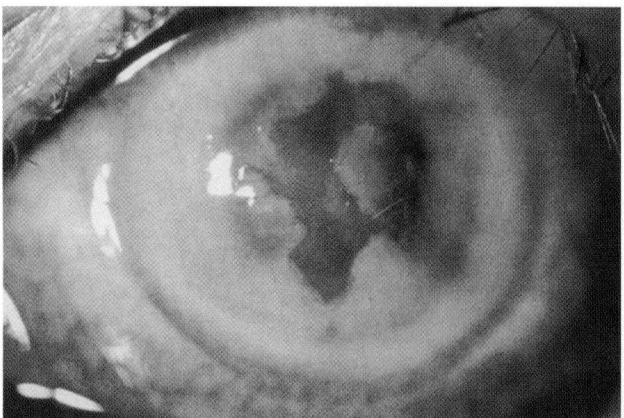

Figure 43   Large geographic ulcer of Herpes simplex keratitis, involving Bowman's membrane and superficial stroma. *Source*: Courtesy of Prof. C. M. Kirkness.

by host cell enzymes. The triphosphate is a potent inhibitor of DNA polymerase and acts as a chain terminator to the growing viral DNA strand. Such drugs are more inhibitory to herpetic than cellular DNA polymerase and preferentially inhibit viral DNA synthesis. They are less toxic than the first generation agents and include acyclovir, bromovinyldeoxyuridine (BVDU), and ethyldeoxy-uridine (EDU). The regimens advocated for these compounds are listed in Table 6. In addition to these agents, human interferon B (leukocyte-derived) and B (fibroblast-derived) and interferon from recombinant DNA sources have also been used clinically.

IDU is more effective than placebo in the treatment of dendritic ulcer, with about 80% healing within two weeks. A similar healing rate has been found with ARA-A, while a 97% healing rate is claimed for F3T. Acyclovir is highly effective against the dendritic ulcer, more effective than IDU, and equally effective as ARA-A and F3T. Since acyclovir is less toxic than any of the first-generation antivirals, it is the treatment of choice for dendritic ulcer, with F3T and ARA-A being alternatives. BVDU and EDU also appear to be highly effective agents. While dendritic keratitis may be self-limiting, with about 26% of placebo-treated cases resolving within two to three weeks, antiviral therapy, combined with a minimal wiping debridement, using a cotton-tipped applicator, will produce a clinical cure in 76% to 100% and shorten the median healing time.

F3T, acyclovir, and ARA-A are all effective for treating the more aggressive geographic ulcer. Where prolonged use is needed, acyclovir is preferable because of its low toxicity. Ganciclovir has become available as a topical preparation and demonstrated a similar or superior therapeutic effect, faster treatment response, and less irritation (keratitis punctata superficialis) in HSV dendritica patients (50).

In disciform keratitis and various forms of stromal keratitis, antivirals are given prophylactically to prevent dendritic lesions while the inflammatory features of the disorder are suppressed with corticocorticosteroids. Disciform keratitis is treated with a combination of acyclovir or F3T (rather than vidarabine), with a tapering dose of topical corticocorticosteroid. A suitable corticosteroid regime is Predsol

Table 6 Antiviral Drugs for Ocular Therapy

Drug availability	Form	Concentration	Frequency
Idoxuridine (IDU)	Ointment	0.5%	5 times daily for 14 days
	Drops	0.1%	1-hourly by day, 2-hourly by night for 14 days
Vidarabine (ARA-A)	Ointment	3.0%	5 times daily for 14 days
Trifluorothymidine (F3T)	Drops	1.0%	1-hourly by day, 2-hourly by night for 14 days
Acyclovir (ACV)	Ointment	3.0%	5 times daily for 14 days
	Tablet	200 mg	400–800 mg 4-hourly for 5 days
	Intravenous		5 mg/kg over 1 hr every 8 hr
Ganciclovir	Drops	0.15%	5 times daily until re-epithelialized, then 3 times daily for 7 days
Bromovinyldeoxyuridine (BVDU)	Drops	0.1%	2-hourly by day
	Ointment	0.5%	
Ethyldeoxyuridine (EDU)	Drops	–	1–2 hourly by day
	Gel		At night
Famciclovir	Tablet	125–250 mg	8-hourly for 5 days
Valaciclovir	Tablet	500 mg	8-hourly Bd

phosphate 0.5% q.d.s., reducing to 0.3%, 0.1%, 0.03%, and 0.01%, according to the response, and weaning finally to a topical nonsteroidal anti-inflammatory drug (NSAID), such as fluoromethalone. Initial therapy with Predsol acetate 1.0% may be required for severe keratouveitis.

In the NEI (HEDS) study of stromal keratitis, patients receiving F3T with a weaning dose of topical corticosteroids (who had not received corticosteroids for at least 10 days before the trial) had a shorter duration and less progression than those receiving F3T with placebo alone (51).

Indolent (metaherpetic) ulcers are not due to active viral disease but to an epithelial surfacing defect. They are treated by lubricants and anticollagenases, and withdrawal of antivirals and preservatives. High water content soft CLs are used for botulinum ptosis, but ultimately penetrating keratoplasty may be required.

Antiviral Prophylaxis Against Herpetic Disease of the Anterior Segment

The frequency of the recurrence rate of dendritic ulcer after a single attack rises after the first attack. The risk of recurrence is increased after corticosteroid use and for this reason it is usual to use topical antivirals prophylactically whenever topical corticocorticosteroids are in use in any but the lowest concentration. Weaning of topical corticosteroids should be carried out in a stepwise fashion under ophthalmic supervision. As the intervals between visits increase, it is worth stepping down the dose one week before a planned visit, to anticipate rebound inflammation prior to further weaning.

In addition to their use during treatment of disciform and stromal keratitis, keratouveitis, trabeculitis, and glaucoma, topical corticosteroids are used in patients undergoing keratoplasty for herpetic keratitis. Although it is feasible to give topical acyclovir to a graft on a long-term basis without much fear of toxicity, the incidence of spontaneous epithelial recurrence during the first postoperative year is low and evidence of benefit is small. Antiviral cover is particularly useful as prophylaxis during the intensive use of topical corticocorticosteroids for allograft rejection, when herpetic recurrence rates can rise from 15% to 32%.

ACV will reach effective aqueous levels in the intact eye after topical use (1.69 μg/mL; range: 0.38–4.87) and has therefore been recommended for the treatment of uveitis, limbitis, and secondary glaucoma. Acyclovir levels of 1.03 μmol/L can be achieved in human aqueous during oral ACV therapy with lower levels (0.31 μmol/L) in cornea. F3T does not penetrate intact epithelium, but effective concentrations may penetrate in the presence of an epithelial lesion. It is current practice to treat patients undergoing keratoplasty for herpetic disease with a full oral dose of acyclovir (800 mg × 5 daily) over the operative period and for two weeks thereafter, followed by a maintenance dose of 400 mg twice daily for 6 to 12 months.

Prophylactic antivirals are also used in patients receiving topical corticosteroids to suppress the inflammatory features of herpetic keratouveitis. Similar considerations to those in corneal graft prophylaxis apply and risks of antiviral drug toxicity again arise because of the prolonged period (weeks or months) of immunosuppressive therapy.

A controlled trial of oral acyclovir in patients receiving topical corticosteroids and trifluridine for stromal keratitis has showed no benefit from this additional antiviral therapy (52). Another controlled trial showed that topical corticosteroid, though significantly better than placebo in reducing persistent or progressive stromal inflammation, had no detrimental effect on visual outcome at 6 months (51). No form of treatment will reduce the frequency of clinical recurrences.

Resistant strains of virus have been identified to IDU, ARA-A, and F3T and have been responsible for clinical disease. Second generation antivirals such as acyclovir, BVDU, and EDU are ineffective against thymidine kinase negative (TK-) HSV. Culture of HSV with acyclovir has permitted the emergence of acyclovir resistant TK- strains within 10 days, and this has raised some concerns as to the long-term expectation of clinical resistance to acyclovir and related drugs. F3T and acyclovir are useful in disease resistant to ARA-A and IDU. BVDU has been used effectively in cases resistant to IDU, ARA-A, and F3T.

Penciclovir (for which the more bioavailable, oral pro-drug is Famciclovir) is available to treat H. simplex virus types 1 and 2 and H. zoster. It is converted in vivo to the triphosphate by virus-induced thymidine kinase and has been shown to be active against acyclovir-resistant H. simplex, which has an altered DNA polymerase.

HSV Infection and Cornea Transplants

Primary transplant failure has been reported to be associated with HSV infection. Indeed, in up to 33% of patients with primary graft failure, HSV type 1 DNA could be detected. In addition, transmission of HSV-1 through transplantation has been confirmed based on genetic characterization of HSV-1 DNA isolated from a donor cornea before and after corneal transplantation (53). Since corneal donor tissue is not routinely screened for the presence of HSV, accidental transfer of the infection in an unaffected recipient is possible but fortunately rare. Clinical and experimental investigations demonstrated that HSV infected transplants undergo a dramatic endothelial cell loss that leads to exclusion from use during inspection in eye banking (54). It is advisable to keep the scleral rim for later analysis in these rare patients with primary graft failure (55,56).

Herpes Zoster Ophthalmicus (VZV)

Involvement of the first division of the Vth cranial nerve by H. zoster virus is associated with a vast array of ocular complications ranging from involvement of the lids in the primary infection to persistent conjunctivitis, keratouveitis, glaucoma, papillitis, ocular nerve palsy, and deep ocular pain. Although the use of topical ocular corticosteroids to suppress the inflammation does not have the dire consequences seen with Herpes simplex eye disease (e.g., induction of dendritic or geographic ulceration), studies have suggested that outcome in those treated with acyclovir alone has been better than in those receiving corticosteroids alone (acyclovir 3% ointment vs. betamethasone 0.1% ointment, five times daily). No recurrence occurred in the acyclovir-treated group, whereas the recurrence rate was 63% in the corticosteroid only group. Such recurrences were more difficult to suppress than the initial disease features. Corneal epithelial disease healed significantly more quickly in the acyclovir patients. Acyclovir has been used intravenously in doses of 5 mg/kg eight-hourly or greater and has been shown in some studies to be effective in improving the healing time of the rash and diminishing the pain associated with the early phase of the disease. However, post-herpetic neuralgia has not been helped and non-dermatological features of the eye disease have not been suppressed. Penciclovir is a newer alternative, as discussed above, and is being marketed for its better effect in treating H. zoster.

There is clinical benefit from oral administration of acyclovir given at 600 mg five times a day. It is well tolerated and reduces the incidence and severity of dendritiform keratopathy, stromal keratitis, and uveitis. Treatment within 72 hours of the onset of skin lesions reduces pain in the acute phase of the disease but not post-herpetic neuralgia. Studies have indicated that doses of 800 mg five times a day are well tolerated and produce higher serum levels.

Adenovirus Keratoconjunctivitis

Adenovirus is a common cause of acute conjunctivitis, presenting either in association with an upper respiratory tract infection, fever, and malaise (pharyngoconjunctival fever), or as part of an outbreak of moderate or severe keratoconjunctivitis (epidemic keratoconjunctivitis). There are 47 serotypes of adenovirus of which half cause ocular infection. However, in one series in London, 92% of ocular adenovirus infections were due to serotypes 3, 4, 7, 8, 10, 15 (15/29), and 19.

Epidemic keratoconjunctivitis is caused mainly by adenoviruses 8, 19, and 37, and gives rise to epidemics within the community or transmitted within the (eye) hospital or outpatient clinics. Onset of conjunctivitis in one eye is followed by signs in the fellow eye within two to seven days. A punctate epithelial keratitis, about four days after onset, is succeeded by a coarse punctate keratitis at one week and then, between the first and second week (later than day 9), by a multifocal, subepithelial keratitis considered to reflect an immune reaction to viral antigen. The keratitis is responsible for symptoms of pain, glare, photophobia, and visual loss, and although the condition may resolve within three to four weeks it may give rise to disability for over a year, sometimes with recurrent episodes of papillary conjunctivitis. Viral shedding has usually subsided by two weeks but there is possible latent retention of virus in lymphoid tissue.

Although antiviral agents, interferon, and antibiotics have not been shown to affect the course of adenoviral keratoconjunctivitis, some amelioration of symptoms has been observed in patients treated with trifluorothymidine. There had been promising results in rabbits with 0.5% cidofovir given bd for seven days, with a decrease in duration and of viral shedding without toxicity (57). However, a clinical trial has shown that cidofovir is of no use in patients, since either the concentration is not sufficient (1%) or toxicity is too high (2%) (58).

Povidone iodine, a broad spectrum antiseptic, has been reported to be effective in patients who presented with classic EKC and were treated with either 10% solution by swab or 5% solution as drops. Marked clinical improvement was observed in >90% of patients within 48 hours (59).

More recently, AdV-infected eyes were topically treated for seven days with 1% N-chlorotaurine eye drops in a prospective double-blind, randomized study. The agent was well tolerated, shortened duration of illness, and might be a useful therapeutic approach in severe epidemic keratoconjunctivitis (60).

The use of topical anti-inflammatory agents has been a matter of debate. It is mainly agreed upon that in the early course of the infection (<2 weeks) corticocorticosteroids are not indicated. Topical corticocorticosteroids can suppress the symptoms and signs in adenovirus keratoconjunctivitis but their use may be followed by a rebound keratitis and they must therefore be used judiciously. Resolution of the infection occurs in most patients within three weeks. Instead, corticosteroids may negatively affect the naturally occurring host response, total antigen load, and persistence of virus. In patients with reduced visual acuity, phototherapeutic keratectomy was effective (61).

While adenovirus infection occurs in the young contact lens wearer, the differential diagnosis for both punctate epitheliopathy and early nummular stromal infiltration is *Acanthamoeba* keratitis, particularly when the infiltration occurs on or before day 8. Thygeson keratitis is another important differential diagnosis, with unknown etiology. Prevention and control of infection in hospitals is considered in Appendix C while laboratory diagnosis is given in Chapters 2 and 3. Spread of infection is by person-to-person, eye contact, and occasionally by the fecal-oral route.

Prevention

Prophylaxis is particularly important in ophthalmic practices, since its role as a vector for spreading the virus is well documented and substantial expenses have been calculated as a result of outbreaks of a preventable disease (62).

Hygienic handrub is an important measure to control adenoviral infection. Unfortunately, most alcoholic or povidone iodine–based solutions that are commonly used in ophthalmic practices are not able to eradicate the virus (63). In one study, only 50% of hands that were contaminated with the virus after tear-hand contact were "clean" following correct hand washing (64). Therefore, it is recommended to use gloves when patients are examined suspected to be infected with adenovirus (65). It also appears that some serotypes of adenoviruses (5,8,15) persist much longer on ocular instruments than previously anticipated. Some isolates survived for four to five weeks following the initial outbreak.

Routine disinfection of ocular instruments, especially tonometers and contact lenses, is essential for safe clinical practice. The effectiveness of different disinfectants has been evaluated. Chlorine-containing agents are best and are able to eliminate adenovirus from a metallic surface within one minute. However, a concentration of 1000 ppm free chlorine is necessary to be effective, whereas most commercial preparations contain only between 25 to 500 ppm chlorine. This is given in more detail in Appendix C.

Other Virus Infections

Measles

Most cases of measles are associated with conjunctivitis. In the West, it causes a mild epithelial keratitis that resolves without sequelae. Keratitis, which occurs in 67% to 100% of cases, is a major cause of blindness in developing nations, presenting as stromal disease with ulceration, when secondary infection from traditional eye medicines and vitamin A deficiency are compounding factors. Although there is no specific antiviral drug available, topical antibiotics and systemic vitamin A supplements will help prevent secondary bacterial disease. Measles immunization has an important role to play in developing countries for preventing blindness.

Mumps

Mumps commonly causes dacryoadenitis and conjunctivitis, but keratitis is rare. It may, however, cause superficial punctate epitheliopathy, nummular anterior stromal keratitis, and unilateral central interstitial keratitis. This is a distinct syndrome, presenting as a profound drop in visual acuity 1 week after the onset of parotitis;

clearing takes place from the periphery within one month, leaving minimal scars and no vascularization. Successful mumps immunization has effectively reduced the incidence of this infection and its ocular complications.

Epstein-Barr Virus

Epstein-Barr virus causes infectious mononucleosis, a common disease among teenagers (the "kissing" disease), when follicular conjunctivitis commonly occurs and can be the presenting sign. It has now been implicated as well as a cause of both epithelial and stromal keratitis (66). It presents with pleomorphic, multifocal, anterior stromal infiltrates with ring opacities in young adults. Treatment has involved topical acyclovir and trifluridine, which have some antiviral activity against Epstein-Barr virus, but not as effectively as for H. simplex, while epithelial keratitis has resolved with topical corticocorticosteroids, suggesting the disease has an underlying immunopathological component. Recent study has found that normal cadaveric ocular tissues, except the optic nerve, gave detection of Epstein-Barr virus DNA genome using PCR and 32P-labelled oligonucleotide probes, although the significance of these findings is not clear (66).

AHC

Acute hemorrhagic conjunctivitis is caused by enterovirus 70 and coxsackie A24v virus. It is self-resolving over several weeks and there are no complications with keratitis.

Vaccinia

Vaccinia virus has been reported as a cause of keratouveitis masquerading as H. simplex, HZV, or *Acanthamoeba* infection. This was further complicated by the infection arising in a contact lens wearer. The clue was recent vaccination against smallpox in a military person.

Prions

Prions are known to exist in the cornea, causing a "silent" infection in patients who later progress to Creutzfeldt-Jacob Disease (CJD) (67). The first recognized case of iatrogenic CJD occurred in 1974 in a 55-year-old woman who had been given a corneal transplant 18 months earlier from a donor later discovered to have died from the disease; the speed of transmission from receiving infected donor tissue to developing clinical disease is notable. Homogenates from the brain of the deceased recipient were shown to be able to transmit spongiform encephalopathy to chimpanzees (68). There is no reliable test available for corneal or systemic infection by prions for potential or cadaveric donors. All known cases are listed in Table 7 with other virus infections following corneal transplantation.

Cataract as a Complication of Severe Microbial Keratitis

Cataract formation can result from severe microbial keratitis alone but is probably enhanced by associated treatment with high doses of topical corticocorticosteroids. Conditions considered to give rise to complicated cataract, which reflects lens changes occurring during the course of ocular disease, include keratitis, scleritis, chronic uveitis, and various other inflammatory disorders. Possible causes include production of toxins by bacteria, especially the proteases and exotoxin A of *P. aeruginosa*, iridocyclitis, and corticocorticosteroid treatment. Each has a

Table 7 Documented Cases of Iatrogenic CJD and Other Virus Infections Following Corneal Transplantation

Year	Author	Country	Outcome	References
1974	Duffy et al.	USA	Creutzfeldt-Jacob disease	(68)
1978	Anonymous	USA	Rabies	(69)
1980	MMWR	France	Rabies	(70)
1981	MMWR	Thailand	Rabies	(71)
1988	Hoft et al.	USA	Hepatitis B	(72)
1996	Javadi et al.	Iran	Rabies	(73)
1997	Hoft et al.	USA	Hepatitis B	(74)
1997	Heckman et al.	Germany	Creutzfeldt-Jacob disease	(75)
2000	Thiel et al.	Germany	Creutzfeldt-Jacob disease	(76)

cataractogenic effect that may be potentiated when occurring together. In addition, there can occasionally be a toxic effect of drugs, such as chlorhexidine for treatment of *Acanthamoeba* keratitis, on the lens epithelium that may disturb lens permeability and homeostasis. Such toxicity is likely to be related to excessive dosage. Fortunately this is rare and not yet recognized for antibiotics.

REFERENCES

1. Tabbara KF, El-Asrar AM, Al-Omar O, et al. Single dose azithromycin in the treatment of trachoma. A randomized controlled study. Ophthalmol 1996; 103:842–46.
2. Mabey DC, Solomon AW, Foster A. Trachoma. Lancet 2003; 362:223–229.
3. Whitty CJ, Glasgow KW, Sadiq ST, et al. Impact of community-based mass treatment for trachoma with oral azithromycin on general morbidity in Gambian children. Pediatr Infect Dis J 1999; 18:955–58.
4. Bowman RJ, Sillah A, Van Dehn C, et al. Operational comparison of single-dose azithromycin and topical tetracycline for trachoma. Investig Ophthalmol Vis Sci 2000; 41:4074–79.
5. Burton MJ, Holland MJ, Faal N, et al. Which members of a community need antibiotics to control trachoma? Conjunctival *Chlamydia trachomatis* infection load in Gambian villages. Investig Ophthalmol Vis Sci 2003; 44:4215–22.
6. Melese M, Chidambaram JD, Alemayehu W, et al. Feasability of eliminating ocular *Chlamydia trachomatis* with repeated mass antibiotic treatments. JAMA 2004; 292: 721–25.
7. Solomon AW, Holland MJ, Alexander ND, et al. Mass treatment with single-dose azithromycin for trachoma. N Eng J Med 2004; 351:1962–71.
8. Francis V (editor). Community Eye Health J 2004; 17:49–68.
9. Parmar P, Salman A, Kalavathy CM, et al. Microbial keratitis at extremes of age. Cornea 2006; 25:153–8.
10. Houang E, Lam D, Seal D, et al. Microbial keratitis in Hong Kong: relationship to climate, environment and contact-lens disinfection. Trans R Soc Trop Med Hyg 2001; 95:361–7.
11. Thomas PA. Current perspective on ophthalmic mycoses. Clin Microbiol Rev 2003; 16: 730–97.
12. Mah-Sadorra JH, Yavuz SG, Najjar DM, et al. Trends in contact lens-related corneal ulcers. Cornea 2005; 24:51–58.
13. Pleyer U, Baykal HE, Rohrbach JM, et al. Cogan-I-Syndrom:Zu oft zu spät erkannt? Ein Beitrag zur Frühdiagnose des Cogan-I-Syndroms. Klin Monatsbl Augenheilk 1995; 207: 3–10.

14. Schaefer F, Bruttin O, Zogafros L, et al. Bacterial keratitis: a prospective clinical and microbiological study. Brit J Ophthalmol 2001; 85:842–47.
15. Erie JC, Nevitt MP, Hodge DO, et al. Incidence of ulcerative keratitis in a defined population from 1950 to 1988. Arch Ophthalmol 1995; 111:1665–71.
16. Cheng KH, Leung SL, Hoekman HW, et al. Incidence of contact-lens-associated microbial keratitis and its related morbidity. Lancet 1999; 354:181–85.
17. Bourcier T, Thomas F, Borderie V, et al. Bacterial keratitis:predisposing factors, clinical and microbiological review of 300 cases. Br J Ophthalmol. 2003; 87:834–38.
18. Lam D, Houang E, Seal D, et al. Incidence and Risk Factors for Microbial Keratitis in Hong Kong: comparison with Europe and North America. Eye 2002; 16:608–618.
19. Sweeney DF (ed). Silicone Hydrogels—Continuous-Wear Contact Lenses. Butterworth Heinemann, London 2004 (ISBN 0-7506-8779-7). 1–318.
20. Beattie TK, Tomlinson A, McFayden AK. Attachment of Acanthamoeba to First and Second-Generation Silicone Hydrogel Contact Lenses. Ophthalmology 2006; 113:117–125.
21. Watt K, Swarbrick HA. Microbial keratitis in orthokeratology: review of the first 50 cases. Eye Contact Lens 2005; 31:201–8.
22. Chang MA, Jain S, Azar DT. Infections following laser in situ keratomileusis: an integration of the published literature. Surv Ophthalmol 2004; 49:269–80.
23. Bennett HGB, Hay J, Seal DV, et al. Antimicrobial management of presumed microbial keratitis: guidelines for treatment of central and peripheral ulcers. Brit J Ophthalmol 1998; 82:137–45.
24. Pfister DR, Cameron JD, Krachmer JH, Holland EJ. Confocal microscopy findings of Acanthamoeba keratitis. Am J Ophthalmol 1996; 121:119–128.
25. Hyndiuk RA, Eiferman RA, Caldwell DR, et al. Comparison of ciprofloxacin ophthalmic solution 0.3% to fortified tobramycin-cefazolin in treating bacterial corneal ulcers. Ophthalmol 1996; 103:1854–63.
26. O'Brien TP, Maguire MG, Fink NE, et al. Efficacy of ofloxacin vs cefazolin and tobramycin in the therapy for bacterial keratitis. Arch Ophthalmol 1995; 113:1257–65.
27. Healy DP, Holland EJ, Nordlund ML, et al. Concentrations of levofloxacin, ofloxacin, and ciprofloxacin in human corneal stromal tissue and aqueous humour after topical administration. Cornea 2004; 23:255–63.
28. Thibodeaux BA, Dajcs JJ, Caballero AR, et al. Quantitative comparison of fluoroquinolone therapies of experimental Gram-negative bacterial keratitis. Curr Eye Res 2004; 28:37–42.
29. Sarayba MA, Shamie N, Reiser BJ, et al. Fluoroquinolone therapy in Mycobacterium chelonae keratitis after lamellar keratectomy. J Cataract Refract Surg 2005; 31:1396–402.
30. Klont RR, Eggink CA, Rijs AJ, et al. Successful treatment of Fusarium keratitis with cornea transplantation and topical and systemic voriconazole. Clin Infect Disease 2005; 40:e110–12.
31. Sponsel W, Chen N, Dang D, et al. Topical voriconazole as a novel treatment for fungal keratitis. Antimicrob Agents Chemother 2006; 50:262–8.
32. Bunya VY, Hammersmith KM, Rapuano CJ, et al. Topical and oral Voriconazole in the treatment of fungal keratitis. Am. J Ophthalmol 2007; 143:151–53.
33. Seal DV, Kirkness CM, Bennett HGB, et al. Population-based cohort study of microbial keratitis in Scotland: incidence and features. Contact Lens & Ant Eye 1999; 22:49–57.
34. Joslin, CE, Tu EY, McMahon TT, et al. Descriptive epidemiology of the Chicago-Area Acanthamoeba keratitis. ARVO 2006 abstract: 5563.
35. Sharma S, Garg P, Rao GN. Patient characteristics, diagnosis and treatment of non-contact lens related Acanthamoeba keratitis. Br J Ophthalmol 2000; 84:1103–1108.
36. Pyott A, Hay J, Seal DV. Acanthamoeba keratitis: first recorded case from a Palestinian patient with trachoma. Brit J Ophthalmol 1996; 80:849.
37. Jin X-Y, Lo S-Y, Zhang W-H. Diagnosis, treatment and prevention of Acanthamoeba. Ophthalmol China 1992; 1:67–71.

38. Ledee DR, Byers TJ, Seal DV, et al. *Acanthamoeba griffini*: molecular characterisation of a new corneal pathogen. Investig Ophthalmol Vis Sci 1996; 37:544–550.

39. Kilvington S, Gray J, Dart N, et al. *Acanthamoeba* keratitis: the role of domestic tap water contamination in the United Kingdom. Investig Ophthalmol Vis Sci 2004; 45: 165–69.

40. Shoff ME, Rogerson A, Seal DV, et al. Prevalence of *Acanthamoeba* and other naked amoebae in South Florida Domestic Water. J Water Health 2007; in press.

41. Kennedy SM, Devine P, Hurley C, et al. Corneal infection associated with *Hartmanella vermiformis* in a contact lens wearer. Lancet 1995; 346:637–38.

42. Aitken D, Hay J, Seal DV, et al. Amebic keratitis in a wearer of disposable contact lenses due to a mixed *Vahlkampfia* and *Hartmanella* infection. Ophthalmol 1996; 103:485–94.

43. Cassella JP, Hay J, Seal DV. Rational drug targeting in *Acanthamoeba* keratitis: implications of host-cell protozoan interaction. Eye 1997; 11:751–54.

44. van Klink F, Taylor WM, Alizadeh H, et al. The role of macrophages in Acanthamoeba keratitis. Investig Ophthalmol Vis Sci 1996; 37:1271–81.

45. Seal DV, Hay J, Kirkness CM, et al. Successful medical therapy of *Acanthamoeba* keratitis with topical chlorhexidine and propamidine. Eye 1996; 10:413–21.

46. Beattie TK, Tomlinson A, Seal DV, et al. Enhanced attachment of *Acanthamoeba* to extended-wear silicone hydrogel contact lenses: a new risk factor for infection? Ophthalmology 2003; 110:765–771.

47. Tomlinson A, Simmons P, Seal DV, et al. Salicylate inhibition of *Acanthamoeba* attachment to contact lenses: a way to reduce the risk of infection. Ophthalmology 2000; 107:112–117.

48. Simmons P, Tomlinson A, Seal DV, et al. Effect of patient wear and extent of protein deposition on adsorption of *Acanthamoeba* to five types of hydrogel contact lenses. Optom and Vis Sci 1996; 73:362–68.

49. Simmons PA, Tomlinson A, Seal DV. The role of *Pseudomonas aeruginosa* biofilm in the attachment of *Acanthamoeba* to four types of hydrogel contact lens materials. Optom Vis Sci 1998; 75:860–66.

50. Hoh HB, Hurley C, Claoue C, et al. Randomised trial of ganciclovir and acyclovir in the treatment of herpes simplex dendritic keratitis: a multicentre study. Br J Ophthalmol 1996; 80:140–3.

51. Wilhelmus KR, Gee L, Hauck WW, et al. Herpetic eye disease study—a controlled trial of topical corticosteroids for H. simplex stromal keratitis. Ophthalmol 1994; 101: 1883–96.

52. Barron BA, Gee L, Hauck WW et al. Herpetic eye disease study—a controlled trial of oral acyclovir for H. simplex stromal keratitis. Ophthalmol 1994; 101:1871–82.

53. Remeijer L, Maertzdorf J, Doornenbal P, et al. Herpes simplex virus 1 transmission through corneal transplantation. Lancet 2001; 357:442.

54. Sengler U, Reinhard T, Adams O, et al. Herpes simplex virus infection in the media of donor corneas during organ culture:frequency and consequences. Eye 2001; 15: 644–47.

55. Robert PY, Adenis JP, Pleyer U. [How "safe" is corneal transplantation? A contribution on the risk of HSV-transmission due to corneal transplantation]. Klin Monatsbl Augenheilkd. 2005; 222:870–3.

56. Cockerham GC, Bijwaard K, Sheng ZM, et al. Primary graft failure: a clinicopathologic and molecular analysis. Ophthalmology 2000; 107:2083–90.

57. Romanowski EG, Roba LA, Wiley LA, et al. The effects of corticosteroids on adenoviral replication. Arch Ophthalmol 1996; 114:481–85.

58. Hillenkamp J, Reinhard T, Ross RS, et al. The effects of cidofovir 1% with and without cyclosporin as a 1% topical treatment of acute adenoviral keratoconjunctivitis: a controlled clinical pilot study. Ophthalmol 2002; 109:845–50.

59. Abel R Jr, Abel AD. Use of povidone-iodine in the treatment of presumptive adenoviral conjunctivitis. Ann Ophthalmol Glaucom 1998; 30:341–343.

60. Teuchner B, Nagl M, Schidlbauer A, et al. Tolerability and efficacy of N-chlorotaurine in epidemic keratoconjunctivitis—a double-blind, randomized, phase-2 clinical trial. J Ocul Pharmacol Ther 2005; 21:157–65.

61. Quentin CD, Tondrow M, Vogel M. Phototherapeutic keratectomy (PTK) after epidemic keratoconjunctivitis. Ophthalmology 1999; 96:92–6.

62. Piednoir E, Bureau-Chalot F, Merle C, et al. Direct costs associated with a nosocomial outbreak of adenoviral conjunctivitis infection in a long-term care institution. Am J Infect Control 2002; 30:407–10.

63. Eggers HJ. Experiments on antiviral activity of hand disinfectants: some theoretical and practical considerations. Int J Med Microbiol 1990; 273:36–51.

64. Jernigan JA, Lowry BS, Hayden FG, et al. Adenovirus type 8 epidemic keratoconjunctivitis in an eye clinic: risk factors and control. J Infect Dis 1993; 167:1307–1313.

65. Gordon R. Prolonged recovery of desiccated adenovirus 5, 8, and 19 in vitro may partially explain clinical epidemics. Ophthalmol 1992; 99:83S.

66. Chodosh J, Gan Y-J, Sixbey JW. Detection of Epstein-Barr virus genome in ocular tissues. Ophthalmol 1996; 103:687–90.

67. Collinge J, Palmer MS. Human prion diseases. In: Collinge J, Palmer MS (eds) Prion Disease. Oxford University Press, Oxford 1997;18–56.

68. Duffy P, Wolf J, Collins G. Possible person-to-person transmission of Creutzfeldt-Jakob disease. New Eng J Med 1974; 290:692.

69. Anonymous. Patient received cornea: rabies case linked to transplant. Am Med News 1978; 21:3.

70. Anonymous. Second rabies death attributed to graft [news]. Aorn J 1980; 31:818.

71. Anonymous. Human-to-human transmission of rabies via corneal transplant—Thailand. MMWR Morb Mortal Wkly Rep 1981; 30:473–74.

72. Hoft RH, Pflugfelder SC, Ullman S, et al. Clinical evidence of hepatitis B transmission resulting from corneal trnasmission. Ophthalmol 1988; 95:162–84.

73. Javadi MA, Favaz A, Mirdehghan SA, et al. Transmission of rabies by corneal graft. Cornea 1996; 15:431–33.

74. Hoft RH, Pflugfelder SC, Forster RK et al. Clinical evidence for hepatitis B transmission resulting from corneal transmission. Cornea 1997; 16:132–37.

75. Heckmann JG, Lang CJ, Petruch F, et al. Transmission of Creutzfeldt-Jacob disease via a corneal transplant. J Neurol Neurosurg Psychiatr 1997; 63:388–90.

76. Thiel HJ, Erb C, Heckmann J, et al. Manifestation of Creutzfeldt-Jacob disease 30 years after corneal transplantation. Klin Monatsbl Augenheilkd 2000; 217:303–07.

SEVEN

Uvea and Retina

INTRODUCTION

Uveitis is a relatively common inflammatory eye disease with an annual incidence between 17 and 52.4 per 100,000 person-years and prevalence between 38 and 370 per 100,000. It is the most common form of inflammatory eye disease and an important cause of visual impairment and blindness. Given that uveitis predominantly affects the young adult population in their most productive years, the personal and socio-economic burden is significant.

The classification of intraocular inflammation is commonly described by anatomical location of the primary site of inflammation. This relatively simple classification system has proven to be beneficial, since it may provide clues on the association with an underlying etiology.

- Anterior uveitis
- Posterior uveitis
- Intermediate uveitis
- Panuveitis

ANTERIOR UVEITIS

Anterior uveitis is the most frequent form of intraocular inflammation that involves the iris and/or ciliary body (Fig. 1). Iritis presents with a gradual onset of moderate ocular pain in the 1st trigeminal nerve distribution, marked tenderness over the globe, tearing, and reduced vision of variable degree. There is conjunctival injection and a characteristic limbal flush due to congestion of perilimbal episcleral vessels. The protein content of the aqueous humor is increased by a rise in vessel permeability and detected as an aqueous flare in the beam of the slit lamp. This may give rise to a secondary open angle glaucoma. A fibrin coagulum may be present in particular in HLA-B27+ individuals. Inflammatory cells circulating in the thermal currents of the aqueous may be deposited in acute cyclitis as keratic precipitates on the corneal endothelium with widespread opacities in the vitreous. These precipitates may damage the corneal endothelium and cause corneal haze due to stromal and epithelial edema.

Precipitates have been described as granulomatous or non-granulomatous and have been considered as a different type of cellular infiltrate. Granulomatous deposits may consist of macrophage and lymphocyte aggregation, whereas lymphocytes and neutrophils are considered non-granulomatous. Pupillary miosis

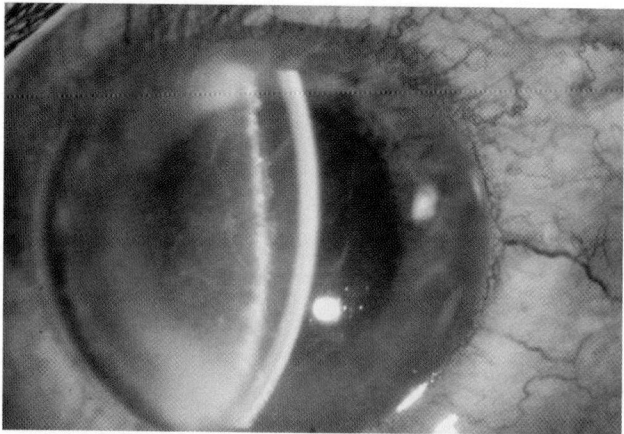

Figure 1   Acute uveitis in an HLA-positive patient due to an autoimmune etiology showing fibrin exudation and ciliary injection.

is due to a sensory reflex and the release of inflammatory mediators, but on pupil dilatation the pupil may become irregular at the site of posterior synechia. A small, irregular pupil on presentation may remain following previous inflammation.

Etiology

The cause of acute uveitis is often not established but patients may be investigated for both infectious and autoimmune etiologies. Anterior uveitis may be part of a systemic illness such as a spondyloarthropathy. It may also arise from an infection such as herpes simplex; be part of an ocular syndrome, such as Fuchs' heterochromic iridocyclitis; be part of trauma, as in cataract surgery (see: endophthalmitis); or result from an idiopathic eye disease with a presumed immune pathogenesis.

Non-infectious Anterior Uveitis

In Europe and North America, up to 50% of patients with anterior uveitis are HLA-B27 positive. The clinical presentation of uveitis in patients associated with ankylosing spondylitis (AS) or reactive arthritis (ReA) is relatively consistent. Most patients present with acute onset of inflammation. The inflammation is active in one eye, but the attacks tend to recur, often in the contralateral eye. Intraocular pressure is often lower in the affected eye.

Infectious Anterior Uveitis

Viral Infections

Herpes Viruses.   Several studies emphasize the role of herpes viruses in the etiology of anterior (and posterior) uveitis. In fact, the number of patients affected by viral infection may be underestimated. In a prospective study of 110 patients analyzing aqueous humor samples, viral infections were present in 15% with a predominance of HSV. In another investigation, intraocular antibody synthesis was positive for HSV in 23.5% of patients with anterior uveitis (1). This high incidence might be biased by patient selection and referral pattern; however, it indicates that herpes viruses are an important cause of anterior uveitis often in the absence of keratitis.

Herpes viruses demonstrate some similar clinical pattern such as unilateral non-granulomatous or granulomatous iritis or iridocyclitis, iris atrophy, increased intra-ocular pressure (IOP), and a high tendency to recur. They have been associated with other entities such as Posner-Schlossman syndrome.

Herpes Simplex Virus (HSV). The condition presents with recurrent non-granulomatous or granulomatous iritis or iridocyclitis in the absence of keratitis. Keratic precipitates and iris atrophy are frequently present. Often, increased IOP is seen caused by trabeculitis. Analysis of aqueous humor by PCR or local antibody synthesis is helpful to differentiate the infectious cause from other diagnoses (Fig. 2).

Varicella Zoster Virus (VZV). Intraocular inflammation caused by VZV can present as zoster sine herpete without cutaneous vesicles or as reactivation in chickenpox or herpes zoster ophthalmicus.

Anterior uveitis caused by zoster sine herpete is infrequently reported and affects otherwise healthy, immunocompetent individuals. Clinical hints for the diagnosis include unilateral non-granulomatous or granulomatous iritis or iridocyclitis, increase of intraocular pressure, iridoplegia, and sectorial iris atrophy. Hypopyon and hyphema may occur (Fig. 3). Histopathology demonstrates vasculitis, necrosis, and lymphocyte infiltration in both iris and ciliary body. Intraocular inflammation may become recurrent and result in posterior synechia, cataract formation, and secondary glaucoma.

More frequently, uveitis can be observed in patients affected by herpes zoster ophthalmicus. It has been reported in approximately 50% of immunocompetent and immunocompromised patients.

The clinical presentation is similar to VZV uveitis sine herpete but is accompanied by cutaneous vesicles. Herpetic keratouveitis either stromal or disciform should be differentiated since it requires additional corticosteroids, whereas isolated anterior uveitis responds to (systemic) antiviral treatment.

Often the diagnosis can be based on clinical grounds, but, in particular, in patients suspicious for "VZV uveitis sine herpete," aqueous or vitreous viral PCR (Chapter 3) or antibody analysis is helpful.

Cytomegalovirus (CMV). Whereas CMV is typically associated with retinitis, recently it also has been shown to be associated with anterior uveitis. Clinical presentation and symptoms are similar to those associated with other viruses of the herpes family. In some patients, multiple, fine stellate-shaped endothelial precipitates

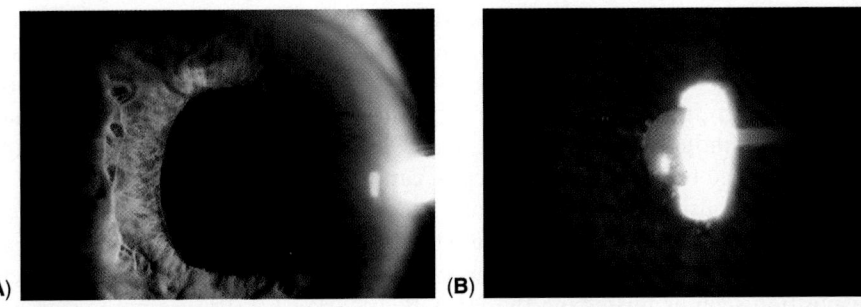

(A)   (B)

Figure 2   (A) Herpetic anterior uveitis with iris atrophy. (B) On transillumination.

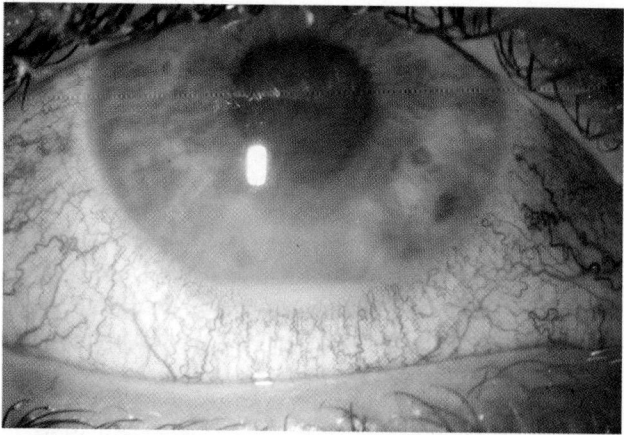

Figure 3   Uveitis due to varicella-zoster virus with hypopyon and massive iris hyperemia.

have been described as characteristic. A spillover from the posterior chamber in CMV retinitis is important to rule out. As in the previous infections, aqueous humor analysis for CMV is helpful to establish the diagnosis.

Human T-Cell Lymphotrophic Virus-1 (HTLV-1).   HTLV infection is related to adult T-cell leukemia and has been reported as an infrequent cause of anterior uveitis. Transmission of the virus may occur via blood transfusion, breast feeding, and sexual contact. Most reports are from Southwest Japan, parts of Europe (Spain), and Central Africa, probably indicating an endemic occurrence. It clinically presents as bilateral mild granulomatous/non-granulomatous anterior uveitis with keratic precipitates and synechia. Often, moderate vitreous inflammation with vascular sheathing and optic disc edema occurs (see below). The diagnosis is based on serological tests for HTLV or PCR analysis of intraocular fluids.

Bacterial Infection

Isolated cases of anterior uveitis due to bacteria are rare. Spontaneous, acute anterior uveitis is usually associated with systemic infection such as syphilis, borreliosis, tuberculosis, cat scratch disease, Whipple's disease, or others.

In these patients it is often not clear whether the infectious cause is directly related to intraocular inflammation or due to an associated immune response. The frequent association of anterior uveitis with gastrointestinal infection with Gram-negative bacteria, such as those of salmonella, and shigella from a recent infection such as diarrhea, in particular in HLA-B27 positive individuals, has raised the question whether it is directly linked or indirectly associated with a low grade infection (2). Although intraocular antibodies have been detected in some patients, no direct evidence for intraocular infection can be provided in most cases (3).

Uveitis can also be due to an immune reaction to circulating microbial antigens such as yersinia, or chlamydia, associated with urethritis. Occasionally it presents following a sore throat caused by *Streptococcus pyogenes* (4) in a patient of genetic predisposition such as possessing the HLA-B27 marker (2). Precise pathogenesis is not clear but may involve molecular mimicry by antigens cross-reacting with ocular tissue. This mechanism can also give rise to arthropathies and may involve deposition of circulating cell wall antigens in tissue. Treatment in the latter case involves anti-inflammatory drugs rather than anti-infectives since the intact microbe is usually eradicated from the patient at the time of presentation. The history of a previous infectious episode should be pursued, however, and a retrospective diagnosis may be possible if there is a suitable serological test for the particular antigen, such as *S. pyogenes* hyaluronic acid or DNAse B.

Spirochetal Infection. Anterior (and more frequently) posterior uveitis (see below) can be caused by spirochetal infection including Treponema and Borrelia species. Severe uveitis can be due to syphilis (5,6) and a serology test for *Treponema pallidum* should always be performed; if positive, systemic treatment is given with penicillin for two weeks. Uveitis also develops in Lyme disease due to the spirochete *Borrelia burgdorferi* (7).

Mycobacterial Infection. Uveitis can also be associated with mycobacterial infection. Active tuberculosis may cause a tuberculous iritis, due to live bacilli within a granulomatous reaction. In the early stages, there are minute, gray and translucent iris nodules and at a later stage a yellowish nodule surrounded by numerous satellite lesions. The presence of keratic precipitates indicates involvement of the ciliary body. Pseudohypopyon composed of caseating material may occur.

If the iritis is exudative rather than granular, then it may be due to an inflammatory response to circulating antigen similar to the phlyctenous response (see Chapter 6).

Initial investigation for pulmonary disease requires microscopy and culture of three consecutive sputum samples for *Mycobacterium tuberculosis*. If these are negative, gastric washings can be particularly useful to identify the mycobacterium. Uveitis should be suspected in all tuberculous patients presenting with ocular pain, decreased vision, and tearing. Systemic treatment for active tuberculosis must be given for a minimum of 6 months, and topical mydriatics and corticosteroids used with care.

Anterior uveitis also accompanies leprosy. Lepromatous leprosy causes more severe disease than the tuberculoid variety; systemic treatment is required. Rare associations with anterior uveitis have been described due to *Bartonella henselae* (cat scratch disease; Fig. 4) and *Tropheryma whippelii* (Whipple's disease). Patients that do not demonstrate an underlying systemic infection should be investigated for an immune defect (e.g., HIV infection).

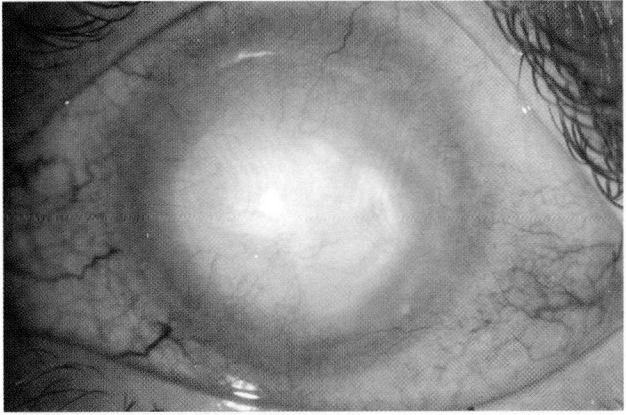

Figure 4 Interstitial keratitis due to *Bartonella henselae* (cat scratch disease).

Parasitic Infection. *Onchocerciasis.* In onchocerciasis, inflammation of part or all of the uvea occurs when microfilaria penetrate the sclera, and can result in blindness from secondary glaucoma, optic atrophy, or phthisis bulbi. For details, see below (posterior uveitis); treatment is discussed in Appendix C.

Treatment

Anti-inflammatory treatment for anterior uveitis includes the use of topical corticosteroids and induction of mydriasis with topical atropine (1%). Atropine paralyzes sphincter pupillae and the ciliary muscle; one drop causes wide dilatation of the pupil in 30 minutes and lasts for three to seven days.

POSTERIOR UVEITIS

Retina and Choroid

Handling of patients with acute posterior uveitis remains a challenge for the ophthalmologist in many respects. Vision might be acutely endangered, there are a great variety of etiologies to be respected, and therapy has to be based mainly on systemic application—with all the risk that might be associated with it. There is considerable overlap in the clinical presentation of posterior uveitis even when the underlying etiology is completely different. In particular, the distinction between an infectious cause and a presumed autoimmune response is important, since it will determine subsequent treatment (Table 1). However, even an experienced ophthalmologist may be confused by certain manifestations such as syphilis ("The Great Mimicker") (Figs. 5 and 6) or intraocular lymphoma, as a masquerade syndrome (8).

Because the clinical reaction to an inflammatory stimulus is limited in posterior uveitis, there is a common pattern that follows the general principles of inflammation:

- Cellular infiltration of the vitreous
- Choroidal

Table 1 Differential Patterns of Posterior Uveitis

Anatomical site	Infectious etiology	Non-infectious etiology
Pars plana	Toxocara	Pars planitis (+/− multiple sclerosis), sarcoidosis
Vitreous infiltration	Toxoplasmosis	Sarcoidosis
	Endophthalmitis	Multifocal choroiditis
	HTLV-1 associated uveitis	Ocular lymphoma
Choroiditis/RPE	Ocular histoplasmosis	"White dot syndromes"
	DUSN	e.g., APMPPE, EMWDS
	Fungal infection	Multifocal choroiditis
	Mycobacterial infection	POHS
Retina	Toxoplasmosis	Behcet's disease
	ARN (acute retina necrosis)	Sarcoidosis
	CMV retinitis	
	Progressive outer retinal necrosis	
Vasculitis (retinal)	Syphilis, borreliosis	Behcet's disease
	Acute retinal necrosis	Sarcoidosis
	CMV	

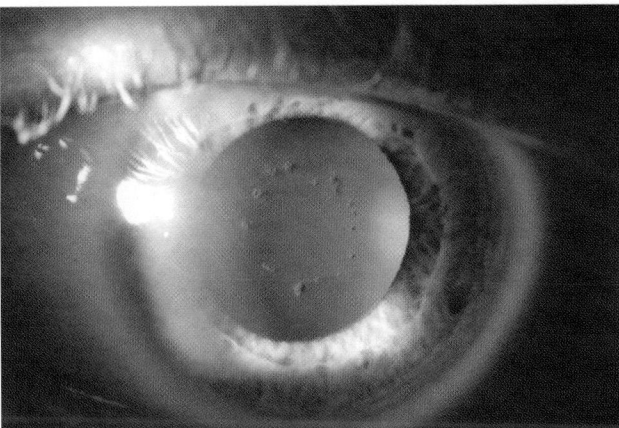

Figure 5 Iritis in secondary syphilis.

- Retinochoroidal inflammation (infiltration)
- Vasculitis
- Exudative changes (either as circumscribed edema—such as macula edema, or papillary edema—or extensive, such as complete exudative retinal detachment in VKH)

Symptoms in patients with posterior uveitis are relatively uniform but may depend on the morphological manifestations, including

- Floaters
- Blurring of vision
- Impaired vision
- Metamorphopsia

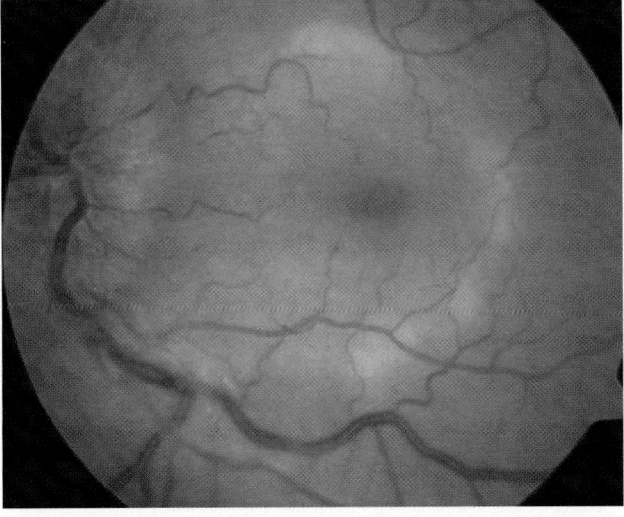

Figure 6 Retinitis in secondary syphilis.

Some symptoms may not be spontaneously reported, such as micropsia, macropsia, or nyktophthalmia, and should be explored by the physician. In some patients with unilateral uveitis, changes of vision may not be noticed unless the second eye becomes involved. Based on these findings, some differential diagnoses can guide further evaluation.

Anatomy and Vascular Supply

The retina, which lines the choroid, has two layers of epithelium. The anterior epithelium of the iris is continuous with the outer epithelial layer of the ciliary body, and this is continued into the pigment epithelium of the retina, a single layer of hexagonal cells lying immediately adjacent to the membrane of Bruch next to the choroid. The posterior epithelium of the iris, which is pigmented, passes into the inner non-pigmented layer of the ciliary body, which changes at the ora serrata into the neurosensory retina. This consists of three strata of cells: (*i*) the visual cells lying externally (rods and cones); (*ii*) a relay of bipolar cells (the outer plexiform layer) and whose synapses form the inner and outer plexiform layers; and (*iii*) a layer of ganglion cells lying internally with, innermost, the nerve fiber layer composed of ganglion cell axons running centrally into the optic nerve. These neural cells are bound together by neuroglia with a supportive and nutritive function.

The arteries of the eye are derived from the ophthalmic artery, which is a branch of the internal carotid. Most of the venous return passes to the cavernous sinus although there are lesser anastomoses within the orbit. The retina is supplied by the central artery, which divides on the surface of the disc into the main retinal trunks. These are end arteries and have no anastomoses at the ora serrata. The uveal tract is supplied by the ciliary arteries.

Posterior uveitis refers to inflammation of the choroid, which extends to the adjacent retina, causing a retinochoroiditis. It may be due to a bacteremia, fungemia, or larva migrans and evolve into endophthalmitis (Chapter 8). More commonly, there is a non-purulent chronic granulomatous reaction. The commonest causes of retinochoroiditis in Europe are due to *Toxoplasma gondii* infection, and larva migrans from *Toxocara canis* infection.

Parasitic Infections

Toxoplasma Retinochoroiditis

The intracellular protozoan parasite, *Toxoplasma gondii*, found in cats, is the commonest cause of posterior uveitis, accounting for approximately 10% of all posterior uveitis cases (9). The infection can be congenital or acquired and is widespread in many parts of the world, with seroconversion in up to 70% (10). Following infection of the fetal retina during intrauterine life, it often presents as delayed retinochoroiditis in the second and third decade of adult life. However, approximately 1% of the population per year acquires toxoplasmosis as a primary infection that may be subclinical or cause a glandular fever-type syndrome. There is now evidence from various outbreaks in children and adults that primary toxoplasmic infection can result in an acute primary choroiditis more frequently than previously considered (11,12).

Recurrent retinochoroiditis usually presents as a yellow inflammatory lesion, resulting in an overlying vitreous haze, at the margin of a pre-existing choroidal scar (Fig. 7).

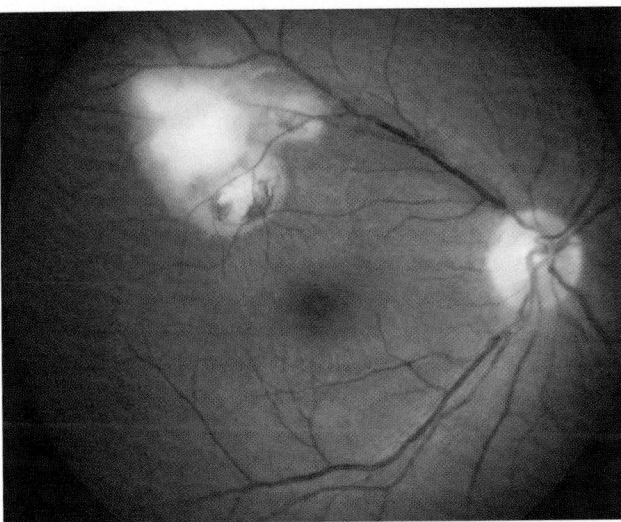

Figure 7   Recurrent retinochoroiditis due to toxoplasma.

Another presentation is a juxtapapillary lesion at the margin of the optic disc, which causes a typical arcuate field defect (13). Small peripheral retinal lesions (Fig. 8) may be allowed to run their course, but lesions near the macula, optic disc, or maculopapular nerve fiber bundle, or those associated with severe vitritis, should be treated. Therapy is directed against both the dividing organism and the inflammatory host response. The problem is complicated by multiplication of the protozoan within "tissue cysts" within cells, which are impervious to drug

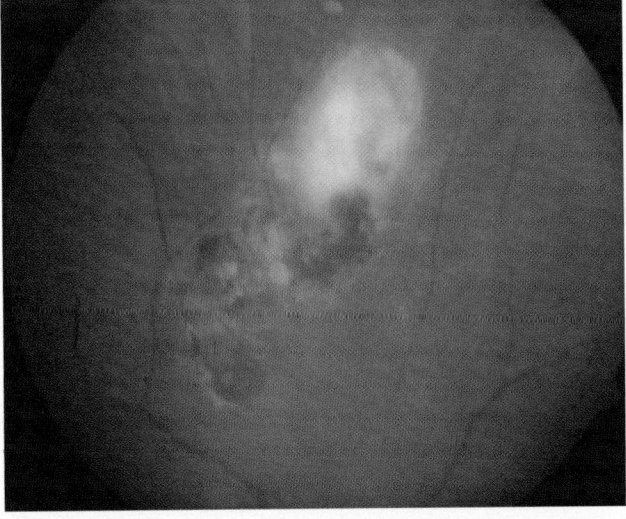

Figure 8   Small peripheral lesions of toxoplasma.

penetration, so that recurrence can always be expected (13). Toxoplasma infection is encountered in immunocompromised patients (see AIDS section below). The differential diagnosis of tuberculosis must always be considered. Atypical cases of toxoplasmosis are more frequently reported at a higher age and may mimic acute retinal necrosis (ARN).

Diagnosis

The diagnosis is clinical, based on the characteristic fundus lesion and often marked by vitreous infiltration. Intraocular detection of toxoplasma in atypical cases is much improved by PCR on a vitreous tap or retinal biopsy or by detecting specific local antibody, using a ratio between anterior chamber (aqueous) and serum levels (14). Direct evidence of *T. gondii* DNA in the intraocular fluids was found more reliable than antibody detection. These techniques have proven more reliable than traditional serology, because of the high seroconversion rate in many parts of the world.

Treatment of Toxoplasma Retinochoroiditis

Indications for treatment are as follows:

- Primary infection during pregnancy
- Congenital infection of the newborn
- Acute retinochoroiditis/reactivation with sight-threatening localization (macula, optic nerve, large peripheral lesion (>3 PD) with dense vitreous infiltration)
- Acute acquired infection by blood transfusion
- Infection in immunocompromised patients

A number of drugs have been used in the treatment of toxoplasmosis, implying that there is no "gold standard." Currently used drugs include the following:
Classic triple therapy:

- Pyrimethamine 100 mg stat then 25 mg/day p.o. for four to six weeks, and
- Sulphadiazine 2 g then 1 g four times/day p.o. for four to six weeks, and
- Folinic acid 3 mg p.o. or i.m. twice a week.

An alternative regimen is

- Clindamycin 300 mg four times/day p.o. for four to six weeks, and
- Sulphadiazine 2 g then 1 g four times/day p.o. for four to six weeks.

Pyrimethamine and sulphadiazine act synergistically to interfere with folic acid synthesis. They should be commenced early in the course of the disease and continued for four to six weeks. Pyrimethamine therapy should be avoided in early pregnancy and monitored closely due to the risk of bone marrow depression. Folinic acid supplements reduce this risk but platelet and white cell counts should be performed weekly.

Clindamycin has also been shown effective in the treatment of ocular toxoplasmosis and has good ocular tissue penetration (see Chapter 4). It does, however, carry the risk of pseudomembranous colitis, although this is small when used on an outpatient basis.

Tetracycline and minocycline may also be effective but have not yet been fully evaluated in clinical trials.

Oral corticosteroid therapy is indicated in vision-threatening disease but should not be used without concurrent specific antiprotozoal therapy or in immunocompromised patients.

Prevention of recurrences has been reported from Brazil using long-term intermittent trimethoprim/sulfamethoxazole (Septrin) treatment (16).

Ocular Toxocara

Toxocara canis and *T. catis* are worms whose natural hosts are the dog or cat, respectively. Man is an accidental host, infected from ingesting the ova from contaminated soil, in whom the larval stage develops, causing visceral and ocular "larva migrans," but adult worms are not found. These larva migrate around the body and occasionally deposit themselves in the central nervous system, including the retina. Here, they can present as a retinochoroiditis or as a granulomatous reaction resembling a possible tumor, for which eyes have been eviscerated in the past (Figs. 9–12) (17–20).

The parasite is widespread and in some tropical countries seroconversion may be found in up to 90%, whereas in Europe the infection level is between 3% and 30%. It has been reported as a cause for posterior uveitis in 3% to 18% of children.

Clinical Presentation. Three major clinical presentations have been described.

▪ Peripheral granuloma: may occur bilaterally, presenting as "pars planitis." Secondary complications may result from traction at the posterior pole.
▪ Posterior pole granuloma: a large white granuloma, often with traction bands to the mid-periphery of the retina, usually accompanied by some cellular vitreous infiltration.

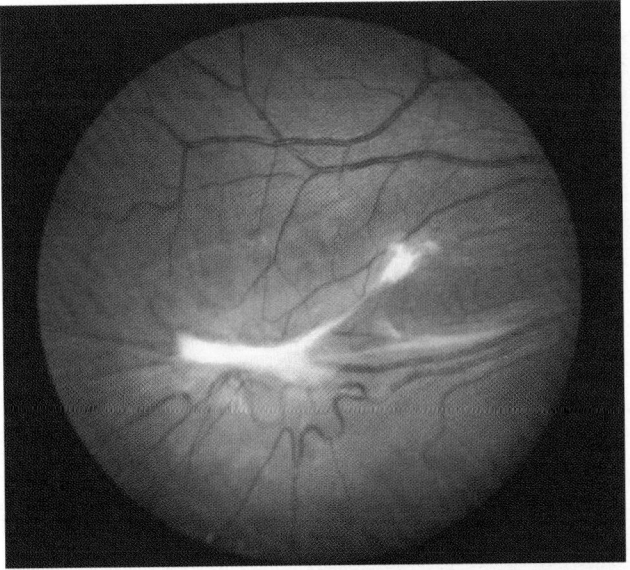

Figure 9 Fundus of a 15- year-old boy showing an epiretinal membrane and puckering due to *Toxocara* (VA 1/15).

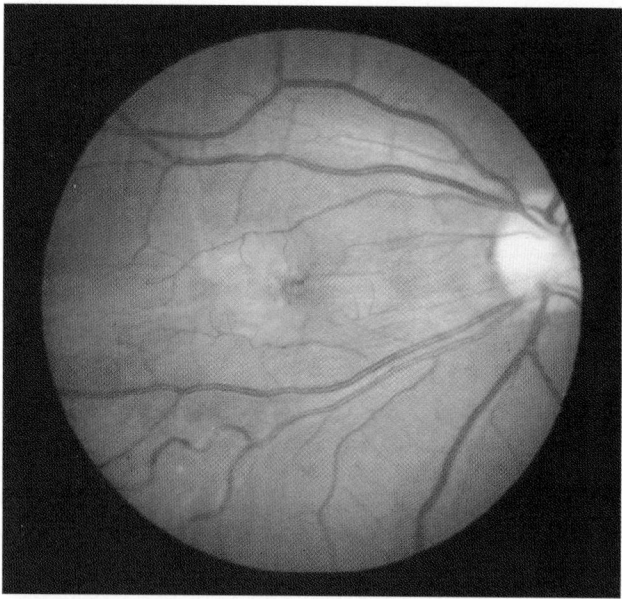

Figure 10 Subretinal membrane due to *Toxocara* (VA 1/10).

■ Endophthalmitis: mainly unilateral presentation with all signs of panuveitis, including hypopyon and severe vitreous infiltration; retinal detachment can be a presenting sign.

Diagnosis.   Serological diagnosis only confirms previous exposure and does not imply whether the retinal lesion is that of toxocariasis or not. Furthermore, the

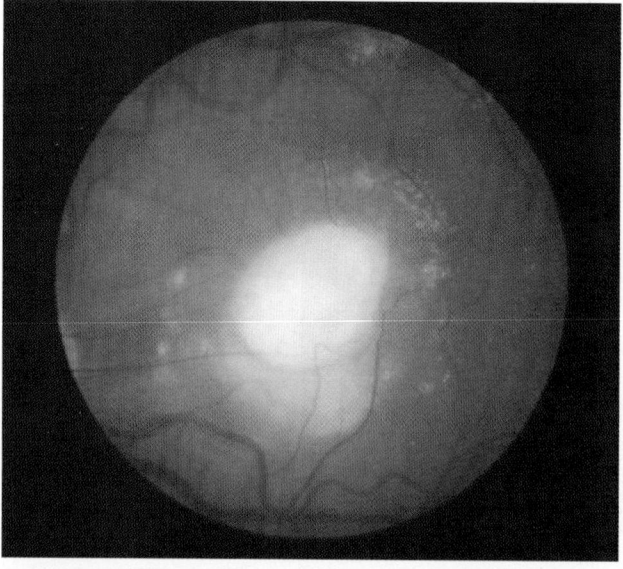

Figure 11 Subretinal granuloma in a 17-year-old due to *Toxocara* (VA 1/25).

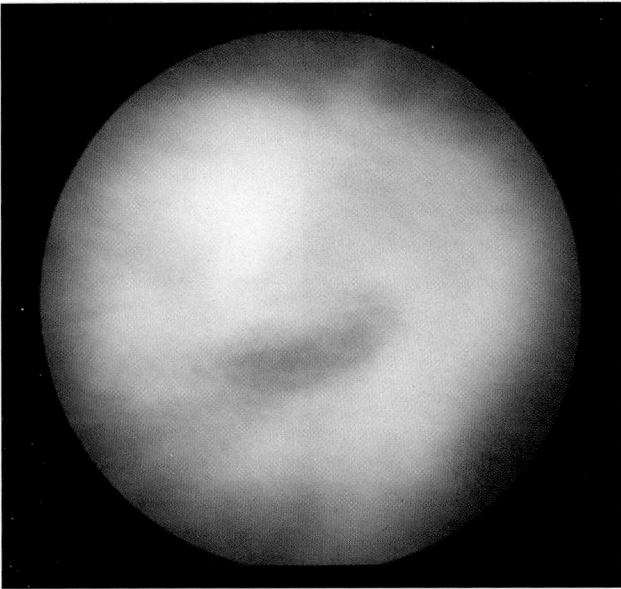

Figure 12   Total ablation of the retina in a 34-year-old due to *Toxocara* (VA NLP).

serological test may be negative when a choroidal lesion is present. Serological diagnosis is thus unreliable and should not be performed. Intraocular diagnosis is preferred. Fine needle biopsy in a reference center with cytology for tumor cells and a test system for Toxocara DNA by PCR is the best approach.

Treatment. If the retinal lesion is peripheral, treatment is conservative or symptomatic, but if it is close to the macula, treatment is warranted. Diethylcarbamazine is given orally for three weeks at 3 mg/kg. There may be symptoms of allergic reaction to the dying larva, for which prednisolone is usually given to suppress the inflammation. When blindness results it is usually unilateral, but bilateral cases are known. Therapy can also be given with albendazole or with a single dose of ivermectin.

Diffuse Unilateral Subacute Neuroretinitis (DUSN)

This relatively rare neuroretinitis is caused by subretinal nematodes. The disease is caused by migration of different worms in the subretinal space, producing the typical "tracks."

In the early stages, there is decreased visual acuity and blurred vision, often developing over the previous year. The condition presents with vasculitis, multiple foci of outer retinitis, meandering pigmentary retinal changes, papillitis, intraretinal hemorrhages, and, later, optic atrophy (Fig. 13). The most common causes are the larva of ascaridoid nematodes including *Toxocara canis* (see above) and the dog hookworm *Ancylostoma caninum*. There are a few reports of trematode infection of the eye, including *Paragonimus* sp. (the lung fluke) in endemic areas such as Thailand, and ocular schistosomiasis may occur. *Alaria* sp., a trematode worm of the small intestine of carnivorous mammals with two intermediate hosts of the snail and the frog, has been reported recently, causing neuroretinitis in two Chinese patients in

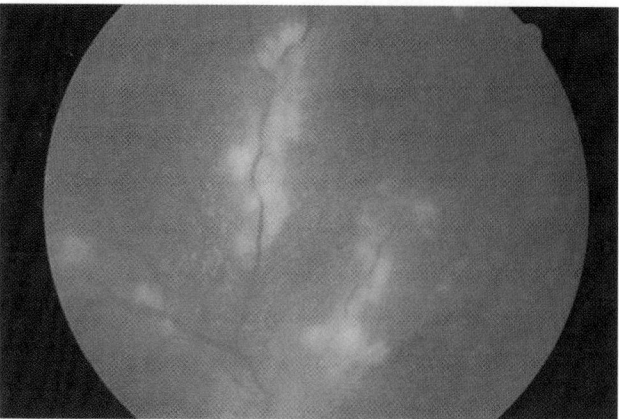

Figure 13 Retinal exudates around vessels in visceral larva migrans (presenting as diffuse unilateral subacute neuroretinitis).

San Francisco (21). The worm has distinctive migratory larva called mesocercaria, which infect the human from ingestion of the paratenic host—commonly the frog, the legs being consumed when undercooked. Drug therapy for nematodes includes thiabendazole and ivermectin and for trematodes includes praziquantel and albendazole. Corticosteroid treatment is beneficial in decreasing the associated inflammation (15). Measuring the aqueous-to-serum antibody ratio to the specific antigen can be useful diagnostically.

Ocular Onchocerciasis

Ocular onchocerciasis, or "river blindness," results from infection with the filarial parasite *Onchocerca volvulus*. The disease is endemic in areas of Africa and Central and South America where it is a major cause of blindness. The ocular manifestations include keratitis, anterior uveitis, glaucoma, chorioretinitis, and optic neuritis. The inflammatory reaction that occurs is due to a cell-mediated immune response to dead microfilaria. In the cornea they can be seen biomicroscopically at the center of ill-defined opacities (Fig. 14) representing a sequence of inflammation and scarring. This is exacerbated by diethylcarbamazine (DEC) treatment.

For several decades DEC and suramin have been used systemically in the treatment of ocular onchocerciasis. Both are effective microfilaricidal drugs, with a positive effect on keratitis and uveitis; they are, however, less beneficial in posterior segment disease. The use of DEC may be followed by a severe systemic reaction, which is largely prevented by the concomitant use of systemic corticosteroids. An appropriate therapeutic regimen has been provided by Taylor and Nutman (22). DEC was also known to precipitate or exacerbate active optic neuritis in some onchocercal patients as part of a general inflammatory reaction (the Mazzotti reaction).

Ivermectin is an important new drug that acts on the adult female worms to inhibit reproduction, so that no new microfilaria are produced for several months. It also kills microfilaria in tissue including skin and eye, slowly eliminating microfilaria from the anterior chamber but offering the advantages of minimal ocular inflammation and much less systemic reaction than DEC or suramin. Ivermectin represents an important advance in the mass therapy of onchocerciasis in endemic

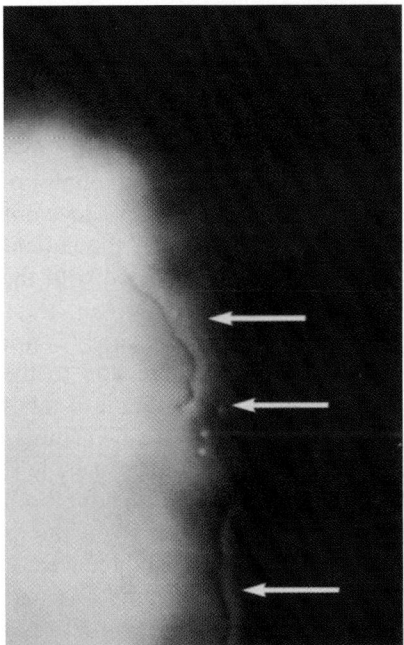

Figure 14 Microfilaria of *Onchocerca volvulus* in the cornea. An immune reaction to the organism has not yet occurred.

areas. It is given as a single dose (12 mg) and has been donated by Merck, Sharpe and Dohme. It has to be given yearly and the eradication program is a continuous one. Ivermectin should not be given to children under five years, to pregnant women, or to patients with other severe infections such as trypanosomiasis. Ivermectin does not appear to precipitate or exacerbate optic neuritis. Single dose treatment is as effective as six months, except for chorioretinal lesions (23).

Bacterial Infections

Whereas borreliosis, or Lyme disease, has probably been overstressed as a cause of posterior uveitis, syphilis has regained importance because of an increasing number of affected sexually active persons. Since late stages of the infection have a major impact on life quality and specific treatment is available, early detection is important.

Both syphilis and borreliosis are caused by spirochetes and share some similarities regarding their presentation as causes of ocular infections. Both may cause interstitial keratitis, anterior and posterior uveitis, and are often accompanied by neural involvement of the peripheral or central nervous system.

Syphilis.   Syphilis is a great mimicker of other diseases that can present with retinochoroiditis in the secondary stage, which can be associated with syphilitic meningitis. Syphilitic neuroretinitis presents as papillitis, with peripapillary flame-shaped hemorrhages, periarteriolar sheathing, and stellate macular exudates (Fig. 15). At this stage, there can be the characteristic macular-papular rash of secondary syphilis with "snail-track" mouth ulcers or the patient may be asymptomatic apart from a loss of visual acuity. Residual changes include optic atrophy. Necrotizing retinitis may occur, associated with panuveitis and retinal vasculitis. Retinitis can become more extensive and bilateral in patients with AIDS (24). Progressive changes can resemble acute retinal necrosis.

Treatment for ocular syphilis should be similar to neurosyphilis and given for two weeks with high-dose intravenous penicillin G (3 MU, four-hourly with normal renal function) or high-dose intramuscular depository penicillin G with oral probenecid. It is essential that the patient's serology be repeated to follow an expected reduction in titer of the nonspecific reagin test. If this reduction does not begin after six weeks, then the penicillin course must be repeated and the patient closely followed up in a specialist clinic. No reduction in titer is expected with the specific treponemal antigen serological tests.

Failure of the standard two-week course to treat the patient adequately is not uncommon, and if the patient absconds from follow-up, then progression to the tertiary stage of neurosyphilis can occur. Besides the classical presentation of tabes dorsalis (shooting pains in the legs), this can present as pseudoretinitis pigmentosa with a progressive decrease in visual acuity and nyctalopia. Findings include a diffuse, granular appearance of the retinal pigment epithelium throughout both fundi with choroidal atrophy. Systemic findings include Argyll-Robertson pupils (small and irregular and reactive to the near response but not to light) and reduced or absent reflexes. Patients may also present with abstruse signs of meningovascular disease, with cranial nerve palsies, or generalized vascular disease with peri- and endarteritis, while a gumma (syphilitic granuloma) can occur anywhere. Congenital syphilitic chorioretinopathy (Fig. 16) persists for life with poor vision, narrowing of retinal vessels, optic disc pallor, and peripheral pseudo-retinitis pigmentosa; iritis and interstitial neuritis may also be present.

Borreliosis (Lyme Disease). Although ocular manifestations are a rare feature of this tick-borne disease, the spirochete (*Borrelia burgdorferi*) invades the eye early and may remain dormant, accounting for both early and late ocular manifestations. A non-specific follicular conjunctivitis occurs in approximately 10% of patients with early Lyme disease, while keratitis occurs within a few months

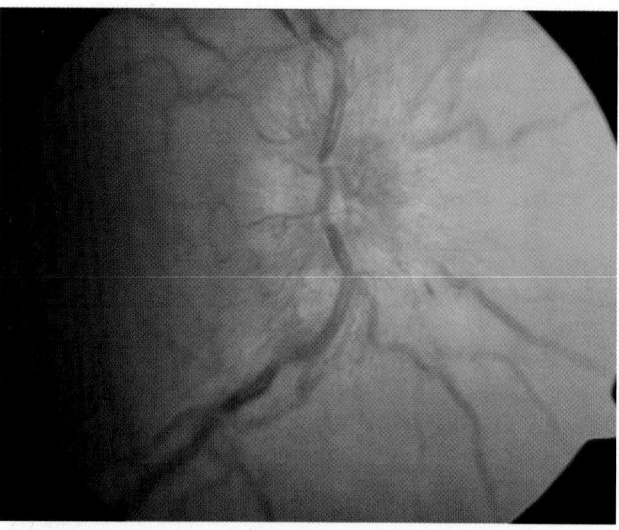

Figure 15   Papillitis in secondary syphilis.

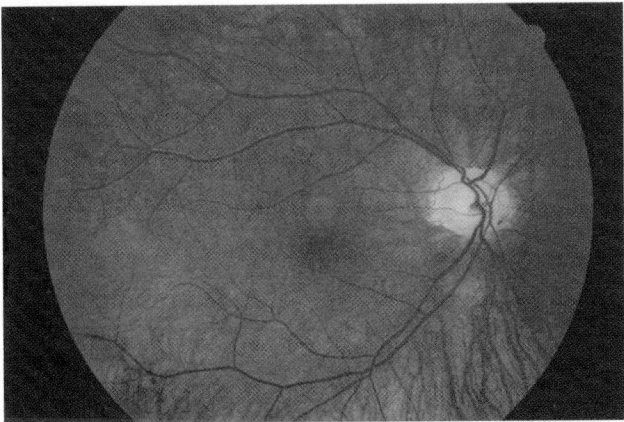

Figure 16   Congenital syphilitic retinopathy.

of onset and is characterized by nummular, interstitial opacities. Inflammatory events include orbital myositis, episcleritis, uveitis, vitritis, and retinal vasculitis. In some cases, when serology is negative, a vitreous tap is required for diagnosis. Neuro-ophthalmic manifestations include bilateral mydriasis, neuroretinitis, pigmentary retinopathy, involvement of multiple cranial nerves, optic atrophy, and disc edema. Seventh cranial nerve paresis can lead to neuroparalytic keratitis. In endemic areas, Lyme disease may be responsible for approximately 25% of new-onset Bell's palsy.

The diagnosis can be difficult and is usually based on a history of exposure within an endemic area, erythema migrans, and a positive serology. Erythema migrans (Fig. 17) has been considered an important clinical sign of Lyme disease but it is present in only about 30% of patients. Response to treatment has also been considered as a diagnostic sign in some patients. Serology in Lyme disease is difficult and the diagnosis should never be solely based on it. Indirect enzyme-linked immunosorbent assay (ELISA) results are often false-positive and should be confirmed by western blotting. In contrast there are patients described where the organism has been cultured from the eye in the presence of negative serology so that seronegativity cannot exclude vitritis due to borreliosis. Antibodies may be measured by ELISA and Western blot. PCR has been used successfully to confirm borreliosis within the vitreous and cerebrospinal fluid (Chapter 3) (25). Interestingly, the bladder has been identified as a reservoir of the organisms, and urine PCR has been found positive in patients with ocular borreliosis (7). Serum reagin tests for syphilis, such as the rapid plasma reagin or VDRL tests, are non-reactive in Lyme borreliosis, but false-positive tests for specific treponemal antigen such as the fluorescent treponemal antibody test can occur.

The use of oral beta-lactam antibacterials [phenoxymethylpenicillin (penicillin V), amoxicillin, cefuroxime axetil] and oral tetracyclines have been recommended as effective first-line treatment modalities for early Lyme disease. Oral macrolides are considered second-line agents, as their clinical efficacy has been less than that of the beta-lactams and tetracyclines. Courses of therapy ranging from 10 to 21 days are supported by the available evidence, although the optimal duration of therapy is unknown.

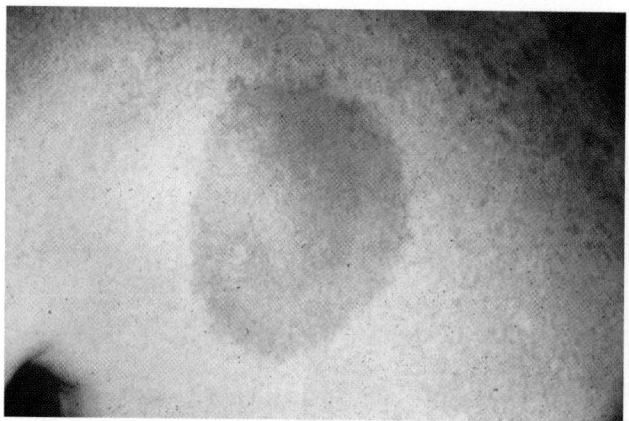

Figure 17   Erythema migrans associated with Lyme disease.

Whipple's Disease. Whipple's disease is a rare systemic disorder caused by *Tropheryma whippelii* with malaise, fever, migrating arthralgias, fatigue, abdominal discomfort, diarrhea, and weight loss. Ocular signs include uveitis, vitritis, and retinal vasculitis. Small bowel biopsy shows characteristic, diastase-resistant, PAS-positive macrophages in the mucosal lamina propria. The etiological agent has been identified as a Gram-positive actinomycete called *Tropheryma whippelii*, but there is still debate on a single or multiple origin. There is some evidence of a predisposing immunodeficiency.

The condition, including its ocular features, may be treated successfully with antibiotics, but relapse is common and it is important to use antibiotics with good penetration of the blood-brain barrier to minimize central nervous system complications. Combination therapy is therefore recommended, e.g., two weeks treatment with parenterally administered streptomycin and benzylpenicillin followed by sulphamethoxazole (800 mg) and trimethoprim (160 mg) orally twice daily for one year. Quinolones may also be effective, as may be Azithromycin.

Brucellosis. Infection arises from unpasteurized milk, with *Brucella abortus* from the cow or *B. melitensis* from the goat. Initially there is an acute phase of generalized fever and systemic disease with leukopenia and a relative lymphocytosis with a slightly enlarged liver and/or spleen. This is followed by a subacute phase with intermittent bouts of low-grade fever when the ocular manifestations can appear with reduced visual acuity from a chorioretinitis. A chronic granulomatous uveitis can also be present with secondary glaucoma. Ocular manifestations are reported as less severe after *B. abortus* than after *B. melitensis* infection, although the former causes more severe systemic symptoms. Typically, the retinal lesion has distinct margins with a hemorrhage around it (Fig. 18); it will contain live bacilli that require antibiotic treatment by the systemic route.

A serological agglutination test with brucella antigen in the third week of illness will confirm the diagnosis with a titer of 1:800 or greater but this can normalize after a further six weeks. Chemotherapy traditionally given involves doxycycline 100 mg twice daily and streptomycin 1 g daily IM. This is not always effective, however, and the later appearance of retinochoroiditis can be treated with ofloxacin 800 mg orally daily and rifampicin 900 mg orally daily. During treatment, serial fundus examinations should be performed weekly and progressive

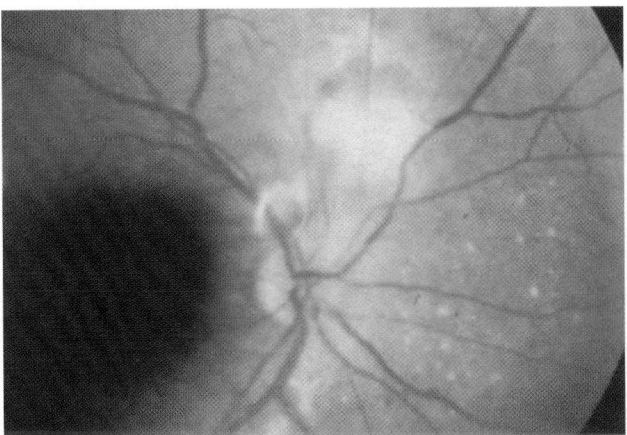

Figure 18 Acute chorioretinitis caused by *Brucella melitensis*.

improvement should occur over six weeks when the choroidal lesion leaves a scar and the retinal lesion disappears; if this does not occur, specialist advice is needed for additional antibiotics. Occasionally, papillitis and retrobulbar neuritis can be present.

Tuberculosis.   Tuberculosis can affect any part of the uveal tract. Tuberculous choroiditis occurs in miliary and chronic disease. Miliary tubercles can be seen especially in tuberculous meningitis, and used to be common at a late stage in children. They appear as round, pale yellow spots and vary in number from 3 to 70. They consist of giant cells around live bacilli. Before chemotherapy, their presence

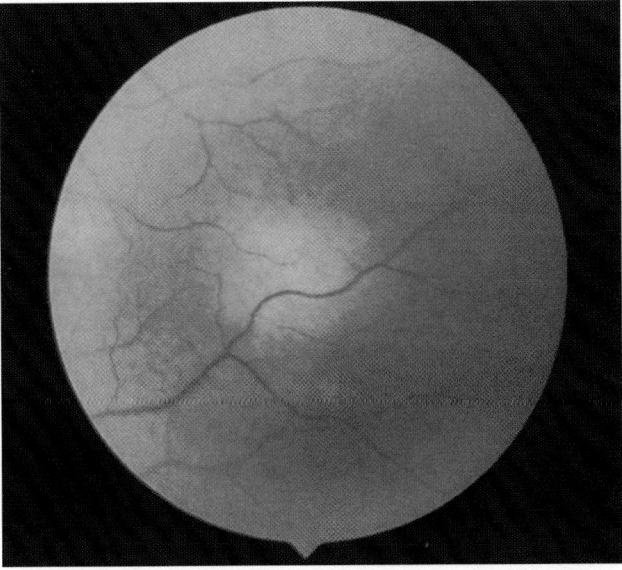

Figure 19 Choroidal tubercle healing on chemotherapy showing a loss of vessels. *Source*: Courtesy of Prof. Susan Lightman.

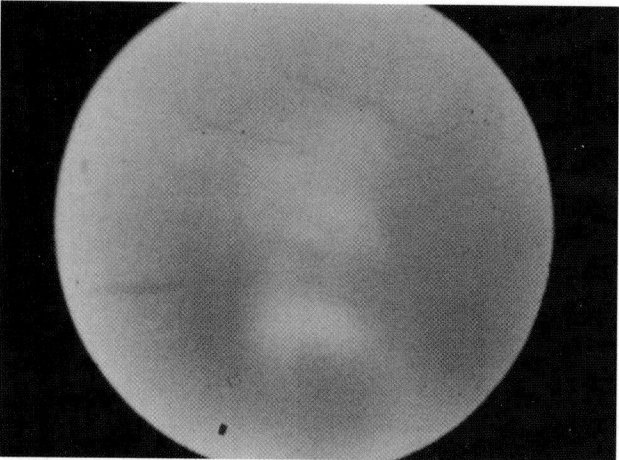

Figure 20 Active tuberculous choroido-retinitis. *Source*: Courtesy of Prof. Susan Lightman.

was a prelude to death, but recovery is now common with triple anti-tuberculous therapy for a minimum of six months. Chronic tuberculosis can occur as a diffuse flat lesion (Fig. 19) if undergoing anti-tuberculous chemotherapy or as raised lesions in active tuberculous chorioretinitis (Fig. 20). The healed lesion (Fig. 21) closely resembles that due to toxoplasmosis. Disseminated choroiditis can occur with extensive development of granulomatous tissue, and (rarely) it can appear as a solitary mass resembling a neoplasm such as a retinoblastoma or pseudoglioma.

Uncommon Bacterial Infections. Posterior uveitis can be caused by a number of unusual and rare bacteria such as *Actinobacillus* sp., *Bacillus cereus*, *Bartonella henselae*, *Listeria monocytogenes* (26), and *Clostridium perfringens*, as well as

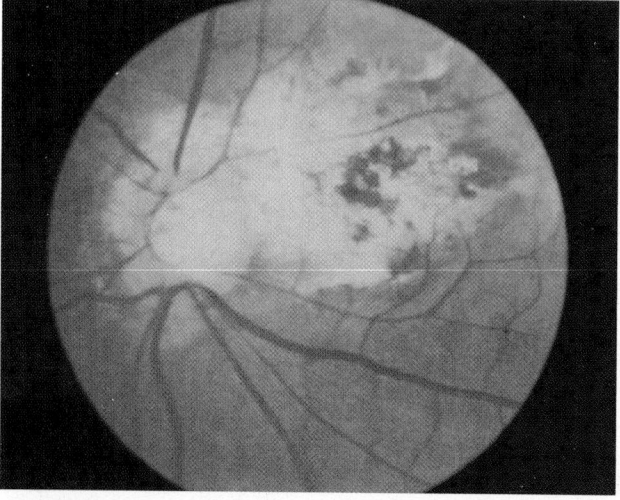

Figure 21 Extensive choroido-retinal scar of healed tuberculosis. *Source*: Courtesy of Prof. Susan Lightman.

Rickettsia and others, especially in immunodeficient patients. They should be suspected in cases of unusual acute or chronic endophthalmitis, often presenting with anterior uveitis as well as multifocal choroiditis and vitreous infiltration. The diagnosis is difficult but is based on isolation of organisms from an anterior chamber and/or vitreous tap or by identification of microbial DNA by PCR (Chapter 3).

Viral Infections

Acute Retinal Necrosis (ARN) Syndrome. Acute necrotizing retinitis is a disorder caused by viruses of the herpes family (VZV>HSV>CMV). A set of criteria for diagnosis has been established that include the typical retinal necrosis features with a characteristic demarcation line. ARN has been described in both immunocompetent and immunosuppressed individuals (27–30). The clinical features differ according to immune status. Herpes zoster and H. simplex have been implicated more frequently in immunocompetent patients. HSV-associated ARN occurs in an earlier age group and is more frequently associated with the HSV-2 type. ARN tends to follow shingles in older patients.

ARN is characterized by a rapid onset of painless loss of vision with areas of necrotizing retinitis, often peripheral, which enlarge and coalesce (Figs. 22–25). There is marked vasculitis, macular edema, and acute optic neuritis. A prodromal phase with flu-like symptoms is common. The anterior segment is usually involved and may even mislead the diagnosis if pupil dilation is not performed and the peripheral lesion overlooked. Mutton fat keratic precipitates, cells in the anterior chamber, and elevated intraocular pressure are common.

Prompt treatment with intravenous acyclovir at 500–800 mg/m^2, eight-hourly for 7 to 14 days, is effective in halting progression of the infection and may prevent involvement of the other eye (29). The outcome, however, is reported as poor.

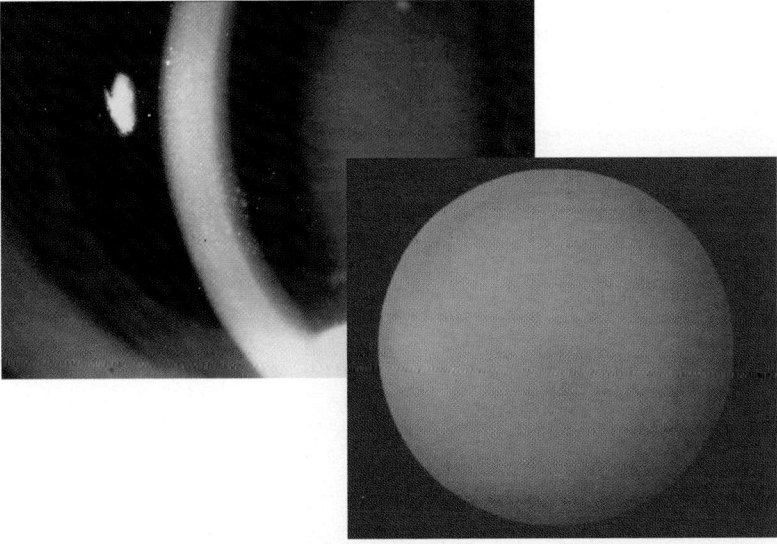

Figure 22 Keratic precipitates (spillover) in ARN syndrome (VZV confirmed) with dense vitreous infiltration.

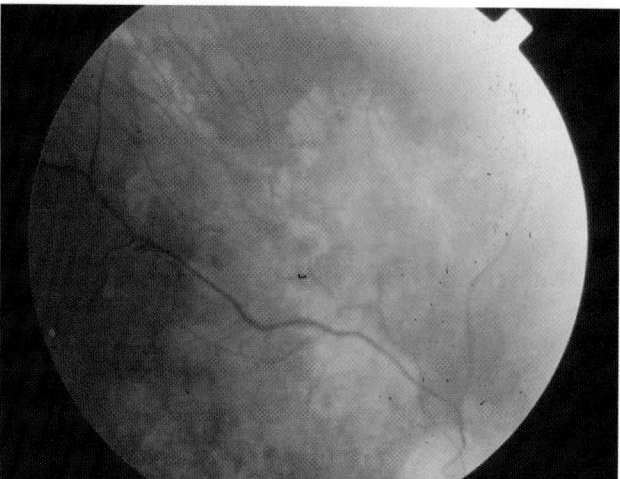

Figure 23 Well-demarcated peripheral lesion of retinal necrosis.

Occlusive vasculitis with optic neuropathy, macular edema, and retinal detachment are frequent and severely limit the outcome. Successful therapy with famciclovir has been reported recently following failed therapy with acyclovir—it may become the antiviral drug of choice for herpetic necrotizing retinitis because it has good bioavailability and is effective against acyclovir-resistant H. simplex. Adjunctive treatment using aspirin or heparin have been suggested. Oil tamponade in eyes undergoing vitrectomy for retinal repair may have a virostatic effect.

Progressive Outer Retinal Necrosis (PORN). A distinct form of retinal necrosis can be found almost exclusively in immunocompromised patients. It is mainly caused by VZV, but other viruses have been implicated, affecting the outer retina. Typical changes are rapidly progressing white lesions at the level of the RPE without hemorrhages or exudates. The diagnosis is based on a typical clinical appearance but

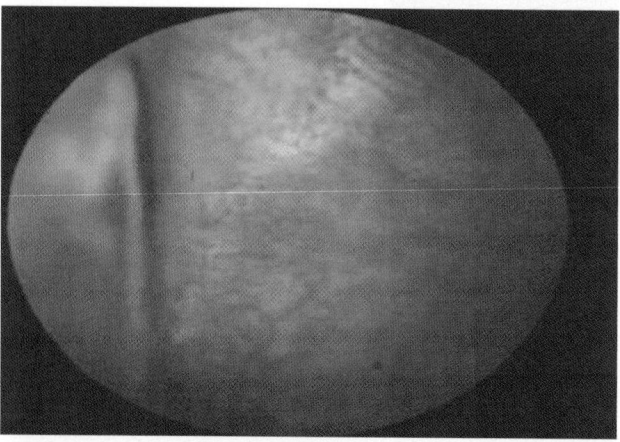

Figure 24 Peripheral retinal lesions in ARN on sceral buckle following retinal detachment.

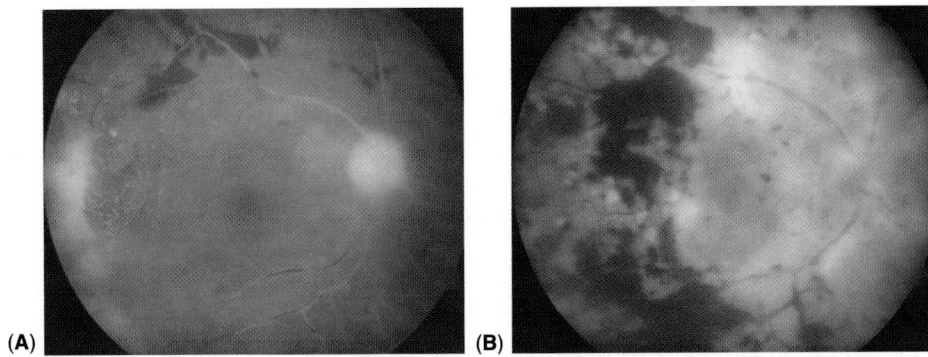

Figure 25 **(A)** Hemorrhagic stage of ARN associated with retinal vasculitis caused by CMV in an immunocompetent patient. **(B)** Angiogram of patient in **(A)**.

may be confirmed by intraocular virus detection (PCR/ELISA). Even when treated early with antivirals (ganciclovir, foscarnet), the prognosis is poor.

Rare Virus Infections. There are a number of cases reported caused by Epstein-Barr, measles, or rubella viruses. In addition, virus vaccines may be related to posterior uveitis in rare instances. Clinical manifestations present commonly as retinitis or chorioretinitis that is often accompanied by retinal vasculitis.

Since most of these virus are widespread, their relation to intraocular inflammation is often difficult to prove and presumptive. Analysis of an intraocular specimen by PCR or ELISA is the commonly suggested diagnostic procedure. Since no specific treatment is available for most of these viruses, symptomatic control of inflammation is mainly achieved with corticosteroids.

Acquired Immunodeficiency Syndrome (AIDS). AIDS is treated nowadays with highly active anti-retroviral therapy (HAART) or by combination therapy between an anti-retroviral drug and a proteinase inhibitor, called cART (31–33). Treatment is given as part of systemic therapy of HIV infection and involves dual therapy with an anti-retroviral drug inhibiting production of reverse transcriptase required for HIV replication, viz. azidothymidine (AZT), and a protease inhibitor viz. indinavir preventing virus particle function (Table. 2) or a combination of two nucleoside analogue inhibitors (viz. emthicitatine and tenofovir) and one non-nucleoside onalogue inhibitor (viz. efavirenz), which has given better results than treatment with one alone.

The HAART regime commenced nearly 10 years ago and has induced a dramatic reduction in the incidence of HIV-related microangiopathy and opportunistic retinal infections, as well as reconstituting the immune system, which can result in uveitis (34). HAART therapy results in the return of full immune function and normal CD4+ lymphocyte counts. Jabs et al. (35) have conducted longitudinal studies and identified a substantial reduction in the incidence of cytomegalovirus (CMV) retinitis from the pre-HAART era. Nevertheless, new cases of CMV retinitis continue to occur (32). For this reason, the considerable amount of work performed with CMV retinitis in AIDS prior to HAART has been retained from the First Edition, as the patient population with CMV retinitis is now too small to justify the large trials of therapy that took place then.

This combined approach (HAART/cART) has prolonged the lifespan of HIV-infected individuals by 10 years and more and increased their immune responsiveness. There is a lack of HIV reproduction in lymphocytes, although the virus is not

Table 2 Effects of Selected Antiviral Drugs

Drug	HIV	CMV	HSV 1/2 HZV	HSV 1/2 IDU-R	HSV 1/2 VID-R	EBV	Adenovirus	Measles[a] Mumps
Inhibitors of viral DNA polymerase								
Idoxuridine	R	R	S	R	?	R	R	R
Vidarabine (adenine arabinoside)	R	R	S	?	R	R	R	R
Trifluridine (trifluorothymidine)	R	R	S	S	?	(S)	R	R
Ganciclovir (Cymevene)	R	S	S	S	S	?	R	R
Foscarnet (Foscavir)	(S)	S	S	S	S	?	S	R
Cidofovir (phase II clinical trials; not for use)	R	S	S	?	?	(S)	S	R
Acyclovir (Zovirax)	R	S	S	S	?	(S)	R	R
Valaciclovir (Valtrex)	R	R	S	S	?	(S)	S	R
Famciclovir[b,c] (Famvir)	R	R	S	S	?	(S)	R	R
Inhibitors of reverse transcriptase (nucleoside analogues)								
Azidothymidine (AZT) (Zidovudine) (Retrovir)	S/(R)							
Didanosine (ddI) (Videx)								
Dideoxycytidine (Zalcitabine) (ddC) (Hivid)		R	R	R	R	R	R	R
Stavudine (d4T) (Zerit)								
Lamivudine (3TC) (Epivir)								
Abacavir (Ziagen)								
Tenofovir (Viread)								
Emtricitabine (Emtriva)								

Inhibitors of reverse transcriptase (non-nucleoside analogues)

Efavirenz (Sustiva)						
Nevirapine (Viramune)						
Delavirdine (Resciptor)	S	R	R	R	R	R

Protease inhibitors

Atazanavir (Reyataz)						
Amprenavir (Agenerare)						
Darunavir (Prezista)						
Lopinavir (Kaletia)						
Tipranavir (Aptivas)						
Saquinavir (Invirase)						
Indinavir (Crixivan)						
Ritonavir (Norvir)	S	R	R	R	R	R
Nelfinavir (Viracept)						

[a] Important effective protection against keratitis due to these viruses can be gained by mass immunization with live attenuated vaccine, with amelioration of disease severity by maintaining adequate levels of vitamin A.

[b] Valaciclovir is the pro-drug for acyclovir absorption is increased five fold.

[c] Famciclovir, the oral pro-drug for penciclovir, is effective against acyclovir-resistant HSV.

HIV: Human immuno-deficiency virus, CMV: Cytomegalovirus, HSV: Herpes *simplex* virus, HZV: Herpes zoster virus, EBV: Epstein-Barr virus

Abbreviations: R, Resistant; S, sensitive; ?, unknown.

eradicated, and the CD4+ Th lymphocyte count rises above $200/mm^3$, at which level opportunistic infection is prevented by the immune response. As such, far fewer patients now present to ophthalmologists with opportunistic infections. The combination chemotherapy has to be maintained for the lifespan of the individual but it does allow them to lead a normal life and be in full-time employment.

In the HAART era, CMV retinitis and other opportunistic infections are associated with intraocular inflammation (34). Interestingly, patients with newly diagnosed CMV retinitis had similar ocular findings to those seen in the pre-HAART era, while CMV retinitis was still the commonest cause of visual acuity loss in AIDS (36,37). Newly diagnosed cataract was another source of visual acuity loss. Hanna et al. (36) found ocular inflammation occurred with a 3.5-fold increased odds with a late HIV diagnosis. Other predictors of ocular inflammation were injection drug use and older age. The authors recommend early HIV testing to improve ocular inflammation morbidity.

Others have studied factors for development of opportunistic infection at high CD4+ cell counts in patients on HAART (38). The CD4 cell count and HIV-RNA level were the best predictors. A Japanese series found ocular complications in one third of patients with AIDS given the HAART regime (39). Immune recovery uveitis presented in AIDS patients who had previous CMV retinitis.

Treatment interruption is common with AIDS patients. The risk of AIDS or death occurring is increased twofold for patients who stop cART for three months or more (40). Those with low CD4+ counts, high viral loads of HIV, or prior AIDS were at greatest risk of death. In the era of HAART there are still visual field defects and evidence of damage to retinal nerve fiber layers in HIV patients using multifocal electroretinogram techniques (41). It can be shown that damage lies in the inner retinal structure while the outer retina is spared.

Human immunodeficiency virus (HIV), which causes AIDS, has been identified in tear fluid, conjunctiva, corneal epithelium, and retina, but the principal ophthalmic manifestations of AIDS relate to the occurrence of florid opportunistic infections and to conjunctival and orbital involvement with Kaposi's sarcoma (Figs. 26, 27) and other neoplasms. Therapy is directed against the relevant organism

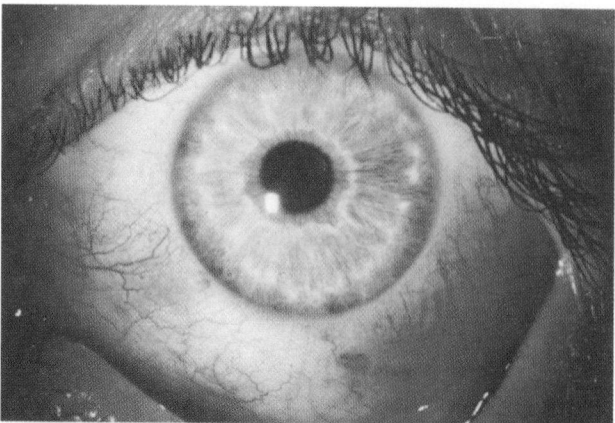

Figure 26 Close up of Kaposi's sarcoma of the conjunctiva in an HIV-positive patient.

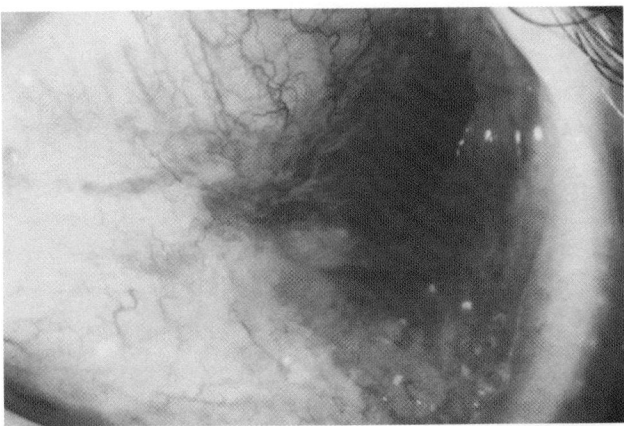

Figure 27 Post-irradiation therapy of Kaposi's sarcoma of the conjunctiva in an HIV-positive patient.

and is generally more intense and prolonged than is required in immunocompetent individuals.

HIV, in the absence of other organisms, can cause a characteristic retinitis due to microvasculopathy of cotton wool spots, hemorrhages ("flame-shaped" in the nerve fiber layer or "dot- or sheet-shaped" deeper in the retina), and microaneurysms (Figs. 28, 29). Macular exudates and edema with optic neuropathy can occur. These forms of HIV retinopathy do not generally interfere with vision.

Opportunistic Infections. These usually involve the posterior segment, but severe keratitis can occur in HIV-infected individuals, particularly in drug addicts in whom *Candida albicans* can cause aggressive keratitis in the absence of predisposing factors with a poor prognosis (42). A bilateral epithelial keratopathy caused by Encephalitozoon has been described that responds to itraconazole, while successful treatment of other patients with microsporidium keratitis has been reported with dibromopropamidine (43). Molluscum contagiosum and multiple verruca of the eyelids, especially among young African children, are common cutaneous manifestations of HIV infection in the tropics.

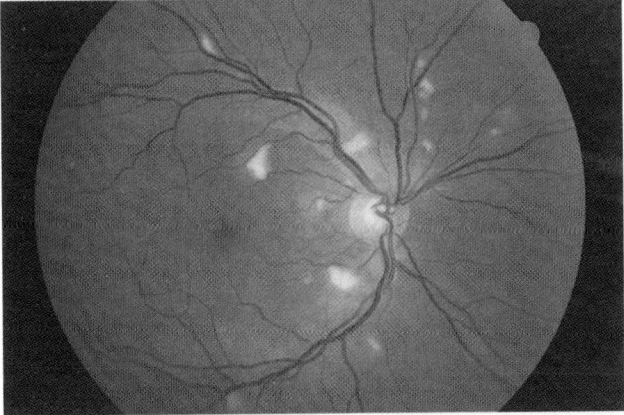

Figure 28 HIV micro-vasculopathy with "cotton wool" spots.

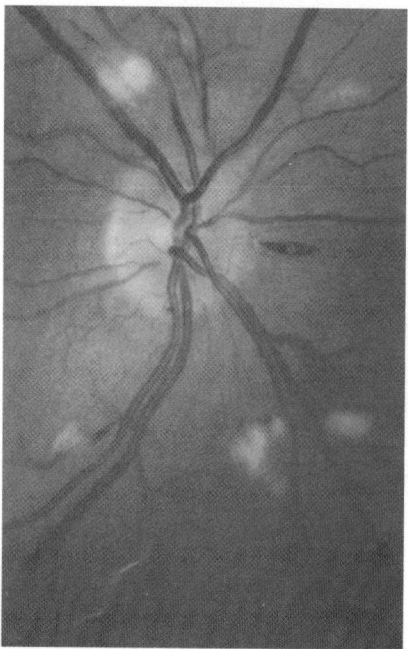

Figure 29 HIV micro-vasculopathy with "flame" and "dot" hemorrhages.

The *Mycobacterium tuberculosis* group, including multiple-resistant strains of *M. avium-intracellulare,* has been recognized increasingly to cause severe systemic infection that can also manifest as chorioretinitis. The presentation is similar to that given above but endophthalmitis can also occur. The infection can be community acquired but has also been recognized in hospital-acquired outbreaks in AIDS wards. Systemic anti-tuberculous treatment is required.

Treponema pallidum infection can coexist with HIV, with similar modes of transmission. Ocular signs present as described above, but in the presence of HIV infection there can be rapid progression to neurosyphilis. The retinitis that occurs can simulate CMV infection or present as a retinal pigment epitheliitis. There may be vitritis alone as a primary manifestation of ocular syphilis in an HIV-positive individual, presenting as a decrease in vision with a dense vitreal haze. Serological testing in HIV infection can be misleading, particularly with a false-negative cardiolipin reagin test result such as VDRL. Syphilis should be remembered as a treatable cause of visual loss in HIV positive patients.

Cytomegalovirus is the commonest opportunistic ocular infection and produces a hemorrhagic necrotizing retinitis (Fig. 30). CMV retinitis is found in about 20% of terminal AIDS patients and is bilateral in about 17%, when the delay from presentation with HIV infection is shorter. Occasionally, H. simplex virus and Epstein-Barr virus, and toxoplasma, may cause a clinically similar retinitis. CMV retinitis may also present in HIV disease in the absence of a severe depletion in the CD4 lymphocyte count (44).

Patients with CMV retinitis complain of visual field defects, flashing lights, and symptoms of "floaters." If the central retina or optic nerve is involved, they complain of a reduction in visual acuity. CMV causes diffuse, full thickness retinitis that is sharply demarcated from normal adjacent retina. It can present as a fulminant

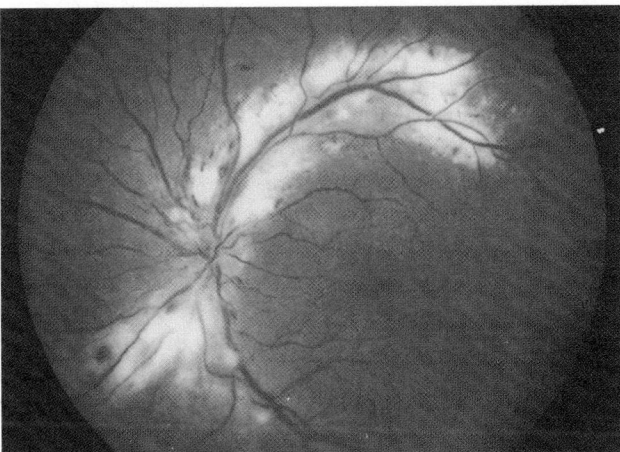

Figure 30 "Pizza pie" retinitis of severe cytomegalovirus infection above and below the optic disc in a patient with AIDS.

retinitis with yellow-white areas of necrosis surrounded by hemorrhages ("pizza pie" retinitis) (Fig. 31) or as a "granular" retinitis, in which hemorrhages are not a feature and may be absent. Both types follow a similar clinical course and may coexist together in the same eye.

In a study of CMV retinitis in AIDS patients, 58% presented with unilateral disease and 15% of these developed contralateral infection despite treatment with ganciclovir. "Smoldering retinitis" was a clinical sign seen in 33% of the patients whose retinitis progressed while receiving ganciclovir; bilateral CMV retinitis was associated with HIV encephalitis (45).

Response to therapy is partly related to immune status. At presentation, CMV retinitis does not frequently pose an immediate threat to vision, but it may do so with development of retinal detachment, in association with peripapillary disease or with involvement of the central retina. Retinal detachment, an important cause of blindness from CMV retinitis, can be treated successfully by vitrectomy, silicone oil,

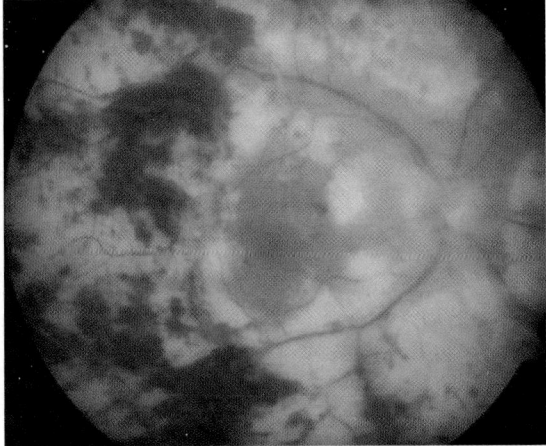

Figure 31 Cytomegalovirus retinitis in a bone marrow transplant recipient.

and endolaser. Elimination of scleral buckling may reduce intraoperative time, patient morbidity, and the risk of an accidental needle stick (46).

CMV retinitis occurring in AIDS patients implies a high risk for the development of CMV encephalitis, particularly when the retinitis involves the peripapillary region. However, in patients with AIDS without CMV retinitis, central nervous system symptoms are more likely due to other causes.

Therapy. Progression of CMV retinitis may be delayed in the short term by intravenous therapy with either ganciclovir or foscarnet, or the combination. Repeated, local, intravitreal therapy is more effective, and particularly valuable when there are no signs of disseminated CMV disease (47,48). The development of slow-release, intraocular implant systems for ganciclovir had provided further improvements in management but are no longer available (49). The prophylactic use of oral ganciclovir in persons with advanced AIDS has been shown to significantly reduce the risk of CMV, but active retinitis can still occur (50).

Parenteral Therapy. The virostatic drugs dihydroxypropoxymethylguanine (DHPG, or ganciclovir), similar in structure to acyclovir, and trisodium phosphonoformate (foscarnet) improve or temporarily stabilize the retinitis. Ganciclovir and foscarnet are equally effective in controlling CMV retinitis, but foscarnet is less well tolerated; neither are virucidal. Ganciclovir is myelotoxic while foscarnet causes renal toxicity. Repeated therapy is indicated because of the high relapse rate. Ganciclovir is given by intravenous infusion over one hour in a dose of 5 mg/kg every 12 hours.

Combined daily therapy of both ganciclovir and foscarnet has been shown more beneficial (51), with prolonged intervals between progressions without increased toxicity. Such combination therapy may halt the progress of peripheral outer retinal necrosis in AIDS patients as well as be effective in treatment of "clinically resistant" CMV retinitis (52,53). Intravenous cidofovir has also been shown to be effective for control of peripheral retinitis (54), with improved results, but symptomatic anterior uveitis can develop in the absence of retinitis, requiring treatment with topical prednisolone (55). Treatment with 5 mg/kg once weekly for two weeks, then 5 mg/kg every other week delayed the progression of retinitis in AIDS patients compared to delay in therapy. Proteinuria (23%) and neutropenia (15%) were possibly related to therapy and may lead to discontinuation of the drug.

Many ophthalmologists in the past found that the expected lifespan of AIDS patients, when they developed CMV retinitis, was limited to approximately six months, but this has improved with combination therapy of reverse transcriptase and virus proteinase inhibitors. Local intraocular therapy may replace systemic therapy.

Intraocular Delivery. An effective treatment of CMV retinitis, which avoids the risk of systemic toxicity, is intravitreal injection of antiviral drugs (56,57). Intravitreal therapy can be given as follows:

- apply a lid speculum and anesthetize the conjunctiva with 4% lidocaine;
- give a subconjunctival injection of 1% lidocaine with epinephrine;
- wipe the area to be injected with 5% povidone iodine (Betadine);
- use calipers to mark a point 4 mm posterior to the corneoscleral limbus;
- use a 27- or 30-gauge, 0.5-inch needle and "tuberculin" syringe to pass through pars plana 4 mm posterior to the limbus with the needle tip aimed at the optic nerve head; and
- inject a volume of 0.05 ml containing the drug.

Intravitreal ganciclovir or foscarnet have been given on a weekly basis with little local ocular complication. These drugs have also been given together for the treatment of clinically resistant retinitis (53). An intravitreal dose of ganciclovir (200–400 mg) was as effective as intravenous therapy. A dose of up to 2 mg (2000 mg) can be given in 0.05 ml and is expected to provide adequate intravitreal levels (0.25–1.22 mg/mL) for up to seven days. Levels at 24 hours have been recorded as 143.4 mg/mL and at 72 hours as 23.4 mg/mL (56). The intravitreal dose of foscarnet is 1.2–2.4 mg in 0.1 mL. Caserta et al. (57) gave 666 intravitreal injections to 49 eyes of 33 patients with a mean number of injections per eye of 14 and a range of 1 to 75. Their frequency of rhegmatogenous retinal detachment due to the fragility of the treated necrotic retina was 14%, without occurring until the 75th injection in one patient, which is similar to that for intravitreal therapy. Other complications included two vitreous hemorrhages and one patient with raised intraocular pressure, adequately managed with topical timolol (beta-blocker) and iopidone.

Intravitreal cidofovir (20 mg/0.1 mL) halted progression of CMV retinitis for 6–10 weeks after a single injection, with a median time to progression of 55 days after the initial injection or 63 days after repeat injections. Patients also received oral probenecid. Intravitreal cidofovir has been used for maintenance treatment of retinitis (58). Cidofovir caused a slight lowering of intraocular pressure and a mild to moderate iritis that responded to treatment.

Retinal detachment from treated CMV retinitis is now managed by instillation of silicone oil tamponade instead of scleral buckling. This is less traumatic for the patient and avoids a possible needle-stick injury to the surgeon. A lower dose of both these antiviral drugs has been given intravitreally to patients whose eyes contain silicone oil (46). Seven patients have received a ganciclovir implant placed into the posterior segment partially filled with silicone oil, although all patients also received intravenous therapy and three had recurrent CMV retinitis (59).

More recently, the development of sustained-release intraocular devices such as Vitrasert have provided the opportunity to deliver controlled amounts of drugs for prolonged periods with minimum local toxicity, confirmed by a clinicopathologic study (60). The median time to progression of CMV retinitis has been recorded as 226 days following a ganciclovir implant delivering 1 mg/hr compared to 15 days for deferred treatment. Complications in 12% of patients have included retinal detachment at three to six months, vitreous hemorrhage that may require a pars plana vitrectomy, endophthalmitis, and cystoid macular edema (61). Another problem with this approach is the development of occult CMV retinitis in the other eye. Oral ganciclovir prophylaxis should be given to prevent this.

The presentation of CMV retinitis with use of combination systemic therapy of reverse transcriptase and protease inhibitors has changed. The epidemiology of CMV retinitis in AIDS patients without use of systemic proteinase inhibitors is well known, when the CD4 Th cell count on presentation is < 50 cells/mm^3, with a risk of recurrence in treated cases of 22% at nine months. For patients receiving systemic proteinase inhibitors, with an initial CD4 cell count of < 50 cells/mm^3 at the beginning of proteinase inhibitor treatment, the risk of developing CMV retinitis was 8% at nine months and the median time to progression was 211 days compared to previous experience without proteinase inhibitors of 50 to 70 days (62). This data suggested that the overall inhibition of HIV activity was associated with reduced expression of active CMV retinitis. This is further corroborated by findings that the CD4 count is not so helpful in predicting disease outcome.

Patients develop more intraocular inflammation associated with active or inactive CMV retinitis than prior to use of protease inhibitors, especially when they recover immune function that reacts to the presence of CMV antigen in the retina (63,64). Furthermore, cystoid macular edema, which had previously only been reported very occasionally with CMV retinitis, has appeared as a cause of visual loss in treated CMV retinitis patients on systemic proteinase inhibitor therapy due to immune function recovery. This is thought to be caused by inflammation-induced blood-retinal barrier breakdown. Topical non-steroidal anti-inflammatory drugs and corticosteroids have been used with benefit.

Other opportunistic infections are seen less frequently in AIDS and present as follows:

Herpes simplex can cause acute retinitis with detachment requiring specific treatment.

The correlation between severe H. zoster infection in young adults and seropositivity for HIV is well established with a high, positive predictive value in Africa.

Herpes zoster virus (HZV) can present as a rapidly progressive necrotizing retinitis and be the initial manifestation of AIDS. A rapid, sensitive PCR-based assay has been developed to detect HZV DNA in vitreous samples of AIDS patients with progressive outer retinal necrosis syndrome to assist in diagnosis (Chapter 3). HZV can also present as severe chronic retinitis, requiring prolonged systemic penciclovir therapy or as recurrent corneal epithelial lesions, which may cause peripheral ulceration but respond well to acyclovir.

Toxoplasma gondii causes retinochoroiditis less commonly than brain involvement. It may cause the presenting symptom, with blurred vision and floaters or pronounced visual loss from macular or papillomacular bundle or optic nerve head involvement. It can still be the initial presentation of AIDS with optic nerve and orbital inflammation (65). The retinochoroiditis is unassociated with a pre-existing retinochoroidal scar, suggesting that the lesions are a manifestation of acquired rather than congenital disease. The presence of IgM antibodies may support this, although antibody levels in AIDS may not reflect the magnitude of disease. Lesions may be single or multifocal, arise in one or both eyes, or consist of massive areas of retinal necrosis. They may resemble those of CMV retinitis and may occur concurrently in the same eye. In comparison, toxoplasmic lesions tend to be thick and opaque, with smooth borders and a relative lack of hemorrhage. A prominent anterior chamber and vitreous reaction may occur.

Treatment of a toxoplasma ocular infection with pyrimethamine, clindamycin, and sulphadiazine is effective in over 75% of patients. Once resolution is observed, maintenance therapy is continued, as relapses occur in the absence of treatment. Corticosteroid treatment is unnecessary and its use has been associated with the development of concomitant CMV retinitis.

Pneumocystis carinii causes pneumonia in up to 80% of AIDS patients. It can cause a choroidopathy by systemic spread from primary lung infection. Multiple yellow placoid fundus lesions are seen. Parenteral pentamidine or co-trimoxazole (Septrin) is required for treatment, but pentamidine should not be given with foscarnet, if there is coexisting CMV infection, due to the risk of profound hypocalcemia.

Candida albicans and *Cryptococcus neoformans* can produce retinal lesions or endophthalmitis, particularly in AIDS patients who are intravenous drug abusers (see below). Cryptococcosis has presented as an inflammatory mass in the iris and as a cloudy choroiditis.

Coccidioidomycosis, endemic in arid regions of the southwestern United States, Mexico, and Central and South America, is an infectious disease caused by the dimorphic soil fungus *Coccidioides immitis*. It presents with multiple, inactive coccidioidal chorioretinal scars in a non-AIDS population which has suffered from systemic coccidioidomycosis. The scars are described as similar to those caused by histoplasmosis. Coccidioidomycosis has become the third most common opportunistic infection in AIDS patients in Arizona, with chorioretinal scars present in 39% of AIDS patients with systemic infection.

Fungal Infections (Refer also to Chapter 9)

Ocular *Candida albicans*. Endogenous or metastatic endophthalmitis, a severe vision-threatening intraocular infection, occurs through hematogenous dissemination from a concurrent infection in the host or an external source such as a catheter or an intravenous needle. The incidence rate of endogenous fungal endophthalmitis has been estimated in the past to range from 9% to 45% of patients with disseminated fungus infection or fungemia (66). However, recent studies have reported an incidence of 2% or 2.8% (67,68).

In an 18- year review between 1982 and 2000 at the Cleveland Clinic, a tertiary care center, the average occurrence for their unit was 1.8 cases per year (69). Ninety percent of patients were diagnosed with medical conditions including diabetes, hypertension, gastrointestinal disorders, cardiac disorders, malignancy, and immunosuppression or else prolonged surgical complications. Acute ocular symptoms, rather than systemic symptoms, were the most common reasons to present to the physician. This situation has now changed, with early aggressive systemic antifungal chemotherapy for either growth of fungi from blood cultures or local sites of fungal infection thought on clinical criteria to be disseminated (68).

Because of the rapid advance of medical technology, a longer lifespan of patients with chronic diseases, and a rising prevalence of long-term intravenous access, the disease may become more common in clinical practice—albeit less common in those with organ transplants and immunosuppression, as the result of early ophthalmological screening of all susceptible patients (67).

Endogenous endophthalmitis is most commonly caused by fungi. Fungal organisms account for more than 50% of all cases of endogenous endophthalmitis. *Candida albicans* is by far the most frequent cause (75% to 80% of fungal cases) followed by *Aspergillus* sp. A review of the literature 20 years ago revealed that endophthalmitis occurred in 5% of patients with candidemia but 10 years previously this rate had reached 78% (8,70).

Aspergillosis is the second most common cause of fungal endophthalmitis, especially in intravenous drug users. Less frequent are other *Candida* sp, *Torulopsis glabrata*, *Cryptococcus neoformans*, *Sporothrix schenckii*, *Scedosporium apiospermum* (*Pseudallescheria boydii*), *Blastomyces dermatitidis*, *Coccidioides immitis*, and mucor. The incidence of endogenous fungal endophthalmitis has increased significantly in recent years. From 1990 to 1999, the incidence of bloodstream infection from *Candida* sp. reported by the Centers for Disease Control and Prevention increased fivefold for large teaching hospitals and tertiary care centers and has continued today.

Risk Factors. Multiple risk factors predispose to the development of endogenous fungal endophthalmitis.

- Many have received intravenous antibiotics. Antibiotic use allows *Candida* superinfection in the gastrointestinal tract and elsewhere.
- Many affected individuals have had major surgery. Surgery in areas such as the gastrointestinal tract enables hematogenous seeding both pre- and postoperatively when tissues such as the intestinal mucosa experience circulation compromise, predisposing to infiltration by *Candida* sp. Additionally, the patient is more likely to undergo prolonged intravenous antibiotic administration postoperatively.
- Many patients have intravenous catheters and/or an intravenous infusion. The intravenous catheter predisposes the patient to infection with *Candida* sp., followed by hematogenous seeding.
- Many patients are either treated with corticosteroids before the development of endophthalmitis or steroids were initiated when the ocular disease was diagnosed.
- Some have received therapy with other immunosuppressive agents before the onset of endophthalmitis. The immune system is a natural defense system against candidal infection; immunosuppressive therapy predisposes the individual (Chapter 1).

Additionally, some patients with systemic candidiasis have impaired host defenses, such as diabetes, malignancy, liver disease (cirrhosis or hepatitis), and alcoholism.

Tanaka et al. retrospectively analyzed predisposing factors for the development of endogenous fungal endophthalmitis in 79 eyes of 46 patients over a 12-year period between 1986 and 1998 (71). Disease onset was 11 days from the initiation of intravenous hyperalimentation. The authors found additional correlations:

- β-D-glucan $\geq 20\,$pg (90%)—a test for deep mycoses
- Fever of unknown origin (76%)
- Male gender (74%)
- Presence of cancer (72%)
- Neutrophils $\leq 500/$mL (67%)—a test for deep mycoses
- Cand-tec $\geq \times 4$ (57%)—a test for deep mycoses

Intravenous drug abuse is associated with the development of *Candida* endophthalmitis. The source of the fungi is either drug paraphernalia or contaminated illegal drugs. Interestingly, pure heroin inhibits the growth of *C. albicans*. The endophthalmitis is characteristically associated with only a transient or no fungal septicemia (candidemia). The patients tend to be young, and if confirmation of systemic infection is needed, the organisms can often be cultured from cutaneous lesions.

Endogenous fungal endophthalmitis also occurs in neonates. Both premature and newborn infants are at high risk for developing endogenous *Candida* endophthalmitis. Factors that place these neonates at increased risk are:

- Newborn leukocytes are less lethal to *C. albicans*
- Deficient host immune system found in newborns
- Inadequate anatomic barriers (skin, mucosa, gastrointestinal mucosa) to resist fungal invasion from surface colonization
- Intravenous hyperalimentation, indwelling catheters, broad-spectrum antibiotic use, and previous surgery, especially abdominal procedures, put the neonate at high risk.

Identification and screening of infants at high risk for developing *Candida* endophthalmitis is recommended. Regular assessment should result in prompt diagnosis and treatment of this infection, leading to good ocular outcomes in these patients.

Factors that alter natural tissue barriers and replace the normal bacterial flora with fungal organisms promote tissue invasion. Factors that promote the growth of *Candida* on mucosal surfaces include

- Corticosteroid use in diabetic individuals, which increases glucose levels, a nutrient important for the growth of *Candida* sp.; and
- Ileus accompanying gastrointestinal disease and surgery, which promotes fungal overgrowth.

Use of broad-spectrum antibiotics can promote *Candida* sp. growth by killing normal bacterial microflora. Other risk factors include use of radiation and chemotherapy in the treatment of malignancies, parental hyperalimentation, indwelling bladder catheters, and debilitation. In an older prospective study, Brooks found that 28% of patients with *Candida* sepsis developed candidal endogenous endophthalmitis (72). These studies included patients with a variety of ocular lesions not necessarily indicating vitreous cavity involvement, but associated with candidiasis. In contrast, in another more recent prospective multi-center study, Donohue et al. found that of 118 patients with candidemia, *Candida* chorioretinitis developed in 9% of patients; however, no patients progressed to *Candida* endophthalmitis (66). This study employed a tighter definition of endophthalmitis as either *Candida* chorioretinitis with extension of the inflammatory process into the vitreous or vitreous abscesses presenting as intravitreal fluff balls.

In all cases of chorioretinitis, treatment with systemic amphotericin B or an imidazole such as fluconazole, itraconazole, or voriconazole results in resolution of the lesions without vitreous involvement. Clinicians need to be aware of the systemic toxic effects of amphotericin, which are well described in the product literature. Risk factors for the development of *Candida* chorioretinitis in their series included multiple positive blood cultures, immunosuppression, fungemia with *C. albicans* versus other species, and visual symptoms. The investigators concluded that frequent ocular examination of patients demonstrating candidemia with these risk factors should result in prompt diagnosis of candidal chorioretinitis. Early treatment of candidal chorioretinitis with an imidazole should keep the infection confined to the posterior uvea, preventing extension of the infection into the vitreous cavity.

Scherer and Lee found a much lower rate of endophthalmitis (2.8%) in patients with nosocomial systemic fungal infections (68). The investigators attributed this low incidence to a recent change in treatment protocol. Previously, ophthalmic consultation was obtained so that a diagnosis of fungal endophthalmitis could be used as an indicator of deep, systemic fungal disease, which would justify treatment with systemic antifungal agents. Physicians were reluctant to start patients on amphotericin B with only positive blood cultures or a presumptive diagnosis of deep tissue fungal infection because some patients were reported to recover spontaneously and because use of this antifungal agent was associated with serious hematological and renal toxicity. With greater physician experience, systemic antifungal therapy is being initiated in the presence of a single positive blood culture. More than 90% of patients in this series were receiving systemic antifungal agents at the time of ocular examination. Screening of susceptible patients at an early stage is also taking place and is well justified. Use of imidazoles for therapy, such as fluconazole, itraconazole, and voriconazole, have been found as effective for treating *Candida albicans* retinochoroiditis as use of amphotericin B.

Pathophysiology. In contrast to exogenous infection where the organisms are introduced directly into the vitreous, pathogens in endogenous endophthalmitis become lodged in the small vessels of the retina or choroid. Retinitis, choroiditis, retinochoroiditis, chorioretinitis, endophthalmitis, and panophthalmitis then result from infectious seeding. Secondary extension into the retina and vitreous may follow with further propagation of the organism. The blood-ocular barrier is broken down by the inflammation elicited by the embolus. The organisms then spread to contiguous perivascular tissue, stimulating a further inflammatory response. The vascular nature of endogenous endophthalmitis renders it more susceptible to systemic antibiotics.

Clinical Manifestations and Organisms. A high degree of suspicion is required to make an early diagnosis of endogenous endophthalmitis in order to maximize the visual outcome. In addition to the ocular findings, it is essential to assess the clinical manifestations of the debilitating conditions, such as endocarditis, which cause the initial infection and allow it to propagate. Therefore, routine vital signs and physical examination are warranted.

Patients commonly complain of eye pain, blurring of vision, ocular discharge, and photophobia. A comprehensive ophthalmic examination including visual acuity, intraocular pressure, pupillary reaction, slit-lamp examination for inflammation, and dilated fundus examination is performed. Early in the disease, Roth's spots (round, white retinal spots surrounded by hemorrhage) and retinal periphlebitis may be seen on fundus examination. Later in the disease, obvious signs such as chemosis, proptosis, and hypopyon may be manifest. Ultrasonography is performed when a clinical examination is inadequate or to supplement clinical findings. Clinicians must be more aware of the need to examine the eyes of extremely ill and sedated patients, such as those in intensive care units.

Fungal endophthalmitis may present acutely or as an indolent course over days to weeks. Individuals with candidal infections may present with high fever, followed several days later by ocular symptoms. Pyrexia of unknown origin may be associated with an occult retinochoroidal fungal infiltrate.

Endophthalmitis

Candidiasis. The first case of endogenous candidal endophthalmitis was described in 1947. Some of the patients may be too ill or sedated to recognize the symptoms, which include floaters, localized scotoma, subtle changes in vision, photophobia, ciliary injection, or ocular pain. Most patients present with macular lesions or significant vitreous involvement. Early lesions start as an inner choroidal focal inflammatory reaction surrounding a small locus of *Candida* cells. A small minority of cases begin with a localized retinitis. The characteristic lesion is described as a creamy-white, round or oval, one-eighth to one-quarter disc diameter in size, circumscribed chorioretinal lesion with overlying vitreous inflammation of varying intensity.

As the infectious focus breaks into the vitreous, it enlarges in a globular fashion with budding extensions. Vitreous extension is accompanied by focal perivascular inflammatory deposits and increasing vitreous haze, with strand-like clusters of inflammatory cells and a diffuse cellular infiltration. Lesions are commonly multifocal and located in both eyes (Figs. 32 and 33). A small focal intraretinal

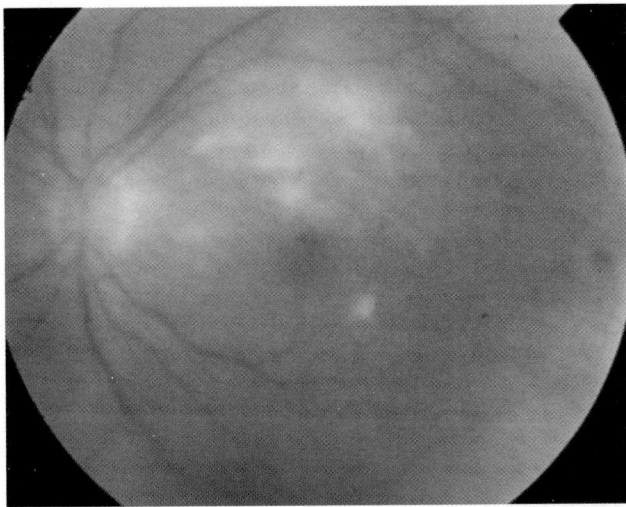

Figure 32 Blurred vision with hazy vitreous with an infiltration around the macular area. Young male who had undergone kidney transplantation five years previously. Chronic immunosuppression with *Candida* gastrointestinal tract infection. Treated with intravitreal amphotericin and intravenous Ambisome. The vitreous biopsy was negative by Gram stain and culture.

hemorrhage can be seen and may appear as a Roth's spot if it surrounds an infectious focus. Focal retinal necrosis and scarring combined with membrane formation and contraction are the major causes of permanent vision loss.

Candidal infection can take days to weeks to produce damage that bacterial organisms cause in hours. Anterior segment inflammation lags behind posterior segment inflammation but mirrors the intensity of the vitreous reaction and progresses to iridocyclitis with synechia, hypopyon, pupillary block, and even a ciliary body abscess.

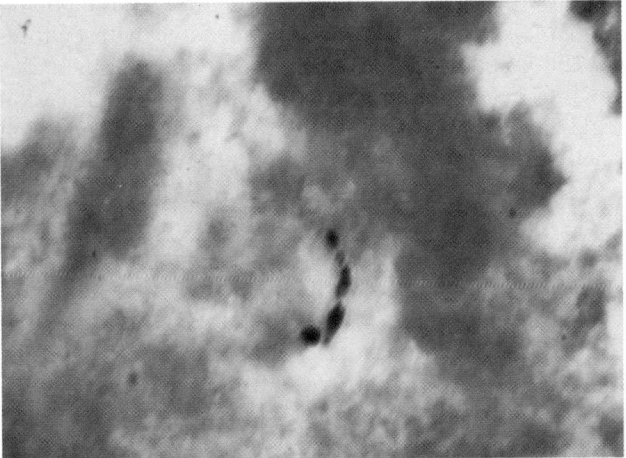

Figure 33 Gram stain from vitreous in a patient with severe bilateral *Candida albicans* endophthalmitis following a contaminated blood transfusion after childbirth.

Histologically, the chorioretinal lesions of early candidal endophthalmitis show an intense localized inflammatory reaction arising in the choroid, composed of suppurative and granulomatous elements surrounding a small nidus of *Candida* cells. From the inner choroid, the inflammatory reaction breaks through Bruch's membrane into the retina to produce a localized microabscess. In the absence of effective treatment, the infectious nidus enlarges to rupture through the internal limiting membrane and "spills" into the vitreous. The characteristic budding yeast and pseudohyphae of *Candida* cells are seen more frequently in the vitreous than the choroid where the organism originated, because the host immune defenses are better able to confine the infectious process. Perivasculitis, papillitis, and localized serous retinal detachment sometimes accompany the spreading infection. In the late stage, there is tractional retinal detachment associated with proliferative vitreoretinopathy and a cyclitic membrane.

In summary, the clinical diagnosis of *Candida* endophthalmitis can be defined as

- chorioretinitis with focal, deep, white infiltrative lesions but with no evidence of direct vitreal involvement except for diffuse vitreal haze, or
- endophthalmitis with extension of the surrounding inflammation into the vitreous or with a vitreous abscess manifesting as "fluff balls," or
- there may be a combination of both.

Neonatal candidemia and end-organ damage, including endophthalmitis, has been the subject of a recent meta-analysis (73).

Aspergillosis.   This ubiquitous fungus is the second most common source of endogenous fungal endophthalmitis, after *Candida* sp. Only five of the more than 900 varieties of Aspergillus cause endogenous endophthalmitis. These include *A. flavus*, *A. fumigatus*, *A. candidus*, *A. niger*, and *A. terreus*. *A. fumigatus* is the most frequently encountered pathogenic species in humans.

Initially, typical signs and symptoms of endophthalmitis occur. Endogenous aspergillus endophthalmitis may be the presenting feature of disseminated aspergillosis or of an endocarditis. The chorioretinitis may manifest as depigmented discrete lesions, Roth's spots, or elevated pale yellow, fluffy lesions. Choroidal and retinal vascular occlusions have been reported. The disease progresses to involve all chambers of the eye and may eventually lead to serous or exudative retinal detachment, choroidal or vitreous abscess formation, and intravitreal granulomata.

In a retrospective analysis of 12 eyes in 10 patients diagnosed with culture-proven endogenous *Aspergillus* endophthalmitis, Weishaar et al. found that there was a characteristic acute onset of intraocular inflammation and a chorioretinal lesion located frequently in the macula (74). Despite resolution of the intraocular infection with vitrectomy and intravitreal amphotericin B, the visual outcome was generally poor, especially with macular lesions. All patients presented with a one- or two-day history of pain, and marked loss of visual acuity; vitritis was present in all 12 eyes. Nine of twelve eyes presented with a macular-involving chorioretinal inflammatory lesion. Four patients had associated pulmonary diseases and were receiving concurrent steroid therapy; six patients had a history of intravenous drug use. Blood cultures and echocardiograms were negative. Management consisted of vitrectomy in 10 of 12 eyes and intravitreal amphotericin B in 11 of 12 eyes. One patient was treated with oral antifungals while six patients completed systemic amphotericin B therapy. In three eyes without central macular involvement, final visual acuities were 20/25 to 20/200. In eight eyes with initial macular involvement, final visual acuities were 20/400 in three eyes and 1/200 or less in another three eyes. Two eyes were enucleated for chronic pain and recurrent retinal detachment.

In a histopathological study examining enucleated eyes secondary to candidal (13 eyes) and aspergillosis (12 eyes) endophthalmitis, Rao and Hidayat found that the vitreous was the primary focus of infection for *Candida* sp., whereas *Aspergillus* sp. preferred the subretinal pigment epithelium area (75). Therefore, vitreous biopsy may not yield positive results in *Aspergillus* endophthalmitis. Retinal and choroidal vessel wall invasion was seen in aspergillosis but not with candidiasis. *Candida* endophthalmitis was noted in patients with a history of gastrointestinal surgery, hyperalimentation, or diabetes mellitus, whereas aspergillosis was present in patients who had undergone organ transplantation or cardiac surgery. Both infectious agents induced suppurative non-granulomatous inflammation. This intraocular infection is usually associated with a high rate of mortality caused by cerebral and cardiac complications (75).

In another series analyzing endogenous fungal endophthalmitis over a 10-year period, Essman et al. found 16 of 20 cases due to *Candida* sp., with the remaining four cases caused by *Aspergillus* sp. (76). Eyes with *Candida* endophthalmitis presented in an indolent fashion with a mean duration from onset of symptoms to initial treatment of 61 days compared to five days with *Aspergillus* endophthalmitis. A final visual acuity of 20/400 or better was achieved in 75%, 80%, and 0% of eyes infected with *C. albicans*, *C. tropicalis,* and *Aspergillus* sp. Four eyes with *C. albicans* endophthalmitis had a final visual acuity of less than 20/400 as a result of retinal detachment. Direct infection of the macula was the cause of the poor final visual acuity in two patients infected with *Aspergillus* sp., while the third patient had a persistent retinal detachment. Only two patients in this series had positive non-ocular cultures. The diagnosis of endogenous fungal endophthalmitis occurred in an outpatient setting for all the patients; the presence of a disseminated focus of infection was found in only two patients (76)

Cryptococcosis. *Cryptococcus neoformans* is a non-mycelial, yeast-like encapsulated fungus with worldwide distribution. The organism is found in bird droppings, contaminated soil, and vegetation. *C. neoformans* is a budding, spore-forming yeast that produces smooth, yellow-tan colonies on culture. Systemic disease is most often acquired through inhalation of aerosolized organisms but direct entry through the digestive tract or skin can occur. Dissemination occurs most frequently to the central nervous system. In 1995, there were fewer than 30 cases of confirmed ocular cryptococcosis in the literature. The incidence of cryptococcosis and meningitis caused by *C. neoformans* in patients with AIDS has accounted for its increased incidence, with a rate of 2% to 9%. Patients may also be immunocompetent.

Systemic features include meningitis, pneumonia, mucocutaneous lesions, multiple skin lesions, pyelonephritis, endocarditis, hepatitis, and prostatitis. Ocular involvement includes sequelae from the systemic involvement as well as the primary infection. For instance, meningitis can cause papilledema, optic atrophy, or extraocular muscle paresis. In one series of 80 patients, Kestelyn et al. found papilledema in 32.5%, abducens nerve palsy in 9%, and optic atrophy in 2.5% of cases (77). The primary infection can cause choroiditis, retinitis, uveitis, inflammatory iris mass, keratitis, conjunctival granuloma, phthisis bulbi, periorbital necrotizing fasciitis, and orbital infection.

Diagnosis is made presumptively in a patient with characteristic clinical findings. In clinically suspicious patients with a negative serum, sputum, cerebrospinal fluid, blood examination, or vitreous biopsy, a vitrectomy may be positive with a fungal stain (Chapter 2) or by PCR (Chapter 3). Therapy depends on the etiology and requires a combination of systemic antifungal agents (amphotericin or imidazoles) and pars plana vitrectomy. If cryptococcal organisms are seen on the India ink stain at the time of diagnostic vitreous tap or vitrectomy, intravitreal amphotericin B should be administered immediately and systemic therapy should be started. Early vitrectomy is recommended if severe vitritis fails to clear or worsens with antifungal chemotherapy.

Pseudallescheriosis. *Pseudallescheria boydii* is a ubiquitous organism found worldwide in soil, stagnant water, and vegetation. This is an opportunistic fungus with low virulence that can imitate aspergillosis both clinically and histologically; thus, it also referred to as the "great imitator." The definitive diagnosis is made on fungal culture, which can differentiate it from *A. fumigatus*. The former appears velvety white and becomes dark gray, whereas the latter appears as green colonies (Chapter 2).

P. boydii is the sexual state of the organism (*Scedosporium apiospermum* is the asexual phase, which is cultured in the laboratory and is the preferred name) and causes invasive infections including sinusitis, otitis, pulmonary infection, prosthetic and native valve endocarditis, prostatitis, meningitis, brain abscess, arthritis, osteomyelitis, and disseminated disease. Eye pain is the most common symptom, followed by decreased or blurred vision, photophobia, and tearing. Visual acuity can range from 20/200 to no light perception. Other signs of endophthalmitis are also present.

Fusarium. *Fusarium* is an aggressive mycelial fungus that causes endophthalmitis that is particularly difficult to treat; it occurs in immunocompetent and -compromised patients. *Fusarium* endophthalmitis usually requires extensive surgery and intravitreal and systemic amphotericin, but may be amphoteniom-resistant, as well as imidazoles, but does not always respond. Sensitivity tests should be conducted in the laboratory.

Histoplasmosis. The dimorphic fungus *Histoplasma capsulatum* is found worldwide in valleys with rivers, but it is endemic in the midwest of the United States ("histoplasmosis belt"). It has also been reported with a similar clinical appearance (presumed ocular histoplasmosis syndrome) in Europe. Residents in those parts of the world may suffer from pulmonary disease and posterior uveitis. The uveitis usually follows a benign, asymptomatic course. Punched-out chorioretinal scars, 0.2–0.7 disc diameter in size, are characteristic for ocular histoplasmosis. An acute change in vision, however, occurs as a consequence of subretinal neovascularization and maculopathy. The prognosis is limited by involution of the neovascular complex that leads to an atrophic, disciform chorioretinal scar.

The diagnosis is mainly made clinically. Since pulmonary infection usually precedes ocular manifestation, a chest X ray may reveal calcified changes. The histoplasma complement fixation test is positive in only 15% of patients. The histoplasma skin test should not be performed anymore since it may reactivate the infection in ocular histoplasmosis.

Antifungal treatment with use of amphotericin B has not been effective in ocular histoplasmosis. Whereas only limited success could be obtained using focal laser coagulation to control SRNVM, photodynamic treatment and anti-VEGF treatment have been reported as successful. This true infection must be distinguished from the pseudo-infection mistakenly labeled ocular histoplasmosis syndrome, which consists of an inflammatory disease mimicking histoplasmosis. Histoplasmosis can become reactivated in AIDS patients.

Laboratory Examination.   The most important laboratory studies are Gram stain and culture of the aqueous and vitreous fluids collected before the institution of antibiotic therapy (Chapter 2). PCR tests should also be applied if possible for DNA of fungal species (Chapter 3). The use of blood culture bottles for the inoculation of ocular specimens is a good idea and increases the yield or chances of a positive culture. Cultures of other sites, including the intravenous catheter tip if present, should be obtained. Immunologic tests for specific bacterial antigens can be performed in patients who have already received antibiotics but the usefulness of these tests has been challenged. Molecular biology with PCR of anterior chamber (aqueous humor) and vitreous samples is producing positive results when traditional tests of stain and culture were negative and is particularly useful (Chapter 3).

Treatment

The outcome of endogenous endophthalmitis is generally disappointing compared with that of exogenous endophthalmitis. Three factors that result in a poor prognosis are:

- Virulent organisms
- Host immunosuppression
- Diagnostic delay

Even with aggressive treatment, vision is limited to counting fingers in about 40% of patients. Given its vascular spread, endogenous endophthalmitis is more responsive to intravenous antibiotics than is exogenous endophthalmitis, which occurs from intraocular injection of pathogens. Systemic medications also treat the source of the organism, thereby reducing the likelihood of invasion to the other eye. Furthermore, the nature of the clinical presentation, as well as the presumed (or confirmed) source of infection, can be used to guide antibiotic choice. The most important therapeutic principle in endophthalmitis is early diagnosis with correct identification of the organism. Sensitivity tests should be performed if possible and therapy with either amphotericin or imidazole instituted. Much better visual recovery takes place with early treatment.

Smiddy suggested that systemic treatment with oral fluconazole is as effective as amphotericin for managing simple choroiditis or minimal endophthalmitis (i.e., vitritis) (78). An alternative imidazole is itraconazole or voriconazole. For persistent or progressive vitritis, he recommends vitrectomy to clear the organism. Instead of intravitreal amphotericin B in conjunction with vitrectomy, the author asserted that an extended course of oral fluconazole following vitrectomy without intravitreal amphotericin B resolved the infection in a vast majority of patients. According to Smiddy, if the macula is spared and preretinal membranes can be effectively removed, visual acuity results can be exceedingly good.

Intravitreal injections have revolutionized the treatment of exogenous endophthalmitis, but their usefulness in endogenous cases is controversial (Chapter 8). Furthermore, in an attempt to reduce inflammation-mediated damage, corticosteroids such as dexamethasone have been administered intravitreally with some controversy (72). Topical steroids have been used empirically in patients with anterior (focal or diffuse) disease to prevent complications such as glaucoma and formation of synechia. Similarly, surgical intervention such as vitrectomy is widely accepted in post-surgical and post-traumatic endophthalmitis; however, its benefits

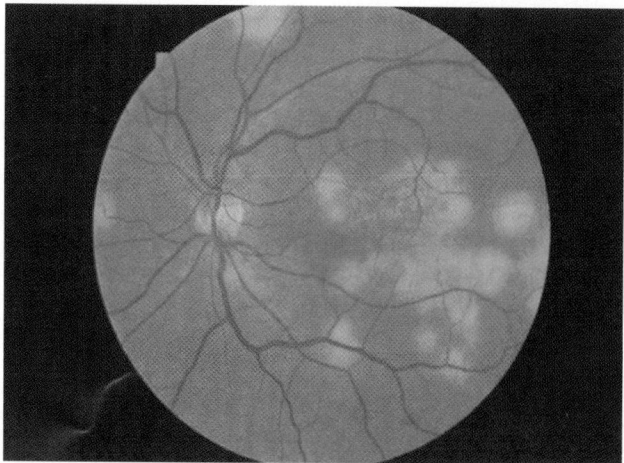

Figure 34 Unilateral APMPPE on presentation.

in endogenous endophthalmitis have been the subject of debate. Surgical intervention is generally recommended for patients infected with especially virulent organisms such as *S. pyogenes* (BHS group A), *S. pneumoniae, St. aureus,* or *Ps. aeruginosa*, in those with visual acuity of 20/400 or less and in cases of pupillary glaucoma from extensive posterior synechia or severe vitreous involvement. Vitrectomy and intravitreal antibiotics may also prevent ocular atrophy or the need for enucleation when the hypopyon level is high and there is evidence of vitreous opacities and retinal detachment on the B-scan examination.

Non-infectious Causes of Posterior Uveitis

Acute Posterior Multifocal Placoid Pigment Epitheliopathy (APMPPE)

The condition presents with a short history of progressive blurring of vision. There are multiple, discrete, cream-colored lesions with ill-defined margins deep within the retina in the posterior pole. It has been associated with a wide range of disorders including Lyme disease, infection by adenovirus, herpes viruses, and swine flu vaccination. It was originally described in association with tuberculosis (refer to Chapter 8) and this reflects current experience (Fig. 34–35). When due to tuberculosis, it may be synonymous with multifocal choroiditis. APMPPE may occur with a closed pulmonary lesion as opposed to miliary disease.

Leber's Idiopathic Stellate Neuroretinitis

This condition presents with sudden loss of vision following a viral-type illness, such as hepatitis A, cat-scratch fever, or leptospirosis and occasionally after other systemic infections including toxocaral disease. It may also occur in the early stage of asymptomatic neurosyphilis, and appropriate tests should always be made. In the first week after the onset of decreased visual acuity there is usually only swelling of the disc to be seen, but one week later a typical macular star (Fig. 36) is present. This is not a maculopathy but is caused by vascular leakage from the optic nerve head with reabsorption of serum when lipid precipitates in a stellate pattern. The swelling

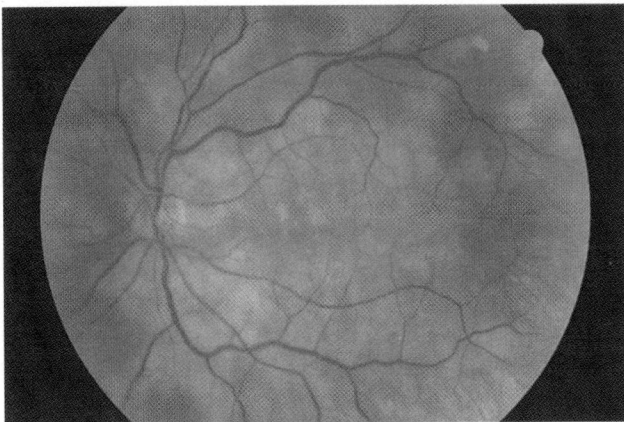

Figure 35 Unilateral APMPPE 3 weeks after presentation, considerably worsened and associated with active pulmonary tuberculosis.

of the optic disc may be mild, segmental, or massive. It is not related to demyelinating disease, but a differential diagnosis for the disc swelling may include leakage from the parafoveal capillaries with diabetes mellitus and hypertension.

There is no specific therapy. The disc swelling begins to decrease after two weeks and resolves within 10 weeks. The macular star begins to decrease after one month but requires up to one year for complete resolution. The prognosis for vision is good but there are exceptions.

Multiple Evanescent White Dot Syndrome

Viral-induced chorioretinitis has been postulated as a cause for some of the transient white dot syndromes. The patient has usually suffered a flu-like illness in the week before presentation. There is an evanescent multifocal chorioretinitis seen as multiple white dots in the paramacular area that can be documented by

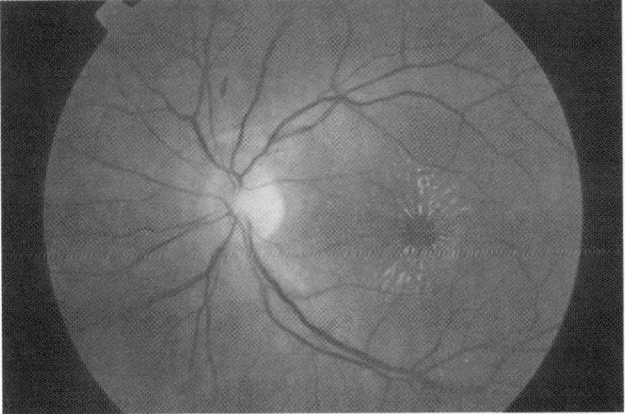

Figure 36 Leber's idiopathic stellate neuroretinitis following a systemic virus infection (non-A hepatitis).

fluorescein angiography at the retinal pigment epithelial layer. The disease is usually unilateral in young adults who present with a sudden loss of vision, decreasing to between 20/20 and 20/300, sometimes with a temporal scotoma. There can be vitreal cells and a blurring of the disc margin. Symptomatic macular lesions have responded to oral corticosteroid therapy. Recovery to normal vision usually occurs after six weeks.

REFERENCES

1. Liekfeld A, Schweig F, Jaeckel C, Wernecke K-D, Hartmann C, Pleyer U. Intraocular antibody production in intraocular inflammation. Graefe's Arch Clin Exp Ophthalmol 2000; 238:222–27.
2. Feltkamp TEW. Are Gram negative bacteria involved in HLA-B27 associated uveitis? Brit J Ophthalmol 1995; 79:718–20.
3. Otasevic l, Walduck A, Meyer TF, Aebischer T, Hartmann C, Orlic N, Pleyer U. Helicobacter pylori infection: a possible risk factor for anterior uveitis? Infection 2005; 33:82–85
4. Benjamin A, Tufail A, Holland GN. Uveitis as the only clinical manifestation of post-streptococcal syndrome. Am J Ophthalmol 1997; 123:258–60.
5. Schürmann D, Bergmann L, Bertelmann E, et al. Early diagnosis of acquired ocular syphilis requires a high index of suspicion and may prevent visual loss. AIDS 1999, 13:623–24.
6. Porstmann A, Marcus U, Pleyer U. Syphilisdiagnose durch den Augenarzt Klin Monatsbl Augenheilkd 2002; 219:349–352.
7. Pleyer U, Priem S, Bergmann L, Burmester G, Hartmann C, Krause A. Detection of *Borrelia burgdorferi* DNA in urine of patients with ocular Lyme borreliosis. Br J Ophthalmol 85:552–555, 2001
8. Peyman G, Lee P, Seal DV. Endophthalmitis: Diagnosis and Management. London & New York: Taylor & Francis, 2004:1–278.
9. Hill D, Dubey JP. Toxoplama gondii: transmission, diagnosis and prevention. Clin Microbiol Infect 2002; 8:634–640.
10. Holland GN. Ocular toxoplasmosis: a global reassessment. Part I: epidemiology and course of disease. Am J Ophthalmol 2003; 136:973–988.
11. Holland GN. Ocular toxoplasmosis: a global reassessment. Part II: disease manifestations and management. Am J Ophthalmol 2004; 137:1–17.
12. Montoya JG, Liesenfeld O. Toxoplasmosis. Lancet. 2004; 363(9425):1965–76.
13. Bosch-Driessen LE, Berendschot TT, Ongkosuwito JV, et al. Ocular toxoplasmosis: clinical features and prognosis of 154 patients. Ophthalmol 2002; 109:869–78.
14. Mahalakshmi B, Therese KL, Madhavan HN, et al. Diagnostic value of specific local antibody production and nucleic acid amplification technique-nested polymerase chain reaction (nPCR) in clinically suspected ocular toxoplasmosis. Ocul Immunol Inflamm 2006; 14:105–112.
15. Wallon M, Kodjikian L, Binquet C, et al. Long-term ocular prognosis in 327 children with congenital toxoplasmosis. Pediatrics 2004; 113:1567–72.
16. Silveira C, Belfort R Jr, Muccioli C, et al. The effect of long-term intermittent trimethoprim/sulfamethoxazole treatment on recurrences of toxoplasmic retinochoroiditis. Am J Ophthalmol 2002; 134:41–46.
17. Bertelmann E, Velhagen KH, Pleyer U, et al. Okuläre Toxocariasis. Diagnostische und therapeutische Optionen. Ophthalmologe 2003; 100:950–54.
18. Campos Junior D, Elefant GR, de Melo E, et al. Frequency of seropositivity to Toxocara canis in children of different socioeconomic strata. Rev Soc Bras Med Trop 2003; 36:509–13.

19. Despommier D. Toxocariasis:Clinical aspects, epidemiology, medical ecology, and molecular aspects. Clin Microbiol Rev 2003; 16:265–72.
20. Stewart JM, Cubillan LD, Cunningham ET. Prevalence, clinical features, and causes of vision loss among patients with ocular toxocariasis. Retina 2005; 25:1005–13.
21. McDonald HR, Kazacos KR, Schatz, et al. Two cases of intraocular infection with *Alaria* mesocercaria (Trematoda). Am J Ophthalmol 1994; 117:447–55.
22. Taylor HR, Nutman TB. Onchocerciasis. In: JS Pepose, GN Holland, KR Wilhelmus (eds). Ocular Infection and Immunity. Mosby Year Book, Chicago: 1996; 1481–504.
23. Mabey D, Whitworth JA, Eckstein M, et al. The effects of multiple doses of Ivermectin on ocular onchocerciasis: a six-year follow-up. Ophthalmol 1996; 103:1001–08.
24. Gaudio PA. Update on ocular syphilis. Curr Opin Ophthalmol 2006; 17:562–66.
25. Schmidt BL. PCR in laboratory diagnosis of human *Borrelia burgdorferi* infections. Clin Microbiol Rev 1997; 10:185–201.
26. Mendez-Hernandez C, Garcia-Feijoo J, Garcia-Sanchez J. *Listeria monocytogenes*-induced endogenous endophthalmitis: bioultrasonic findings. Am J Ophthalmol 2004; 137:579–81.
27. Lau CH, Missotten T, Salzmann J, et al. Acute Retinal Necrosis Features, Management, and Outcomes. Ophthalmology 2007; 114:756–62.
28. Gargiulo F, De Francesco MA, Nascimbeni G, et al. Polymerase chain reaction as a rapid diagnostic tool for therapy of acute retinal necrosis syndrome. J Med Virol. 2003; 69:397–400.
29. Aizman A, Johnson MW, Elner SG. Treatment of acute retinal necrosis syndrome with oral antiviral medications. Ophthalmology. 2007; 114:307–12.
30. Balansard B, Bodaghi B, Cassoux N, et al. Necrotising retinopathies simulating acute retinal necrosis syndrome. Brit J Ophthalmol 2005; 89:96–101.
31. Wood E, Hogg R, Yip B, et al. Superior virological response to boosted protease-inhibitor-based highly active anti-retroviral therapy in an observational treatment program. HIV Med 2007; 8:80–85.
32. Jabs DA, van Natta ML, Holbrook JT, et al. Logitudinal study of the ocular complications of AIDS 1. Ocular diagnoses at enrollment. Ophthalmol 2007; 114:780–6.
33. Bannister WP, Ruiz L, Loveday C, et al. HIV-1 subtypes and response to combination antiretroviral therapy in Europe. Antivir Ther 2006; 11:707–15.
34. Accorinti M, Pirraglia MP, Corradi R, et al. Changing patterns of ocular manifestations in HIV seropositive patients treated with HAART. Eur J Ophthalmol 2006; 16:728–32.
35. Jabs DA, van Natta ML, Holbrook JT, et al. Logitudinal study of the ocular complications of AIDS 2. Ocular examination results at enrollment. Ophthalmol 2007; 114:787–93.
36. Hanna DB, Gupta LS, Jones LE, et al. AIDS-defining opportunistic illnesses in the HAART era in New York City. AIDS Care 2007; 19:264–72.
37. Thorne JE, Holbrook JT, Jabs DA, et al. Effect of cytomegalovirus retinitis on the risk of visual acuity loss among patients with AIDS. Ophthalmol 2007; 114:591–8.
38. Podiekareva D, Mocroft A, Dragsted UB, et al. Factors associated with the development of opportunistic infections in HIV-1 infected adults with high CD4+ cell counts: a EuroSIDA study. J Infect Dis 2006; 194:633–41.
39. Uemura A, Yashiro S, Takeda N, et al. [Ocular complications in patients with HIV infection]. Nippon Ganka Gakkai Zasshi 2006, 110.698–702.
40. Holkmann OC, Mocroft A, Kirk O, et al. Interruption of combination antiretroviral therapy and risk of clinical disease progression to AIDS or death. HIV Med 2007; 8:96–104.
41. Flakenstein I, Kozak I, Kayikcioglu O, et al. Assessment of retinal function in patients with HIV without infectious retinitis by multifocal electroretinogram and automated perimetry. Retina 2006; 26:928–34.
42. Hemady RK. Microbial keratitis in patients infected with the human immunodeficiency virus. Ophthalmol 1995; 102:1026–30.

43. Rastrelli PD, Didier E, Yee RW. Microsporidial keratitis. Ophthalmol Clin N Am 1994; 7:635–40.

44. Baglivo E, Dosso A, Leuenberger PM. Cytomegalovirus retinitis in an AIDS patient without severe depletion in CD4 cell count. Brit J Ophthalmol 1995; 79:962–63.

45. Faber DW, Wiley CA, Lynn GB, et al. Role of HIV and CMV in the pathogenesis of reitinitis and retinal vasculopathy in AIDS patients. Investig Ophthalmol Vis Sci 1992; 33:2345–33.

46. Garcia RF, Flores-Aguila M, Quiceno JI. Results of rhegmatogenous retinal detachment repair in CMV retinitis with and without scleral buckling. Ophthalmol 1995; 102:236–45.

47. Badouin C, Chassain C. Treatment of CMV retinitis in AIDS patients using intra-vitreal injections of highly concentrated ganciclovir. Ophthalmologica 1996; 210:329–35.

48. Hodge WG, Lalonde RG, Samplais J, et al. Once weekly intra-ocular injections of ganciclovir for maintenance therapy of CMV retinitis: clinical and ocular outcome. J Infect Dis 1996; 174:393–406.

49. Marx JL, Kapusta MA, Patel SS, et al. Use of the ganciclovir implant in the treatment of recurrent CMV retinitis. Arch Ophthalmol 1996; 114:815–20.

50. Spector SA, McKinley GF, Lalezari JP, et al. Oral ganciclovir for the prevention of cytomegalovirus disease in persons with AIDS. New Eng J Med 1996; 334:1491–7.

51. Weinberg DV, Murphy R, Naughton K. Combined daily therapy with intravenous ganciclovir and foscarnet for patients with recurrent CMV retinitis. Am J Ophthalmol 1994; 117:776–82.

52. Kuppermann BD, Flores-Aguilar M, Quicenco JI, et al. Combination ganciclovir and foscarnet in treatment of clinically resistant CMV retinitis in patients with AIDS. Arch Ophthalmol 1993; 111:1359–66.

53. Desatnik HR, Forster RE, Lowder CY. Treatment of clinically-resistant CMV retinitis with combined intravitreal injections of ganciclovir and foscarnet. Am J Ophthalmol 1996; 122:121–3.

54. Lalezari JP, Stugg RJ, Kupperman BD, et al. IV cidofovir for peripheral CMV retinitis in patients with AIDS. Ann Int Med 1997; 126:257–63; 264–74.

55. Madreperla SA, Bardenstein DS, Salata R, et al. Anterior uveitis associated with treatment of CMV retinitis in AIDS patients by intravenous cidofovir. Investig Ophthalmol Vis Sci 1997; 38 (suppl.): abstract 3425.

56. Morlet N, Young S, Naidoo D, et al. High dose intravitreal ganciclovir for CMV retinitis: a shelf life and cost comparison study. Brit J Ophthalmol 1995; 79:752–55.

57. Caserta FP, Goldstein GA, Gramates PH, et al. Does intravitreal therapy for CMV retinitis increase the risk of retinal detachment? Investig Ophthalmol Vis Sci 1997; 38 (suppl.): abstract 3414.

58. Rahal FM, Arevalo F, Munguia D, et al. Intravitreal cidofovir for the maintenance treatment of CMV retinitis. Ophthalmol 1996; 103:1078–83.

59. Chong LP, Marx JL, Thach Ab, et al. Recurrence rate of CMV retinitis in eyes with the gangciclovir implant and silicone oil. Investig Ophthalmol Vis Sci 1997; 38 (suppl.): abstract 3417.

60. Charles NC, Steiner G. Ganciclovir intraocular implant—a clinicopathologic study. Ophthalmol 1996; 103:416–21.

61. Lim JI, Wolitz RA, Dowling AH, et al. Visual and anatomic outcomes associated with posterior segment complications after ganciclovir implant procedures in AIDS and patients with CMV retinitis. Am J Ophthalmol 1999; 127: 288–93.

62. Labetoulle M, Goujard C, Frau E, et al. Cytomegalovirus vetinitis in advanced HIV-infected patients treated with putease inhibitors: incidence and outcome over 2 years. J Acquir Immune Defic Syndr 1999; 22: 228–34.

63. Morrison VL, Kozak I, LaBree LD. Intravitreal triamcinolone acetonide for the treatment of immune recovery uveitis macular edema. Ophthalmol 2007; 114:334–9.

64. Baker ML, Allen P, Shortt J, et al. Immune recovery uveitis in an HIV-negative individual. Clin Exp Ophthalmol 2007; 35:189–90.

65. Lee MW, Fong KS, Hsu LY, et al. Optic nerve toxoplasmosis and orbital inflammation as initial presentation of AIDS. Graefes Arch Clin Exp Ophthalmol 2006; 244:1542–44.

66. Donahue SP, Greven CM, Zuravleff JJ, et al. Intraocular candidiasis in patients with candidemia. Clinical implications derived from a prospective multicenter study. Ophthalmol 1994; 101:1302–09.

67. Feman SS, Nichols JC, Chung SM, Theobald TA. Endophthalmitis in patients with disseminated fungal disease. Trans Am Ophthalmol Soc 2002; 100:67–70; discussion 70–1.

68. Scherer WJ, Lee K. Implications of early systemic therapy on the incidence of endogenous fungal endophthalmitis. Ophthalmology 1997; 104:1593–8.

69. Binder MI, Chua J, Kaiser PK, Procop GW, Isada CM. Endogenous endophthalmitis: an 18-year review of culture-positive cases at a tertiary care center. Medicine (Baltimore) 2003; 82:97–105.

70. Harvey RL, Myers JP. 1987. Nosocomial fungemia in a large community teaching hospital. Arch Intern Med 1987; 147:2117–20.

71. Tanaka M, Kobayashi Y, Takebayashi H, Kiyokawa M, Qiu H. Analysis of predisposing clinical and laboratory findings for the development of endogenous fungal endophthalmitis. A retrospective 12-year study of 79 eyes of 46 patients. Retina 2001; 21:203–9.

72. Brooks RG. Prospective study of *Candida* endophthalmitis in hospitalized patients with candidemia. Arch Intern Med 1989; 149:2226–8.

73. Benjamin DK, Poole C, Steinbach WJ, Rowen JL, Walsh TJ. Neonatal candidemia and end-organ damage: a critical appraisal of the literature using meta-analytic techniques. Pediatrics 2003; 112:634–40.

74. Weishaar PD, Flynn HW Jr, Murray TG, et al. Endogenous Aspergillus endophthalmitis. Clinical features and treatment outcomes. Ophthalmology 1998; 105:57–65.

75. Rao NA, Hidayat AA. Endogenous mycotic endophthalmitis: variations in clinical and histopathologic changes in candidiasis compared with aspergillosis. Am J Ophthalmol 2001; 132:244–51.

76. Essman TF, Flynn HW Jr, Smiddy WE, et al. Treatment outcomes in a 10-year study of endogenous fungal endophthalmitis. Ophthalmic Surg Lasers 1997; 28:185–94.

77. Kestelyn P, Taelman H, Bogaerts J. Ophthalmic manifestations of infections with *Cryptococcus neoformans* in patients with Acquired Immune Deficiency Syndrome. Am J Ophthalmol 1993; 116:721–7.

78. Smiddy WE. Treatment outcomes of endogenous fungal endophthalmitis. Curr Opin Ophthalmol 1998; 9:66–70.

EIGHT

Endophthalmitis Including Prevention and Trauma

INTRODUCTION

Endophthalmitis implies infection of the vitreous compartment together with the retinal and uveal coats of the eye. It may present as endogenous or as exogenous infection, involving either intraocular surgery (such as cataract, glaucoma, or occasionally squint) when pathogens harbored on the lids and conjunctival sac will determine the infecting organism(s) or following penetrating injury to the eye. The formation of thin-walled drainage blebs after glaucoma surgery using mitomycin C may predispose to late infection. Bacterial and fungal endophthalmitis may also follow penetrating keratoplasty.

Endogenous Bacterial Endophthalmitis

Endogenous or metastatic endophthalmitis, a severe vision-threatening intraocular infection, occurs through bloodstream spread from a concurrent infection in the host or an external source such as a catheter or an intravenous line. Endogenous bacterial infection is relatively rare, accounting for 2% to 8% of all cases (1,2). Many patients were diagnosed with medical conditions including diabetes, hypertension, gastro-intestinal disorders, cardiac disorders, malignancy, and immunosuppression or else prolonged surgical complications (2). Systemic symptoms rather than acute ocular symptoms were the most common reasons to present to the physician and many cases were initially misdiagnosed.

Jackson et al. recently reviewed 267 reported cases of endogenous bacterial endophthalmitis and also presented a 17-year prospective series (2). Blood cultures were the most frequent means for establishing the infective cause. The most common Gram-positive bacteria were *Staphylococcus aureus*, group B streptococci, *Streptococcus pneumoniae* (3), and *Listeria monocytogenes*. Gram-negative bacteria included *Klebsiella* sp., *Escherichia coli*, *Pseudomonas aeruginosa*, and *Neisseria meningitidis*. Gram-positive bacteria were mostly involved in North America and Europe, with Gram-negative bacteria in East Asia, which is similar to the situation with bacterial keratitis (Chapter 6). The visual outcome was poor, resulting frequently in a blind eye. In addition, the overall mortality rate for these patients was 5%. The outcome of endogenous bacterial endophthalmitis is not considered to have improved over the last 55 years, presenting a challenge to today's clinicians to use new and better techniques and drugs (refer to flow charts below). Recently reported results for the effectiveness of vitrectomy and intravitreal antibiotics in this group to save useful vision are encouraging (1).

Endogenous bacterial endophthalmitis is bilateral in approximately 14% to 25% of cases. In bilateral infection, simultaneous ocular involvement is the rule; however, one eye is characteristically more severely affected than the other eye. Delayed involvement of the second eye can occur even in patients already being treated with systemic antibiotics. The right eye is involved twice as often as the left, probably because of this eye's proximity and more direct blood flow from the right carotid artery. There is no gender predisposition.

Endogenous Fungal Endophthalmitis

The incidence rate of endogenous fungal endophthalmitis has been estimated in the past to range from 9% to 45% of patients with disseminated fungus infection or fungemia (4). However, recent studies have reported an incidence of 2% or 2.8% (Chapter 7) (4).

This situation has now changed for *Candida albicans* (Chapters 7 and 9), with early aggressive systemic antifungal chemotherapy for either growth of fungi from blood cultures or local sites of fungal infection thought on clinical criteria to be disseminated. Because of the rapid advance of medical technology, a longer lifespan of patients with chronic diseases, and a rising prevalence of long-term intravenous access, *C. albicans* retinochoroiditis may become more common in clinical practice albeit less common in those with organ transplants and immunosuppression as the result of early ophthalmological screening of all susceptible patients.

Endogenous endophthalmitis is most commonly caused by fungi. *Candida albicans* is by far the most frequent cause (75% to 80% of fungal cases), followed by *Aspergillus* sp. Endophthalmitis occurs in 5% to 78% of patients with candidemia (4); however, this probably reflects the catchment population of the reporting hospital, particularly if it has an organ transplant unit with highly immunosuppressed patients. *C. albicans* retinochoroiditis is fully described in Chapters 7 and 9.

Aspergillosis is the second most common cause of fungal endophthalmitis, especially in intravenous drug users. Less frequent are other *Candida* sp., *Torulopsis glabrata*, *Cryptococcus neoformans*, *Sporothrix schenckii*, *Scedosporium apiospermum* (*Pseudallescheria boydii*), *Blastomyces dermatitidis*, *Coccidioides immitis*, and mucor. These fungi are described in Chapter 9.

The most important host sources of infectious agents include endocarditis, the gastrointestinal tract, genitourinary tract, skin and wound infection, pulmonary infections, meningitis, and septic arthritis. Other predisposing factors include invasive procedures, such as hemodialysis, bladder catheterization, gastrointestinal endoscopy, total parenteral nutrition, chemotherapy, dental procedures, and intravenous drug abuse.

Management of Endogenous Endophthalmitis

The patient requires immediate investigation (see Flow Chart 1 below) with blood cultures, and anterior chamber and vitreous taps and possibly vitrectomy (1). Gram stain and semi-quantitative culture of aqueous and vitreous should be performed (Chapter 2). Samples should be kept frozen for polymerase chain reaction (PCR) if no organism is identified (Chapter 3). Isolation of any bacterial colonies on direct

Flow Chart 1   Diagnostic Guidelines for Acute Endophthalmitis

Observe the patient for:

- pain
- blurring or loss of vision, which may appear as a darkened image due to developing vitritis, down to perception of light
- swollen lids
- inflamed or edematous conjunctiva
- discharge into conjunctiva
- corneal edema sometimes with an filtrate or ring abscess
- cloudy anterior chamber with cells, hypopyon, or fibrin clot
- afferent pupillary defect
- vitreous clouding (vitritis) from inflammation precluding a view of the retinal arterioles
- involvement of posterior segment with retinitis, and/or retinal periphlebitis, retinal edema, and papillary edema
- absent red reflex when the vitreous is viewed through the pupil may be a poor guide to the state of the vitreous, which may be most opaque anteriorly where the inflammatory process has begun. If the pupil is observed while transilluminating the sclera, the red reflex may become apparent and can then form a better guide to control of the disease

↓

Check B-scan ultrasonography for vitritis and retinal detachment—this is a useful adjunct for the clinical evaluation of infectious endophthalmitis, especially in an eye with opaque media

↓

MAKE A CLINICAL DIAGNOSIS OF ENDOPHTHALMITIS
(with photography if possible)

↓

THIS IS A MEDICAL EMERGENCY!

↓

PERFORM AN INTRAVITREAL TAP WITHIN ONE HOUR!

- perform an anterior chamber tap for microbiology [Gram stain, culture, and PCR[a] (Chapter 3)]; either perform the vitreous tap in the Operating Theater using a vitrector or phako/vitrector or use a portable vitrector (Becton Dickinson (BD) Visitrec 5100) in the outpatient clinic (refer to text)
- collect samples of aqueous and vitreous for microbiology (Gram stain, culture, and PCR[a])
- inject empirical choice of antibiotics[b] (refer to text and flow diagram below) **AND** unpreserved dexamethasone (400 µg in 0.1 mL) into vitreous with separate syringes and needles

[a] Expert PCR (polymerase chain reaction) for bacteria and fungi causing acute and chronic endophthalmitis is available from Dr. Udo Reischl, Institute for Medical Microbiology and Hygiene, University Hospital, 93053 Regensburg, Germany (udo.reischl@klinik.uni- regensburg.de) or from Prof. Jorge Alio and Dr. Consuelo Ferrer, VISSUM-Instituto Oftalmologico de Alicante, 03016 Alicante, Spain (cferrer@vissum.com) at a cost of 250 EUR per sample. Collect samples of one drop of aqueous and one drop of vitreous, each placed in a separate sterile Eppendorf plastic tube. Specimens should be stored at + 4°C for up to 24 hours and sent by next-day courier service to the laboratory or stored at -20°C for longer periods. Do not send by courier over a weekend. It is advised to contact the units by e-mail before sending any samples.

[b] ALWAYS have a chosen empirical regime of antibiotics ready in advance for intravitreal use in a clinic or operating theater setting. Have instructions prepared for making up correct dilutions (see text) and have necessary sterile equipment (bottles and syringes) available.

inoculation of agar plates cultured aerobically, microaerophilically, or anaerobically should be deemed indicative of culture-positive endophthalmitis. A combination of intravitreal antibiotics should be injected, such as vancomycin (1 mg) or cephazolin (or ceftazidime) (2 mg) plus amikacin (0.4 mg) [or gentamicin (0.2 mg)]; 0.1 mL of each of the two chosen antibiotics should be injected in separate syringes intravitreally after the tap has been performed (see text below). An alternative, less toxic combination is intravitreal vancomycin plus ceftazidime, so avoiding use of an aminoglycoside (gentamicin or amikacin). Systemic antibiotics should be given accordingly to target the focus of infection such as endocarditis in association with an infectious disease physician. The vitreoretinal surgeon should be consulted urgently about the merits or advantages to be gained by early vitrectomy, either partial or full (1). Discussion on further management including surgery is given below. Treatment of retinochoroiditis due to *Candia albicans* is given in Chapters 7 and 9.

Exogenous Bacterial Endophthalmitis

The most common presentation today is following cataract (phacoemulsification) surgery for which the incidence and other details are reviewed below. It may also follow glaucoma and, occasionally, squint surgery or trauma with a foreign body.

Symptoms of acute endophthalmitis include pain, swollen lids, conjunctival discharge, blurring of vision, and loss of vision (Figs. 1–5). Signs of acute endophthalmitis include swollen lids, conjunctival and corneal edema, reduced visual acuity down to perception of light, a turbid anterior chamber with cells, hypopyon or fibrin clot, and/or vitreous clouding precluding a view of the retinal vessels; often no view of the posterior segment is possible. However, it is worth stressing that loss of the red reflex when the vitreous is viewed through the pupil may be a poor guide to the state of the vitreous, which may be most opaque anteriorly where the inflammatory process has begun. If the pupil is observed while transilluminating the sclera, the red reflex may become apparent and can then form a better guide to

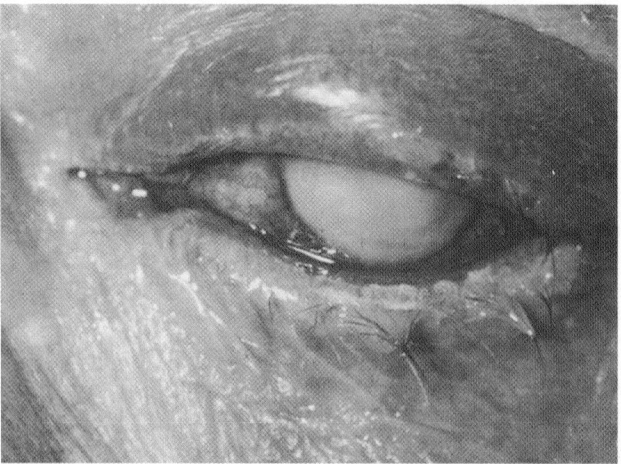

Figure 1   Acute suppurative endophthalmitis due to *Streptococcus pyogenes* occurring 3 days after cataract surgery, presenting with a swollen lid, fully opaque cornea with severe limbitis, no visual perception and much pain.

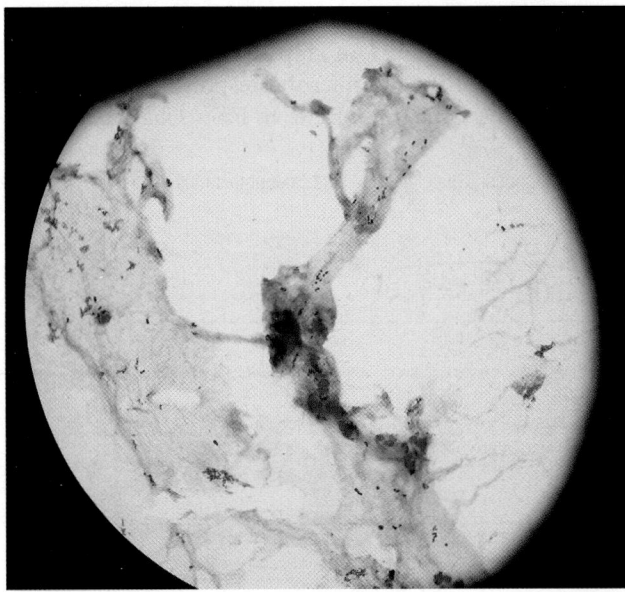

Figure 2   Gram stain of vitreous showing streptococci and pus cells.

control of the disease. Ocular echography is a useful adjunct for the clinical evaluation of infectious endophthalmitis, especially in eyes with opaque media.

Patients may present within 6 weeks of cataract or other ophthalmic surgery with reduced visual acuity and signs of inflammation in the anterior chamber that have been

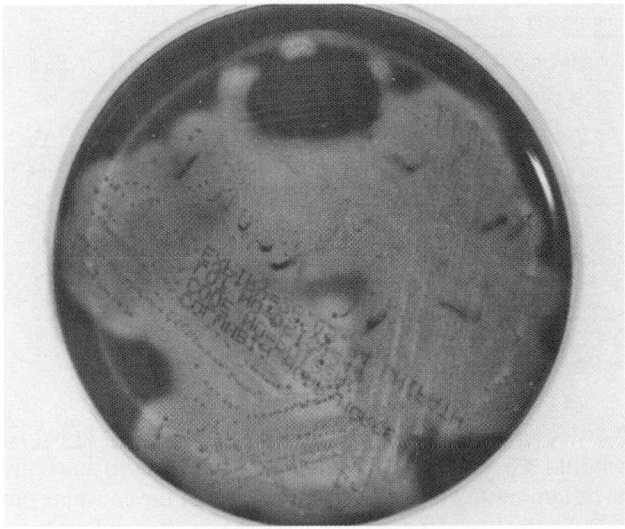

Figure 3   Blood agar plate culture of vitreous sample from the patient in Fig. 1 showing a heavy growth of beta-haemolytic streptococci (*Streptococcus pyogenes* Lancefield group A) with a pool of inhibition at the site of plate inoculation due to a low concentration of vancomycin present from intravenous therapy, ineffective as the antibiotic did not penetrate the posterior segment of the eye in sufficient concentration.

managed with corticosteroids. They should be managed as for acute endophthalmitis. If vitrectomy is not indicated, an anterior chamber (AC) tap and a vitreous tap will be required to confirm infection and to identify the microbe responsible; antibiotics should be instilled into the vitreous at the same time (Refer to Flow Chart 2).

Flow Chart 2   Diagnostic Guidelines for Late Endophthalmitis

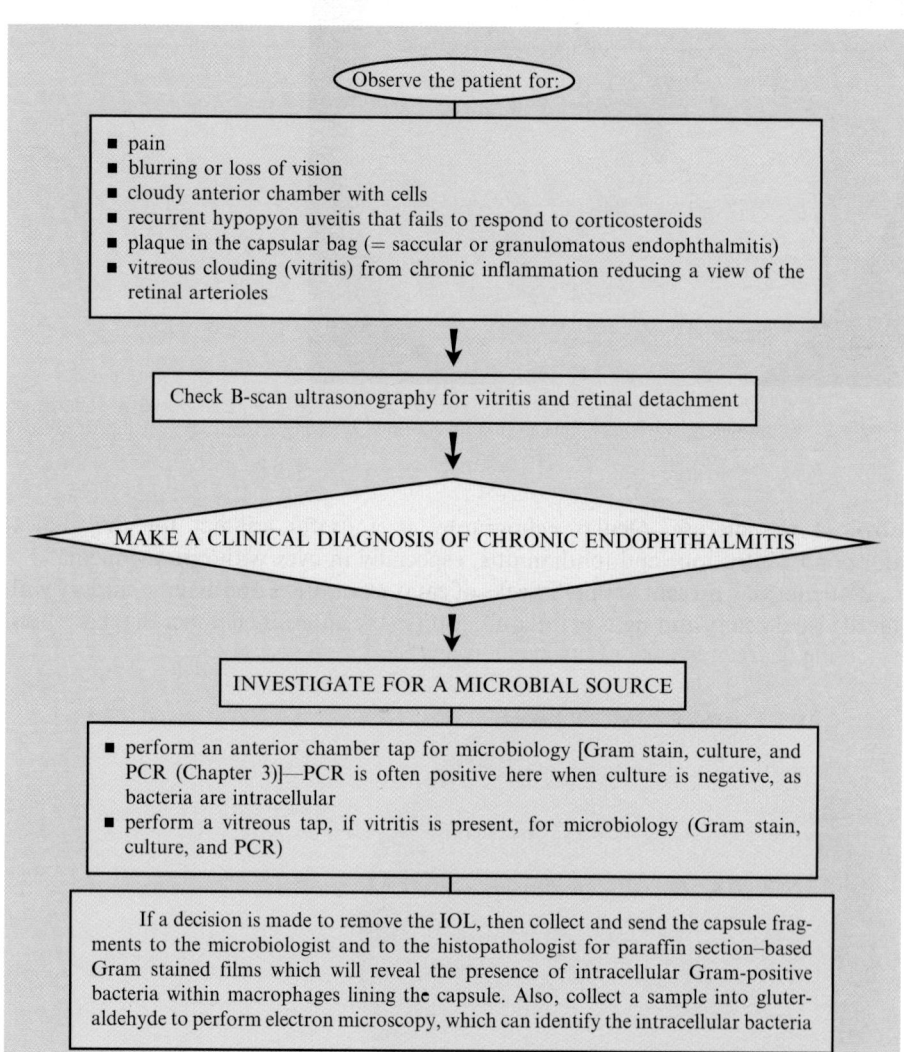

Patients presenting late after surgery with hypopyon uveitis that has failed to respond to corticosteroids should be investigated for chronic infection, and the possibility considered that the inflammation is due to infection within the capsular sac with *P. acnes* (Fig. 6), other propionibacteria, or corynebacteria (Fig. 7), or coagulase-negative staphylococci (CNS) (Figs. 8, 9) (4). This can be established, even at a late stage after 1 year or more, by an AC and vitreous tap followed by Gram stain, culture, and PCR (see Chapter 3); both samples should be collected, as only one or other may be culture-positive. There may be insufficient numbers of organisms present in the sample to yield a positive Gram stain in the presence of a positive culture; laboratories should

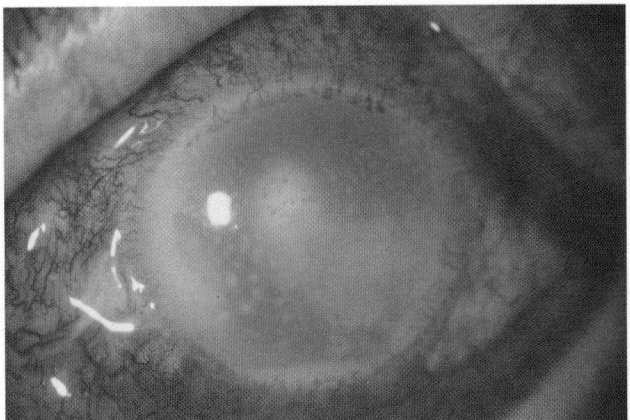

Figure 4   Acute suppurative endophthalmitis due to *Staphylococcus aureus*.

be aware that *P. acnes* can appear as a Gram-variable coccobacillus when stained from the AC or vitreous. To isolate *P. acnes*, there must be strict anaerobic growth conditions with incubation for at least 14 days. The PCR technique (see below and Chapter 3) has been found more sensitive to identify bacteria within these specimens than culture. There may be viable *P. acnes* organisms sequestered within macrophages within lens capsular remnants when surgical removal may be required with intraocular lens exchange to prevent recurrent or persistent endophthalmitis. The lens fragment should be processed for histology when a Gram stain on the fixed tissue can demonstrate Gram-positive cocci within the macrophages (4).

A medical emergency exists as soon as the clinical diagnosis of acute bacterial endophthalmitis is made. There is a need to perform an anterior chamber tap and a vitreous tap within one hour of clinical diagnosis and to instill intravitreal antibiotics. The acute bacterial process is very unforgiving, giving rise to massive inflammation within the

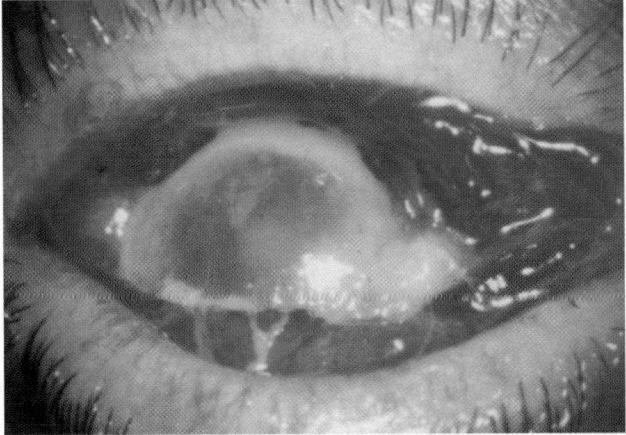

Figure 5   Excessive conjunctival chemosis, corneal edema, infiltration, and exudation with ring abscess, anterior chamber reaction, and papillary membrane formation—acute postoperative endophthalmitis due to *Pseudomonas aeruginosa* at day 4.

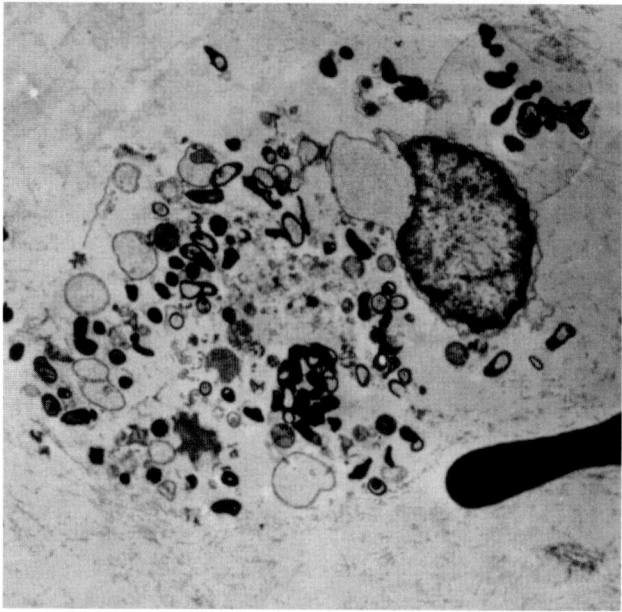

Figure 6 Electron micrograph shows *Propionibacterium acnes* in a macrophage. *P. acnes* was cultured from an eye with late-onset chronic plaque-type endophthalmitis.

posterior segment with a breakdown of the blood-ocular barrier. To save vision the acute bacterial process must be stopped but also the acute inflammation that it has caused must be minimized, hence the current opinion that dexamethasone 400 µg in 0.1 mL should be injected intravitreally at the same time. The full arguments for and against this approach, based on a scientific review, are considered by Peyman, Lee, and Seal (4).

The question arises how to carry out the urgent taps and antibiotic instillation. In some hospitals it is not possible to take the patient to the operating theatre at any given time because of its use by others. It is *not* advisable to keep the patient waiting for more than three hours, because the chances of retaining reasonable vision

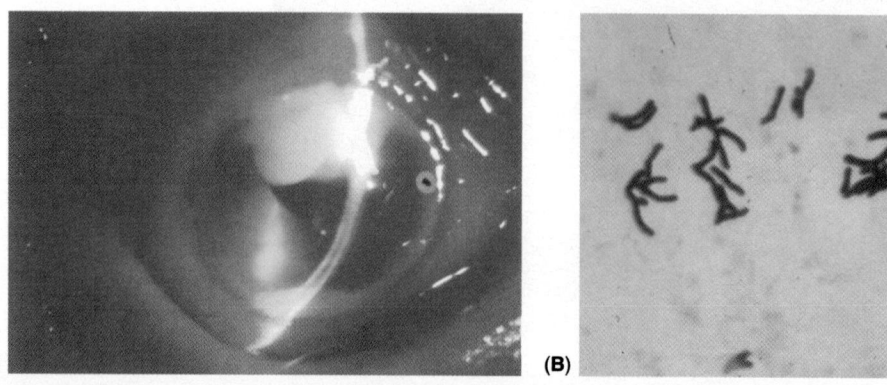

(A)   (B)

Figure 7 (**A**) Postoperative endophthalmitis due to *Corynebacterium macginleyi*. (**B**) Gram stain of *Corynebacterium macginleyi* (Courtesy of Prof. J. Alio).

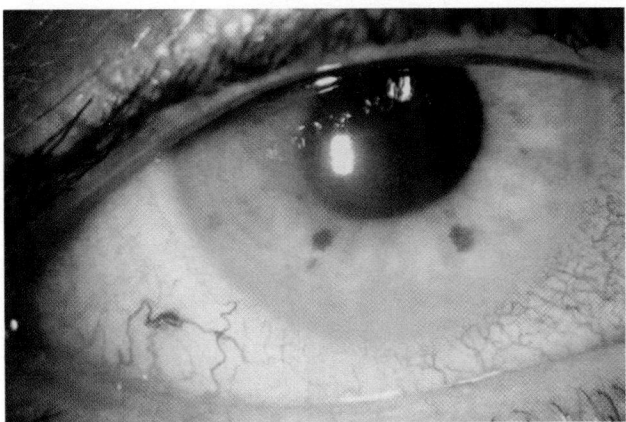

Figure 8 Chronic low-grade endophthalmitis following cataract surgery, with steroid-suppressed hypopyon, due to coagulase-negative staphylococci (CNS), with maintenance of a clear cornea and partially reduced visual perception.

diminish quickly. For this reason, a portable vitrector has been developed by Peyman and Visitec™ to make an intravitreal tap within the hospital ward or outpatient clinic and to instill intravitreal antibiotics and dexamethasone at the same time. This procedure is quite safe to perform in the outpatient clinic providing that antiseptic prophylaxis is used with povidone iodine. Jager et al. (5) found no statistical difference in acquired endophthalmitis rates after vitrectomy between the two sites, operating theater and outpatients (clinic).

The portable vitrector consists of a battery box, with off/on and cutting speed switches, and a hand-held vitrector (Fig. 10). The vitrector has a 23-G guillotine probe, for use through a sclerotomy puncture site. In addition, a 20-G bevel needle can be placed on it, to allow direct puncture and use at the pars plana. The microguillentine cutter is controlled with a switch on the handpiece. There is also an irrigating and aspiration module, to allow collection of the vitreous sample (Fig. 10).

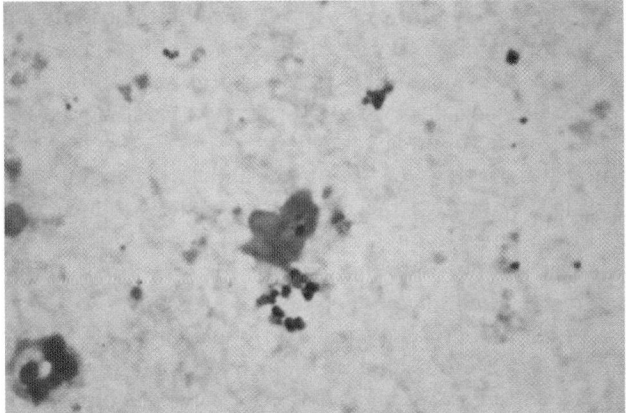

Figure 9 Gram stain of vitreous from a vitrectomy tap of the patient in Fig. 8 showing small numbers of Gram-positive cocci in characteristic tetrads of coagulase-negative staphylococci, with few red-staining polymorphonuclear leucocytes.

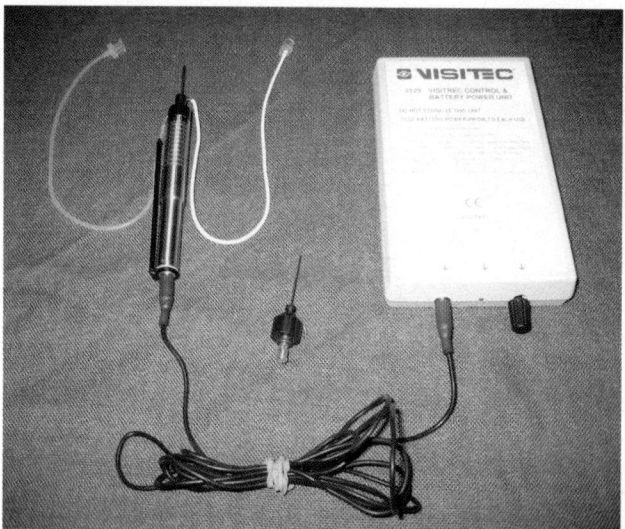

Figure 10   Portable vitrector with tubings. Becton Dickinson Visitrec Vitrectomy System 5100 Visitrec Surgical Vitrectomy Unit marketed by Visitec, Waterloo Industrial Estate, Bidford-on-Avon, Warwickshire, U.K.

The portable vitrector is designed for sample collection only and not for use for a therapeutic vitrectomy. When a partial or full vitrectomy is required, this needs use of the phako/vitrector equipment in the operating theatre.

Antibiotics (refer below) are instilled with separate syringes and 25-G needles for each drug, either injecting directly through the pars plana or by injecting through the sclerotomy wound if present. The scleral wound is usually self-sealing and does not need to be sutured.

The patient is supine (lying flat). The eye is prepared with 5% povidone iodine and then anesthetized by topical, peri-, or retro-bulbar anesthetic. A lid speculum is inserted and an anterior chamber tap is made for Gram stain, culture, and PCR.

For the portable vitrector, the biopsy probe is inserted through the pars plana transconjunctivally (Fig. 11). The probe has a MVR blade at its tip so no incision is necessary. The single instrument provides one-step sclerotomy, cutting, and aspiration. A mini-core vitrectomy can also be performed with slow manual aspiration of up to 1 cc vitreous for microbiology and PCR.

The cutting probe is removed from the eye. Antibiotics and corticosteroids are injected with a 30-mm (30-G) needle through the *same* incision site with *separate* syringes and needles. The incision is small enough not to need suture closure.

Identification of the pathogen in infectious endophthalmitis is rational, as this allows targeted antibiotic therapy. It should be done as soon as possible, *and within one hour*, after the diagnosis is made. Microscopy results are available after a few minutes, pathogen culture results after 24 hours, and resistance testing results after 6 to 10 hours when the RAST method is used, or after 24 to 48 hours with conventional methods. PCR can provide a result within six hours.

The highest rate of pathogen identification is obtained with microscopic and microbiological processing of vitreous material, obtained either using the vitrectomy cutter before switching on the irrigation or as an aspirate.

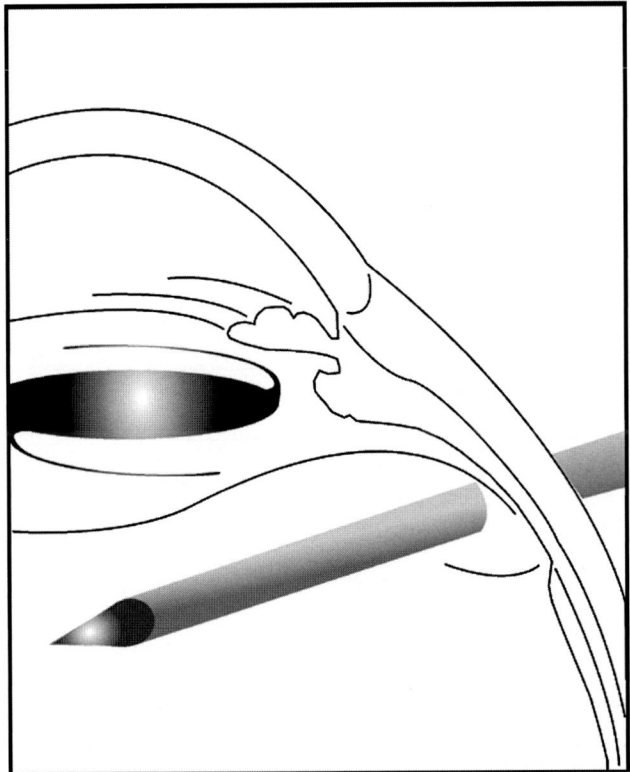

Figure 11 Site of insertion of vitrector guillotine. *Source*: Courtesy of European Society of Cataract and Refractive Surgeons.

Pathogen identification from the anterior chamber is less successful and also potentially contaminated by the vitrectomy cassette. Conjunctival and corneal swabs are pointless, as the correlation with the microorganisms isolated is too low. The culture media should be inoculated directly in the operating theatre or clinic. Transport of material on cotton buds produces a high loss of pathogens and reduces the detection rate, which ideally is about 90%.

The PCR method offers much improved pathogen detection especially in the case of chronic endophthalmitis with low pathogen counts (Chapter 3). However, the increased risk of contamination because of the high sensitivity of the method, the lack of resistance testing, and the partial lack of general quality standards have limited its routine use so far. PCR has been extensively evaluated in the multi-center European prophylaxis study of postoperative endophthalmitis following cataract surgery and found useful identifying 6 out of 20 pathogens causing endophthalmitis that were Gram-stain and culture-negative (6).

TREATMENT

Endogenous Endophthalmitis

This is considered above.

Flow Chart 3 Treatment Guidelines for Acute Endophthalmitis (presumed and not proven)

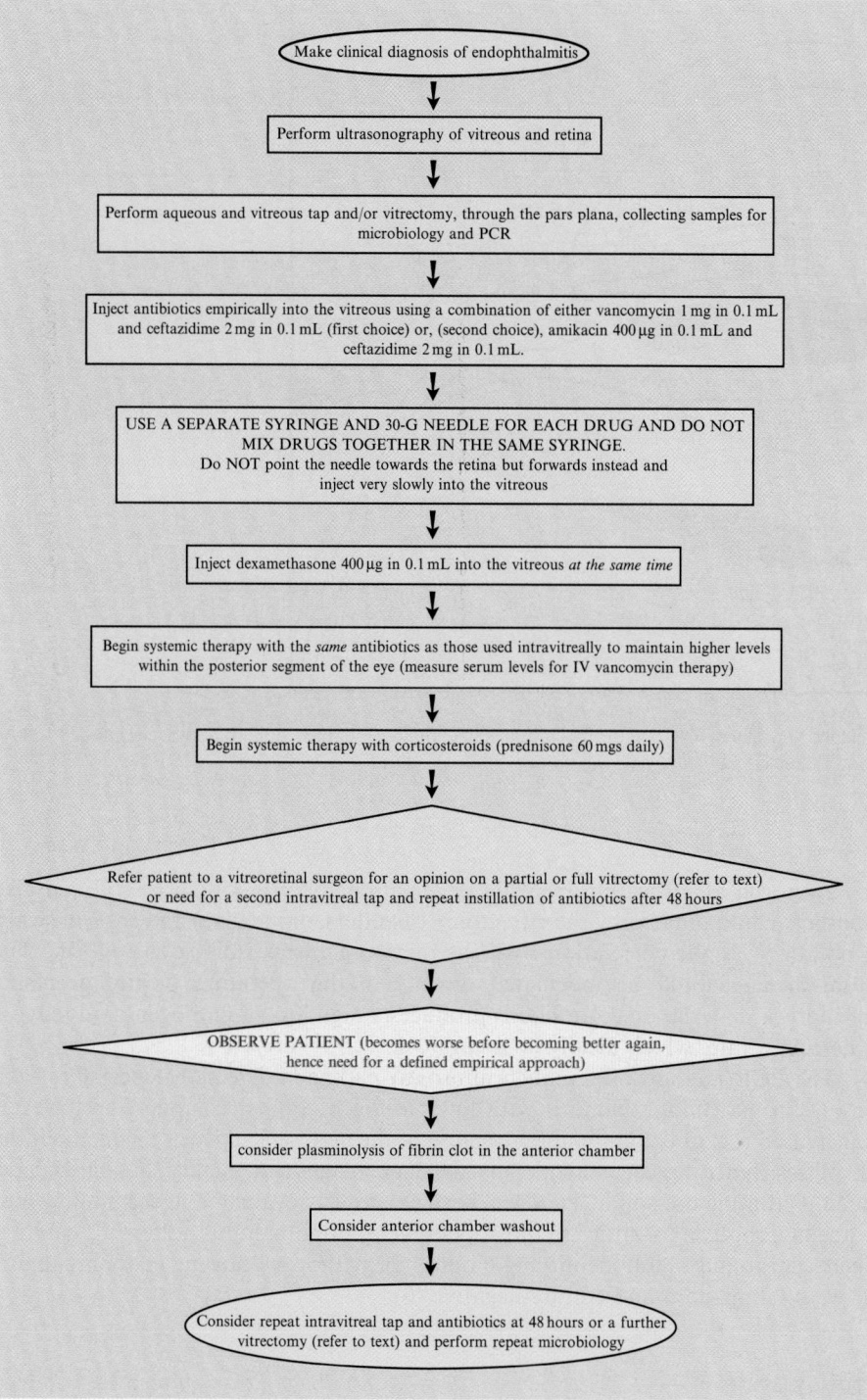

Make clinical diagnosis of endophthalmitis

↓

Perform ultrasonography of vitreous and retina

↓

Perform aqueous and vitreous tap and/or vitrectomy, through the pars plana, collecting samples for microbiology and PCR

↓

Inject antibiotics empirically into the vitreous using a combination of either vancomycin 1 mg in 0.1 mL and ceftazidime 2 mg in 0.1 mL (first choice) or, (second choice), amikacin 400 µg in 0.1 mL and ceftazidime 2 mg in 0.1 mL.

↓

USE A SEPARATE SYRINGE AND 30-G NEEDLE FOR EACH DRUG AND DO NOT MIX DRUGS TOGETHER IN THE SAME SYRINGE.
Do NOT point the needle towards the retina but forwards instead and inject very slowly into the vitreous

↓

Inject dexamethasone 400 µg in 0.1 mL into the vitreous *at the same time*

↓

Begin systemic therapy with the *same* antibiotics as those used intravitreally to maintain higher levels within the posterior segment of the eye (measure serum levels for IV vancomycin therapy)

↓

Begin systemic therapy with corticosteroids (prednisone 60 mgs daily)

↓

Refer patient to a vitreoretinal surgeon for an opinion on a partial or full vitrectomy (refer to text) or need for a second intravitreal tap and repeat instillation of antibiotics after 48 hours

↓

OBSERVE PATIENT (becomes worse before becoming better again, hence need for a defined empirical approach)

↓

consider plasminolysis of fibrin clot in the anterior chamber

↓

Consider anterior chamber washout

↓

Consider repeat intravitreal tap and antibiotics at 48 hours or a further vitrectomy (refer to text) and perform repeat microbiology

Flow Chart 4   Treatment Guidelines for Chronic Endophthalmitis
(presumed and not proven)

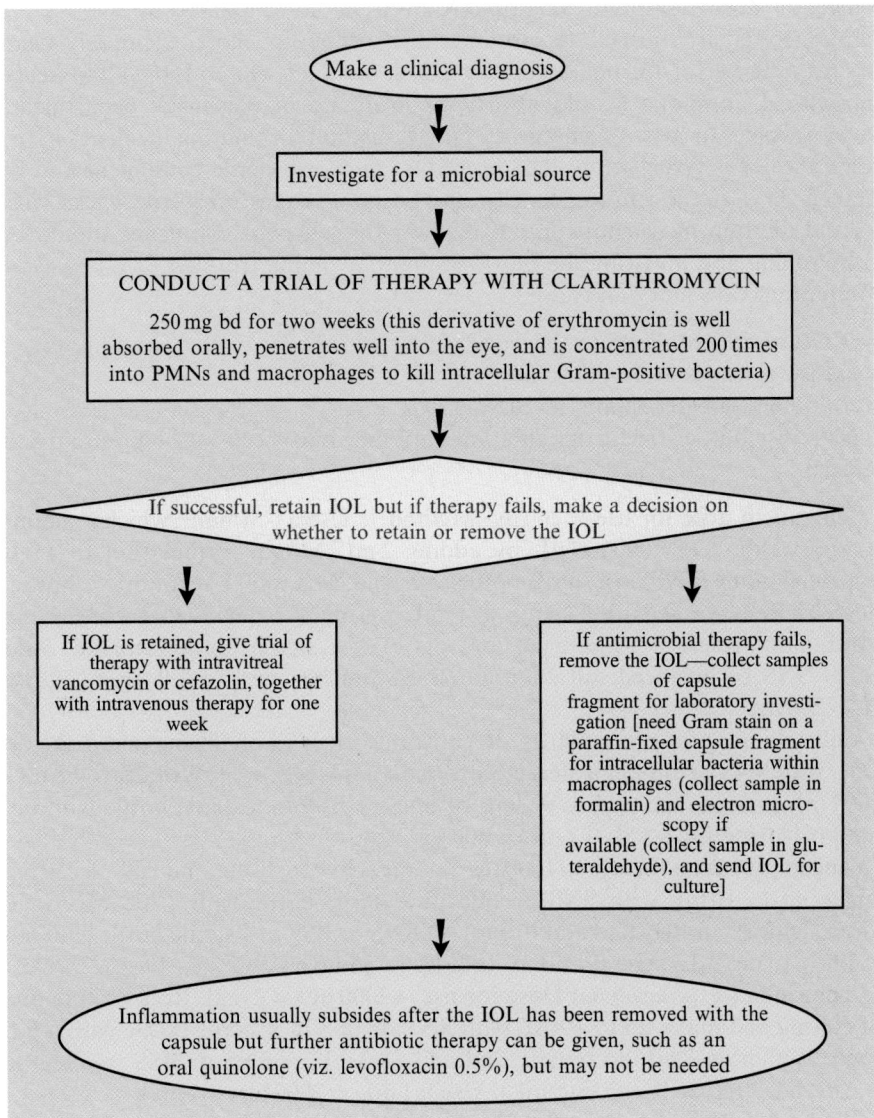

Make a clinical diagnosis

Investigate for a microbial source

CONDUCT A TRIAL OF THERAPY WITH CLARITHROMYCIN

250 mg bd for two weeks (this derivative of erythromycin is well
absorbed orally, penetrates well into the eye, and is concentrated 200 times
into PMNs and macrophages to kill intracellular Gram-positive bacteria)

If successful, retain IOL but if therapy fails, make a decision on
whether to retain or remove the IOL

If IOL is retained, give trial of
therapy with intravitreal
vancomycin or cefazolin, together
with intravenous therapy for one
week

If antimicrobial therapy fails,
remove the IOL—collect samples
of capsule
fragment for laboratory investigation [need Gram stain on a
paraffin-fixed capsule fragment
for intracellular bacteria within
macrophages (collect sample in
formalin) and electron microscopy if
available (collect sample in gluteraldehyde), and send IOL for
culture]

Inflammation usually subsides after the IOL has been removed with the
capsule but further antibiotic therapy can be given, such as an
oral quinolone (viz. levofloxacin 0.5%), but may not be needed

Exogenous Endophthalmitis

Bacterial endophthalmitis is treated by a combination of intravitreal and adjunctive
systemic antibiotic therapy. Subconjunctival therapy is a less satisfactory alternative to
the intravitreal route. An antibiotic combination is injected intravitreally and repeated,
as required according to the clinical response, at intervals of 48 to 72 hours depending
on the persistence of the drugs selected within the eye (see Chapter 4). Intravitreal
antibiotic doses are scaled down to avoid retinal toxicity, but the margin for error
between chemotherapy and toxicity is narrow for the aminoglycosides (for gentamicin,
200 mg is effective but 400 mg is toxic, causing macular infarction); thus, the total dose
injected needs to be highly accurate. For this reason, a combination of intravitreal

vancomycin (for Gram-positive bacteria) and ceftazidime (for Gram-negative bacteria) is now favored. In addition, it is now advised that if a full vitrectomy is performed, the dose of intravitreal antibiotic given should be reduced by 50%.

Pharmacists are trained to produce drug dilutions more accurately than doctors (expected error for pharmacists $+/-5\%$, for doctors $+/-150\%$) and hence the intravitreal antibiotic should always be made up in pharmacy departments whenever possible. In cases of emergency only, a method for diluting the drugs in the operating theater is given below. The procedure must use sterile equipment and be undertaken on a sterile surface; ideally, the hospital makes up sterile packs with bottles and dilution instructions in advance for this purpose. All drugs should be mixed by rolling and inverting the bottle 15 times, avoiding frothing.

Important do's and don'ts are:

- *Never* return diluted drugs to the same vial or syringe for further dilution;
- Do *not* use syringes more than once;
- *Never* dilute at greater than 1 in 20; and
- Do *not* re-use bottles for further dilution until they have been washed, rinsed, and sterilized.

Gentamicin dose for use $= 200\,\mu g$. Method 1: Use a "Minim" which contains $3000\,\mu g/mL$. Dilute to $2000\,\mu g/mL$ by adding $2\,mL$ Minim formulation to $1\,mL$ sterile normal saline (SNS) in a sterile bottle with a lid. Mix well. Use $0.1\,mL = 200\,\mu g$. Method 2: Remove $0.5\,mL$, using a 1-mL syringe, from a vial containing $40\,mg/mL$ non-preserved gentamicin and place in a sterile bottle with lid. Add $9.5\,mL$ of SNS or balanced salt solution (BSS) and mix well ($= 2.0\,mg/mL$). Use $0.1\,mL = 200\,\mu g$.

Amikacin dose for use $= 400\,\mu g$. Reconstitute one vial of $500\,mg$ and make up to $10\,mL$ with SNS or BSS in a sterile bottle with a lid. Mix well. Withdraw $0.8\,mL$, using a $1\,mL$ syringe, and add to $9.2\,mL$ of SNS or BSS in a sterile bottle with lid. Mix well ($= 4.0\,mg/mL$). Use $0.1\,mL = 400\,\mu g$.

Vancomycin dose for use $= 1000\,\mu g$ ($= 1\,mg$). Reconstitute one vial of $250\,mg$ and make up to $10\,mL$ with SNS or BSS in a sterile bottle with a lid. Mix well. Withdraw $2\,mL$ accurately and add to $3\,mL$ of SNS or BSS in a sterile bottle with lid. Mix well ($= 10\,mg/mL$). Use $0.1\,mL = 1000\,\mu g$ ($= 1\,mg$).

Cephazolin (*or ceftazidime*) Dose for use $= 2000\,\mu g$ ($= 2\,mg$). Reconstitute one vial of $500\,mg$ and make up to $10\,mL$ with SNS or BSS in a sterile bottle with a lid. Mix well. Withdraw $2\,mL$ accurately and add to $3\,mL$ of SNS or BSS in a sterile bottle with lid. Mix well ($= 20\,mg/mL$). Use $0.1\,mL = 2000\,\mu g$ ($= 2\,mg$).

Clindamycin dose for use $= 1000\,\mu g$ ($= 1\,mg$). Transfer the contents of a 2-mL ampoule, containing $300\,mg$, to a sterile bottle and add $1\,mL$ SNS or BSS, replace lid and mix well. Withdraw $1\,mL$, using a 1-mL syringe, and add to $9\,mL$ of SNS or BSS in a sterile bottle with a lid. Mix well ($= 10\,mg/mL$). Use $0.1\,mL = 1000\,\mu g$ ($= 1\,mg$).

Amphotericin dose for use $= 5\,\mu g$ (but $10\,\mu g$ has been used). Reconstitute a $50\,mg$ vial with $10\,mL$ water for injection. Withdraw $1\,mL$, using a 1-mL syringe, and add to $9\,mL$ of water for injection in a sterile bottle with lid. Mix well. Withdraw $1\,mL$ of this dilution, using a 1-mL syringe, and add to $9\,mL$ of Dextrose 5% in a sterile bottle with a lid, to complete a dilution of $1/100$. Mix well ($= 50\,\mu g/mL$). Use $0.1\,mL = 5\,\mu g$.

Miconazole dose for use $= 10\,\mu g$. Withdraw $1\,mL$, using a 1-mL syringe, from an ampoule of IV miconazole at $10\,mg/mL$ and add to $9\,mL$ SNS or BSS in a sterile bottle with lid. Mix well. Withdraw $1\,mL$, using a 1-mL syringe, and add to $9\,mL$ SNS or BSS in a sterile bottle with a lid. Mix well ($= 100\,\mu g/mL$). Use $0.1\,mL = 10\,\mu g$.

For Spain and Portugal, use Braun's 50-mL plastic bottles of saline for drug dilution

The following method can be used with Braun's 50-mL bottles of saline, suitable for drug dilution with two rubber injection ports:

Vancomycin – mix vial of 500 mg with 5 mL saline withdrawn from Braun 50-mL bottle, shake well and then return to Braun bottle. Dilution gives 10 mg/mL (dose of 0.1 mL contains 1 mg).

Ceftazidime – mix vial of 1 g (1000 mg) with 5 mL saline withdrawn from Braun 50-mL bottle, shake well and then return to Braun bottle. Dilution gives 20 mg/mL (dose of 0.1 mL contains 2 mg).

Amikacin – add 2 mL saline to vial of 500 mg. Shake well and remove 0.8 mL (= 200 mg) to Braun 50 mL bottle and shake well. Dilution gives 4 mg/mL (dose of 0.1 mL contains 400 µg).

How to Give the Antibiotics

Up to 0.1 mL can be lost in the hub of the needle when drugs are diluted or made up for injection into the eye. Always draw up sufficient drug to fill a 1-mL syringe to at least the halfway mark (0.5 mL) and expel 0.1 mL accurately from the syringe. Never expel the drug from the 0.1-mL mark, because of errors with the hub. An antibiotic combination is injected intravitreally and repeated as necessary according to the clinical response at intervals of 48 to 72 hours, depending on the persistence of the drugs selected within the eye (Chapter 4; see also Flow Charts 3 and 4).

Adjunctive Routes

Systemic Administration. According to the randomized, multi-center Endophthalmitis Vitrectomy Study (EVS), systemic antibiotics do not appear to have any effect on the course and outcome of endophthalmitis after cataract operations (7). However, the study design used different drugs systemically (amikacin and ceftazidime) from those used intravitreally (vancomycin and ceftazidime), which does not contribute towards maintaining effective antibiotic levels within the eye (Chapter 4). 38% of eyes with endophthalmitis demonstrated Gram-positive cocci, against which ceftazidime is only slightly active, whereas vancomycin would have been much more effective. Adjunctive systemic antibiotic therapy is recommended with the *same* antibiotics as those given intravitreally for management of acute bacterial endophthalmitis (Chapter 4).

Vancomycin provides good cover for Gram-positive bacteria including methicillin-resistant *Staphylococcus auresus* (MRSA). Ceftazidime is used to cover the Gram-negative spectrum. Moxifloxacin, levofloxacin, and imipenem are effective against Gram-positive and -negative bacteria but have not yet been fully evaluated for intravitreal use and should not be used at present.

Clindamycin, vancomycin, or cefuroxime are effective for *Propionibacterium acnes* endophthalmitis. However, this must often be preceded by surgery, combined with intravitreal antibiotic injection.

For fungal infection, systemic amphotericin should be used systemically if it has been given by the intravitreal route. It is relatively toxic, but safe with experience, and the help of an infectious disease specialist should be found. In addition, systemic imidazoles are often given but are more likely to be effective for systemic infection than for the endophthalmitis. Voriconazole or fluconazole (for *Candida albicans*) or itraconazole (for other *Candida* species, *Aspergillus* or *Cryptococcus*) can be given but may not be as effective as amphotericin (see Chapters 4, 7, and 9). 5-fluorocytosine can be used for *Candida albicans*. Fusarium endophthalmitis is particularly difficult to treat, requiring both surgical removal where possible and chemotherapy (intravitreal, AC instillation or wash-outs, topical and systemic routes); it is usually but not always sensitive to amphotericin, and combination therapy is often given with imidazoles (Chapter 9).

Anti-inflammatory Therapy

Effective anti-inflammatory therapy, e.g., with corticosteroids, is rational in order (*i*) to limit tissue destruction by infiltrating leukocytes, (*ii*) to stem the effect of antigens and highly inflammatory cell walls released by bacterial disintegration after administration of antibiotics, and (*iii*) to diminish the toxic effects of intraocular cytokines. Intravitreal dexamethasone injection (400 μg in 0.1 mL) at the end of the vitrectomy leads under antimicrobial therapy to a more rapid subsidence of the intraocular inflammation, but without improving the long-term functional outcome, although this finding may reflect studies in which dexamethasone or corticosteroid have been given later rather than initially with the first injection of antibiotics.

Oral administration of prednisolone (1 mg/kg body weight) a day after intravitreal antibiotic therapy with or without vitrectomy has not shown any negative effect on the course of infection in bacterial endophthalmitis. There have also been case reports of systemic steroid administration with mycotic infections, likewise without adverse effects. 200 mg prednisolone is often rationally given systemically in parallel with intravenous antibiotics; however, there are no published studies on this subject.

To summarize, not only the microbes but also their interplay with the immune mechanisms are important in the outcome of endophthalmitis. Cell walls of dead bacteria, especially those of streptococci, and including those recently killed by antibiotics, are highly inflammatory. As a direct consequence, anti-inflammatory treatment with intravitreal dexamethasone (400 μg in 0.1 mL) should be given along with specific intravitreal antimicrobial therapy. This may not be advisable, however, in fungal endophthalmitis, and a decision for their use should be judged against the degree of inflammation and the virulence, and the sensitivity test result if known, of the infecting fungus (Chapter 9). In addition, surgical removal of a high bacterial load in the vitreous can also be important to save vision, by removing the main source of the acute inflammatory effect.

Exogenous Fungal Endophthalmitis

Exogenous fungal endophthalmitis (see also Chapter 9) is invariably sight-threatening unless managed aggressively by surgery and with antifungal chemotherapy. Although rare in Europe, usually being associated with trauma, it can follow cataract and other surgery in rural settings in tropical and hot countries. The filamentous fungus, *Fusarium*, and to a lesser extent *Aspergillus*, cause infection following trauma with soil contamination. It is also recognized as a complication of failed therapy of contact lens-associated *Fusarium* keratitis (Appendix D).

Effective therapy requires mycological identification and, preferably, information about drug sensitivity (Chapter 2). The number of drugs available for ocular use is limited, not only by problems of local and systemic toxicity, but by poor solubility or ocular penetration characteristics. Because of the infrequency of oculomycosis in Europe, no commercial ocular antifungal preparations are available. In the United States and India, pimaricin (5% natamycin) is available.

Effective antifungal drugs include the polyene antibiotics (amphotericin B, natamycin), the cytostatic 5-fluorocytosine (flucytosine, 5-FC), and the imidazoles (thiabendazole, miconazole, ketoconazole, clotrimazole, econazole, fluconazole, itraconazole, voriconazole, and posaconazole). These are all reviewed, with advantages and disadvantages, in Chapter 9 and Appendix A.

Amphotericin B is the only fungicidal antibiotic, depending on concentration achieved, and is active against a wide range of fungi including *Aspergillus* sp., *Fusarium* sp. (but not always), and *Candida* sp. It may be given topically as a drop (0.05–0.5%), subconjunctivally, and intravitreally. It is toxic by any of these routes but the use of topical preparations at concentrations of 0.15% or less will minimize toxicity, which relates in part to the presence of deoxycholate in the parenteral preparation. In addition to intravitreal therapy, amphotericin is given parenterally by slow intravenous infusion in the management of endophthalmitis, often on a background of more widespread systemic infection. Renal (low potassium) and hematological status must be kept under surveillance.

For fungal endophthalmitis, amphotericin (5 mg) or miconazole (10 mg) is usually required intravitreally with a vitrectomy. Amphotericin B can be administered intravenously combined with oral flucytosine (100–200 mg/kg per day) for severe *Candida* endophthalmitis associated with retinochoroiditis. For *Candida* retinochoroiditis without endophthalmitis, treatment is effective with systemic ketoconazole (200–400 mg/day), fluconazole, or voriconazole. 5-FC can be added as well.

5-FC is only active against *Candida* sp. It is well absorbed by the oral route and achieves high blood and tissue levels. It has been used effectively in the treatment of *Candida* endophthalmitis, in combination with systemic or intravitreal amphotericin B or fluconazole to prevent the otherwise rapid emergence of resistant strains.

Endophthalmitis Following Phacoemulsification Cataract Surgery

At the start of the 20th century (c. 1910), the incidence of endophthalmitis after cataract operations was 10%. In the period of extra capsular cataract extraction (ECCE) and improved hygiene conditions (1970–1990), the infection rate fell to 0.12% in Europe and to 0.072% in the United States. However, since the introduction of phacoemulsification and clear cornea incisions, the retrospective data with phacoemulsification has increased to between 0.1% and 0.5% (8).

The clear cornea incision is thought to have contributed to the increase in the number of endophthalmitis cases following phacoemulsification surgery (8–10). Taban et al. (8) performed a meta-analysis of 215 studies that addressed post-cataract surgical endophthalmitis that met his selection criteria. A total of 3,140,650 cataract extractions were pooled from ECCE and phacoemulsification surgery, giving an overall incidence of 0.128% for postoperative endophthalmitis. He found this incidence varied with time from 0.265% in 2000/2003, 0.087% in the 1990s, 0.158% in the 1980s, to 0.327% in the 1970s. He found the clear corneal incision (CCI) of phacoemulsification to be a risk factor between 1992 and 2003 with an increased rate of 0.189% compared to 0.074% for scleral tunnel (ST) incision. However, Taban reviewed the limitations of his meta-analysis study depending mostly on retrospective studies with limited statistical power. He commented on the paucity of prospective randomized case-controlled studies.

There is a range of technical factors relating to the cataract operation that influence the risk of endophthalmitis. With regard to the incision, leak-proof closure plays an important role. However, surgeons are reluctant to use a stitch! When the clear cornea incision (CCI) was first used, the data regarding the incidence of infection tended to be poor.

Following phacoemulsification cataract surgery, patients can present: (*i*) acutely, with fulminant endophthalmitis within 10 days of surgery often due to *Staphylococcus aureus, Streptococcus pyogenes, Streptococcus pneumoniae,* or alpha-hemolytic strepto-cocci of the 'viridans' group (*S. suis, S. mitis, S. oralis, S. sanguis, S. salivarius*) during which vision can be lost over 12 hours (3,7,11,12); (*ii*) subacutely, with infection at up to six weeks after surgery, often due to CNS, and with vision reduced but retained—it is these patients who were mainly investigated in the recent EVS Group (7) when full vitrectomy was shown to be of benefit only if vision was reduced to light perception; or (*iii*) with chronic low-grade uveitis, often due to *Propionibacterium acnes* and occasionally coagulase-negative staphylococci (4,13–15). Patients present initially as hypopyon uveitis within six weeks of surgery, which fails to respond to corticosteroids and needs eventual vitrectomy with intravitreal antibiotics and often the removal as well of the intraocular lens (IOL). In chronic saccular endophthalmitis, there is granulomatous inflammation and characteristic white capsular plaque (Fig. 12) (4,13,14). A "trial of therapy" with clarithromycin or azithromycin should be conducted as the drug penetrates well into cells and tissue (Chapter 4), and *P. acnes* is very sensitive to it (MIC90 0.03 mg/L) (Fig. 13) (13,15,16). In addition, culture-negative endophthal-mitis has been found to respond better if the patient is treated with clarithromycin than

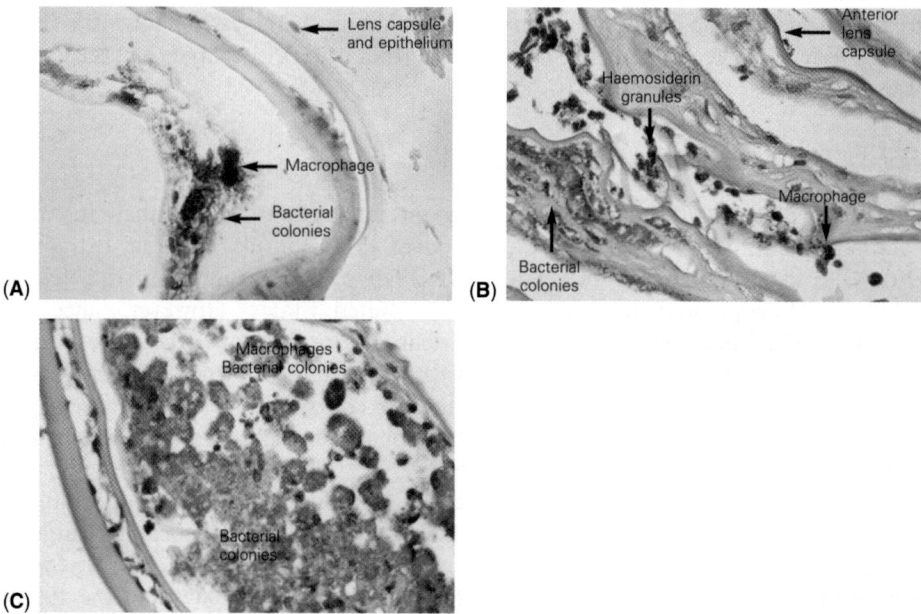

Figure 12   (A–C) Hematoxylin and eosin stain of the capsule fragment removed surgically from three patients with chronic "saccular" endophthalmitis showing macrophages and bacterial colonies. (Courtesy of Drs. Abreu and Cordoves)

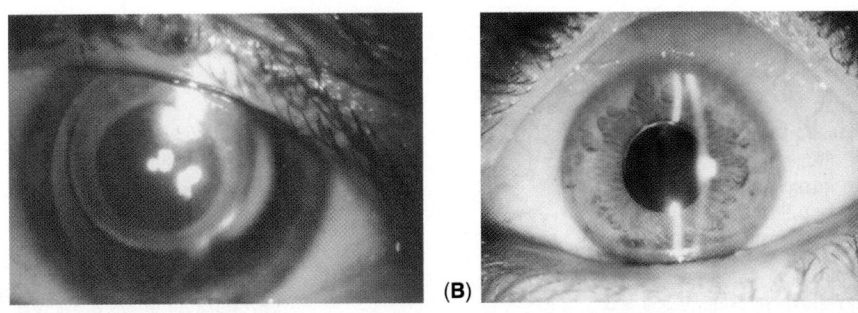

Figure 13 **(A–B)** Saccular plaque-type endophthalmitis before **(A)** and after **(B)** effective clarithromycin therapy.

without this drug (16). If PCR testing is available, some of these culture-negative cases can be expected to become positive (Chapter 3).

Prophylaxis and Risk Factor Analysis of the 16,000 Patient Multi-Center European Study of Postoperative Endophthalmitis (6,12,17)

Study Design

All patients taking part in the study during 2003 to 2006 underwent phacoemulsification cataract surgery with placement of an intraocular lens as a single procedure without additional surgery. All patients were operated on within a modern theater with assisted ventilation and full sterile and aseptic techniques. Patients were recruited with an Information Sheet in their own language and had to conform to both inclusion and exclusion criteria listed in the protocol (17). Twenty-four hospitals took part in the study in nine countries: Austria (one), Belgium (five), England (eight), Germany (one), Italy (two), Poland (one), Portugal (one), Spain (four), and Turkey (one).

The study was set up as a factorial design to test for the effects of two treatment interventions (17):

- Cefuroxime injected into the anterior chamber at the end of surgery as 1 mg in 0.1 mL saline (0.9%); and
- Levofloxacin 0.5% administered one drop one hour before surgery, 1 drop 30 minutes before surgery and three drops at five minute intervals immediately following surgery.

Consequently, there were four prophylaxis groups (Table 1), three (B to D) with perioperative antibiotics (levofloxacin and/or cefuroxime) and one (A) without (placebo control group) forced by the randomization process to be of approximately equal sizes so that the study was balanced. Small number randomization was used in blocks of 12, each containing three patients for each of the four study groups.

All patients received povidone-iodine 5% (Betadine) as one drop into the conjunctival sac and onto the cornea for a minimum of three minutes before surgery. In order to achieve this time interval, all clinicians had to change their practice so that the povidone iodine was first given outside the operating theatre. Further 5% povidone iodine was given inside the operating theatre when the surrounding skin was also prepared with 10% povidone iodine. In addition all patients were given levofloxacin 0.5% eye drops (Oftaquix) every eight hours for six days starting the day *after* surgery.

Table 1   Factorial Design of Multi-Center European Study of Prophylaxis of Postoperative Endophthalmitis Following Phacoemulsification Surgery

Group A	Group B
Placebo vehicle drops × 5[a]	**Placebo vehicle drops** × 5[a]
Povidone iodine 5%	Povidone iodine 5%
No intracameral injection	**Intracameral cefuroxime 1 mg**
Post-op Levofloxacin 0.5%	Post-op Levofloxacin 0.5%
(from post-op day 1 to 6)	(from post-op day 1 to 6)
Group C	**Group D**
Levofloxacin drops 0.5% × 5[a]	**Levofloxacin drops 0.5%** × 5[a]
Povidone iodine 5%	Povidone iodine 5%
No intra-cameral injection	**Intra-cameral cefuroxime 1 mg**
Post-op Levofloxacin 0.5%	Post-op Levofloxacin 0.5%
(from post-op day 1 to 6)	(from post-op day 1 to 6)

[a] One drop one hour before surgery, one drop one half-hour before surgery, one drop immediately post-operation, one drop five minutes later and one drop five minutes later again.

Case Definitions

A diagnosis of presumed endophthalmitis was made for any patient presenting within six weeks after operation with pain and loss of vision thought to be due to infection. Samples of aqueous and vitreous were collected from these patients for investigation using Gram staining, culture, and PCR testing using two reference molecular biology laboratories (Flow Chart 1). Gram staining and culture were performed locally in each hospital microbiology laboratory. If any one of these investigations produced a positive result, the case was classified as a *proven* case of infective endophthalmitis; otherwise the case was classified as *presumed unproven* endophthalmitis. Each unproven case was carefully considered if there was evidence of the Toxic Anterior Segment Syndrome (TASS) (see below) or non-infective uveitis when the case was removed from the data set.

Results

Endophthalmitis Cases.   Results for identification of bacteria in proven cases within the three prophylaxis and one placebo control groups are given in Table 2, together with groupings of the unproven cases.

    Statistical Analysis of the Study Data.   Of a total of 16,603 patients recruited, a small number were lost to follow-up (2.2%). The final 'per protocol' data set consisted of 15,971 patients, and the majority of patients were females (58%). The median age of females was 75 and males 73 years. An additional 240 patients were included in the "intent to treat" data set (12). The incidence rates in each of the four groups together with the overall numbers are given in Table 3.

    Factors found to be significantly associated, by logistic regression analysis, with presumed endophthalmitis for the per protocol group were "site of incision," with patients receiving the clear cornea procedure 5.75 times (95% CI 1.30 to 25.39) more likely to experience endophthalmitis than patients receiving scleral tunnel; "cefuroxime injection," with patients not receiving cefuroxime 4.76 times (95% CI 1.81 to 12.54) more likely to experience endophthalmitis than patients receiving cefuroxime; "any surgical complications," with patients experiencing complications at time of surgery 4.04 times (95% CI 1.19 to 13.76) more likely to experience endophthalmitis; and "IOL optic material," with patients receiving a silicone lens

Table 2 Bacteriology Results of all Endophthalmitis Cases for Patients Recruited to the European Study (n = 29)

Group A	Group B
3 Coagulase-negative staphylococcus (CNS)[a]	2 Coagulase-negative staphylococcus (CNS)
2 *Streptococcus pneumoniae*	1 Non-Proven
Streptococcus salivarius	
Streptococcus suis	
Propionibacterium acnes	
Mixed *Staphylococcus aureus*, CNS, *Propionibacterium acnes*	
Mixed *Streptococcus mitis*, CNS	
4 Non-Proven	n = 3
n = 14	
Group C	**Group D**
Staphylococcus aureus	
3 Coagulase-negative staphylococcus (CNS)	1 Coagulase-negative staphylococcus (CNS)
Streptococcus salivarius	
Streptococcus oralis	1 Non-Proven
Streptococcus salivarius	
3 Non-proven	n = 2
n = 10	

[a] One case of CNS in Group A was not included in the per protocol analysis given in the text (n = 28) but was included in the intention-to-treat analysis as published (n = 29) (12).

3.29 times (95% CI 1.52 to 7.09) more likely to experience endophthalmitis than patients receiving an acrylic (or other material) lens. None of patient age or gender, or the use of perioperative levofloxacin eye drops as used in the study, was found to have a significant association with presumed endophthalmitis. Surgeon experience level was associated with a p value of 0.061, so that it just failed to be significant at the traditional 5% level, but it was notable, as the result suggested that the more experienced surgeon is more likely to be associated with endophthalmitis cases but is, of course, more likely to be operating on the more complicated cases!

When analysis is restricted to proven endophthalmitis for the per protocol cases, similar risk factors are obtained. However, "any surgical complications" was

Table 3 Presumed and Proven Patient Numbers and Incidence Rates in Each of the Four Groups Within the European Study Based on the Per-Protocol Analysis

Group A	Group B
Number of patients 3990	Number of patients 3997
Incidence rates (%)	**Incidence rates (%)**
Presumed: 0.33 (95% CI 0.17 and 0.56)	Presumed: 0.075 (95% CI 0.02 and 0.22)
Proven: 0.23 (95% CI 0.10 and 0.43)	Proven: 0.05 (95% CI 0.006 and 0.18)
Group C	**Group D**
Number of patients 3984	Number of patients 4000
Incidence rates (%)	**Incidence rates (%)**
Presumed: 0.25 (95% CI 0.12 and 0.46)	Presumed: 0.05 (95% CI 0.001 and 0.18)
Proven: 0.18 (95% CI 0.07 and 0.36)	Proven: 0.025 (95% CI 0.001 and 0.14)

no longer a key risk factor and most odds ratios were higher. "Patient gender" was associated with a p value of 0.052, suggesting that male patients may be more at risk of endophthalmitis than female patients, with an odds ratio of 2.53 (95% CI 0.99 to 6.47). Analysis conducted on the "intention-to-treat" basis has been considered as well when it included one extra case due to CNS in group A (placebo control) (12).

Conclusions. The intracameral injection of cefuroxime has been proven to give a statistically significant beneficial effect in reducing the risk of postoperative infective endophthalmitis following phacoemulsification cataract surgery by 4.8 times for the presumed cases [5 in cefuroxime group (incidence 0.062%), 23 in non-cefuroxime group (incidence 0.288%)], Fisher's exact test p value 0.00053, and, for the proven group [3 in cefuroxime group (incidence 0.038%), 16 in non-cefuroxime group (incidence 0.21%)], Fisher's exact test p value 0.0025.

Unfortunately, however, cefuroxime is not currently licensed for this use within the EU, but clinicians in some, but not all, countries may use it on an off-label or named-patient basis in discussion with their hospital pharmacist; cefuroxime needs to be carefully diluted for the intracameral injection, maintaining sterility at 1 mg in 0.1 mL.

Topical levofloxacin given at the time of surgery, as perioperative and delayed postoperative prophylaxis, did not give a statistically significant reduction in endophthalmitis rates. As all patients received postoperative levofloxacin starting 18 hours after surgery, as part of the study design, it is not possible to ascertain what the endophthalmitis rate would have been with more intensive postoperative use. It is now known that the postoperative dose gives a high aqueous level of 4.4 µg/mL at a peak of one hour but will reduce to low levels within four hours (18). Postoperative levofloxacin drop prophylaxis thus needs to be given two-hourly postoperatively in the recovery area then two-hourly until night-time to maintain effective bactericidal levels in the AC. Thereafter, levofloxacin should be given six-hourly for two weeks.

In Sweden, the details of every cataract operation, including associated complications such as endophthalmitis, are recorded into a national register. All patients for the last five years have received prophylaxis with aqueous chlorhexidine 0.2% preoperatively and intracameral cefuroxime 1 mg in 0.1 mL perioperatively (19) with no use of a postoperative antibiotic. The incidence of endophthalmitis during the period between 2002 and 2004, based on a sample of 225,400 operations, was 0.05% (19). There was a difference in bacteria isolated between the European and Scandinavian studies, when there were isolates of enterococci, resistant to cefuroxime, and Gram-negative enterobacteriaceae, mostly sensitive to cefuroxime, present in the Swedish study that did not occur in the cefuroxime arm of the European study.

The data showed that the clear cornea (CC) site of incision in phacoemulsification surgery was associated with a significantly higher risk of endophthalmitis than the scleral tunnel incision with a binary logistic regression model, results for presumed cases (5.75 times greater for CC, 95% confidence intervals 1.3–25.39, p value 0.021), and for proven cases (7.39 times greater for CC, 95% confidence intervals 0.096–56.9, p value 0.055). The significance of the "site of incision" risk factor was not unequivocally proven, however, by the European study because it was confounded with a possible center effect. Only two of the 24 centers in the study used the scleral tunnel approach for all their surgery, with 396 and 2503 cases; other centers contributed just 120 cases. Taban et al.'s (8) meta-analysis looked at clear corneal incisions for phacoemulsification in the United States and Europe and noted that 92% of American surgeons did not practice suture closure at the end of their cataract procedures. Taban's conclusion was that the upward trend in endophthal-mitis frequency coincides temporally with the development of sutureless clear corneal incisions (8) and that view is supported by the European study. McCulley (personal communication) has emphasized that if all surgeons were to put just one suture into the clear corneal wound, it would become watertight and result in a much lower rate of endophthalmitis. Wallin et al. conducted a large retrospective cohort study of cataract surgery in their institute and found, with sutureless surgery, that surgical complications and wound leak on the first postoperative day were most strongly associated with postoperative endophthalmitis (20).

The integrity of clear corneal incision is critical. Up to 80% of post-cataract surgery endophthalmitis cases are associated with wound defects, gaping, and leakage. Wound integrity varies as a function of intraocular pressure which itself varies with blinking. There are large fluctuations in IOP in the postoperative period. Optical coherence tomography has also shown variations in gaping of unhealed wounds. India ink migrates from the ocular surface into the stroma through the wound and into the anterior chamber. The clear cornea wound is expected to take up to one week minimum for the epithelial surface to regenerate to become watertight. This is why a postoperative antibiotic drop, such as a topical quinolone viz. levofloxacin 0.5%, is needed to stop bacteria entering into the anterior chamber in the first few days after surgery. Our advice is to foster careful wound construction with a minimum tolerance for wound leakage, the placement of sutures whenever necessary or possible, and continued vigilance in the surveillance of infection.

An important finding of the European multi-center study was that silicone IOLs had a higher risk of postoperative endophthalmitis than acrylic IOLs by a factor of 3.29. For presumed cases of endophthalmitis, the incidence was 13/4070 (1 in 313) for the silicone IOLs and 15/11,873 (1 in 791) for the acrylic and others (p 0.017). For proven cases of endophthalmitis, the incidence was 10/4073 (1 in 407) for the silicone IOLs and 9/11,879 (1 in 1,320) for the acrylic and others (p 0.03). Similar findings for an increased risk for silicone IOLs were found in a retrospective case-control study in 2004 from Singapore (21). These results will have an important impact on future materials used for IOL manufacture. It has been known since the 1990s that *Staphylococcus epidermidis* adheres to polypropylene haptics more easily than to polymethyl methacrylate (PMMA), increasing the risk of endophthalmitis by a factor of 4.5 times. Furthermore, hydrophilic heparin-coated IOLs have been shown to demonstrate lower adherence for staphylococcal bacteria.

Analysis of Streptococcal Cases. An anomaly arose in the European study with finding that 8 out of 20 proven cases were due to streptococci and that seven out of eight of these were in England and 1 in lzmir,Turkey—none were found in continental Europe. An analysis is given in Table 4 when the difference is found to be highly significant (p = 0.0004). This implies that England had a particular problem with postoperative endophthalmitis due to streptococci (*S. pneumoniae* x2, á-hemolytic streptococci x5) not found elsewhere except Turkey (*S. mitis* x1), probably associated with the 'English' disease of chronic bronchitis often caused by streptococci. This in turn means that intracameral cefuroxime is of more value in England but of less value in continental Europe, as 8 out of 17 proven isolates in the non-cefuroxime group (Groups A and C) were due to these streptococci. Furthermore, only three patients with postoperative endophthalmitis lost all useful vision and they were all in the streptococcal group; none lost vision and all but one had a good recovery in the continental European group due predominantly to CNS. For this reason we recommend that use of intracameral cefuroxime be

Table 4 Streptococcal Cases of Postoperative Endophthalmitis (figures used from the per-protocol group) in the European Study

Endophthalmitis cases	Place recruited	Total number recruited	Incidence rate
7	England	5484	1 : 783
1	Turkey	859	1 : 859
0	Continental Europe	9628	0 : 9628
		Total 15,971	

Note: Statistical difference between England/Turkey and Continental Europe using Fisher's exact test (http://www.matforsk.no/ola/fisher.htm), because of the small proportions, is p = 0.0004.

made in the United Kingdom but that its beneficial value will be less in continental Europe.

If a unit has a low or very low rate of postoperative endophthalmitis, then we advise to continue with existing antibiotic prophylaxis or to introduce intensive quinolone drops viz. levofloxacin 0.5% that should be given pre-, peri-, and immediately postoperatively as described above. If a unit has post-operative endophthalmitis problems, then intra-cameral prophylaxis with cefuroxime or vancomycin is needed.

Seasonality of Cases. Seasonality was investigated (Table 5) but no significant difference could be found for operations between winter and summer months, although a trend towards more cases presenting in summer months was present.

Intraoperative Prophylaxis

Addition of Antibiotic to Irrigating Solutions in Cataract Surgery

According to questionnaire reporting surveys in various countries, antibiotics are used in the irrigating solution by approximately 60% of responding cataract surgeons, but only 65% replied to the survey (4). This low response rate is unreliable and open to bias (Chapter 10). While it is suggested that the addition of antibiotics to the irrigating solution has a protective effect, it has not been possible to reduce the incidence of endophthalmitis in any prospective scientific study (22). All information on the incidence of endophthalmitis comes either from retrospective studies or from studies of antibiotic use where there was no control group.

Table 5 Seasonality of Postoperative Endophthalmitis [figures taken from per-protocol group (n = 15,971)]

Number of operations:
Winter (October 1st to March 31st) = 8430
Summer (April 1st to September 30th) = 7541
Number of cases of post-operative endophthalmitis:
Winter = 10 Summer = 18
Incidence rate of endophthalmitis:
Winter = 1 : 843 Summer = 1 : 419

Note: The study was run for two years and three months between October 2003 and January 6th, 2006. Statistical difference using Fisher's exact test (http://www.matforsk.no/ola/fisher.htm), because of the small proportions, is p= 0.087.

Anterior chamber contamination at the end of a cataract operation varies between 0% (0 of 98 eyes) and an extreme figure of 43% (13 of 30 eyes), but even with a greater number of investigated eyes it is between 0.18% (1 of 552) and 13.7% (98 of 700) (4). Whether the reduction in the contamination rate from 12 of 100 to 5 of 100 when vancomycin is used in the irrigation solution and from 22 of 110 to 3 of 110 with vancomycin/gentamicin is meaningful remains doubtful, especially as no difference was found in another corresponding study (8 of 190 control, 9 of 182 vancomycin) (4).

In any case, the onset of action of various antibiotics in vitro is observed only after three to four hours and full activity only occurs after about one day. In animal studies of pars plana vitrectomy, antibiotic prophylaxis can be established only for low but not for moderate numbers of bacteria.

There is also the risk of overdose (aminoglycoside retinal toxicity) and the danger of developing resistance, which is disturbing particularly with the reserve antibiotic vancomycin. Relevant scientific organizations and authors therefore advise against giving prophylactic antibiotics in the irrigating solutions or call this in question, especially as a benefit has not so far been proven (Centers for Disease Control, Atlanta, in 1995 and the American Academy of Ophthalmology in 1999) regarding vancomycin.

Subconjunctival Antibiotic Injection Prophylaxis

This technique has been much used over the last 30 years, especially in the United Kingdom, but probably has little prophylactic effect on the prevention of endophthalmitis (4). While there has been no formal study to establish the technique, there have been plenty of studies, including the EVS, in which patients who develop endophthalmitis have received a prophylactic subconjunctival injection. One of the reasons, though, that this might have occurred is that many surgeons used gentamicin, which has no antibiotic effect on streptococci and *Propionibacterium acnes*. Researchers have investigated the pharmacokinetics of cefuroxime and found that when 125 mg is given by the subconjunctival route, levels reached 20 µg/mL in the anterior chamber, which are far lower than those (3000 µg/mL) which occur when injected by the intracameral route (4). However, one recent retrospective uncontrolled study from Canada found one case of endophthalmitis in 8856 surgeries using subconjunctival antibiotics and nine cases of endophthalmitis out of 5030 surgeries not using them (23). While this was statistically significant (p < 0.009), the study was open to bias as regards surgeon, incision, and type of patient; such data needs to be repeated in a controlled prospective study. In contrast, the effect of withdrawing subconjunctival cefuroxime, which resulted in an increased rate of postoperative endophthalmitis, is described in Chapter 10.

Postoperative Prophylaxis

In order to minimize the risk of infection, particularly after clear cornea incisions until wound healing is secure, use of the preoperatively applied topical antibiotic drop (such as a quinolone viz. levofloxacin 0.5%) is recommended for two weeks but is not proven. Longer application (more than two weeks) is discouraged unless there are other medical reasons for it. Postoperative antibiotic drops should be applied two-hourly on the day of operation and six-hourly thereafter, from first waking in the morning to just before sleeping, with one drop well placed in the conjunctival sac.

Surgical Management of Acute Postoperative Bacterial Endophthalmitis— Diagnostic and Therapeutic Vitrectomy

Treatment of acute postoperative endophthalmitis should be immediate (within a few hours) with a "complete" three-port pars plana vitrectomy by a vitreoretinal surgeon. First, the infusion port is inserted through the pars plana, 3.5 mm from the limbus, but is *not turned on*. The vitreous cutter is inserted through a separate 3.5-mm sclerotomy and directly visualized through the pupil. A handheld syringe is attached to the aspirating line and the surgical assistant aspirates while the surgeon activates the cutter until the eye visibly softens and the cutter is disappearing from view. The infusion is turned on to reform the globe and the cutter removed. The syringe and the tubing now contain 1–2 mL of infected but undiluted vitreous and the two together are promptly sent to the laboratory for immediate Gram stain, culture, and PCR analysis.

The microbiologist has previously been informed the sample is en route.

The vitreous cutter is now connected to the machine for aspiration control and a light pipe is inserted through the pars plana. A standard three-port vitrectomy is performed within the limits of visualization. It is useful to perform a posterior capsulotomy with the cutter and aspirate the fibrin and pus from the anterior chamber and intraocular lens surface. This procedure not only improves visualization but permits flow through the entire eye and facilitates recovery.

Caution must be exercised against too-aggressive surgery. These eyes have concomitant retinal vasculitis and retinal edema and the inadvertent creation of a retinal break can be catastrophic. Iatrogenic retinal detachment in eyes with acute bacterial endophthalmitis is comparable to that of AIDS.

Once the vitrectomy is as complete as possible, the intravitreal antibiotics (see text) are injected. Note that the dose is reduced by 50% if a full vitrectomy has been performed. This injection should occur *slowly*, over minutes, and the needle pointed away from the macula. Separate syringes and separate needles are used through an existing entry site. Intravitreal dexamethasone (preservative-free) is then injected.

The procedure is performed under general, peribulbar, or retrobulbar anesthesia but not topical. General anesthesia is advantageous because the patients are usually old, frail, sick and in pain. Furthermore, the eyes are inflamed, hyperemic, and bleed.

Whilst the above procedure achieves immediate diagnostic and therapeutic vitrectomy and reduces the need for re-operation, it is often not possible due to the lack of a vitreoretinal surgeon and a vitreoretinal operating room. The duty surgeon frequently does not have the required skill.

As time is of the essence, an alternative approach may be the practical option. The basic fundamental requirement is the intravitreal injection of the antibiotics. This should be preceded by vitreous biopsy. Simple aspiration with a needle is frequently unsuccessful unless the needle fortuitously enters the syneretic vitreous cavity; infected vitreous cannot be aspirated with a syringe and needle. Every duty surgeon must be taught to perform a vitreous biopsy with a vitreous cutter. The simplest technique is with the portable vitrector with an MVR blade at its tip as marketed by Visitec (Becton Dickinson Visitrec Vitrectomy System 5100 Visitrec™ Surgical Vitrectomy Unit) (Figure 10). Following the sampling, antibiotics and corticosteroids are injected through the sclerotomy as above. The incision does not require suture closure and no conjunctival incision is necessary.

This has the advantage of time over completeness but it ignores the fundamental surgical principle of, "Where there is pus, release it." It provides a smaller sample but permits the earlier injection of intravitreal antibiotics and earlier microbiology. It also buys time pending the availability of a vitreoretinal surgeon and vitreoretinal operating room.

Epidemiology of Outbreaks of Endophthalmitis Following Cataract Surgery

For detailed discussion on how to investigate an outbreak of cataract surgery–associated postoperative endophthalmitis, the reader is referred to Chapter 10.

Other Types of Postoperative Endophthalmitis

Late cases of endophthalmitis after cataract operation are the second commonest form of endophthalmitis accounting for 20% to 30% of cases; the symptoms are milder and *Propionibacterium acnes* has been identified as the principal pathogen. The difficulty of culturing the principal pathogen, which is often enclosed in the synechized capsular sac, and the high rate of recurrence, which can only be reduced by vitrectomy, possibly combined with posterior capsulectomy, is problematic. A further advantage of vitrectomy is that adequate material for culturing the causative organism can be obtained but capsular bag material is needed as well (see Flow Chart 3). Early vitrectomy is advisable but a trial of therapy should be given first with clarithromycin 250 mgs bd which can be effective without surgery because the drug is well absorbed and concentrated x200 into macrophages and other cells (Chapter 4) (4,13–15).

Bleb-associated endophthalmitis usually follows a chronic course. The commonest pathogens are *Streptococcus* sp. and Gram-negative bacteria, especially *Haemophilus influenzae*. The visual prognosis is usually poor and demands aggressive therapy, consisting of immediate vitrectomy and intravitreal antibiotics. Systemic antibiotics should also be given.

Limitations of Endophthalmitis Vitrectomy Study (EVS) Study

The EVS conclusions refer mostly to subacute postoperative endophthalmitis due to CNS (80% of cases) and the advice and conclusions overall do not relate to acute pyogenic pathogens such as *Staphylococcus aureus*, *Streptococcus pyogenes*, and *Streptococcus pneumoniae*. The conclusions also do not apply to late postoperative (saccular), bleb-induced, posttraumatic, or endogenous endophthalmitis. These forms of endophthalmitis have a more aggressive bacterial spectrum and therefore require different operative techniques (vitrectomy), and both intravitreal and systemic antibiotics. The principles of management are the same for posttraumatic and endogenous endophthalmitis, as for acute postoperative endophthalmitis, but the visual outcome is poorer.

According to the EVS, conducted in the United States, patients with acute endophthalmitis after a cataract operation with an initial vision of hand movements or better should be treated by vitreous biopsy and intravitreal antibiotics (7). For patients whose vision consisted only of light perception, immediate vitrectomy was recommended. This advice, however, was based on the selection of patients admitted to the study, when 80% of cases were due to CNS. The advice does not reflect the management of acute streptococcal endophthalmitis, where there is good reason to proceed with an immediate vitrectomy in order to remove the highly inflammatory bacterial cell walls from within the vitreous milieu. Retrospective studies have shown that affected patients can profit from early vitrectomy.

Follow-up EVS analyses showed differences between diabetics and non-diabetics. Diabetics with a visual acuity of hand movements or better obtained vision of 20/40 more often (57%) by vitrectomy than after biopsy (40%), but the results ultimately were not statistically significant because of the low number of diabetic participants in the study.

Endophthalmitis Associated with Microbial Keratitis

Occasionally, severe microbial keratitis progresses to endophthalmitis such as with *Fusarium* sp. (Appendix D and Chapter 9). This devastating complication, with a poor prognosis, can occur in patients with Sjögren's syndrome, with use of topical corticosteroids, generalized immune suppression or dysfunction, lack of an intact posterior capsule (following cataract surgery) and corneal trauma. Fungi are the most frequently reported organisms, of which *Fusarium* sp. are the commonest, while bacterial species include *Mycobacterium chelonae*, *Nocardia* sp., *Staphylococcus aureus*, streptococci and coliforms as well as Capnocytophaga. Treatment requires radical surgery and intravitreal antibiotics, guided by the organisms seen on Gram and Ziehl-Neelsen stains of anterior chamber and vitreal taps.

Differentiation of Postoperative Endophthalmitis—the TASS Syndrome

The toxic anterior segment syndrome (TASS) presents acutely 12 to 48 hours after surgery with pain and blurred vision. There is diffuse corneal edema, classically from limbus to limbus, with endothelial cell damage. There is a small hypopyon with possible iritis that may result in iris atrophy.

TASS never occurs as a single case—usually as three, four, or five cases presenting closely to each other within a single unit (24). It is a toxic reaction and not infective, but an infective endophthalmitis is an initial differential diagnosis. It is a worrying situation for a clinician.

TASS can be due to a variety of stimuli including bacterial endotoxin (lipopolysaccharide cell wall of Gram-negative bacteria) from water within an ultrasound bath for cleaning instruments or even from contaminated but sterile water used to make steam in an autoclave. Such an event, due to the latter cause, took place once during the European multi-center study and involved one study patient and two other patients within 36 hours, all of whom made a good recovery with corticosteroids.

TASS may also be due to ophthalmic viscoelastic devices that have become denatured, the wrong concentration of antibiotics used in the BSS irrigating solution during phacoemulsification cataract surgery, use of drugs containing preservatives, BSS made up at the wrong pH, or ethylene oxide residue left in plastics.

If an outbreak occurs of several cases, then the surgeon must stop operating and investigate for the source of the problem. Checks should be made on how the instruments have been cleaned and sterilized, what type of water was used for cleaning and autoclaving, and if reusable phako tips and cannulae were in use. Samples should be collected for endotoxin assay from the various tanks of water used for cleaning and sterilizing instruments. Dilutions of drugs and their pH should be checked.

Treatment is given with corticosteroids, which can be used aggressively once infection is excluded by making an anterior chamber tap for microscopy and culture and PCR testing if available (Chapters 2, 3). If no recovery occurs in the cornea within 6 weeks then it is not likely to happen.

POSTTRAUMATIC ENDOPHTHALMITIS

Posttraumatic endophthalmitis, along with postoperative endophthalmitis, is the second commonest form of endophthalmitis. The incidence of endophthalmitis after perforating injury is between 2% and 17%.

Trauma due to an intraocular foreign body involves a greater risk of endophthalmitis than trauma without a foreign body. The signs of infection usually occur early, but are often masked by the posttraumatic reactions of the injured tissue (25,26).

An exact history (e.g., "did the accident happen in the country or in the city?," type of foreign body, symptoms) enables an early diagnosis to be made. In rural districts, the occurrence of posttraumatic endophthalmitis was reported in 30% of 80 patients after an injury. In contrast, posttraumatic endophthalmitis occurred in 11% of 204 patients in non-rural districts.

The start, course and symptoms of endophthalmitis after trauma are very varied, corresponding to the causative organisms. The initial symptoms are usually pain, intraocular inflammation, hypopyon, and vitreous clouding. Similar to postoperative endophthalmitis, two thirds of the bacteria in posttraumatic endophthalmitis are Gram-positive and 10% to 15% are Gram-negative. In contrast to postoperative endophthalmitis, virulent *Bacillus* species are the commonest pathogens in posttraumatic endophthalmitis. They were isolated in 20% of all cases. In the rural population, they are also found in 42% of cases of posttraumatic endophthalmitis. They are thus the second commonest cause of all cases of endophthalmitis. Most *Bacillus* infections are associated with intraocular foreign bodies. Infections that are caused by *Bacillus* species usually commence with rapid loss of vision together with severe pain. *Bacillus* sp. are resistant to penicillin and cephotospoxins, except *B. anthrasis* but are sensitive to gentamicin and vancomycin (refer to Chapter 4).

Other bacteria responsible for the remaining 50% of cases include *Staphylococcus aureus*, streptococci, coliforms, and, occasionally, *Clostridium perfringens* or *C. bifermentans,* the latter being associated with soil contamination of a foreign body injury (4).

Fungi are the causative organisms in 10% to 15% of cases of endophthalmitis after trauma. Fungal endophthalmitis usually commences only weeks to months after the injury. While mixed infections tend to be rarer in postoperative endophthalmitis, they have been isolated in 42% of the trauma-associated cases of endophthalmitis.

Compared to postoperative endophthalmitis, the prognosis of posttraumatic endophthalmitis is usually poor. This is due to a spectrum of more virulent pathogens, to mixed infections, to the degree of tissue injury caused by the preceding trauma and to the *failure* to instill prophylactic intravitreal antibiotics at the time of surgery. While final vision of 20/400 or better occurs in 37% to 42% of cases of postoperative endophthalmitis, patients with posttraumatic endophthalmitis achieve a final vision of 20/400 or better in only 9% to 50% with the remaining losing all vision (4).

Prevention of Endophthalmitis Associated with an Intraocular Foreign Body (IOFB)

Intraocular penetration of a projectile into the posterior segment of the eye is a medical emergency, especially metal splinters arising from hammering farmyard equipment or dirty or soil-contaminated items and machinery. A 20-year review found that 8% of patients with an IOFB developed endophthalmitis, of whom half lost all light perception (4). While antibiotic prophylaxis has been advocated for the last 20 years for IOFB removal, this is not universally practiced. Some ophthalmologists guess which fragments are contaminated with bacteria and give antibiotics to only these patients but this is no longer justifiable; all patients with an IOFB require antibiotic prophylaxis.

Bacillus spp. are the most virulent pathogens carried by an IOFB, and are responsible for 50% of cases of endophthalmitis, most of which result in severe visual loss. *Staphylococcus aureus*, coliforms, streptococci, and, occasionally, *Clostridium perfringens* can also cause sight-threatening endophthalmitis. If trauma takes place in a rural area, there is more likely to be a polymicrobial infection. Because magnet removal is crude, a track of necrotic tissue is left behind which is seeded with bacteria. It is for this reason that we advocate that antibiotic prophylaxis be used for all IOFB removals.

The following regimen is suggested (*cephalosporins* are excluded because *Bacillus* spp. produce beta-lactamases that inactivate them):

- Intravitreal amikacin 400 µg (or gentamicin 200 µg) plus vancomycin 1000 µg (= 1 mg) or clindamycin 1000 µg (= 1 mg) or, if the intravitreal route is not possible or appropriate, give subconjunctival gentamicin 40 mg plus clindamycin 34 mg;
- Topical gentamicin (forte) 15 mg/mL plus clindamycin 20 mg/mL;
- For severe injuries, give intravenous therapy with the same drugs as given intravitreally for three days—give adequate dosage for weight but assay to avoid systemic toxicity, especially for renal and VIII nerve function.

If the patient presents early with good vision and the IOFB is recognized and treated as above, then the chances of endophthalmitis are reduced; without an X ray the IOFB can be missed or dismissed as non-penetrating or the cause of a possible corneal abrasion. If the IOFB is missed, and the patient returns after 24 hours with loss of vision and pain, then it is a medical emergency. To save vision, the IOFB needs prompt removal with the instillation of intravitreal antibiotics—any delay or dependence on intravenous antibiotics alone will contribute to a blind eye (25).

REFERENCES

1. Zhang YQ, Wang WJ. Treatment outcomes after pars plana vitrectomy for endogenous endophthalmitis. Retina 2005; 25:746-50.
2. Jackson TL, Eykyn S, Graham EM et al. Endogenous bacterial endophthalmitis: a 17-year prospective series and review of 267 reported cases. Surv Ophthalmol 2003; 48:403–23.
3. Miller JJ, Scott IU, Flynn HW et al. Endophthalmitis caused by *Streptococcus pneumoniae*. Am J Ophthalmol 2004; 138:231–36.
4. Peyman G, Lee P, Seal DV. Endophthalmitis: Diagnosis and Management. Taylor & Francis, London & New York: 2004:1–278.

5. Jager RD, Aiello LP, Patel SC et al. Risks of intravitreous injection: a comprehensive review. Retina 2004; 24:676–98.
6. Seal DV, Alio J, Ferrer C et al. Laboratory management of endophthalmitis comparison of microbiology and molecular biology methods in the European ESCRS multicentre study and appropriate chemotherapy. Paper prepared, 2007.
7. Endophthalmitis Vitrectomy Study Group. Results of the EVS study: a randomised trial of immediate vitrectomy and of intravenous antibiotics for the treatment of post-operative bacterial endophthalmitis. Arch Ophthalmol 1995; 113:1479–96.
8. Taban M, Behrens A, Newcomb RL et al. Acute endophthalmitis following cataract surgery. A systematic review of the literature. Arch Ophthalmol 2005; 123: 613–20.
9. Nichamin LD, Chang DF, Johnson SH et al. What is the association between clear corneal cataract incisions and postoperative endophthalmitis ? J Cataract Refract Surg 2006; 32:1556–59.
10. Nagaki Y, Hayasaka S, Kadoi C et al. Bacterial endophthalmitis after small-incision cataract surgery. Effect of incision placement and intraocular lens type. J Cataract Refract Surg 2003; 29:20–26.
11. Verbraeken H. Treatment of post-operative endophthalmitis. Ophthalmologica 1995; 209:165–71.
12. ESCRS Endophthalmitis Study Group. Prophylaxis of postoperative endophthalmitis following cataract surgery: results of the ESCRS multicentre study and identification of risk factors. JCRS 2007; 33:978–88.
13. Abreu JA, Cordovés L. Chronic or saccular endophthalmitis: diagnosis and management. J Cataract Refract Surg 2001; 27:650–51.
14. Warheker PT, Gupta SR, Mansfield DC, Seal DV, Lee WR. Post-operative saccular endophthalmitis caused by macrophage-associated staphylococci. Eye 1998; 12: 1019–1021.
15. Warheker PT, Gupta SR, Mansfield DC, Seal DV. Successful treatment of saccular endophthalmitis with clarithromycin. Eye 1998; 12:1017–1019.
16. Okhravi N, Guests S, Matheson MM et al. Assessment of the effect of oral clarithromycin on visual outcome following presumed bacterial endophthalmitis. Curr Eye Res 2000; 21:691–702.
17. Seal DV, Barry P, Gettinby G et al. ESCRS study of prophylaxis of postoperative endophthalmitis after cataract surgery. Case for a European multicenter study. J Cataract Refract Surgery 2006; 32:396–406.
18. Sundelin K, Stenevi U, Seal D et al. Enhanced anterior chamber penetration of topical levofloxacin 0.5%. Paper prepared, 2007.
19. Lundstrom M, Wejde G, Stenevi U et al. Endophthalmitis after cataract surgery. A nationwide prospective study evaluating incidence in relation to incision type and location. Ophthalmol 2007; 114:866–70.
20. Wallin T, Parker J, Jin Y et al. Cohort study of 27 cases of endophthalmitis at a single institution. J Cataract Refract Surg 2005; 31:735–41.
21. Wong TY, Chee S-P. Risk factors of acute endophthalmitis after cataract extraction: a case-control study in Asian eyes. Brit J Ophthalmol 2004; 88:29–31.
22. Ciulla TA, Starr MB, Masket S. Bacterial endophthalmitis prophylaxis for cataract surgery: an evidence based update. Ophthalmol 2002; 109:13–24.
23. Colleaux KM, Hamilton WK. Effect of prophylactic antibiotics and incision type on the incidence of endophthalmitis after cataract surgery. Can J Ophthalmol 2000; 35: 373–78.
24. Mamalis N, Edelhauser H, Dawson DG et al. Toxic Anterior Segment Syndrome. J Cataract Refract Surg 2006; 32:324–32.
25. Spoor TC. An Atlas of Ophthalmic Trauma. Martin Dunitz, London 1997, pp 1–207.
26. Peyman GA, Meffert SA, Conway MD, Chou F. Vitreoretinal Surgical Techniques. Martin Dunitz, London 2001, pp. 1–605.

NINE

Tropical Ophthalmomycoses

Philip Thomas
Joseph Eye Hospital, Tiruchirrapalli, Tamil Nadu, India

INTRODUCTION

Fungal ocular infections, principally mycotic keratitis and conjunctivitis, are common in the tropics, as compared with temperate climates. Some (*Lasiodiplodia theobromae*, *Colletotrichum* species) are unique to tropical and subtropical regions (1). This chapter highlights important aspects of the ophthalmomycoses, with emphasis on the problems experienced in tropical countries. Actinomycetales (*Actinomyces, Nocardia*), although branching Gram-positive bacteria, and not fungi, are also considered since they are important causes of tropical ocular infections.

Fungal infections of the eye are of two basic types: (*i*) those involving the orbit and (*ii*) those involving the globe, which are more common; those involving the globe can again be subdivided into external ocular infections and intraocular infections. The fungi that have been implicated worldwide as pathogens from more than one ocular site are listed in Table 1 (1–3).

FUNGAL INFECTIONS OF THE ORBIT

The principal routes of infection in orbital mycoses are by inoculation (trauma or surgery), by extension of a panophthalmitis, or by spread of infection from the paranasal sinuses (4). Rhino-orbital-cerebral zygomycosis (rhinocerebral zygomycosis, mucormycosis) and sino-orbital aspergillosis are the commonest orbital mycoses (1).

Rhino-Orbital-Cerebral Zygomycosis

This is a potentially lethal, acute fungal infection of the paranasal sinuses and orbit. *Rhizopus* spp., especially *Rhizopus arrhizus*, are the principal causes; less common causes include *Absidia corymbifera*, *Apophysomyces elegans*, and other Mucorales, which abound in the environment. These organisms proliferate within the nasal turbinates and sinuses, penetrate the muscular walls of the arteries, and spread by vascular and direct extension to the orbit. Diabetic ketoacidosis is a key predisposing factor, but renal failure, lymphoproliferative disorders, malignancies, hepatic and other gastrointestinal diseases, severe burns and malnutrition, fulminant diarrhea in small children, and administration of desferrioxamine may also predispose to the

(Text continues on page 277)

Table 1 Sites of Fungal Infection[a]

Fungus	Lids	Conjunc.	Cornea[b]	Lacrimal system[c]	Sclera	Intraocular	Orbit	Other sites
Filamentous and dimorphic fungi								
Acremonium			•✓	◉ ?	◉ ?	•✓		
Alternaria	◉ ?	•✓	•✓	◉ ?	•✓	◉ ?	•✓	
Aspergillus	•✓		•✓	◉ ?		•✓	•✓	Optic nerve •✓
Bipolaris/Drechslera			•✓			•✓	•✓	
Blastomyces	•✓	◉ ?	•✓	◉ ?	◉ ?	•✓	◉ ?	Uvea •✓
Cladosporium			◉ ?			◉ ?		
Chrysonilia (formerly *Neurospora*)	•✓	•✓	◉ ?			◉ ?		
Coccidioides	•✓		◉ ?		◉ ?	•✓	◉ ?	Uvea •✓
Curvularia			•✓			•✓	◉ ?	
Dermatophytes (*Microsporum Trichophyton*)	◉ ?	◉ ?	◉ ?	◉ ?			◉ ?	

Exophiala

Fusarium

Fonsecaea

Graphium (see *Scedosporium apiospermum*)

Histoplasma

Lasiodiplodia

Microspheropsis

Ovadendron

Paecilomyces

Paracoccidioides

Penicillium

Phialophora

Phycomycetes (zygomycetes)[d]

(Continued)

Table 1 Sites of Fungal Infection[a] (*Continued*)

				Site of infection				
Fungus	**Lids**	**Conjunc.**	**Cornea[b]**	**Lacrimal system[c]**	**Sclera**	**Intraocular**	**Orbit**	**Other sites**
Scedosporium apiospermum			•√		•√	•√	•√	
Scedosporium prolificans		●?	•√	●?	●?			
Scopulariopsis				●?		●?		
Sporothrix	●?	●?	●?	●?	●?	•√	•√	Uvea •√
Sphaeropsis			•√			•√		
Volutella			●?	●?		•√		
Yeasts and nonfilamentous fungi								
Malassezia	●?	●?	●?	●?		•√		
Candida	●?	●?	•√	•√	•√	•√	●?	Uvea •√
Cryptococcus	●?	•√	•√		●?	•√	•√	Uvea •√
Rhodotorula	●?		●?	•√				

Trichosporon

Geotrichum

Torulopsis

Branching bacteria
(Actinomycetales)
Actinomyces
Nocardia

Fungi-like organisms
Pythium
Pneumocystis carinii

Rhinosporidium

Uvea

[a] Fungi listed have been reported to cause infection in more than one ocular tissue.
[b] Comprises lacrimal gland, lacrimal passage, and lacrimal sac.
[c] A more detailed list of fungi implicated in keratitis is given in Table 2.
[d] Phycomycetes (zygomycetes) causes acute orbital cellulitis (mucormycosis); other fungi causing orbital infections usually cause chronic cellulitis.

Table 2   Reported Etiological Agents in Mycotic Keratitis[a]

Hyaline Filamentous Fungi

Genus	Species
Acremonium	A. Atrogriseum, A. curvum,[b] A. kiliense, A. potronii, A. recifei,[c] Acremonium species[c]
Arthrographis	A. kalrae
Aspergillus	A. clavatus,[b] A. fischerianus, A. flavipes, A. flavus, A. glaucus[b], A. fumigatus, A. janus[c], A. niger, A. terreus, A. nidulans[c], A. oryzae[b], A. wentii[c]
Beauveria	B. bassiana
Cephaliophora	C. irregularis
Chrysonilia	C. sitophila[c] (formerly Neurospora sitophila)
Chrysosporium	C. parvum[c]
Cylindrocarpon	C. lichenicola (C. tonkinense)
Diplosporium	Diplosporium species[d]
Engyodontium	E. alba (formerly Beauveria alba)
Epidermophyton	Epidermophyton species[d]
Fusarium F. aquaeductum[b]	F. dimerum, F. oxysporum, F. solani, F. verticilloides (F. moniliforme), F. nivale[c], F. subglutinans, F. ventricosum
Glenospora	G. graphii[d]
Metarhizium	M. anisopliae
Microsporum	Microsporum species[c], M. canis[b]
Myrathecum	Myrathecum species[d]
O. vadendron	O. sulphureo-ochraceum[b]
Paecilomyces	P. farcinosus, P. lilacinus, P. variotii[b]
Penicillium	P. citrinum, P. expansum
Rhizoctonia	Rhizoctonia species
Sarcopodium	S. oculorum
Scedosporium	S. apiospermum (reported as Pseudallescheria boydii; previously Allescheria boydii, Petriellidium boydii, Monosporium apiospermum)
Scopulariopsis	S. brevicaulis
Tritirachium	T. oryzae
Ustilago	Ustilago speciesd
Verticillium	V. searrae[c], Verticillium species[b]

Phaeoid Hyphomycetes

Genus	Species
Alternaria	A. alternata, A. infectoria,[c] Alternaria spp.[c]
Aureobasidium	A. pullulans[c]
Bipolaris	B. hawaiiensis, B. spicifera (formerly Drechslera)
Cladosporium	C. cladosporioides[c]
Curvularia	C. brachyspora, C. geniculata, C. lunata, C. pallescens,[b] C. senegalensis, C. verruculosa[c]
Dichotomophthoropsis	D. nymphearum, D. portulacae[b]
Doratomyces	D. stemonitis[b]
Exophiala	E. jeanselmei var. dermatitidis, E. jeanselmei var. jeanselmei[c]
Exserohilum	E. rostratum, E. longirostratum
Fonsecaea	F. pedrosoi
Lecytophora	L. mutabilis[c]
Phaeoisaria	P. clematitidis
Phaeotrichoconis	P. crotalariae
Phialophora	P. bubakii[b], P. verrucosa
Tetraploa	T. aristata

Phaeoid Sphaerosidales

Genus	Species
Colletotrichum	C. capsici[c], C. coccodes[c], C. dematium[c], C. graminicola, C. gloenosporioides, Colletotrichum state of Glomerulla cingulata

(Continued)

Table 2 Reported Etiological Agents in Mycotic Keratitis[a] (*Continued*)

Lasiodiplodia	*L. theobromae*
Microsphaeropsis	*M. olivacea*[c]
Phoma	*P. oculo-hominis*[c], *Phoma species*
Sphaeropsis	*S. subglobosa*

Yeast and Yeast-like Fungi

Genus	**Species**
Candida	*C. albicans, C. famata*[b]*, C. glabrata*[c]*, C. guilliermondii, C. krusei, C. parapsilosis, C. tropicalis*[c]
Cryptococcus	*C. laurentii, C. neoformans*[c]
Geotrichum	*G. camdidum*[d]
Malassezia	*M. furfur*[c]
Rhodotorula	*R. glutinis,*[c] *R. rubra,*[c] *Rhodotorula species*
Rhodosporidum	*R. toruloides*[d]

Dimorphic Fungi

Genus	**Species**
Blastomyces	*B. dermatitidis*
Coccidioides	*C. immitis*[b]
Paracoccidioides	*P. brasiliensis*
Sporothrix	*S. schenckii*[b]

Other Fungi

Genus	**Species**
Absidia	*A. corymbifera*
Chlamydoabsidia	*C. padenii*[b]
Pythium	*P. insidiosum*

Newly Reported Agents
Ulocladium atrum (75)
Scytalidium sp. (76)
Blastoschizomyces capitatus (77)

[a] Based on Thomas and Geraldine (2005).
[b] Listed as cause of mycotic keratitis by Wilson and Ajello (1998).
[c] Deemed uncertain by Thomas and Geraldine (2005).
[d] Uncertain.

disease (5). Patients frequently present with acute lid edema, proptosis, restricted ocular motility, and diminished consciousness; black necrotic eschars and black pus formation in the nose and palate are characteristic, while a cherry-red spot (due to occlusion of the central retinal or ophthalmic artery) and X-ray evidence of extensive bony destruction are suggestive of the condition (5,6).

Magnetic resonance imaging and computerized tomography are important aids in establishing an anatomic diagnosis in suspected rhinocerebral fungal infections. Prompt laboratory diagnosis is essential and may be life-saving (see Chapters 2 and 3) (3). Species of Mucorales are readily detected in fixed tissue sections by Gomori methenamine silver (GMS), hematoxylin-eosin (H&E), or periodic acid-Schiff (PAS) stains, appearing as distinctive, aseptate (or sparsely septate) hyphae (which often appear vacuolated and bizarre) with right-angle branching, usually in the interior of blood vessels. The muscles, optic nerve sheath, and even the bone may be infiltrated.

Treatment consists of intravenous amphotericin B (see Appendix A), drainage and curettage of affected sinuses, aggressive surgical debridement of necrotic tissue from orbit/nose, and control of the underlying disease (e.g., diabetic ketoacidosis); hyperbaric oxygen therapy may also be beneficial (5). Amphotericin B has been administered in different ways, in addition to conventional intravenous dosing, in an effort to improve the outcome of this condition, which normally results in extreme morbidity; these have included irrigation and packing of the involved orbit and sinuses, use of a percutaneous catheter, and adjunctive local nebulized and intraventricular amphotericin B. New formulations of amphotericin B, such as amphotericin B lipid complex or colloidal dispersion, and liposomal amphotericin, have also been tried, as reviewed by Thomas (3). There is anecdotal evidence of the efficacy of some azoles in treatment of this condition, including posaconazole for the treatment of periorbital zygomycosis due to *Rhizopus* spp. (7). Even after resolution of the acute phase of the disease, treatment is normally continued for two to three weeks.

There is insufficient data to facilitate evaluation of the treatment of orbital mucormycosis, and no standard of care currently exists to guide physicians as to when exenteration may benefit a mucormycosis patient. A recently published review noted that patients with mucormycosis who were older than 46 years old, and who had frontal sinus involvement and fever, were less likely to survive compared with patients without these conditions; also, patients treated with amphotericin B and those with diabetes mellitus were more than four times more likely to survive compared with patients without these conditions; febrile exenterated patients were significantly more likely to survive compared with febrile non-exenterated patients (8).

Basidiobolus and *Conidiobolus*, which cause subcutaneous zygomycoses in the tropics, have hitherto not been implicated as causes of rhino-orbital-cerebral zygomycosis. However, species of *Aspergillus, Scedosporium apiospermum*, and different genera and species of phaeohyphomycetes, may rarely cause acute orbital infection (3). Orbital actinomycosis has been reported to present as painful ophthalmoplegia, with associated actinomycotic lesions in the lung, and to resolve completely with penicillin therapy (9).

Chronic Orbital Cellulitis

The chronic invasive form of fungal sinusitis, which results in rhino-orbital-cerebral disease, manifests as a slowly destructive disease process that may occur in immunocompromised patients or non-immunocompromised individuals. Due to various fungi (Table 1), this is characterized by pain, slowly progressive proptosis, limitation of ocular motility, and reduced vision. *Aspergillus* species are important causes of chronic orbital cellulitis.

Chronic Orbital Aspergillosis

Chronic orbital aspergillosis, which is described as being more common in hot, humid climates, may refer to the chronic invasive disease (associated with granulomatous inflammation and fibrosis) occurring in immunocompromised persons aged 40 to 60 years, or to chronic invasive disease that occurs in non-immunocompromised patients; predisposing factors include alcoholism, high-dose corticosteroid therapy, insulin-dependent diabetes mellitus, and AIDS. Chronic presentations of orbital aspergillosis are more common than the acute variety and tend to be more localized; manifestations include aspergilloma, dacryocystitis, nerve tumors, and postoperative periorbital swelling. The presenting complaints in chronic disease can be non-specific. Chronic sino-orbital aspergillosis may present subacutely in elderly diabetic individuals or in those on immunosuppressive therapy (10). Thus, neutropenic or otherwise immunocompromised persons are at risk to first develop chronic disease and then fulminant aspergillosis. Principal steps in laboratory diagnosis are outlined in Chapter 2.

To successfully manage invasive sino-orbital aspergillosis, as in patients with AIDS, aggressive surgical debridement and intravenous amphotericin B are needed; however, these measures notwithstanding, the prognosis is usually very poor in invasive disease, with mortality approaching 75% to 80%, primarily due to intracranial extension of the disease process, and most patients succumbing to their illness within a few months. The prognosis may be better if the person is immunocompetent and has only had the disease for a short while, and if protease inhibitors are used (11). The prognosis is better with therapy of the more localized forms of orbital aspergillosis; conservative orbital debridement and intravenous and local administration of amphotericin B may be effective, especially in affected individuals with reversible immunosuppression and good preoperative visual activities. There is anecdotal evidence of the efficacy of amphotericin B colloidal dispersion and itraconazole in therapy of sino-orbital aspergillosis (3), but a recent paper reported that this condition did not respond satisfactorily to surgical debridement and medical management with amphotericin B and itraconazole (10).

Chronic Rhino-Orbital-Cerebral Zygomycosis

In the chronic presentation of rhinocerebral zygomycosis, which is less common than the acute form, the disease is indolent and slowly progressive over weeks to months. The outcome of the disease is also better than that of the acute variety.

Chronic Orbital Infections Due to Other Fungi and Actinomycetes

Chronic invasive orbital infection may also be caused by other species of hyaline filamentous fungi (e.g., *Paecilomyces lilacinus, Scedosporium apiospermum*), phaeohyphomycetes (*Bipolaris* species) and dimorphic fungi (3) and also Actinomycetales (Table 1). Affected individuals may present with nasal congestion, post-nasal drip, and chronic sinus pain, or with slowly progressive unilateral proptosis with decreased vision; other orbital inflammatory signs may be absent. The prognosis in this variety of disease is better than that in acute fulminant disease; however, if intraorbital and intracranial extension occurs, significant morbidity may result. Treatment is by surgical debridement and systemic antifungal therapy; significantly, antifungals other than amphotericin B may play an important role in such infections. Orbital infections with *Actinomyces* or *Nocardia* should be treated with penicillin, rifampin, or vancomycin; combination therapy should be used with amikacin for nocardiosis.

Chronic Localized Noninvasive Fungal Infections

A chronic, localized, noninvasive fungal infection or "fungus ball" may occur in an orbit where exenteration (removal of the entire contents of the orbit) has been performed or in the orbital prosthesis used to fill an exenterated socket.

Allergic Fungal Rhinosinusitis with Orbital Involvement

Allergic fungal sinusitis or allergic fungal rhinosinusitis refers to a chronic sinusitis in immunocompetent, atopic individuals where an infecting fungus does not invade the surrounding tissue, but serves as an allergen resulting in the formation of a thick (peanut butter-like) tenacious, green, or brown "allergic" mucin. The fungi

implicated include species of *Aspergillus* and various phaeohyphomycetes (*Bipolaris,
Exserohilum, Curvularia,* and *Alternaria*). Up to 17% of patients with allergic fungal
rhinosinusitis may experience orbital symptoms. The affected individual may
experience nasal obstruction or rhinorrhea, visual loss or diplopia, proptosis,
epiphora, cranial nerve palsies, and facial deformity. Allergic fungal rhinosinusitis is
a unique subset of fungal sino-orbital disease that differs from invasive sino-orbital
disease in many respects, such as the occurrence of a predominantly eosinophilic
response of the sinus mucosa and characteristic mucoid exudate containing fungal
hyphae, peripheral blood eosinophilia, serum precipitins against the fungus involved,
elevated serum total and fungus-specific IgE and IgG concentrations, and an
immediate hypersensitivity response to the concerned fungal antigen.
Immunotherapy has been advised for such patients, since it appears to significantly
alter their clinical course. Patients with the condition should ideally undergo
conservative but thorough endoscopic surgical exenteration of the lesion, adjuvant
therapy with perioperative systemic corticosteroids and topical corticosteroids,
irrigation and debridement as necessary after surgery, and immunotherapy with all
relevant fungal and non-fungal antigens. Allergic fungal rhinosinusitis does not
require aggressive surgical debridement or intravenous amphotericin B.

EXTRAOCULAR FUNGAL INFECTIONS

Mycotic Infections of Eyelids

Infections of the eyelids are caused almost exclusively by bacteria, particularly
Staphylococcus aureus. Several fungi have been isolated from eyelid lesions (Table 1);
however, the significance of some of the reported isolates is uncertain (1).

Etiological Agents of Mycotic Eyelid Infections

Cryptococcus neoformans has been implicated in necrotizing fasciitis of the eyelids
and periorbital area following trivial injury, and in nodular and ulcerative "sentinel"
lesions of the eyelid in disseminated cryptococcosis in a patient with AIDS (12).
Candida spp. have been isolated from eyelid cultures of 12% of patients with chronic
severe ulcerative blepharitis, with most of the patients also having atopic dermatitis;
however, *Candida* eyelid lesions or colonization probably arises due to spread from a
septic focus following use of broad-spectrum antibacterials or immunosuppressive
agents (13). Ulceration begins at the base of an eyelash; small granulomata manifest
at its edge, and vesicles and pustules may also be present.

   Malassezia furfur (formerly *Pityrosporum orbiculare* and *Pityrosporum ovale*) can
cause pityriasis versicolor (a chronic mild skin infection) around the eyebrows and
eyelids; in addition, *Malassezia* spp. have been found associated with more than
90 percent of patients with active seborrheic or mixed seborrheic/staphylococcal
blepharitis. Eyelid lesions may also be caused by dermatophytes, starting as
erythematous scaly papules that slowly enlarge; healing simultaneously occurs in the
central, paler area; when species of *Microsporum* and *Trichophyton* are involved, there
may be destruction of the eyelashes.

   Among the dimorphic fungi, *Blastomyces dermatitidis* and *Paracoccidioides
brasiliensis* have been reported to cause eyelid infection. *B. dermatitidis* infections,
which follow contiguous spread from adjacent lesions or hematogenous dissemination, initially manifest as small abscesses around the eyelashes and later as
granulomatous ulcers with thick crusts and an underlying purplish discoloration

of the skin; these lesions heal with severe cicatrization and ectropion formation. Conjunctival lesions generally occur due to contiguous spread from eyelid lesions, but may also occur as separate entities. In paracoccidioidomycosis, an eyelid lesion (alone or in association with corneal and conjunctival lesions) starts as a papule, adjacent to the lid border, which grows and ulcerates in the center; the base of the ulceration reveals fine hemorrhagic punctate and elevated, thickened and hardened borders; there may be loss of eyelashes.

Rhinosporidiosis of the lid margins is a rare occurrence.

Diagnosis

Lesions due to certain dermatophytes may fluoresce when exposed to a Wood's (UV) lamp. Otherwise, the usual method of diagnosis of mycoses of the eyelids is by demonstrating fungal hyphae or yeast cells in a 10% potassium hydroxide (KOH) mount or a Gram-stained smear of scrapes from the eyelids.

Therapy

Although several therapeutic regimens have been recommended for mycoses of the eyelids (13), the basis for some of these recommendations is unclear.

Superficial infections of the eyelid. For dermatophytic infections, orally administered griseofulvin (0.5 to 1 G/day in adults, 10 mg/kg BW/day in children, in a single or divided doses until the infection subsides and for two weeks thereafter) or local antifungal drugs (active ingredients being combinations of fatty acids and salicylic acid) probably suffice. Oral itraconazole is also efficacious. In seborrheic blepharitis due to *Malassezia* spp., a combination of topical 2% ketoconazole cream with lid hygiene has been reported to yield better than results than placebo and lid hygiene alone. Oral ketoconazole (100 mg/day) with topical miconazole ointment for six weeks has been used to successfully treat blepharitis associated with *Candida* spp., while oral fluconazole (150 mg) once or twice weekly for two to three months may also be given (13). In many instances, *Candida* infections resolve with topical therapy alone [examples include the use of ointments of nystatin (100,000 units/g), ketoconazole (2%), miconazole (2%), clotrimazole (1%), or econazole (1%)]. However, lesions of the eyelid due to *Rhinosporidium seeberi*, which are rare, always require excision.

Deep infections of the eyelid. These may occur as a consequence of infections due to the thermally dimorphic fungi or of systemic infections due to species of *Aspergillus*, *Candida*, or *Cryptococcus*. Such infections must be treated with drugs administered systemically; intravenous amphotericin B is usually the treatment of choice, although oral ketoconazole, itraconazole, or fluconazole may be considered in milder infections. A combination of surgical excision and intravenous amphotericin B was reported to control lesions of the eyelid in a patient with disseminated infection due to *Cryptococcus neoformans* var. *neoformans* (12). Preseptal cellulitis due to a *Trichophyton* sp. was found to have completely resolved following two courses of oral itraconazole (100 mg per day) (13). Verrucous lesions of the eyelid due to *Blastomyces dermatitidis* have been reported to resolve with a combination of antifungals (potassium iodide, intravenous amphotericin B) and surgery; more recently, oral fluconazole (150 mg) once or twice weekly for two to three months has been recommended. Papular lesions of the eyelids due to *Paragcoccidioides brasiliensis* have been found to respond to intravenous amphotericin B, alone or in combination with oral ketoconazole, without any surgery being required.

Mycotic Dacryocanaliculitis

Dacryocanaliculitis constitutes 2% of diseases of the lacrimal passages. *Actinomyces* spp. are the most frequent causes, but *Nocardia* spp. are sometimes encountered as well. Fungi such as *Alternaria* sp., *Aspergillus fumigatus*, and other *Aspergillus* spp., *Candida* sp., dermatophytes, *Fusarium* sp., *Penicillium* sp., *Scopulariopsis* sp., and *Sporothrix schenckii* have also been reported as causes of canaliculitis (Table 1); however, the significance of some of these isolates is uncertain (1). In addition, most of the clinical features described, as well as the surgical procedures recommended, are for dacryocanaliculitis in general, and are not necessarily specific for fungal infection (1).

Characteristically, only one canaliculus (usually the inferior one) and only on one side is affected. Persistent unilateral watering, associated with itching, is the frequent presenting complaint. Clinical features include: conjunctival inflammation, usually localized to the internal angle and sometimes accompanied by formation of follicles; reddening and swelling of the canaliculus, with the opening being dilated and the edges elevated and inflamed; extrusion of white, yellow, or brown concretions after applying pressure on the canaliculus; patency of the lacrimal passages and no preauricular lymph gland enlargement. Local environmental factors in the canaliculus, such as stasis arising out of congenital diverticula, may predispose to these infections.

Curettage is apparently the most effective treatment for chronic canaliculitis in general; this probably applies to specific actinomycotic and fungal causes of canaliculitis as well. The procedure is performed following application of topical and local anesthesia. Essentially, the punctum is dilated and a small curette is introduced. Occurrence of extensive ectasia and retention of concretions in diverticuli requires a canaliculotomy. Following curettage, the canaliculus is thoroughly irrigated to remove any remaining fragments; unrecognized pockets of retained debris are removed to ensure that the distal part of the drainage system is patent.

Infections caused by actinomycetes can usually be cleared by dilatation and curettage of the canaliculi, while irrigating the canaliculus with penicillin G (1,000,000 U/mL irrigant; 60 000 U/mL drops) is also useful. Many broad-spectrum antibiotics, Lincomycin, and sulfonamides (including 10% sulfacetamide drops) are also effective against *Actinomyces*. Infections due to fungi are usually satisfactorily treated by topical administration of 5% natamycin or by topical application and local syringing of the canaliculi and sac with amphotericin B solution (1.5–8 mg/mL) or nystatin solution (25,000 to 100,000 U/mL); however, the basis for these recommendations is not clear. If medical therapy fails, surgery is required. Canaliculotomy is done and all material removed is stained and cultured; the canaliculus is then syringed daily with the solutions referred to above. The canaliculus should be reconstructed, for which silicone intubation may be needed. Medical therapy alone may yield a cure in only about 10% of patients with canaliculitis; however, canaliculotomy may yield successful results in as many as 80% of patients. Thus, surgical treatment of canaliculitis, in combination with medical therapy, is probably the best line of management of canaliculitis.

Mycotic Dacryocystitis

Dacryocystitis, an infection of the lacrimal sac (the most common infection of the entire lacrimal apparatus), generally originates from the stasis accompanying obstruction of the nasolacrimal duct, which, in turn, may arise due to several possible causes. While dacryocystitis may be acute or chronic, fungi are generally not implicated in acute dacryocystitis. Chronic dacryocystitis is a sequel to partial or complete obstruction at a single site within the lacrimal sac or within the nasolacrimal duct, and infection generally follows obstruction, and does not cause it. In acquired dacryocystitis, infection is generally due to bacteria (especially aerobic and facultative anaerobic bacteria), with fungi accounting for a mere 5% of infections.

Fungi implicated as causes of dacryocystitis include *Acremonium* sp., *Aspergillus* sp., *Candida* sp., *Paecilomyces* sp., *Rhinosporidium seeberi*, dermato-phytes, and *Sporothrix schenckii* (1); however, the significance of some of these species as actual causes of dacryocystitis is uncertain. Infections due to *S. schenckii* and *Acremonium* spp. present as chronic suppurative dacryocystitis, with preauri-cular and submaxillary lymphadenitis and possible abscess formation. Species of *Aspergillus*, *Candida* and *Paecilomyces*, *Rhinosporidium seeberi* and the dermato-phytes may cause chronic granulomatous dacryocystitis. With partial or complete obstruction of the nasolacrimal duct, a laminated concretion (dacryolith) may develop in the lacrimal sac.

Epiphora is frequently the only clinical finding in patients with chronic dacryocystitis, but lid edema, extreme tenderness of the lacrimal sac, conjunctival injection, and a swelling in the medial canthus may also occur; pressure over the area usually results in a purulent discharge through the lower punctum. Tabbara (14) reported *A. fumigatus* colonization of punctal plugs in two patients who were using these for dry eye syndrome. Rhinosporidiosis of the lacrimal sac is known to manifest as bloodstained epiphora (due to the extreme fragility of the lesion) in affected individuals. Samples to be collected to establish the diagnosis are outlined in Chapter 2.

Rhinosporidiosis of the lacrimal sac is managed by dacryocystectomy (*in toto* removal of the sac). Other forms of dacryocystitis are usually managed by dacryocystorhinostomy (patency of the nasolacrimal duct is restored without removing the sac) and removal of dacryoliths, if present. Topical antifungals, probing, and syringing may be useful in congenital mycotic dacryocystitis. With regard to specific fungal causes of dacryocystitis, *C. albicans* infection has been reported to resolve completely following surgery, alone or in combination with topical miconazole and natamycin therapy, while dacryocystitis following lacrimal sac plugging due to *A. fumigatus* was relieved by just removing the plug after opening the lacrimal sac. To prevent fungal colonization of the inserter hole of a punctual plug used to retain tears in a dry eye syndrome, such plugs should be removed prior to refractive or intraocular procedures, and in individuals receiving topical corticosteroids, should be redesigned to avoid the presence of an inserter hole (14). Minimally invasive, transnasal dacryocystorhinostomy surgery employing endoscopic and laser technologies has recently been introduced; postoperative infection is prevented by intraoperative or postoperative antifungal therapy.

Mycotic Dacryoadenitis

Acute fungal infections of the lacrimal gland are rare; zygomycetes can cause a necrotizing dacryoadenitis in the setting of contiguous sino-orbital disease. Chronic granulomatous dacryoadenitis can be caused by certain filamentous fungi and by *Nocardia* (Table 1). Treatment consists of the use of systemic antifungal agents.

Mycotic Conjunctivitis

Fungi are usually transient inhabitants of the normal conjunctival sac but may be stimulated to produce pathogenic effects following the unrestricted use of topical antibacterials or corticosteroids. Severe necrotizing granulomatous conjunctivitis due to *Coccidioides immitis* has been reported in a patient who had received intensive corticosteroid therapy by various routes. Acute conjunctivitis due to fungi appears to be uncommon. The principal routes of inoculation are airborne, fomites, hand-to-eye contact, and spread from the ocular adnexa; endogenous (i.e., hematogenous) conjunctivitis is uncommon.

The clinical manifestations of mycotic conjunctivitis depend on the fungi involved. The following have been described: simple acute or subacute superficial epithelial conjunctivitis due to *Candida* spp. (purulent infection), dermatophytic fungi (chronic lesions), and *Malassezia* spp. (catarrhal conjunctivitis); nodular conjunctivitis, with associated deep lesions and local lymphadenopathy, due to *S. schenckii*; follicular conjunctivitis due to the dimorphic fungi *B. dermatitidis*, *C. immitis*, or *P. brasiliensis*; and chronic conjunctivitis with black conjunctival secretions due to *A. niger*.

Rhinosporidiosis is a chronic infection of the mucous membranes caused by *Rhinosporidium seeberi*, which was once believed to be a fungus. Most reported ocular lesions due to rhinosporidiosis have occurred in hot, dry, climatic regions, with the occasional case from temperate zones. While nasal lesions predominate in endemic areas, a predominance of ocular lesions is believed to indicate an epidemic. Ocular rhinosporidiosis most frequently manifests as single or multiple polypoidal outgrowths of the palpebral conjunctiva, the outgrowths being pink or red, granular or lobulated (occasionally flattened), sessile or stalked, and attached to the upper or lower fornix or the tarsal conjunctiva. The rare occurrence of conjunctival rhinosporidiosis with associated scleral melting and staphyloma formation was recently reported in three patients in India, with the lesions manifesting as grey-white spherules without polyps. The occurrence of rhinosporidiosis of the lid margins, canaliculus, and lacrimal sac has been described earlier. These lesions usually cause no discomfort to the patient but there may be increased lacrimation, discharge, tenderness of lids, and photophobia. Diagnosis depends on direct examination and

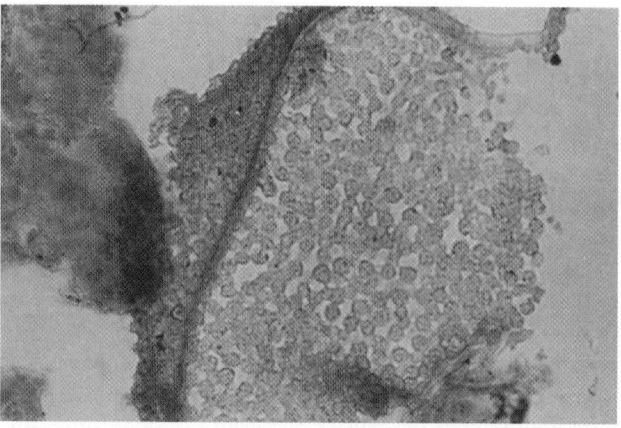

Figure 1 Rhinosporidiosis of the conjunctiva.

histopathology (Fig. 1). Distinctive spherical bodies (spherules or sporangia), which vary in size from 6 to 30 μm, can be demonstrated by simple hematoxylin and eosin staining. Although the presence of such well-defined spherical bodies of varying size in a rather dense stroma covered by hyperplastic epithelium is a distinctive feature, these structures need to be differentiated from the spherules of *Coccidioides immitis*. Serological tests have not been found useful for the diagnosis of the condition.

Treatment of mycotic conjunctivitis can be difficult. Lesions due to *Actinomycetes* generally respond to penicillin, sulphonamides, or broad-spectrum antibiotics. For non-granulomatous (superficial) conjunctivitis, topical miconazole 2%, natamycin 5%, and clotrimazole 2% are effective. For granulomatous (deep or proliferative) conjunctivitis, systemic therapy with oral ketoconazole and fluconazole has been found efficacious. Amphotericin B drops can be used against many fungal infections, while nystatin solution or ointment is also effective, especially against *Candida* sp. No drug treatment has proven effective for ocular rhinosporidiosis. This condition is treated by surgical excision alone.

FUNGAL INFECTIONS OF THE GLOBE

Mycotic Scleritis

Although fungal lesions of the sclera are rare, several fungi have been implicated in scleritis (Table 1). Scleritis may result from a cell-mediated or delayed hypersensitivity reaction (Chapter 1) or from microbial invasion or from a combination of both. Scleritis may be endogenous or exogenous in origin. Exogenous infections may spread from contiguous structures (cornea, conjunctiva) or may be secondary to trauma or surgery. Scleritis has been reported arising from spread of infection from keratitis due to *Absidia corymbifera*, *Acremonium* sp., and *Lasiodiplodia theobromae*. Similarly, scleritis due to fungi (*Aspergillus* sp. or *S. schenckii*) or *Nocardia* may follow ocular trauma. A unique subset of microbial scleritis following ocular surgical procedures is being increasingly reported. Scleritis following cataract surgery involving filamentous bacteria such as Nocardia and fungi such as *Paecilomyces* species and *A. flavus* has been reported. Beta-irradiation therapy, or, rarely, thiotepa, following pterygium excision may damage the anterior sclera, predisposing the eye to fungal infection. Scleritis may occur relatively soon after surgery (two to six weeks) or after an interval of 3 to 10 years. Laboratory examination is essential to identify the causative organism (Chapters 2 and 3).

The outcome of corneoscleritis due to *S. prolificans* and *S. apiospermum* is varied. Mycotic scleritis due to species of *Candida* or filamentous fungi (except *Fusarium*) may be treated by subconjunctival miconazole (and intravenous miconazole, if necessary). Additional remedies include oral administration of fluconazole, flucytosine, itraconazole or ketoconazole, intravenous amphotericin B and topical miconazole, or natamycin. *S. schenckii* infection has been successfully treated with oral potassium iodide (50 mg/drop), 10 drops thrice daily, slowly increasing to 24 drops thrice daily. Cryotherapy and dura mater grafting were found to control a case of posttraumatic scleritis due to *A. fumigatus*, which had deteriorated in spite of fluconazole and amphotericin B therapy. Oral itraconazole therapy has been found useful in *A. flavus* scleritis, which worsened with ketoconazole and amphotericin B therapy. Chung et al. (15) reported a case of *Paecilomyces lilacinus* scleritis with secondary keratitis, which had been initially diagnosed as immune-based scleritis, in an 82-year-old male; the fungal infection was predisposed to by oral cyclophosphamide. The infection ultimately resolved following therapy with topical natamycin and fluconazole, oral itraconazole, intracameral amphotericin B, and debridement of necrotic tissue, but without useful vision.

Mycotic Keratitis (Keratomycosis)

Mycotic keratitis is a suppurative, usually ulcerative, fungal infection of the cornea. It is thought responsible for more than 50% of all cases of ocular mycoses and of all patients with culture-proven microbial keratitis, especially in tropical and subtropical environments. This has important financial implications since fungal keratitis is more expensive to treat than herpetic keratitis, bacterial keratitis, and culture-negative keratitis (16). Fungal keratitis occurs more frequently in tropical countries (17), where it can often be found as part of a mixed infection with bacteria, and in the hot southern states of the United States (18). Although a high incidence of mycotic keratitis can be expected in countries with similar annual rainfall and temperature range, this is not always the case, and also appears to depend on the extent of urbanization (19).

Demographic Features

There are two basic types of mycotic keratitis: keratitis due to filamentous fungi and keratitis due to yeast-like and related fungi. The proportion of corneal ulcers caused by filamentous fungi tends to increase towards tropical latitudes, whereas in more temperate climates, fungal ulcers appear to be uncommon and to be more frequently associated with *Candida* species than filamentous fungi (17,19).

Age may influence the etiological agent and outcome of therapy in infectious keratitis. When patients were categorized into three age-based groups, namely, pediatric (≤16 years), elderly (≥65 years), and control (17 to 64 years), fungal keratitis was found to occur significantly less frequently in the pediatric group compared with the other groups; polymicrobial infections were less frequent in controls (5%) than in the other groups (≥20%) (20). Filamentous fungal keratitis usually occurs in healthy young males engaged in agricultural or other outdoor work, such as onion harvesters in Taiwan (21).

The source of infection is usually organic material; these fungi do not penetrate an intact epithelium and invasion is secondary to trauma, which is the key predisposing factor, occurring in 40% to 60% of patients (22) immunological incompetence and prior administration of corticosteroids or antibiotics are apparently not prerequisites for corneal invasion. Trauma to the cornea predisposes to ulceration of the corneal epithelium; breaching of the epithelium permits invasion of the corneal stroma and deeper part of the cornea by bacteria and fungi. Traumatizing agents of plant or animal origin (even dust particles) either directly implant fungal conidia in the corneal stroma or abrade the epithelium, permitting invasion. Similarly, contact lenses are now recognized as predisposing to *Fusarium* keratitis (Appendix D).

Etiological Agents and Risk Factors

Species of *Fusarium, Aspergillus, Curvularia*, and other dematiaceous hyphomycetes, *Acremonium* and *Penicillium*, predominate, but many other species have been implicated (Table 2) strains of *F. solani* causing human disease, including keratitis, may have a tropical or subtropical (e.g., southern United States) distribution. *Scedosporium apiospermum* and its teleomorph *Pseudallescheria boydii*, *Lasiodiplodia theobromae*, *Colletotrichum capsici*, *Cephaliophora irregularis*, and *Pythium insidiosum* have mainly (or only) been isolated from tropical and subtropical zones. It has long been considered that environmental factors, such as the wind, temperature, and rainfall, influence the occurrence of filamentous fungal keratitis. The relative prevalence of filamentous fungal keratitis has been found to increase toward tropical latitudes, possibly due to the influence of wind, temperature, and rainfall (17,19). Similarly, *Curvularia* keratitis along the Gulf of Mexico was found to cluster during the hotter, moister summer months (the mean temperature being $75°F \pm 9°F$, and relative humidity averaging $73\% \pm 13\%$), possibly reflecting the increase in airborne *Curvularia* spores during these months (23).

Pre-existing "allergic" conjunctivitis and the use of soft contact lenses have been reported as predisposing causes for *Fusarium* keratitis; in fact, *Fusarium* spp. have been cultured from such lenses during use (Appendix D). Full details of a major outbreak of contact lens–associated *Fusarium* keratitis in 2005 and 2006 are provided in Appendix D. Traditional eye medicines are frequently used in the tropics for various eye ailments, but it is uncertain to what extent these contribute to pathogenesis of mycotic keratitis.

In patients sustaining ocular trauma as a risk factor, regional variations in infecting organisms, even within defined age groups, can be discerned. In Australia, filamentous fungi accounted for 3.2% of keratitis following ocular trauma (24) but for 32.7% of keratitis in India (20). In contrast, fungi were notable by their absence in a cohort study of microbial keratitis in Hong Kong (25), in particular for contact lens wearers, perhaps because of urbanization or lens hygiene practiced (26). This interesting finding was not confirmed recently, however, when Hong Kong reported cases of *Fusarium* causing contact lens–associated keratitis as part of an outbreak associated with ReNu MoistureLoc disinfecting solution (Appendix D).

In infectious keratitis following laser-assisted in situ keratomileusis (LASIK), various fungi as well as *Nocardia* species have been implicated (3,27,28)

In keratitis due to *Candida albicans* and related fungi, ocular surface problems (e.g., dry eye, defective lid closure) or a systemic condition (e.g., diabetes mellitus, immunosuppression) predisposes to infection. In a recent report from New York, where *C. albicans* was the most frequent isolate in mycotic keratitis, seropositivity to the human immunodeficiency virus (HIV) was implicated as the most frequent risk factor (18).

Diagnosis

A rapid and accurate diagnosis of mycotic keratitis improves the chances of a complete recovery, especially in the tropics, where patients delay their arrival at a hospital. A suggested strategy for management of a patient with suspected fungal keratitis is provided (Fig. 2).

Clinical Features

Symptoms are similar to those of other types of keratitis and include redness, tearing, pain, sensitivity to light, discharge, decreased vision, and a white corneal infiltrate; symptoms are usually more prolonged in duration (10 days or more) in mycotic keratitis. Filamentous fungal keratitis may involve any area of the cornea, and typically exhibits the following features: firm (sometimes dry) elevated slough;

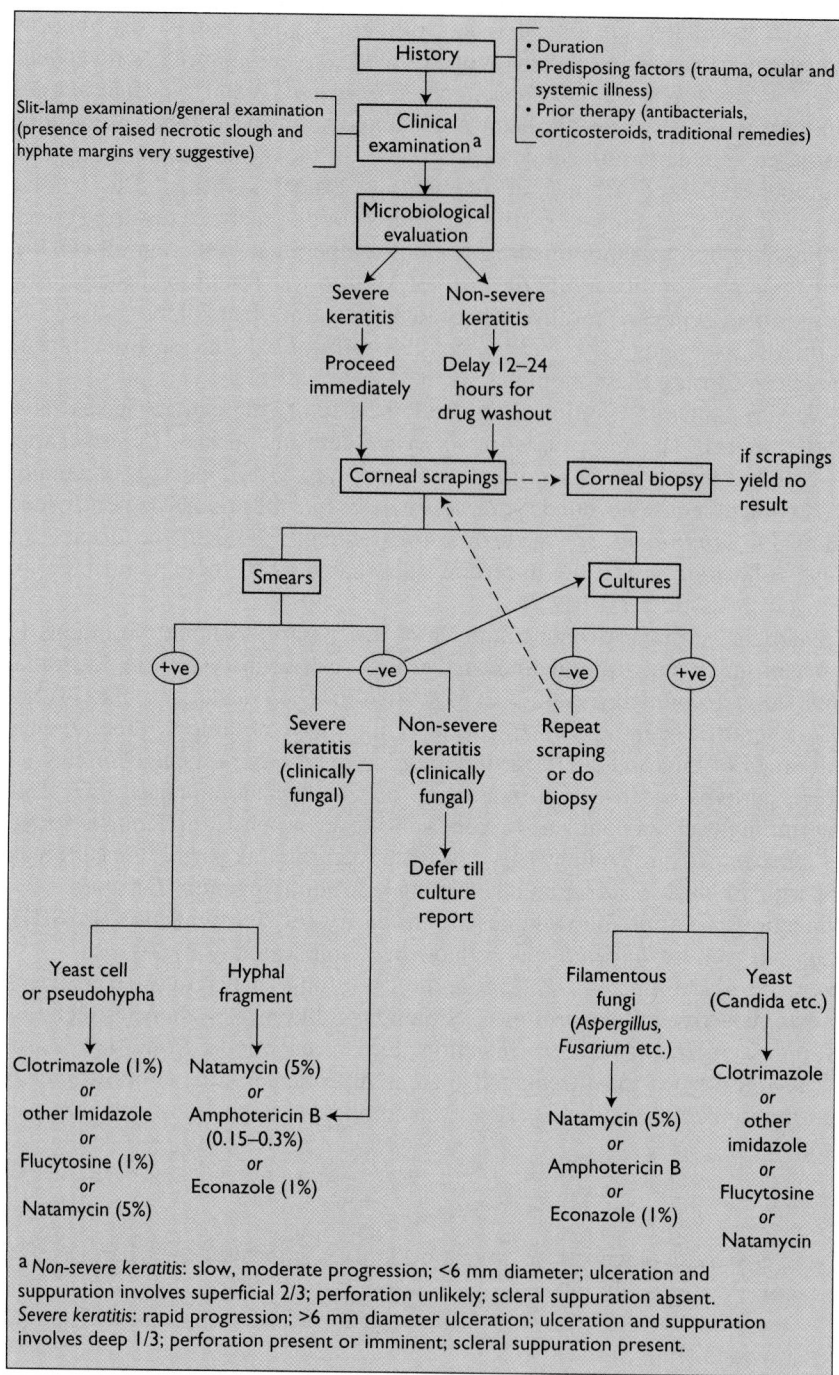

Figure 2 Flow chart for investigation and management of suspected mycotic keratitis.

hyphate lines extending beyond the ulcer edge into the normal cornea; multifocal granular (or feathery) grey-white satellite stromal infiltrates; "immune ring"; minimal cellular infiltration in the adjacent stroma and mild iritis. Of these signs, serrated margins, raised slough, and color other than yellow have been reported to be independently associated with fungal keratitis in a logistic regression model, the probability of fungal infection being 63% if one clinical feature is present and increasing to 83% if all three features are present (29).

Each case of fungal keratitis, while exhibiting these basic features, may differ from others, depending on the etiological agent. Thus, *Fusarium solani* keratitis can completely destroy an eye in a few weeks, since the infection is usually severe and perforation, deep extension, and malignant glaucoma may supervene. *Aspergillus* keratitis is usually less severe, not so rapidly progressive, and more amenable to therapy. In keratitis due to dematiaceous hyphomycetes (e.g., *Curvularia*, *Bipolaris*, *Exserohilum*), there is usually a low-grade, smoldering keratitis, with minimal structural alteration, lasting for weeks or months; the necrotic slough is frequently pigmented and simple debridement may suffice. However, if such dematiaceous fungal ulcers are neglected or augmented by corticosteroids, they may become deep, with intraocular extension or descemetocele formation supervening; one dematiaceous hyphomycete in particular, *Phialophora verrucosa*, is a recognized pathogen causing chromomycosis and requires prompt therapy. *Lasiodiplodia theobromae* resembles *F. solani* in causing a very severe type of keratitis clinically (Fig. 3). It differs, however, in being capable of causing progressive, prolonged keratitis in rabbits without using corticosteroids; this virulence is thought to originate from the production of collagenase by *L. theobromae* and its relative resistance to azole antifungals (3).

Although *Aureobasidium pullulans* is reported to be an infrequent cause of keratitis, this fungus was found to account for 12.5% of all fungal isolates from mycotic keratitis patients in a Nepali hospital (30); importantly, these ulcers showed negligible improvement with topical natamycin therapy and required either topical fluconazole or topical itraconazole in all cases, along with systemic intravenous fluconazole in eight patients, resulting in a favorable response in 22 of 25 patients. Keratitis due to *Scedosporium boydii* (*Pseudallescheria boydii*) resembles other types of keratitis in predisposing factors and clinical features but is usually not amenable to amphotericin B therapy; in vitro studies have revealed that *P. boydii* is susceptible to miconazole but resistant to amphotericin B, itraconazole, and fluconazole (31). *Pythium insidiosum* has been found to cause a very severe form of keratitis in Thailand, which is unresponsive to medical therapy and which frequently requires keratoplasty or evisceration. Chronic, severe, filamentous fungal keratitis may resemble bacterial suppuration and may involve the entire cornea.

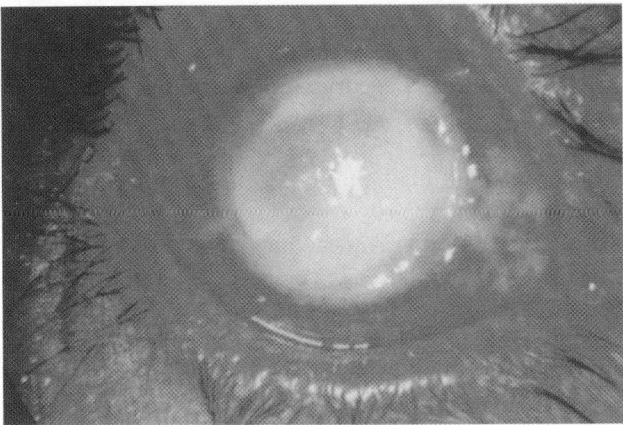

Figure 3 Clinical picture of severe keratitis due to *Lasiodiplodia theobromae*.

Noninvasive Diagnostic Techniques

In recent years, there has been an attempt to use noninvasive means to diagnose infectious keratitis. One such modality is the confocal microscope, which allows in vivo examination of the cornea. First-generation confocal microscopes have yielded to more advanced versions such as the tandem scanning confocal microscope and the Heidelberg Retina Tomograph II (HRT-II) with cornea module.

Impression cytology is another example of a noninvasive, relatively easy to perform, reliable technique used for diagnosis of ocular diseases, including infectious keratitis. A cellulose acetate filter is applied to the ocular surface to remove the most superficial layers of the ocular surface epithelium, the cells obtained then being subjected to histological, immunohistological, or molecular analysis; deeper cells can also be accessed by repeated application over the same site (32)

Microbiological Investigations

Microbiological investigations should always be performed in cases of suspected microbial keratitis. Ideally, topical antibiotics should be withheld until material has been collected for these tests; prior antimicrobial use is frequently encountered. However, prior use of topical antibiotics may only delay the time taken to grow organisms in culture without affecting culture positivity rates (17,33). It is possible to establish an effective and relatively inexpensive ocular microbiological service in a tropical setting. A suggested sequence for microbiological evaluation is provided in Chapter 2.

Conventionally, infectious keratitis is diagnosed by obtaining samples from the infected cornea; scrapings are the commonest samples, but sometimes a biopsy or, in the case of LASIK, material from the stromal bed after lifting the flap may be needed. Material from the anterior chamber or from a corneal endothelial plaque may also be collected (3,27,28). The base and edges of the ulcer should be scraped several times to obtain as much material as possible for microscopy and culture; the blade or spatula may be reused if a sterile medium is streaked but must be changed (the spatula can be flamed) if a smear has been made on an unsterile slide. Corneal biopsy may have to be performed where scrapings yield negative results; aqueous may also have to be obtained from the anterior chamber. The material thus obtained is used for microscopic examination, using various stains, or is inoculated onto appropriate culture media (Chapter 2).

An incubation temperature of 30°C and the use of liquid-shake cultures help in isolation of ocular fungi. Fungal growth usually occurs within three to four days, but culture media may require incubation for up to four to six weeks. Sham control cultures should also be maintained to ensure that no contamination is occurring, either from the environment, media, or contralateral eye. Multiple similar or diverse pathogens may sometimes occur simultaneously or sequentially in an ocular infection. A recent study (34) established criteria to define a bacterial coinfection culture-proven fungal keratitis; these criteria reflect the bacterial load in the wound but give relatively less weight to the virulence of the organism.

Direct microscopic examination of corneal scrapings permits a rapid and cheap presumptive diagnosis of mycotic keratitis. The techniques used have several advantages and disadvantages (Chapter 2) with a wet preparation (potassium hydroxide or lactophenol cotton blue) and a Gram-stained fixed smear or fixed smear for staining by special fungal stains [Giemsa, PAS, Grocott GMS, calcafluor white (CFW)]. The corneal material should be spread out as thinly as possible on the slides so as to facilitate visualization of the fungal hyphae (Figs. 4 and 5). It may be possible to identify a particular fungal genus from its appearance in corneal scrapings.

Histopathological Studies

These are useful in diagnosis of mycotic keratitis since contamination is avoided, tissue penetration can be gauged, and the outcome of surgical procedures can be anticipated. Fungal structures in corneal tissues can be stained by PAS, GMS, CFW, or fluorescein-conjugated lectins (FCL) (3). The inflammatory cellular reaction is usually less marked in fungal than in bacterial keratitis, and filamentous fungi are usually found deep in, and arranged parallel to, corneal stromal lamellae (35). Histomorphologic features of important ocular fungal pathogens are provided (see Chapter 2).

Molecular Methods of Diagnosis

Detection of fungal DNA and identification of the fungal species using corneal scrape material from patients with keratitis is fully described in Chapter 3. A sensitive and rapid polymerase chain reaction (PCR)-based method based on single-stranded conformation polymorphism was recently described for diagnosis of fungal keratitis (36). In a related study on patients with mycotic keratitis, a greater frequency of positive results was obtained by culture than by PCR amplification and sequencing of the internal transcribed spacers (ITSs) of the rDNA regions; however, using the ITS-based molecular approach, specific identification of the fungal pathogens involved was possible within 24 hours (37).

Antifungal Susceptibility Testing

Laboratory estimation of antifungal sensitivities may aid rational specific antifungal chemotherapy, but most laboratories do not routinely perform such tests (see *Antifungal Susceptibility Testing* in Chapter 2). The real value of in vitro antifungal susceptibility testing lies not in predicting responses in individual cases but in providing important baseline data of the spectrum of activity of antifungal compounds against ocular fungal isolates. Xuguang et al. (38) analyzed the

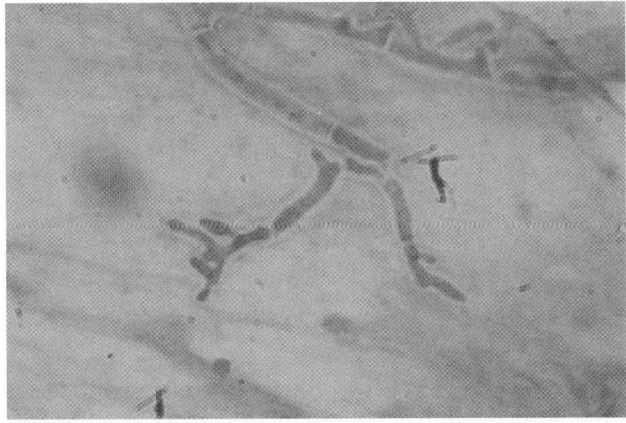

Figure 4   Gram stain of fixed corneal scraping showing branching mycelium.

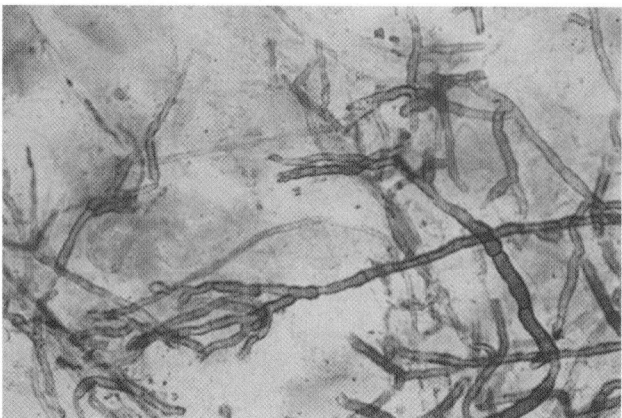

Figure 5 Wet preparation of corneal scraping showing septate hyphae of *Curvularia* sp. stained with lactophenol cotton blue.

antifungal susceptibility in vitro of 681 ocular fungal isolates, more than 80% of which had been isolated from corneal specimens; *Fusarium* species accounted for about 58% and *Aspergillus* species for 17% of positive cultures. Overall, 89% of isolates were susceptible in vitro to natamycin, 69% to terbinafine, 38% to itraconazole, and 15% to fluconazole; 93% of *Fusarium* isolates were susceptible to natamycin and 92% of *Aspergillus* isolates to itraconazole. Ozdek et al. (39) reported the interesting finding that current commercial topical preparations of moxifloxacin and gatifloxacin demonstrated a definite in vitro (in microtube dilution tests) antifungal activity against ocular *Candida* species; they conceded, however, that clinical efficacy of these agents remains unproven.

Management

Mycotic keratitis is managed medically or surgically. Medical therapy consists of non-specific measures and the use of specific antifungal agents. Cycloplegics are used to relieve the iridocyclitis that usually accompanies mycotic keratitis; broad-spectrum antibacterials may be needed to combat secondary bacterial infection.

Antifungal agents that are useful in treatment of mycotic keratitis include (Fig. 2):

- Topical: natamycin (5%), amphotericin B (0.15–0.3%), flucytosine (1%), thiabendazole (4%), econazole (1%), clotrimazole (1%), miconazole (1%), ketoconazole (1–2%), itraconazole (1%), fluconazole (1%), voriconazole or silver sulphadiazine (1%);
- Subconjunctival: miconazole (10 mg in 0.5 mL); amphotericin B (5 mg) in desperate cases;
- Intravenous: amphotericin B, miconazole (600–1200 mg/day), voriconazole: and
- Oral: ketoconazole (200–600 mg/day), itraconazole (100–200 mg/day), fluconazole (50–200 mg/day), or voriconazole.

More details are provided in Appendix A. Since the imidazole drugs (clotrimazole, etc.) only inhibit growth of the fungus, host defense mechanisms must eradicate the organism and treatment is usually prolonged. Only the polyene drugs (amphotericin and natamycin) are fungicidal.

The selection of an antifungal agent for therapy necessarily depends on its easy availability and on other criteria. Treatment with an antifungal agent may be initiated on the basis of direct microscopy findings alone if these are clear-cut and consistent with the clinical evaluation; otherwise, therapy should be withheld while awaiting the results of culture (3). Topical natamycin (5%) is usually chosen as initial therapy for superficial keratomycoses, regardless of whether septate hyphae or yeast cells are seen by direct microscopy; additional antifungal agents (e.g., amphotericin B, miconazole, one of the oral imidazoles) are added for deep corneal infections. The initial antifungal agent may also be chosen depending on whether yeast cells or hyphae are seen by microscopy; if hyphae are definitely seen by microscopy, topical natamycin (5%) is the drug of choice (or 0.15% amphotericin B if natamycin is not available) (3); if yeasts or pseudohyphae are seen, topical 0.15% amphotericin B or topical 1% fluconazole (or other azole). Once the organism has been identified by culture (Fig. 6), the therapeutic regimen may be modified, in particular by in vitro susceptibility tests.

For topical therapy, most workers advise hourly application around the clock for several days and the dosage is then gradually reduced. This can be given every 15 minutes, first day; every 30 minutes, first night; every 15 to 30 minutes, second to the fourth day; one- to four-hourly, second to the fourth night; every 60 minutes, 5th to the 20th day; four- to eight-hourly, 20th to the 90th day. A loading-dose therapeutic regimen has also been suggested, where five separate one-drop doses are given one minute apart (between doses, the patient keeps the eye shut) to achieve a very high corneal drug concentration within 30 minutes; after this, the drops are administered every hour.

Filamentous fungal keratitis is notoriously difficult to treat despite the use of topical and systemic antifungal agents (40). Many patients require adjuvant surgery, ranging from recurrent corneal debridement to corneal transplantation, but the visual outcome is often dismal. Few prospective studies have evaluated the effectiveness of different therapeutic approaches for fungal keratitis (3). In a study in India, 36 (72%) of 50 patients showed a favorable response to primary topical 5% natamycin therapy, while 30 (60%) of 50 patients responded favorably to primary

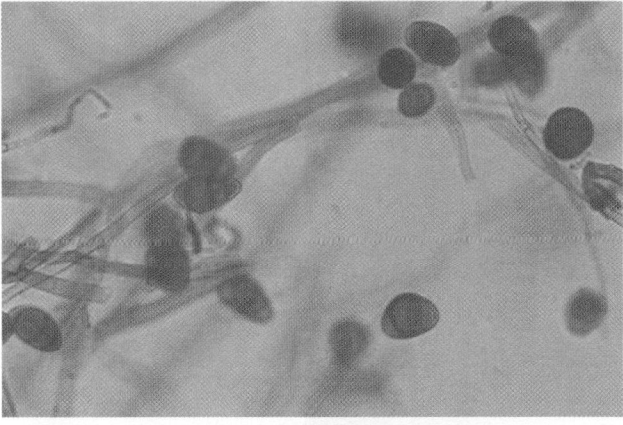

Figure 6 High power view of *Curvularia* sp. showing characteristic brown conidia with transverse septa.

topical 1% itraconazole therapy; species of *Fusarium*, *Aspergillus*, and *Curvularia* were the principal isolates. In patients with *Fusarium* keratitis, 79% responded favorably to natamycin and 44% to itraconazole, a significant difference; no such difference was observed in *Aspergillus* or *Curvularia* keratitis. A major addition to the therapeutic armamentarium for treatment of fungal keratitis is voriconazole, given topically or by other routes (41,42). Bunya et al. (43) used voriconazole topically for nine patients and orally for eight patients in the treatment of fungal keratitis due to filamentous fungi in six and *C. albicans* in three that had been refractory to standard antifungal therapy; of seven patients treated with topical therapy and followed, five conditions healed. Another promising agent is posaconazole, particularly in *Fusarium* keratitis and/or endophthalmitis refractory to other systemic, intracameral, and topical antifungal agents, including systemic and or topical voriconazole (44). In addition, new ways of administering established drugs have been tried. Combined intrastromal (5 µg per 0.1 mL) and intravitreal amphotericin B led to eradication of recurrent fungal keratitis in one patient (45). Subconjunctival fluconazole (0.5–1.0 mL of a 2% solution) was effective in treating 12 of 13 patients with fungal keratitis in one study (46) and led to eradication of infection in 54% of patients with filamentous fungal keratitis in another study (47). Topical fluconazole, alone or in combination with oral ketoconazole, was found effective in culture-proven filamentous fungal keratitis in yet another study (48).

Fungal keratitis (Figs. 7–12) usually responds slowly over weeks to antifungal therapy. Clinical signs of improvement of a fungal corneal ulcer include diminution of pain, decrease in size of infiltrate, disappearance of satellite lesions, rounding out of the feathery margins of the ulcer, and hyperplastic masses or fibrous sheets in the region of healing fungal lesions. Signs of topical antifungal agent toxicity (conjunctival chemosis and injection, recurrent corneal epithelial erosions) should also be observed for. Negative scrapings during treatment do not always indicate that the fungus has been eradicated, since it may become deep-seated; hence, therapy should be maintained for at least six weeks.

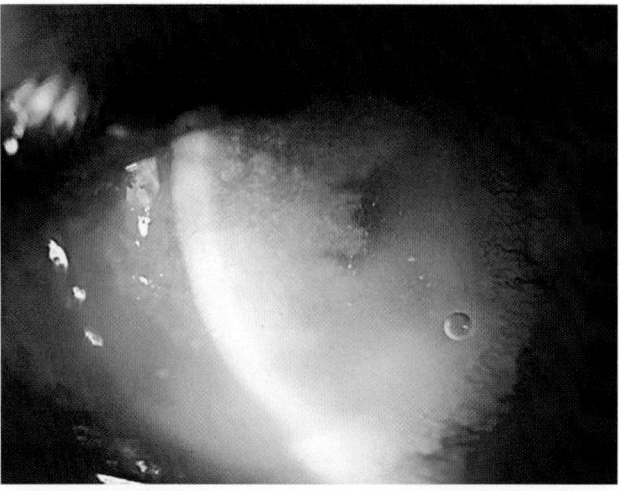

Figure 7   Keratitis due to *Alternaria alternata*. *Source*: Courtesy of J. Alio and C. Ferrer.

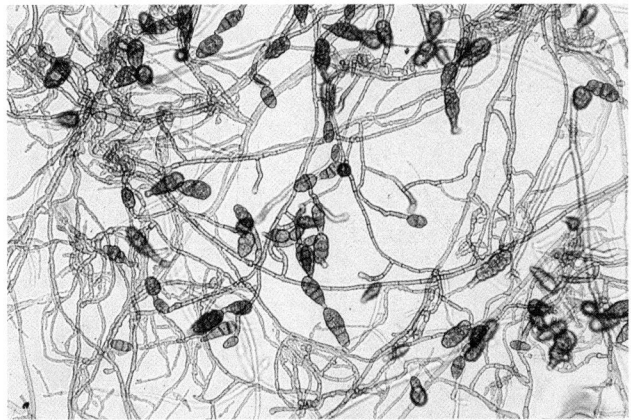

Figure 8 Light microscopy of a culture of *Alternaria alternata.* Showing branching mycelia and characterstic conidia with transverse septa. *Source*: Courtesy of J. Alio and C. Ferrer.

Patients who seem to respond most poorly to medical antifungal therapy are those with deep corneal infection, perhaps due to limited penetration by antifungals, and those who have received corticosteroids before the diagnosis of mycotic keratitis. Fungal growth in tissue is aided by corticosteroids, which mitigates against their use, either alone or in combination with antifungal agents, in managing mycotic keratitis. Sympathetic ophthalmia following severe fungal keratitis has been reported (49).

Methods to enhance the efficacy of antifungal agents by administration of several compounds simultaneously require careful study, since there is the risk of antagonism by combining certain antifungals, for example, amphotericin B and the imidazoles.

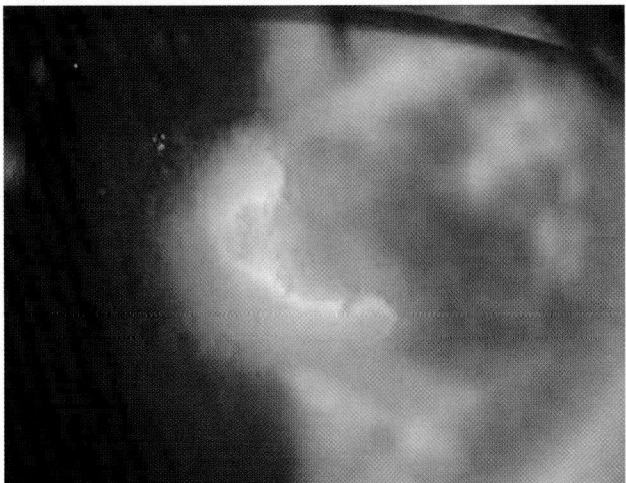

Figure 9 Keratitis due to *Fusarium* sp. associated with contact lens wear. *Source*: Courtesy of J. Alio and C. Ferrer.

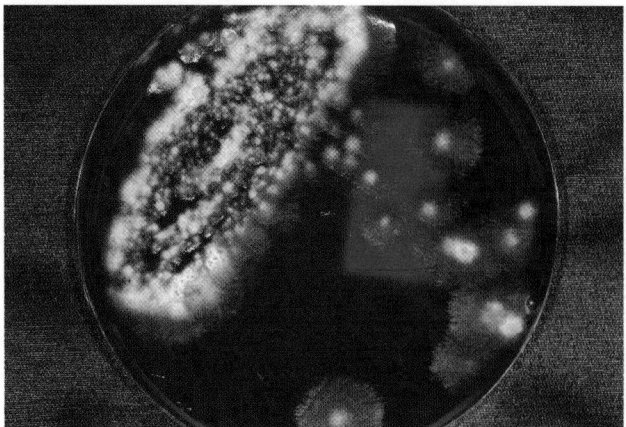

Figure 10   Plate culture of *Fusarium dimerum* isolated from a contact lens. *Source*: Courtesy of J. Alio and C. Ferrer.

Therapeutic surgery may be required for clinical cases of mycotic keratitis that respond poorly, or not at all, to medical therapy, or where perforation or descemetocele formation is imminent. However, every effort should be made to prolong medical therapy for the maximum duration possible, to render the infecting fungus non-viable prior to surgery and therein to improve the outcome. Surgery attempts to remove antigenic and/or infectious elements and also necrotic tissue and other debris that may hinder complete healing of the lesion. Therapeutic photo-kerato-plasty (PKP) seems to be effective in the treatment of fungal keratitis with corneal perforation; in one study, in spite of the occurrence of complications such as graft rejection, graft ulceration, recurrence of fungal infection, complicated cataract, and secondary glaucoma, as many as 85% of grafts remained clear at final follow-up, and 88.5% had improved visual acuity (50). Microbial infections within corneal grafts can occur anytime in the postoperative course and are associated with broken sutures and the use of topical corticosteroids; fungal species alone may account for

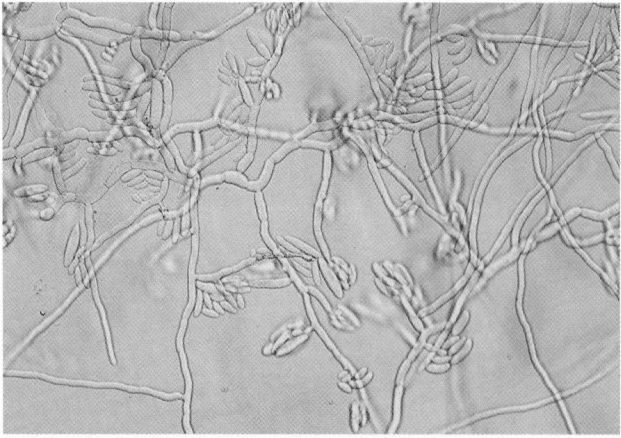

Figure 11   Light microscopy of a culture of *Fusarium* sp. Showing branching mycelia and conidia. *Source*: Courtesy of J. Alio and C. Ferrer.

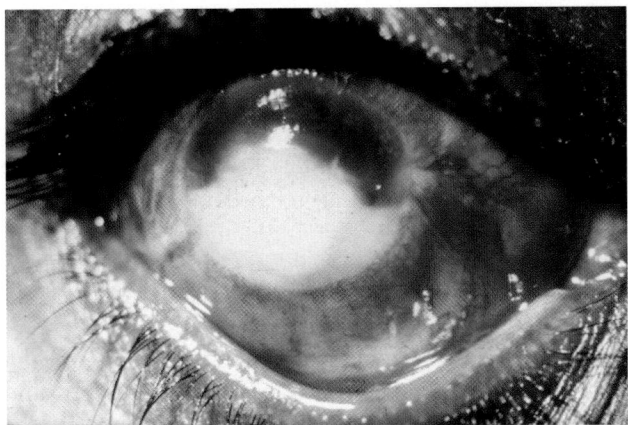

Figure 12 *Fusarium* keratitis in a patient from Bangledesh. *Source*: Courtesy of Dr. Elizabeth Wright.

almost 20% of such infections, and for another 27% of infections in association with bacteria (51).

Amniotic membrane transplantation (AMT) can also be used to treat acute, culture-proven fungal keratitis, with complete epithelialization being achieved in as many as 75% of patients with active disease and in all patients with inactive disease, provided antifungal agents are administered throughout the entire duration of hospitalization and secondary infection does not occur (52).

Prevention

Since trauma is the most important risk factor for mycotic keratitis, prompt treatment of corneal injury may prevent the infection. This hypothesis has proven correct in two recent studies at the village level, one in Burma and the other in India. In the Burma study, patients presenting with posttraumatic corneal abrasion were treated with an ointment containing 1% chloramphenicol and 1% clotrimazole ointment three times daily for three days (53). In the Indian study, almost 99% of individuals with posttraumatic corneal abrasions were found to heal without complications, irrespective of whether they had received a 1% chloramphenicol and 1% clotrimazole ointment or 1% chloramphenicol and a placebo ointment three times a day for three days (54). Thus, both fungal and bacterial ulcers that occur after traumatic corneal abrasions seem to be effectively prevented in a village setting using only antibiotic prophylaxis.

INTRAOCULAR MYCOSES

Endophthalmitis can be defined as an inflammatory process that involves the ocular media and adjacent structures; this inflammatory process may be localized to specific tissues within the eye (e.g., iris, choroid) or may involve the intraocular contents in a generalized fashion For the purposes of simplicity, these various manifestations are grouped together as "intraocular mycoses" in this chapter.

Many fungal species have been isolated from intraocular mycoses (Table 1). Fungal endophthalmitis may arise from endogenous or exogenous sources.

Intraocular Mycoses of Endogenous Origin

Fungi are the most important causes of endogenous endophthalmitis (Chapter 7). A diagnosis of endogenous fungal endophthalmitis points to the possible presence of undiagnosed immunosuppressive or other predisposing factors; an extraocular mycotic infection should also be looked for. Endogenous fungal endophthalmitis usually starts as a chorioretinitis, slowly evolving to become a generalized, sometimes purulent, endophthalmitis or even a panophthalmitis. The common causes of endogenous fungal endophthalmitis include *Candida* (especially *C. albicans* and *C. parapsilosis*), *Aspergillus*, *Cryptococcus*, and *Fusarium*. They are associated with intravenous drug addiction and contaminated intravenous lines.

Intraocular *Candida* infections represent an important clinical problem owing to the potential for visual loss, which can be bilateral. *Candida* chorioretinitis and endophthalmitis are complications of systemic candidiasis, with extension of the fungal pathogens to the uvea and retina (Chapter 7). The course of *C. albicans* endophthalmitis may be so severe as to require therapy with all available antifungal drugs such as fluconazole, liposomal amphotericin B, caspofungin, and voriconazole (55). In patients with a history of injection drug use, persisting intravenous drug or substitution therapy abuse must be considered. Quick diagnosis and adequate treatment can prevent the development of widespread *C. albicans* endophthalmitis, which has a poor visual prognosis (56). Early administration of antifungal and surgical therapy is crucial for achieving good functional results (57). Fluconazole has been effective against *C. albicans* in various animal models. However, despite its clinical relevance, there have been no controlled trials on different treatment regimens in *Candida* endophthalmitis (8); such cases may be resistant to azoles (58).

Andreola et al. (59) described the occurrence of multifocal choroiditis as a consequence of disseminated *Cryptococcus neoformans* infection in a patient who was HIV-positive. Multifocal choroiditis in *C. neoformans* infection is a rare ophthalmic manifestation. The recognition of this condition by ophthalmologists can help physicians to diagnose a disseminated and fatal disease.

Endogenous *Aspergillus* endophthalmitis is a rare complication of invasive aspergillosis, occurring most often in transplant patients who are on vigorous immunosuppressive therapy. Treatment generally includes pars plana vitrectomy, intravitreal amphotericin B, systemic amphotericin B, and oral inridazoles; however, the prognosis remains grave, and the risk of mortality is high. Amphotericin B is currently the only antifungal agent approved for injection into the vitreous, although intravitreal miconazole and voriconazole have also been used successfully. However, it may cause focal retinal necrosis even at low concentrations; moreover, a variety of fungal species, particularly *Aspergillus*, has shown resistance to amphotericin B (60).

In endogenous endophthalmitis, there is usually some immunocompromise or debilitation or a lesion predisposing to the ocular infection. However, de Silva et al. (61) recently reported the occurrence of isolated endogenous endophthalmitis due to *Nocardia* spp. in an immunocompetent adult.

Intraocular Mycoses of Exogenous Origin

Exogenous sources of fungal endophthalmitis include extension of a fungal corneal ulcer and surgical or non-surgical trauma. In addition, outbreaks of fungal endophthalmitis have occurred following use of contaminated ophthalmic irrigating solutions, intraocular lenses, surgical equipment, or the operating theater air supply (62).

Intraocular mycoses may follow non-surgical trauma (usually perforating injuries of the globe) resulting in endophthalmitis or panophthalmitis. *Aspergillus* (especially *A. fumigatus*), *F. solani*, *Sporothrix*, and *P. verrucosa* are among the most important fungi implicated (1).

Postsurgical fungal endophthalmitis most frequently presents following cataract surgery (1); the principal fungal pathogen has been *Aspergillus* but other fungi have also been implicated (Table 1). A retrospective study of the clinical profile of 28 patients, with fungal infection following cataract surgery in India, revealed that 29% presented with keratitis alone, 36% with keratitis and endophthalmitis, and 35% with endophthalmitis only. The mean duration after cataract surgery (68% ECCE, 32% ICCE) was 48 days. The principal predisposing factors were topical corticosteroids (63%), diabetes mellitus (43%), characteristic suppurative infiltration along the surgical wound in the cornea (64%), and anterior vitreous exudates (16%). *Aspergillus* spp. were isolated in 83% and *Lasiodiplodia theobromae* in 6%, with no yeasts isolated. If postsurgical endophthalmitis due to fungi develops, possible contamination of irrigating solutions, equipment, or theater air should be investigated. There have been outbreaks of *Paecilomyces lilacinus* endophthalmitis due to contaminated intraocular lenses and intraocular lens storage solutions and *Candida parapsilosis* endophthalmitis due to contaminated intraocular irrigation solutions (62).

Endophthalmitis may follow a deep fungal keratitis, where the fungus penetrates an intact Descemet's membrane. Fungi implicated in such infections include *Aspergillus*, *Fusarium*, *Acremonium*, *L. theobromae*, *S. apiospermum*, and *Curvularia*. Hassan and Wilhelmus (63) made the interesting observation that, compared with other post-keratoplasty endophthalmitis, candidal endophthalmitis after corneal transplantation occurred significantly more often when outdoor afternoon temperatures exceeded 75°F at the source eye bank than on cooler days; they however acknowledged that a chain of causation that links the donor's environment to the recipient's infection remains to be elucidated.

Candida glabrata is a rare cause of endophthalmitis after penetrating keratoplasty, and adequate therapy for the condition is still to be conclusively defined. Typically, the intraocular infection is noticed within the first two weeks; however, a hyperacute onset (within 10 hours of keratoplasty) has recently been described (64). Such cases of *C. glabrata* endophthalmitis may not respond well to even prolonged therapy with fluconazole and corticosteroids, in which case topical and intracameral application of amphotericin B in combination with topical prednisolone may result in dramatic improvement.

Exogenous postoperative fungal endophthalmitis usually presents between two and four weeks (or later) following surgery. Persistent iritis may be the only presenting sign. In other types of exogenous fungal endophthalmitis the incubation period is usually long—not less than three weeks and sometimes several months. The development and progression of infection are slow, the infection generally remaining localized to the anterior uvea for a long time. Symptoms are similar to those seen in bacterial endophthalmitis, but less marked. Fungus may grow over the iris surface or intraocular lens, and a dense pupillary membrane may also develop, resulting in reduction of vision to bare perception of light; the eye is filled with a granulomatous endophthalmitis mass. Whitish "puff-balls" and vitreous strands are the cardinal signs of fungal infection.

A review of patients with post-cataract surgery endophthalmitis due to *Aspergillus* revealed that 40% of patients were immunocompromised, the mean number of days between cataract surgery and diagnosis with endophthalmitis was 29 (range 10 to 62 days), and 60% of eyes were enucleated despite a variety of treatments (65). Interestingly, the first case of postoperative endophthalmitis due to *Aspergillus ustus*, a species that has only rarely been implicated in human disease, occurred six weeks after cataract surgery in a medically controlled diabetic patient (66). The endophthalmitis did not improve in spite of surgery including complete vitrectomy with removal of the capsular bag and the intraocular lens and vigorous systemic (itraconazole and caspofungin) and intravitreal (amphotericin B and caspofungin) antifungal therapy, necessitating enucleation. Thus, post-cataract surgery endophthalmitis due to *Aspergillus* species should be considered in the differential diagnosis of early onset endophthalmitis following cataract surgery since visual outcomes are generally poor and enucleation is common in these patients.

Keratitis due to *Fusarium* sp. associated with contact lens wear has been found to progress to endophthalmitis in a few patients (see Appendix D) (67).

Management of mycotic endophthalmitis requires identification of the responsible organism and prompt institution of appropriate therapy; suggested guidelines for laboratory evaluation are provided in Chapters 2 and 3. The inoculated culture media are examined daily for two weeks or longer to detect fungal growth, incubation is at 30°C (Chapter 2). The culture is deemed positive if the same organism grows on more than one medium, or if confluent growth occurs on one or more solid media at the inoculation site. Growth in only one liquid medium or scanty growth on one solid medium is of uncertain significance. Up to 65% of vitreous specimens from eyes with a clinical diagnosis of endophthalmitis may be culture-positive but use of PCR can be expected to increase the diagnostic rate to 85% (Chapter 3).

The relatively low yield of positive results with culture has prompted increasing use of molecular methods for diagnosis of fungal endophthalmitis. Use of a primer that targets the conserved 18S rRNA gene sequences of fungi for amplification of fungal DNA in 50 µL samples of vitreous resulted in an almost fourfold increase in identification of patients with fungal endophthalmitis; positive results were generated within 14 hours (see Chapter 3) (68). In another study, conventional methods of diagnosis were compared with PCR for detection of fungal DNA using panfungal primers (ITS1 and ITS4) for diagnosis of fungal etiology; culture was positive in four patients, of which PCR was positive in three, and PCR was positive in an additional three culture-negative cases. All six patients, who were positive for PCR, showed clinical improvement after a full course of antifungal therapy (69).

With the exception of endogenous *Candida* endophthalmitis, antifungal drugs have generally not been found effective against postoperative fungal endophthalmitis, perhaps because there is usually a delay in diagnosis, and because less information is available on pharmacokinetics and sensitivity testing of preferred antifungal agents (refer to Chapter 4). Amphotericin B has been the mainstay of therapy for many years because it can be given by multiple routes (intravenous, subconjunctival, intravitreal, intracameral) and because it is active against *Candida*, *Cryptococcus*, and dimorphic fungi.

For Candida infections, the combined use of topical 5% natamycin or 0.15% amphotericin, oral flucytosine, intravitreal amphotericin B, and, if warranted intravenous amphotericin B is recommended. For systemic therapy, alternatives are oral ketoconazole, itraconazole, fluconazole, or voriconazole. For intravitreal injection, amphotericin B 5 µg (dose = 0.1 mL of a 50-µg/mL solution) or miconazole 10 µg (dose = 0.1 mL of a 100-µg/mL solution) (Appendix A) should be used following the intravitreal tap for laboratory diagnosis. Velpandian et al. (70) observed that plain fluconazole at a concentration of 100 microg and above caused retinal changes, with disorganization of the photoreceptor outer segments, whereas a liposome formulation of fluconazole (200 microg/0.1 mL) did not cause any significant microscopic changes to the retina.

Voriconazole exerts potent activity against a broad spectrum of yeasts and molds. Studies have shown that voriconazole is more active than amphotericin B against filamentous fungi such as *Aspergillus* species. Hence, currently, voriconazole has replaced amphotericin as the treatment of choice for systemic *Aspergillus* infections, and it is also active against *Fusarium*. It achieves good aqueous and vitreous levels with oral administration (53% and 38% of plasma levels, respectively) (71). A recent report described voriconazole treatment of five patients with *Candida* endophthalmitis, three of who also received caspofungin (72). Filamentous fungal infections are more difficult to cure than *Candida* infections, but voriconazole has been used to treat endophthalmitis attributable to *Paecilomyces lilacinus*, *Scedosporium apiospermum*, and *Lecytophora mutabilis*. Durand et al. (73) reported successful treatment of exogenous *Fusarium* and *Aspergillus* endophthalmitis with voriconazole. Inflammation associated with post-cataract *Fusarium moniliforme* endophthalmitis that had persisted despite removal of the intraocular lens, three vitrectomies, and five intravitreal injections of amphotericin resolved with improvement of vision after six months of oral voriconazole. Post-cataract intraocular inflammation, diagnosed as *Aspergillus fumigatus* endophthalmitis after three vitrectomies and removal of the intraocular lens, initially responded to intravitreal amphotericin and systemic voriconazole, but showed early signs of recurrence after a week; improvement was noted after adding intravenous caspofungin. When caspofungin was continued for 6 weeks and voriconazole for six months, vision improved markedly. Thus, voriconazole appears to be a promising new therapy for *Fusarium* and *Aspergillus* endophthalmitis.

Caspofungin may act synergistically with voriconazole in treating *Aspergillus* endophthalmitis. Kramer et al. (60) used voriconazole 10µg for intravitreal injection (dose = 0.1 mL of a 100-µg/mL solution) into a human eye (the first such description) in a patient with endogenous *Aspergillus* endophthalmitis. This was combined with pars plana vitrectomy, after the isolation of *A. terreus* in the vitreous sample. Previous treatment modalities, including vitrectomy with repeated intravitreal amphotericin B and systemic voriconazole, had failed to prevent deterioration. Sen et al. (74) also reported the efficacy of intravitreal voriconazole in a series of five patients with culture-proven fungal endophthalmitis, whose isolates had been found resistant to conventional antifungal agents.

Caspofungin is the first antifungal of the echinocandin class. It exhibits major inhibitory activity against *Candida*, including azole-resistant species (78). It acts by noncompetitive inhibition of b-(1,3)–D-glucan synthesis in fungal cell walls (78). Topical caspofungin has been shown to be effective in experimental *Candida* keratitis (79). Against the filamentous fungi, caspofungin exhibits excellent in vitro activity against *Aspergillus* spp., but limited activity against *Fusarium* spp; the resistance of *F. solani* to caspofungin is partly due to the resistance of the enzyme encoding b-(1,3)–D-glucan synthase to this antifungal drug (80). However, Ozturk et al. (81) reported that topically applied 1% caspofungin was effectove against experimental *F. solani* in rabbits; caspofungin keratitis and suppressed fungal growth in culture (according to these authors, caspofungin was comparable to amphotericin B in the treatment of experimental *F. solani* keratitis).

Unlike the situation in bacterial endophthalmitis, corticosteroids are not used in proven cases of fungal endophthalmitis since the common antifungal drugs are of low potency. Suppression of immunological reactions by corticosteroids (see Chapter 1) leads to unhindered growth of fungi. In all forms of endophthalmitis, vitrectomy is an essential surgical adjunct.

ACKNOWLEDGMENTS

I thank Dr P. Geraldine and Dr. J. Kaliamurthy for their assistance with this manuscript.

REFERENCE

1. Thomas PA, Geraldine P. Oculomycosis. In: Topley and Wilson's Microbiology and Microbial Infections, vol. 5 (Medical Mycology). 10th edition, Ed. Merz WG. Hay RJ. Hodder Arnold, London, 2005. pp. 273–344.
2. Wilson LA, Ajello L. Agents of oculomycosis: fungal infections of the eye. In: Topley and Wilson's Microbiology and Microbial Infections, vol. 4. 9th edition, Ed. Collier L, Balows A and Sussman M. Arnold, London, 1998. pp. 525–567.
3. Thomas PA. Current perspectives on ophthalmic mycoses. Clin Microbiol Rev 2003; 16: 730–797.
4. Klotz SA, Penn CC, Negvesky GJ, et al. Fungal and parasitic infections of the eye. Clin Microbiol Rev 2000; 13:662–85.
5. Ribes JA, Vanover-Sams CL, Baker DJ. Zygomycetes in human disease. Clin Microbiol Rev 2000; 13:236–301.
6. Fairley C, Sullivan TJ, Bartley P, et al. Survival after rhino-orbital-cerebral mucormycosis in an immunocompetent patient. Ophthalmol 2000; 107:555–58.
7. Rutar T, Cockerham KP. Periorbital zygomycosis (mucormycosis) treated with posaconazole. Am J Ophthalmol 2006; 142:187–88.
8. Hargrove RN, Wesley RE, Klippenstein KA, et al. Indications for orbital exenteration in mucormycosis. Ophthal Plast Reconstr Surg. 2006; 22:286–91.
9. Pagliani L, Campi L, Cavallini GM. Orbital actinomycosis associated with painful ophthalmoplegia. Actinomycosis of the orbit. Ophthalmologica 2006; 220:201–05.
10. Sharada DM, Arunkumar G, Vandana KE, et al. Sino-orbital aspergillosis in a diabetic patient. Ind J Med Microbiol 2006; 24:138–40.
11. Johnson TE, Casiano RR, Kronish JW, et al. Sino-orbital aspergillosis in acquired immunodeficiency syndrome. Arch Ophthalmol 1999; 117:57–64.
12. Coccia L, Calista D, Boschini A. Eyelid nodule: a sentinel lesion of disseminated cryptococcus in a patient with acquired immunodeficiency syndrome Arch Ophthalmol 1999; 117:271–72.
13. Velazquez AJ, Goldstein MH, Driebe WT. Preseptal cellulitis caused by Trichophyton (ringworm). Cornea 2002; 21:312–14.
14. Tabbara KF. *Aspergillus fumigatus* colonization of punctal plugs. Am J Ophthalmol 2007; 143:180–01.
15. Chung PC, Lin HC, Hwang YS, et al. *Paecilomyces lilacinus* scleritis with secondary keratitis. Cornea. 2007; 26:232–34.
16. Keay L, Edwards K, Naduvilath T, et al. Microbial keratitis: predisposing factors and morbidity. Ophthalmol 2006; 113:109–16.

17. Leck AK, Thomas PA, Hagan M, et al. Aetiology of suppurative corneal ulcers in Ghana and south India, and epidemiology of fungal keratitis. Brit J Ophthalmol 2002; 86: 1211–15.
18. Ritterband DC, Seedor JA, Shah MK, et al. Fungal keratitis at the New York Eye and Ear Infirmary. Cornea. 2006; 25:264–67.
19. Houang E, Lam D, Fan D, Seal D. Microbial Keratitis in Hong Kong—relationship to climate, environment and contact lens disinfection. Trans Roy Soc Trop Med & Hyg 2001; 95:361–67.
20. Parmar P, Salman A, Kalavathy CM, et al. Microbial keratitis at extremes of age. Cornea 2006; 25:153–58.
21. Lin SH, Lin CP, Wang HZ, et al. Fungal corneal ulcers of onion harvesters in southern Taiwan. Occup Environ Med 1999; 56:423–25.
22. Wong T-Y, Ng T-P, Fong K-S, et al. Risk factors and clinical outcome between fungal and bacterial keratitis. A comparative study. CLAO J 1997; 23:275–81.
23. Wilhelmus KR. Climatology of dematiaceous fungal keratitis. Am J Ophthalmol 2005; 140:1156–57.
24. Keay L, Edwards K, Naduvilath T, et al. Factors affecting the morbidity of contact lens related microbial keratitis—a population study. Investig Ophthalmol Vis Sci 2006; 47: 4302–08.
25. Lam D, Houang E, Seal D, et al. Incidence and Risk Factors for Microbial Keratitis in Hong Kong: comparison with Europe and North America. Eye 2002; 16:608–18.
26. Fan D, Lam D, Houang E, et al. Health Belief and Health Practice in Contact Lens Wear – a dichotomy ? CLAO Journal 2002; 28:36–39.
27. Patel NR, Reidy JJ, Gonzalez-Fernandez F. *Nocardia* keratitis after laser in situ keratomileusis: clinicopathologic correlation. J Cataract Refract Surg 2005; 31:2012–15.
28. Patel SR, Hammersmith KM, Rapuano CJ, et al. *Exophiala dermatitidis* keratitis after laser in situ keratomileusis. J Cataract Refract Surg 2006; 32:681–84.
29. Thomas PA, Leck AK, Myatt M. Characteristic clinical features as an aid to the diagnosis of suppurative keratitis caused by filamentous fungi. Brit J Ophthalmol 2005; 89:1554–58.
30. Panda A, Das H, Deb M, Khanal B, Kumar S. Aureobasidium pullulans keratitis. Clin Experiment Ophthalmol 2006; 34:260–64.
31. Bradley JC, Hirsch BA, Kimbrough RC 3rd, McCartney DL. *Pseudallescheria boydii* keratitis. Scand J Infect Dis 2006; 38:1101–03.
32. Singh R, Joseph A, Umapathy T, et al. Impression cytology of the ocular surface. Brit J Ophthalmol 2005; 89:1655–59.
33. Marangon FB, Miller D, Alfonso EC. Impact of prior therapy on the recovery and frequency of corneal pathogens. Cornea 2004; 23:158–64.
34. Pate JC, Jones DB, Wilhelmus KR. Prevalence and spectrum of bacterial co-infection during fungal keratitis. Brit J Ophthalmol 2006; 90:289–92.
35. Vemuganti GK, Garg P, Gopinathan U, et al. Evaluation of agent and host factors in progression of mycotic keratitis: a histologic and microbiologic study of 167 corneal ulcers. Ophthalmol 2002; 109:1538–46.
36. Kumar M, Mishra NK, Shukla PK. Sensitive and rapid polymerase chain reaction based diagnosis of mycotic keratitis through single-stranded conformation polymorphism. Am J Ophthalmol 2005; 140:851–57.
37. Mancini N, Perotti M, Ossi CM, et al. Rapid molecular identification of fungal pathogens in corneal samples from suspected keratomycosis cases. J Med Microbiol. 2006; 55:1505–09.
38. Xuguang S, Zhixin W, Zhiqun W, et al. Ocular fungal isolates and antifungal susceptibility in northern China. Am J Ophthalmol 2007; 143:131–33.
39. Ozdek SC, Miller D, Flynn PM, et al. In vitro antifungal activity of the fourth generation fluoroquinolones against Candida isolates from human ocular infections. Ocul Immunol Inflamm 2006; 14:347–51.

40. Malecha MA, Tarigopula S, Malecha MJ. Successful treatment of *Paecilomyces lilacinus* keratitis in a patient with a history of herpes simplex virus keratitis. Cornea. 2006; 25: 1240–42.

41. Sponsel W, Chen N, Dang D, et al. Topical voriconazole as a novel treatment for fungal keratitis. Antimicrob Ag Chemother 2006; 50:262–68.

42. Ozbek Z, Kang S, Sivalingam J, et al. Voriconazole in the management of Alternaria keratitis. Cornea 2006; 25:242–244.

43. Bunya VY, Hammersmith KM, Rapuano CJ, Ayres BD, Cohen EJ. Topical and oral voriconazole in the treatment of fungal keratitis. Am J Ophthalmol. 2007 Jan; 143(1): 151–3.

44. Tu EY, McCartney DL, Beatty RF. Successful treatment of resistant ocular fusariosis with posaconazole (SCH-56592). Am J Ophthalmol 2007; 143:222–27.

45. Garcia-Valenzuela E, Song CD. Intracorneal injection of amphotericin B for recurrent fungal keratitis and endophthalmitis. Arch Ophthalmol 2005; 123:1721–23.

46. Yilmaz S, Maden A. Severe fungal keratitis treated with subconjunctival fluconazole. Am J Ophthalmol 2005; 140:454–58.

47. Dev S, Rajaraman R, Raghavan A. Severe fungal keratitis treated with subconjunctival fluconazole. Am J Ophthalmol 2006; 141:783.

48. Sonego-Krone S, Sanchez-Di Martino D, et al. Clinical results of topical fluconazole for the treatment of filamentous fungal keratitis. Graefe's Arch Clin Exp Ophthalmol 2006; 244:782–87.

49. Buller AJ, Doris JP, Bonshek R, Brahma AK, Jones NP. Sympathetic ophthalmia following severe fungal keratitis. Eye 2006; 20:1306–07.

50. Xie L, Zhai H, Shi W. Penetrating keratoplasty for corneal perforations in fungal keratitis. Cornea 2007; 26:158–62.

51. Wright TM, Afshari NA. Microbial keratitis following corneal transplantation. Am J Ophthalmol. 2006; 142(6):1061–2.

52. Chen H-C, Tan H-Y, Hsiao C-H, et al. Amniotic membrane transplantation for persistent corneal ulcers and perforations in acute fungal keratitis. Cornea 2006; 25: 564–572.

53. Maung N, Thant CC, Srinivasan M, et al. Corneal ulceration in South-East Asia. II: A strategy for the prevention of fungal keratitis at the village level in Burma. Brit J Ophthalmol 2006; 90:968–70.

54. Srinivasan M, Upadhyay MP, Priyadarsini B, et al. Corneal ulceration in south-east Asia III: Prevention of fungal keratitis at the village level in south India using topical antibiotics. Brit J Ophthalmol 2006; 90:1472–75.

55. Manfredi R, Sabbatani S. Severe *Candida albicans* panophthalmitis treated with all available and potentially effective antifungal drugs: fluconazole, liposomal amphotericin B, caspofungin, and voriconazole. Scand J Infect Dis 2006; 38:950–51.

56. Sallam A, Lynn W, McCluskey P, et al. Endogenous *candida* endophthalmitis. Expert Rev Anti Infect Ther 2006; 4:675–85.

57. Trpin S, Gracner T, Pahor D, et al. Phacoemulsification in isolated endogenous *Candida albicans* anterior uveitis with lens abscess in an intravenous methadone user. J Cataract Refract Surg 2006; 32:1581–83.

58. Osthoff M, Hilge R, Schulze-Dobold C, et al. Endogenous endophthalmitis with azole-resistant *Candida albicans*—Case report and review of the literature. Infection. 2006; 34: 285–88.

59. Andreola C, Ribeiro MP, de Carli CR, et al. Multifocal choroiditis in disseminated *Cryptococcus neoformans* infection. Am J Ophthalmol 2006; 142:346–48.

60. Kramer M, Kramer MR, Blau H, et al. Intravitreal voriconazole for the treatment of endogenous *Aspergillus* endophthalmitis. Ophthalmol 2006; 113:1184–86.

61. de Silva T, Evans C, Mudhar HS, et al. Isolated endogenous endophthalmitis secondary to *Nocardia spp* in an immunocompetent adult. J Clin Pathol 2006; 59:1226.

62. Peyman, G, Lee, P, Seal DV. Endophthalmitis: Diagnosis and Management. London & New York: Taylor & Francis, 2004:1–278.

63. Hassan SS, Wilhelmus KR. Ecologic effects on eye banking. Am J Ophthalmol. 2006; 142:1062–64.

64. Grueb M, Rohrbach JM, Zierhut M. Amphotericin B in the therapy of *Candida glabrata* endophthalmitis after penetrating keratoplasty. Cornea 2006; 25:1243–44.

65. Callanan D, Scott IU, Murray TG, et al. Early onset endophthalmitis caused by Aspergillus species following cataract surgery. Am J Ophthalmol 2006; 142:509–11.

66. Yildiran ST, Mutlu FM, Saracli MA, et al. Fungal endophthalmitis caused by *Aspergillus ustus* in a patient following cataract surgery. Med Mycol 2006; 44:665–69.

67. Rosenberg KD, Flynn HW Jr, Alfonso EC, et al. *Fusarium* endophthalmitis following keratitis associated with contact lenses. Ophthalmic Surg Lasers Imaging 2006; 37: 310–13.

68. Varghese B, Rodrigues C, Deshmukh M, et al. Broad-range bacterial and fungal DNA amplification on vitreous humor from suspected endophthalmitis patients. Mol Diagn Ther 2006; 10:319–26.

69. Tarai B, Gupta A, Ray P, et al. Polymerase chain reaction for early diagnosis of postoperative fungal endophthalmitis. Indian J Med Res 2006; 123:671–78.

70. Velpandian T, Narayanan K, Nag TC, et al. Retinal toxicity of intravitreally injected plain and liposome formulation of fluconazole in rabbit eye. Indian J Ophthalmol 2006; 54:237–40.

71. Hariprasad SM, Mieler WF, Holz ER, et al. Determination of vitreous, aqueous and plasma concentration of orally administered voriconazole in humans. Arch Ophthalmol 2004; 122:42–47.

72. Breit SM, Hariprasad SM, Mieler WF, et al.. Management of endogenous fungal endophthalmitis with voriconazole and caspofungin. Am J Ophthalmol 2005; 139: 135–40.

73. Durand ML, Kim IK, D'Amico M, et al. Successful treatment of *Fusarium* endophthalmitis with voriconazole and *Aspergillus* endophthalmitis with voriconazole plus caspofungin. Am J Ophthalmol 2005; 140:552–54.

74. Sen P, Gopal L, Sen PR. Intravitreal voriconazole for drug-resistant fungal endophthalmitis: case series. Retina. 2006; 26:935–39.

75. Badenoch PR, Halliday CL, Ellis DH, et al. *Ulocladium atrum* Keratitis. J Clin Microbiol 2006; 44:1190–93.

76. Farjo QA, Farjo RS, Farjo AA. *Scytalidium* keratitis: case report in a human eye. Cornea 2006; 25:1231–33.

77. Levy J, Benharroch D, Peled N. *Blastoschizomyces capitatus* keratitis and melting in a corneal graft. Can J Ophthalmol 2006; 41:772–74.

78. Cornely OA, Schmitz K, Aisenbrey S. The first echinocandin: caspofungin. Mycoses 2002; 45: 56–60.

79. Goldblum D, Fruech BE, Sarra GM, et al. Topical caspofungin for treatment of deratitis caused by Candida albicans in a rabbit model. Antimicrob Agents Chemother. 2005; 49: 1359–63.

80. Ha YS, Covert SF, Momany M FsFKS1, the 1,3-beta-glucan synthase from the caspofungin-resistant fungus *Fusarium* solani. Eukaryot Cell. 2006; 5: 1036–42.

81. Özturk F, Yavaz GF, Kusbeci T, et al. Efficacy of topical caspofungin in experimental *Fusarium* keratitis. Cornea 2007; 26: 726–8.

TEN

Epidemiology

INTRODUCTION

In order to investigate risk factors for infectious diseases, a basic understanding of epidemiological methods is required. This is important to avoid misinterpretation when comparing the results of one study with another. For example, crude rates should not be used to compare populations of different structure without definition. Similarly, it is not appropriate to draw population-related conclusions from studies of limited size and power; this is always dangerous! Definitions can be summarized as follows:

- *Epidemiology*—The study of disease in relation to populations. All findings must relate to a defined population. Epidemiology relates the pattern of disease to the population in which it occurs and requires study of both diseased and healthy persons. For epidemiology, incidence and prevalence rates form the basis for comparison between defined population groups.

- A *population*—for epidemiology is a defined group of people including both diseased and healthy subjects, about whose health some statement is to be made.

- *Group orientation*—Clinical appearances determine decisions about individual patients. Epidemiological observations determine decisions about groups and are the basis for preventive medicine.

- *Incidence*—The proportion (1 in 1000) or percentage (%) with *new* disease, in a defined group occurring within a given time period. It is annualized unless otherwise expressed, but may be stated over any time period.

- *Prevalence*—The proportion or percentage with disease (including both *old* and *new*) of a defined group occurring at or over a specific time. This may take place over a day, week, or month, etc.

- *Error*—In epidemiology, statistical inference is valid only if the sample is truly representative of the population, and thus random, accepting that each individual has defined criteria. In addition, the control groups must have no inherent bias. One way of achieving this is to choose a sample at random from the whole population, but this is difficult without a sampling frame, which is a list of the whole population! Hence, a selection has to be made and this is where error frequently arises.

- *Implications of variability*—in epidemiology include random error and biased error:

  In *random* error, individuals are apt to be wrongly assessed or misclassified—this is serious in clinical practice but is not so serious in epidemiology, which is concerned with group decisions.

  In *biased* error, the wrong groups are selected (usually the controls), which then distorts comparisons—*an epidemiological disaster that cannot be corrected by statistics!* Using hospital-based instead of community-based controls, for patients admitted to the hospital from the community, often causes bias and should be avoided.

  In epidemiology, computed information is usually used. This can have errors from:

- Imprecision—Information correctly entered into the computer but the data is wrong
- Unreliability—Correct data entered into the computer incorrectly

  Disease outcome has to be assessed accurately. Clinical experience alone cannot predict prognosis and outcome. Important causes of bias that frequently occur are bias in the selection of cases, bias in the choice of medical therapy or surgical technique, and bias due to incomplete follow-up with unconfirmed assumptions about final outcomes.

  Within a drug trial, sources of error that must be avoided include misallocation of interventions (wrong drug given to the wrong person) and poor compliance with drop administration (failure to give the drug as expected). These two sources of error should be prevented by careful study design and protocol enforcement. Poor compliance means that the expected effect will not be realized.

RETROSPECTIVE AND PROSPECTIVE STUDIES

Epidemiological methods for retrospective or prospective surveys can be listed as follows:

1. Descriptive Study—This lists out a number of cases of infection, such as post-cataract surgery endophthalmitis, in a particular place over a time period and is often retrospective. The observations may suggest a variety of different causes for increased or decreased infection, such as the introduction of a foldable sterile intra-ocular lens (IOL) inserted through a sterile injector (1). The authors identified the power limitations of their study but considered their findings warranted further analysis in a large multi-center trial. This has now been performed in a study of 16,000 patients in 24 operating centers and this risk factor was *not* found to be significant (2), illustrating the danger of extrapolation from small single unit studies.

  A descriptive study can be a useful start to investigate a particular problem, such as endophthalmitis following cataract surgery, but does not record the incidence or prevalence in the population at large. The assessment of risk factors is open to bias by the population attending the particular hospital and the surgical techniques used, i.e., the sample is not representative of the population at large. This type of study does provide a useful basis, however, on which to plan a case-control or cohort study. It is also cheap to perform. If the population is known from whom the cases have been collected, it may be possible to calculate an approximate incidence figure from a large retrospective study such as that performed by Javitt et al. in the United States (1991 Medicare Study) (3). The authors used cases of postoperative infection reported to the U.S. Medicare Funding Organization, as the numerator, and the total number of cataract surgeries that had been paid for as the denominator. The overall incidence of endophthalmitis following extra capsular cataract extraction (ECCE) was 0.12%.

In another retrospective study, Aarburg et al. in 1998 reviewed the incidence of acute culture-proven postoperative (ECCE and phaco) endophthalmitis, occurring within six weeks of surgery, in their hospital over the previous 10 years from 1984 to 1994 (4) when a rate was found of 0.082% (34 out of 41,654 operations); five patients were not included who were culture-negative, being 12.8% of the total number of patients with presumed clinical endophthalmitis. The incidence was higher after secondary IOL implantation. In Holland, a descriptive study based on a questionnaire survey in 2000 (see Use of Questionnaire Reporting Surveys below), estimated the incidence of endophthalmitis after cataract surgery at 0.11% but there was only a 51% response to the survey (5); the authors attempted to estimate bias by under-reporting and calculated an expected nationwide incidence of 0.15%.

Other descriptive studies have suggested that the clear cornea incision used in phacoemulsification cataract surgery might be associated with postoperative endophthalmitis (6,7), with incidence rates of 0.13% (five cases out of 3886) and 0.66% (six cases out of 912) for clear cornea incisions and 0.05% (five cases out of 10,000) and 0.25% (one case out of 397) for scleral tunnel incisions, respectively, but these results were not statistically significant. In addition, these studies were retrospective and not randomized and so were open to bias in the choice of the incision by the surgeon. The two studies have also produced very different incidence rates. Such studies lack statistical power to draw population-based conclusions but can be improved by comparison between and within each other using a meta-analysis technique (8).

Taban has performed such a meta-analysis (8). He decided to compare postoperative rates of endophthalmitis following cataract surgery with those following penetrating keratoplasty (PK) (corneal graft) as a control group, assuming that changes in infection rates associated with PK would reflect the changing practices and drugs used for antibiotic prophylaxis (8). He obtained 5000 papers on cataract surgery and 1900 reports on PK and selected 215 papers on cataract surgery, including three million patients, and 66 studies on PK, including over 90,000 operations, that fulfilled the study's selection criteria; they were considered to be most representative of all the studies found. The overall endophthalmitis rate was three times higher for PK (0.38%) than for cataract surgery (0.13%).

Taban (8) compared the endophthalmitis rate for cataract surgery before and after 1992, when Howard Fine introduced the clear cornea incision, and pre- and post-2000, by which time many cataract surgeons, perhaps half of them, were practicing with it. His analysis showed a correlation between increased use of the clear cornea incision and increased risk of endophthalmitis. This correlation was evident both qualitatively, using a weighted linear regression analysis and determination of the slope coefficient of plots of infection rates over time, as well as quantitatively, based on a relative risk comparison of pooled estimates of endophthalmitis incidence rates between study periods and of different cataract incision techniques. There was a gradual trend downwards of post-cataract surgery endophthalmitis rates between 1961 and 1991, which was reversed between 1992 and 2003. In contrast, the opposite pattern was seen in the slope coefficients for the graphs of endophthalmitis following PK over time.

For the period 2000 to 2003, the rate of endophthalmitis after cataract surgery and use of the clear cornea incision reached 0.27%, representing a 2.5 times increase over earlier years that was statistically significant (8). However, there was no significant difference found in the rates of endophthalmitis after cataract surgery with the limbal and scleral tunnel approaches for the same period.

Taban commented that his findings needed to be confirmed in a large scale, prospective trial with defined inclusion criteria, and accepted that there was bias resulting from the review of studies submitted or accepted for publication that can potentially affect the results of a meta-analysis, which is inevitably retrospective. It can at best produce a hypothesis for further testing. This hypothesis includes the suggestion of poorer stability of the clear cornea wound than that of the scleral tunnel.

While these approaches within a heath care system or specific hospital provide a guide-line figure for historical incidence rates, it is unreliable for recording late-presenting cases of endophthalmitis and others that may be missed, such as hypopyon uveitis where the diagnosis is of postoperative inflammation, which may itself be due to capsule bag (or saccular) infection with *Propionibacterium acnes* (Chapter 8). In addition, care must be taken in presenting results of descriptive studies as they are derived from the population sampled, which may not be representative of the population as a whole.

2. Case-Control Study—This approach compares those with and without the disease who are suitably matched. The study should include only incident cases of disease, i.e., *new* cases of infection that develop during the study and not chronic cases of long duration before the study started. Case-control studies are usually prospective but can be retrospective as well. Diagnostic criteria must be well defined and, for cases of infection, should be Gram- or other stain-positive and/or culture-positive or involve microbial antigen recognition (antibody-based or DNA by a PCR technique, Chapter 3). Patients who have developed a disease, viz. postoperative endophthalmitis, are compared with controls that have not developed the disease when risk factors can be identified reliably (9,10). This permits estimation of odds ratio statistics but not the population attributable risk (11).

Allowance must be made for potential confounding factors. The most important point to note is that the controls *must* match the patients, so that, for example, if the patients enter the hospital from a wide community then the controls must be selected likewise. If those selected as controls are all from the hospital or its local area, there will be potentially serious bias. Controls are best matched to each patient for age, sex, and location of residence, but some studies, such as those of contact lens wearers, require controls to be matched as well for educational status, income, and type of housing (12). Using a case-control study in this way allowed cigarette smoking to be identified as a risk factor for contact lens wear (CLW)-associated microbial keratitis (12).

Use of a case-control study to compare different groups of people is effective for investigating different risk factors within a defined group, and not too costly, such as CLW patients developing *Fusarium* keratitis with a particular multi-purpose storage solution versus CLW controls without keratitis (13). In this study, Chang et al. (13) showed that the topping up of contact lens storage cases with a multi-purpose solution (ReNu with MoistureLoc) was significantly more frequent in patients who developed keratitis with *Fusarium* spp. than in those who did not do so. Furthermore, this demonstrated that use of old solution in storage cases that had not been air dried while the lenses were in use was a significant risk factor. However, it does *not* identify the frequency of the infection in the population, i.e., neither the number of new cases (incidence) nor the proportion present at any time (prevalence) is measured.

If risk factors of postoperative endophthalmitis are predicted from a case-control study, such as those listed by Menikoff et al. (9) and Montan et al. (10), including rupture of the posterior capsule to give communication with the vitreous cavity, use of IOLs with haptics made of polypropylene, immunosuppressive treatment, wound abnormality, and the use of IOLs without a heparinized surface, then the findings need corroboration in a population-based study. A population-based study is needed because of confounding factors that include possible bias both in the controls selected and in the population of patients being investigated. In some studies this may relate to geographical location (high or low altitude), climate (humid or dry and hot, warm, or cold), and the insect population. It is dangerous to extrapolate the findings of a case-control study in one place to the population at large—the results only apply to the precise population studied and other inferences can be most unreliable. For example, fungal infections are more common in a hot, humid rural environment than in a hot urban, high-density, built-up environment such as Kowloon and Shatin, Hong Kong (14).

A case-control study is best conducted prospectively with random selection of patients and controls for the procedures to be used. For example, Nagaki et al. (15) compared rates of postoperative endophthalmitis following phacoemulsification surgery between those who were operated on through a clear cornea incision with those who were operated on through a scleral tunnel in a prospectively planned study in their hospital. They found that there was a postoperative endophthalmitis rate of 0.29% [11 presumed (nine culture-proven) out of 3831 eyes] for the clear temporal cornea incision group and 0.05% [four presumed (four culture-proven) out of 7764 eyes] for those who had surgery via a superior scleral tunnel approach (groups B and C). The relative risk of postoperative endophthalmitis proven by culture in the scleral tunnel group was 4.6 times lower than the clear cornea group ($p = 0.037$). Their findings suggest that the clear cornea incision is a risk factor for postoperative infection. A similar finding (5.9 times lower, $p = 0.02$) has been found in the European multi-center endophthalmitis prophylaxis study (2). They also compared an acrylic IOL (Alcon, MA60BM) with a silicone IOL (Allergan, SI -40NB) and found no significant difference with use of these two materials for influencing the postoperative endophthalmitis rate (15). However, as the authors only had approximately 4000 patients in each of three groups, the statistical power of their study was limited so that their results, although of great interest, required confirmation in a larger multi-center trial. In the European multi-center endophthalmitis prophylaxis study (2), use of the silicone IOL was statistically significant as a risk factor, compared to the acrylic IOL, for postoperative endophthalmitis at $p < 0.003$.

The application of a case-control study is also given in the section below, Statistical Methods to Analyze Clusters of Cases of Endophthalmitis Following Cataract Surgery.

3. Longitudinal Study—This examines associations between exposure to suspected causes of disease and morbidity (the actual disease). The simplest type of longitudinal study is the cohort study, when those subjects exposed to one or to a number of risk factors and those not exposed are followed up prospectively in a defined community-based population study over a period of time. This measures the incidence of disease in each group in the community and can therefore measure the relative risk of the different exposures, together with the population attributable risk (11). Cohort studies are time-consuming and costly and are therefore reserved for testing precisely formulated hypotheses that have been previously explored by descriptive and/or case-control studies, which are used for initial exploratory investigations.

A cohort study concentrates on risk factors related to the development of disease. The length of study depends on the number of new cases of disease and on the duration of the induction period between risk and the occurrence of the disease; sufficiently large numbers are needed to gain statistical significance. Retrospective cohort studies may be possible if the necessary data has already been collected from the population.

The European endophthalmitis prophylaxis study was a prospective, randomized, multi-center and partially placebo-controlled study with a classic Fisher 2×2 factorial design (2,16) similar to that used for the Endophthalmitis Vitrectomy Study (EVS) (17). This study compared the use of perioperative 0.9% saline drops (placebo vehicle control group) with topical 0.5% levofloxacin drops and use of intracameral cefuroxime (1 mg in 0.1 mL) versus no injection, and the combination of each. It started in October 2003 and finished in January 2006 (2). The study took place in 24 operating centers in eight European countries and Turkey. It involved 16,000 patients having phacoemulsification cataract surgery randomized into blocks of 12, with each of four treatment regimes replicating three times within each block. Data collection was electronic and included full surgical details to assess the effect of incision and IOL type on the postoperative endophthalmitis rate as well as many other details. It gave a reliable figure for proven endophthalmitis rates following phacoemulsification surgery of 0.038% with intracameral cefuroxime at the end of surgery and 0.21% without this prophylaxis (p = 0.005). An anomaly arose, however, with the geographical distribution of the cases, which is reviewed in Chapter 8.

4. Cross-Sectional Study—This measures the prevalence of any disease condition, functional impairment, or health outcomes in a population at a point in time or over a short period. The prevalence is the proportion afflicted by that defined clinical state at any one point in time. Prevalence is estimated using sample surveys, which can introduce bias as discussed below.

Sampling involves selection of a representative fraction of the population. For example, it can be used to assess the prevalence (that is, the proportion) of ischemic heart disease in different workers in a factory in order to alter their risk of myocardial infarction by reducing the stress of those with the highest prevalence figure. The cross-sectional study considers *existing* disease, as contrasted with the *development* of new disease in a cohort study. Care is required in detecting known existing cases since they may not be representative of all cases of the disease, i.e., in the example above, detection of myocardial insufficiency is dependent on the test method used and does not necessarily predict those who will develop acute disease, in this example, myocardial infarction. Furthermore, cases of chronic disease of long duration are over-represented in a cross-sectional study.

A cross-sectional study may be used to initiate a cohort study by defining the population at risk, such as the type of contact lens worn or ocular trauma suffered prior to the development of microbial keratitis. It does not replace the cohort study, which measures the incidence and risk factors of *new* cases of the disease under study.

USE OF QUESTIONNAIRE REPORTING SURVEYS

Current standards of practice for cataract surgery in the aging population involves the full range of choice from non-use of antibiotics to their application pre-, intra-, and postoperatively, delivered either intra- or extraocularly. There have been questionnaire reporting surveys of antibiotic use for prophylaxis of endophthalmitis following cataract surgery in Germany (18), the United States (19), and Australia (20) in the last 5 years, but only approximately 66% of surgeons polled by mail returned their questionnaire, with the exception of the Australians (89%). Two of these three surveys did suggest that subconjunctival antibiotics as commonly given immediately after cataract surgery did *not* significantly reduce the incidence of endophthalmitis, further corroborated by the evidence-based review of Ciulla et al. (21). This finding of no evidence to support the use of subconjunctival antibiotics in preventing endophthalmitis is disputed by the findings of Colleaux et al. (6) in their retrospective review, which found 1 case out of 8856 surgeries (0.011% incidence rate) using subconjunctival antibiotics and 9 out of 5030 surgeries (0.179% incidence rate) in those not using subconjunctival antibiotics. This was statistically significant ($p < 0.009$) but open to bias as regards the type of incision used and type of patient included, as it was not a prospective randomized study.

These questionnaire surveys also suggested that the use of intraocular antibiotics during surgery, given as a supplement in the irrigation fluid, often with vancomycin (20 mg/L) and/or gentamicin (8 mg/L), produced the lowest (unproven) incidence rates of postoperative endophthalmitis. The Australian survey found an incidence rate of postoperative endophthalmitis of 0.11%, no relationship with antibiotic prophylaxis but that more experienced surgeons had fewer cases; none of these inferences were upheld in the European multi-center endophthalmitis prophylaxis study (2).

The danger with a low response rate to a survey, where the sample surveyed has been carefully chosen to be representative of the population under study, is that it may introduce *non-response* bias (22). An absurd but extreme example of this would be to conduct a survey investigating death rates during a year and to ask the question "Are you still alive?" All those responding would say "Yes," but it would not be safe to infer from this questionnaire survey that the death rate was zero!

A less extreme example would be to ask "Have you ever used illicit drugs?" It is very likely that those people who have made use of the drugs will be less willing to respond affirmatively and more likely to lie than those people who have not used them. Non-response bias will reinforce the bias already present because of the tendency to lie.

Questionnaire surveys distributed by mail do not usually provide a response rate above 66%. Without some check of the likely effect of non-response bias, the results of such surveys will remain questionable in many cases. One way to validate such a survey is to choose a manageably small but representative sample of all those surveyed who failed to respond initially and to obtain responses from them. From the results of this exercise, an estimate of the non-response bias in the full survey can be made and used to correct the crude results. In addition, this estimate of the non-response bias will itself be subject to random sampling error as well as non-response bias, which will lead to further loss of precision in the final corrected estimates from the full survey. Wormald (22) has commented that the minimum response rate should be 80% and that all non-responders needed to be pursued. He also states that the survey organizer should be aware of the characteristics of those who persistently fail to respond and of the bias that this introduces to the survey.

Any result from a questionnaire reporting survey with only a 66% response rate, that has not been corrected as discussed above, which gives a "significant" result at the 95% level (p <0.05) should be regarded as highly suspect. This situation arises because of the chance that the non-responders would, if included, give results that would mean that this significance level would not be achieved. A 99% significance level (p <0.01) should be considered the minimum level to be acceptable from an uncorrected questionnaire reporting survey with only a 66% response rate. Even then, the study should have included a poll of non-responders. If the non-response bias remains unknown, then the results of a questionnaire reporting survey with only a 66% response rate can be considered interesting but not statistically or epidemiologically valid.

IS NO EVIDENCE OF A BENEFICIAL EFFECT AS GOOD AS EVIDENCE?

Low rates of postoperative endophthalmitis of 0.05% have been produced in Sweden in 225,000 patients using cefuroxime 1 mg in 0.1 mL as an intracameral injection into the anterior chamber at the end of cataract surgery (National Swedish Cataract Registry) (23). The study for the introduction of intracameral cefuroxime was a retrospective analysis, which was not randomized or placebo-controlled, but the recent study has been prospective (23). As there had been no reliable prospective randomized placebo-controlled study, these figures for prophylaxis by an intraocular antibiotic and those from other studies, including subconjunctival use of gentamicin and cefazolin (6), were considered unreliable in a recent review of evidence-based medicine on this subject (21). The study by Colleaux et al. (6) was the first to find a difference between no use and subconjunctival use of antibiotics for cataract surgery with postoperative endophthalmitis rates of 0.179% and 0.011%, respectively, but the size of the study was limited to justify these figures without doubt.

Based on the lack of evidence for the beneficial effect of prophylactic antibiotics as described above and from "apparent" results of the questionnaire reporting surveys, as well as the experience of patients developing postoperative endophthalmitis who had received subconjunctival injections of antibiotics, one surgeon recently withdrew the routine use of prophylactic subconjunctival cefuroxime 125 mg given at the end of surgery (24); in addition, no preoperative or postoperative antibiotics were used but povidone iodine prophylaxis was given at the time of surgery to disinfect the ocular surface. The result was seven cases of postoperative cataract surgery endophthalmitis out of 427 patients (1 case per 61 patients) compared to zero out of 1073 patients operated on by colleagues in the same unit who continued giving subconjunctival injections of cefuroxime immediately after surgery.

The surgeon stopped operating, reviewed surgical technique, and then restarted with reintroduction of subconjunctival cefuroxime, after which there were no further cases of postoperative endophthalmitis in the next 1350 operations (24). This interesting situation demonstrates that while it is possible to provide evidence for non-effectiveness of prophylactic antibiotics from a questionnaire reporting survey with a response rate of 66% or a literature review of evidence-based medicine (21), it does not mean that such antibiotic prophylaxis is of *no* benefit. The likely problem here was the lack of reliability of the questionnaire reporting survey as an epidemiological tool, as described above. In addition, the variable effect of surgical technique for a clear cornea incision or scleral tunnel approach was not properly considered in the evidence-based review (21), because of a lack of relevant studies on which the evidence-based review depended; this is the weak link in the latter type of study.

It has since been suggested that the clear cornea incision used for phacoemulsification is more likely to be associated with postoperative endophthalmitis (2,8). This type of incision will benefit from prophylactic antibiotics more than the scleral tunnel approach, with topical use recommended for one week postoperatively while the corneal wound is healing to prevent bacterial egress into the anterior chamber. This example emphasizes the need to constantly evaluate epidemiological and study-based data for inherent errors either in the study design or the methods being used, or their application in practice which will produce misleading results for others to copy. Clinicians should always question studies that show no benefit of a particular technique—it may be the study and not the technique that is at fault!

ASSESSMENT OF STATISTICAL ASSOCIATIONS

When a significant result is obtained, consideration should always be given to the following:

1. Was the result due to bias? Have the wrong controls been selected? Have hospital controls been used for community patients? Is there bias in patent selection or in patient referral to the hospital? Is there bias in the selection of the controls? In a questionnaire study, is there non-response bias?
2. Could the result be due to chance? Always calculate the confidence intervals and consider the degree of statistical significance. Statistics define the degree of chance but do not change it.
3. Are there any unsuspected confounding factors at play? This may involve unsuspected secondary microbial contamination, such as bacterial contamination of sterile instruments perhaps because they are not being washed properly. Always consider where and why patients become infected. Is the organism or the host at fault?
4. Does the significant level calculated represent a real effect for a risk factor or a confounding one that may be associated with the infection but is not the primary cause of it? Have the correct statistical methods been used? Should the data have been stratified before analysis? Is there a better statistical model by which to analyze the results of the study? Are the results suitable for analysis by multivariate or logistical regression analysis? Would Beysian methods be more applicable than frequentist ones?

STATISTICAL METHODS TO ANALYZE CLUSTERS OF CASES OF ENDOPHTHALMITIS FOLLOWING CATARACT SURGERY

An increasing number of cases of postoperative endophthalmitis after cataract surgery is a worrying situation for a surgeon. There is an urgent need to establish if the numbers of cases are due to chance or to the development of an outbreak due to a particular cause, such as a sterilization failure. Samples of aqueous humor or vitreous should be collected for both microbiology and molecular biology and the organism identified if at all possible (Chapters 2 and 3). Typically, different types of coagulase-negative staphylococci, a *S. aureus* and a streptococcus, maybe isolated from such patients (Chapter 8).

Statistical analysis is needed of the number of cases of endophthalmitis for the number of operations performed. There is a long-standing argument for a frequentist (25) versus a Bayesian (26–30) approach. In either case, comparison is made with the expected postoperative endophthalmitis rate experienced by others, which may itself vary by technique and country. As the chance of infection/no infection is binomial, the classic approach has been to use the Poisson distribution (approximation to the binomial) and to use it to calculate the probability of the infection rate under question being statistically significantly different (i.e., the degree of chance) from the expected norm (25).

To make this simpler, five tables have been constructed for postoperative endophthalmitis rates of 0.05%, 0.1%, 0.2%, 0.3%, and 0.5%, for between 50 and 5000 operations and for one to six cases of postoperative endophthalmitis. The number of cases of postoperative endophthalmitis for the number of operated patients can then be referred to in the appropriate table (Table 1 to 5) and the probability found (p value) for their occurrence being due to chance. A figure of 0.05 (5% chance) is just significant, 0.02 (2% chance) more so, and 0.01 (1% chance) or less is highly significant and suggests a problem of an autoclave or other sterility failure with bacterial contamination of instruments as one possibility.

A new approach to investigate the problem of ongoing or increasing numbers of postoperative endophthalmitis cases uses "traffic lights" as targets (25). The traffic light system is based on the statistical theory above, which states that if an event is

Table 1   Probabilities of Observing X Cases (or More) During N Operations, Where Cases are Expected to Occur with a Poisson Frequency at a Rate of Five Per 10,000 Operations (1:2000 or 0.05%)

No. of operations	Cases observed (X or more)					
	$X \geq 1$	$X \geq 2$	$X \geq 3$	$X \geq 4$	$X \geq 5$	$X \geq 6$
50	0.0247	0.0003	0.0000	0.0000	0.0000	0.0000
100	0.0488	0.0012	0.0000	0.0000	0.0000	0.0000
150	0.0723	0.0027	0.0001	0.0000	0.0000	0.0000
200	0.0952	0.0047	0.0002	0.0000	0.0000	0.0000
250	0.1175	0.0072	0.0003	0.0000	0.0000	0.0000
300	0.1393	0.0102	0.0005	0.0000	0.0000	0.0000
350	0.1605	0.0136	0.0008	0.0000	0.0000	0.0000
400	0.1813	0.0175	0.0011	0.0001	0.0000	0.0000
450	0.2015	0.0218	0.0016	0.0001	0.0000	0.0000
500	0.2212	0.0265	0.0022	0.0001	0.0000	0.0000
550	0.2404	0.0315	0.0028	0.0002	0.0000	0.0000
600	0.2592	0.0369	0.0036	0.0003	0.0000	0.0000
700	0.2953	0.0487	0.0055	0.0005	0.0000	0.0000
800	0.3297	0.0616	0.0079	0.0008	0.0001	0.0000
900	0.3624	0.0754	0.0109	0.0012	0.0001	0.0000
1000	0.3935	0.0902	0.0144	0.0018	0.0002	0.0000
2000	0.6321	0.2642	0.0803	0.0190	0.0037	0.0006
3000	0.7769	0.4422	0.1912	0.0656	0.0186	0.0045
4000	0.8647	0.5940	0.3233	0.1429	0.0527	0.0166
5000	0.9179	0.7127	0.4562	0.2424	0.1088	0.0420

Source: Based on Ref. 25.

Table 2 Probabilities of Observing X Cases (or More) During N Operations, Where Cases are Expected to Occur with a Poisson Frequency at a Rate of 10 per 10,000 Operations (1:1000 or 0.1%)

No. of operations	Cases observed (X or more)					
	X ≥ 1	X ≥ 2	X ≥ 3	X ≥ 4	X ≥ 5	X ≥ 6
50	0.0488	0.0012	0.0000	0.0000	0.0000	0.0000
100	0.0952	0.0047	0.0002	0.0000	0.0000	0.0000
150	0.1393	0.0102	0.0005	0.0000	0.0000	0.0000
200	0.1813	0.0175	0.0011	0.0001	0.0000	0.0000
250	0.2212	0.0265	0.0022	0.0001	0.0000	0.0000
300	0.2592	0.0369	0.0036	0.0003	0.0000	0.0000
350	0.2953	0.0487	0.0055	0.0005	0.0000	0.0000
400	0.3297	0.0616	0.0079	0.0008	0.0001	0.0000
450	0.3624	0.0754	0.0109	0.0012	0.0001	0.0000
500	0.3935	0.0902	0.0144	0.0018	0.0002	0.0000
550	0.4231	0.1057	0.0185	0.0025	0.0003	0.0000
600	0.4512	0.1219	0.0231	0.0034	0.0004	0.0000
700	0.5034	0.1558	0.0341	0.0058	0.0008	0.0001
800	0.5507	0.1912	0.0474	0.0091	0.0014	0.0002
900	0.5934	0.2275	0.0629	0.0135	0.0023	0.0003
1000	0.6321	0.2642	0.0803	0.0190	0.0037	0.0006
2000	0.8647	0.5940	0.3233	0.1429	0.0527	0.0166
3000	0.9502	0.8009	0.5768	0.3528	0.1847	0.0839
4000	0.9817	0.9084	0.7619	0.5665	0.3712	0.2149
5000	0.9933	0.9596	0.8753	0.7350	0.5595	0.3840

Source: Based on Ref. 25.

Table 3 Probabilities of Observing X Cases (or More) During N Operations, Where Cases are Expected to Occur with a Poisson Frequency at a Rate of 20 per 10,000 Operations (1:500 or 0.2%)

No. of operations	Cases observed (X or more)					
	X ≥ 1	X ≥ 2	X ≥ 3	X ≥ 4	X ≥ 5	X ≥ 6
50	0.0952	0.0047	0.0002	0.0000	0.0000	0.0000
100	0.1813	0.0175	0.0011	0.0001	0.0000	0.0000
150	0.2592	0.0369	0.0036	0.0003	0.0000	0.0000
200	0.3297	0.0616	0.0079	0.0008	0.0001	0.0000
250	0.3935	0.0902	0.0144	0.0018	0.0002	0.0000
300	0.4512	0.1219	0.0231	0.0034	0.0004	0.0000
350	0.5034	0.1558	0.0341	0.0058	0.0008	0.0001
400	0.5507	0.1912	0.0474	0.0091	0.0014	0.0002
450	0.5934	0.2275	0.0629	0.0135	0.0023	0.0003
500	0.6321	0.2642	0.0803	0.0190	0.0037	0.0006
550	0.6671	0.3010	0.0996	0.0257	0.0054	0.0010
600	0.6988	0.3374	0.1205	0.0338	0.0077	0.0015
700	0.7534	0.4082	0.1665	0.0537	0.0143	0.0032
800	0.7981	0.4751	0.2166	0.0788	0.0237	0.0060
900	0.8347	0.5372	0.2694	0.1087	0.0364	0.0104
1000	0.8647	0.5940	0.3233	0.1429	0.0527	0.0166
2000	0.9817	0.9084	0.7619	0.5665	0.3712	0.2149
3000	0.9975	0.9826	0.9380	0.8488	0.7149	0.5543
4000	0.9997	0.9970	0.9862	0.9576	0.9004	0.8088
5000	1.0000	0.9995	0.9972	0.9897	0.9707	0.9329

Source: Based on Ref. 25.

Table 4 Probabilities of Observing X Cases (or More) During N Operations, Where Cases are Expected to Occur with a Poisson Frequency at a Rate of 30 per 10,000 Operations (1:333 or 0.3%)

No. of operations	Cases observed (X or more)					
	X ≥1	X ≥2	X ≥3	X ≥4	X ≥5	X ≥6
50	0.1393	0.0102	0.0005	0.0000	0.0000	0.0000
100	0.2592	0.0369	0.0036	0.0003	0.0000	0.0000
150	0.3624	0.0754	0.0109	0.0012	0.0001	0.0000
200	0.4512	0.1219	0.0231	0.0034	0.0004	0.0000
250	0.5276	0.1734	0.0405	0.0073	0.0011	0.0001
300	0.5934	0.2275	0.0629	0.0135	0.0023	0.0003
350	0.6501	0.2826	0.0897	0.0222	0.0045	0.0008
400	0.6988	0.3374	0.1205	0.0338	0.0077	0.0015
450	0.7408	0.3908	0.1546	0.0482	0.0124	0.0027
500	0.7769	0.4422	0.1912	0.0656	0.0186	0.0045
550	0.8080	0.4911	0.2296	0.0859	0.0265	0.0070
600	0.8347	0.5372	0.2694	0.1087	0.0364	0.0104
700	0.8775	0.6204	0.3504	0.1614	0.0621	0.0204
800	0.9093	0.6916	0.4303	0.2213	0.0959	0.0357
900	0.9328	0.7513	0.5064	0.2859	0.1371	0.0567
1000	0.9502	0.8009	0.5768	0.3528	0.1847	0.0839
2000	0.9975	0.9826	0.9380	0.8488	0.7149	0.5543
3000	0.9999	0.9988	0.9938	0.9788	0.9450	0.8843
4000	1.0000	0.9999	0.9995	0.9977	0.9924	0.9797
5000	1.0000	1.0000	1.0000	0.9998	0.9991	0.9972

Source: Based on Ref. 25.

occurring randomly at a specified rate, then the probability of observing a given number of cases is defined by the Poisson distribution. The practical application involves referring to one of Tables 1 to 5, according to the expected postoperative infection rate, and stopping surgery when a threshold has been reached for the probability being due to chance of less than 2%. Investigation should take place of the infection control procedures and sterilization of instruments, consideration be given to use of prophylactic antibiotics (Chapter 8), and then surgery is recommended. If further cases occur, then the traffic light system is applied again and surgery continues until the likelihood of its occurrence being due to chance is less than 2%, again by referring to the tables. In this way the decision to stop and start surgery can be based on probability levels that exclude cases that might be due to random chance.

An additional approach used to investigate the above study (25) was to perform a case-control study by comparing the 11 cases of postoperative endophthalmitis with 49 controls, randomly selected from the possible 1212 patients who had similar operations in the unit but did not develop postoperative endophthalmitis. A series of univariate analyses was carried out using Fisher's Exact Test followed by stepwise logistic regression using both forward and backward conditional methods to examine the independent effect of each variable on endophthalmitis. These analyses showed that there was a higher risk of endophthalmitis associated with being female and with having a vitrectomy or a previous history of respiratory disease. A similar predominance of female patients was found by Montan et al. in the Swedish prophylaxis study using intracameral cefuroxime (31) and for proven cases in the European multi-center endophthalmitis prophylaxis study (p=0.035) (2).

An alternative approach is to use Bayesian methodology (26–30). This involves establishing a prior, a likelihood, and a posterior. For a prior to be taken seriously, its

Table 5 Probabilities of Observing X Cases (or More) During N Operations, Where Cases are Expected to Occur with a Poisson Frequency at a Rate of 50 per 10,000 Operations (1:200 or 0.5%)

No. of Operations	Cases observed (X or more)					
	X ≥ 1	X ≥ 2	X ≥ 3	X ≥ 4	X ≥ 5	X ≥ 6
50	0.2212	0.0265	0.0022	0.0001	0.0000	0.0000
100	0.3935	0.0902	0.0144	0.0018	0.0002	0.0000
150	0.5276	0.1734	0.0405	0.0073	0.0011	0.0001
200	0.6321	0.2642	0.0803	0.0190	0.0037	0.0006
250	0.7135	0.3554	0.1315	0.0383	0.0091	0.0018
300	0.7769	0.4422	0.1912	0.0656	0.0186	0.0045
350	0.8262	0.5221	0.2560	0.1008	0.0329	0.0091
400	0.8647	0.5940	0.3233	0.1429	0.0527	0.0166
450	0.8946	0.6575	0.3907	0.1906	0.0780	0.0274
500	0.9179	0.7127	0.4562	0.2424	0.1088	0.0420
550	0.9361	0.7603	0.5185	0.2970	0.1446	0.0608
600	0.9502	0.8009	0.5768	0.3528	0.1847	0.0839
700	0.9698	0.8641	0.6792	0.4634	0.2746	0.1424
800	0.9817	0.9084	0.7619	0.5665	0.3712	0.2149
900	0.9889	0.9389	0.8264	0.6577	0.4679	0.2971
1000	0.9933	0.9596	0.8753	0.7350	0.5595	0.3840
2000	1.0000	0.9995	0.9972	0.9897	0.9707	0.9329
3000	1.0000	1.0000	1.0000	0.9998	0.9991	0.9972
4000	1.0000	1.0000	1.0000	1.0000	1.0000	0.9999
5000	1.0000	1.0000	1.0000	1.0000	1.0000	1.0000

Source: Based on Ref. 25.

evidential basis must be explicitly given (26). As well as defining their prior, investigators should quote briefly, with some technical detail of the algebra involved, what analytical form was used for the prior distribution. They should choose, for example, from a Weibull distribution or a Gamma distribution or other possibilities. Bayesian methods are appropriate for use here but they involve a good knowledge of mathematics. However, these methods, if properly applied, have a logical consistency about them that, with proper grounds on which to base a prior, may be superior to the traditional frequentist approach, but mathematical details must be explicitly given.

The recent outbreak study in Scotland used the traditional frequentist approach (25) while that in England used the currently more fashionable Bayesian approach (24). The authors of the latter study criticized the use of the Poisson approximation to the binomial distribution used in the former study but freely admitted that use of this approximation is satisfactory. They found that use of the Bayesian method for the exact distribution led to the same conclusion as using the Poisson approximation. The use of the Poisson approximation has allowed us to calculate values in Tables 1 to 5, which lets the reader quickly establish whether their cases are likely due to an outbreak when surgery should stop for assessment (25). The Bayesian/frequentist debate is a long-standing one and not easy to resolve.

REFERENCES

1. Mayer E, Cadman D, Ewings P, et al. A 10 year study of cataracts surgery and endophthalmitis in a single eye unit: injectable lenses lower the incidence of endophthalmitis. Br J Ophth 2003; 87:867–869.

2. ESCRS Endophthalmitis Study Group. Prophylaxis of post-operative endophthalmitis following cataract surgery: results of the ESCRS multi-centre study and identification of risk factors. J Cat Refract Surg 2007; 33:978–88.

3. Javitt JC, Vitale S, Conner JK. National outcomes of cataract extraction. Endophthalmitis following inpatient surgery. Arch Ophthalmol 1991; 109:1985–89.

4. Aaberg TM, Flynn HW, Schiffman J, Newton J. Nosocomial acute-onset postoperative endophthalmitis survey. A 10-year review of Incidence and Outcomes. Ophthalmol 1998; 105:1004–1010.

5. Versteegh MFL, van Rij G. Incidence of endophthalmitis after cataract surgery in the Netherlands. Documenta Ophthalmologica 2000; 100:1–6.

6. Colleaux KM, Hamilton WK. Effect of prophylactic antibiotics and incision type on the incidence of endophthalmitis after cataract surgery. Can J Ophthalmol 2000; 35:373–378.

7. Thurber LT, Chalfin S, Kavanagh JT. Incidence of post-operative endophthalmitis in clear corneal and scleral tunnel incisions for phacoemulsification surgery. Presented as an ARVO abstract (no. 189), 2003.

8. Taban M, Behens A, Newcomb RL, et al. Acute endophthalmitis following cataract surgery. A systematic review of the literature. Arch Opthalmol 2005; 123:613–20.

9. Menikoff JA, Speaker MG, Marmor M. A case-control study of risk factors for postoperative endophthalmitis. Ophthalmol 1991; 98:1761–68.

10. Montan PG, Koranyi G, Setterquist HE, Stridh A, Philipson BT, Wiklund K. Endophthalmitis after cataract surgery: risk factors relating to technique and events of the operation and patient history. Ophthalmol 1998; 105:2171–2177.

11. Friedman GD. Primer of Epidemiology. 4th edition. McGraw-Hill, New York: 1994.

12. Lam D, Houang E, Lyon D, Fan D, Wong E, Seal D. Incidence and Risk Factors for Microbial Keratitis in Hong Kong: comparison with Europe and North America. Eye 2002; 16 (5):608–618.

13. Chang DC, Grant GB, O'Donnell KO, et al. Multi-state outbreak of *Fusarium* keratitis associated with use of a contact lens solution. JAMA 2006; 296:953–963.

14. Houang E, Lam D, Fan D, Seal D. Microbial Keratitis in Hong Kong—relationship to climate, environment and contact lens disinfection. Trans Roy Soc Trop Med & Hyg 2001; 95:361–67.

15. Nagaki Y, Hayasaka S, Kadoi C, et al. Bacterial endophthalmitis after small-incision cataract surgery. Effect of incision placement and intraocular lens type. J Cataract Refract Surg. 2003; 29:20–26.

16. Fisher, R. A. The design of experiments. 1st ed. London: Oliver & Boyd, 1935.

17. Endophthalmitis Vitrectomy Study Group. Endophthalmitis Vitrectomy Study. Arch Ophthalmol 1995; 113:1479–1496 & Am J Ophthalmol 1996; 122:1–17.

18. Schmitz S, Dick HB, Krummenauer F. Endophthalmitis in Cataract Surgery. Results of a German Survey. Ophthalmology 1999; 106 (10):1869–76.

19. Masket S. Questionnaire Survey of Cataract Surgery Practice in the USA. Presented to the Annual Meeting of the ASCRS, 1998 (Ocular Surgery News).

20. Morlet N, Gatus B, Coroneo M. Patterns of peri-operative prophylaxis for cataract surgery: a survey of Australian ophthalmologists. Austr New Zealand J Ophthalmol 1998; 26:5–12.

21. Ciulla TA, Starr MB, Masket S. Bacterial Endophthalmitis Prophylaxis for Cataract Surgery. An evidence-based update. Ophthalmol 2002; 109:13–24.

22. Wormald R. Assessing the prevalence of eye disease in the community. Eye 1995; 9:674–76.

23. Lundstrom M, Wejde G, Stenevi U, et al. Endophthalmitis after cataract surgery. A nationwide prospective study evaluating incidence in relation to incision type and location. Ophthalmol 2007; 114:866–70.

24. Mandal K, Hildreth T, Farrow M, Allen D. An investigation into post-operative endophthalmitis and lessons learned. J Cat Refract Surg 2004; 30:1960–65.

25. Allardice GM, Wright EM, Peterson M, Miller JM. A statistical approach to an outbreak of endophthalmitis following cataract surgery at a hospital in the West of Scotland. J Hosp Inf 2001; 48:1–7.
26. Spieglhalter DJ, Myles JP, Jones DR, Abrams KR. Bayesian methods in Health Technology Assessment: a review. Health Technol Assess 2000; 4 (38):1–130.
27. Spiegelhalter DJ, Myles JP, Jones DR, Abrams KR. An introduction to Bayesian methods in health technology assessment. Brit Med J 1999; 319:508–512.
28. Bland JM, Altman DG. Bayesians and frequentists. Brit Med J 1998; 317:1151.
29. Matthews RAJ. Why should clinicians care about Bayesian methods? J Statist Planning & Inference 2001; 94:43–58; discussion 59–71.
30. Lee PM. Bayesian Statistics: an introduction. 2nd edition. Arnold, London: 1997.
31. Montan PG, Wejde G, Koranyi G, Rylander M. Prophylactic intra-cameral cefuroxime. Efficacy in preventing endophthalmis following cataract surgery. J Cat Refract Surg 2002; 28:977–982.

APPENDIX A

Modes of Delivery and Formulations of Antibiotics and Antifungal Drugs

Many modes of delivery discussed here are outside the specifications of the product license for the agent and are therefore used at the clinician's responsibility.

TOPICAL PREPARATIONS

Drops and ointments are the standard means of administering antibiotics to the surface of the eye either for prophylaxis or treatment (Tables 1 and 2). Ointments prolong contact time and therefore permit less frequent instillation and are less likely to be washed out of the eye. In recent years, the preparation of fortified antibiotic eye drops has been advocated for the successful treatment of suppurative keratitis. Fortified drops are prepared by combining commercially available parenteral preparations with artificial tear preparations or sterile water to widen the range and concentration of agents used.

For example, to produce drops containing cefuroxime 5% (50 mgs/mL) or gentamicin 1.5% (15 mgs/mL) for immediate use, the vial of antibiotic for injection is opened under aseptic conditions and appropriate dilutions made into sterile bottles with sterile water for injection. The drops should be given to the patient to commence immediate therapy and then stored at $+4°C$ (or refrigerator temperature) between doses. The drops should be used within one week, in order to maintain stability and activity of the antibiotic and to avoid extraneous organisms causing contamination. If bulk production is required, this should be performed by a professional pharmacy department when the solution will be filtered to maintain sterility and preservatives considered to prevent contamination; in addition, stability of the antibiotic must be considered, which is more crucial for cefuroxime (or other penicillins and cephalosporins) than for gentamicin, which is stable.

PERIOCULAR INJECTION

Subconjunctival delivery involves the injection of 0.25–1.0 mL of antibiotic solution deep to the conjunctiva. There is some leakage of antibiotic back into the conjunctival sac, but the bolus chiefly acts as a depot for diffusion that will produce transient high levels of antibiotic in cornea, sclera, anterior chamber and choroid, and aqueous and, to a lesser and variable extent, the vitreous. Vitreous levels are lower because of the absorption of drug into the choroidal and retinal circulations, and because of the natural barriers to penetration into the vitreous across the retina (see earlier). Appropriate doses for subconjunctival injection are given in Table 3.

Table 1 Selected Topical Antimicrobial Drops: Commercially Available and Fortified Hospital-Produced Preparations

Eye drops	Fortified[a] preparation	Commercial
Antibacterial		
Amikacin	25–50 mg/mL	
Bacitracin (not in U.K.)	10,000 U/mL	
Cefazolin	50 mg/mL	
Ceftazidime	50 mg/mL	
Cefuroxime	50 mg/mL	
Cephalothin	50 mg/mL	
Chloramphenicol[b]	5 mg/mL	
Ciprofloxacin		3 mg/mL
Gentamicin	15 mg/mL	3 mg/mL
Framycetin		5 mg/mL
Fusidic acid (gel basis; Fucithalmic)		10 mg/mL
Levofloxacin		5 mg/mL
Neomycin		5 mg/mL
Ofloxacin		3 mg/mL
Oxacillin	66 mg/mL	
Penicillin G	5000 U/mL (0.3%)	
Piperacillin	50 mg/mL	
Propamidine isethionate	1 mg/mL	
Sulphacetamide		100–300 mg/mL
Tetracycline		10 mg/mL (oil vehicle)
Ticarcillin	50 mg/mL	
Tobramycin	15 mg/mL	3 mg/mL
Vancomycin	50 mg/mL	
Combinations		
Neosporin		
Polymyxin B		5000 U/mL
Gramicidin		25 U/mL
Neomycin		2.5 mg/mL
Polytrim		
Polymyxin B		10,000 U/mL
Trimethoprim		1 mg/mL
Antifungal		
Amphotericin[c]	1.5–3.0 mg/mL	
Clotrimazole	1% in arachis oil	
Econazole	1% in arachis oil	
Fluconazole	1% in arachis oil	
Flucytosine[c]	1%	
Ketoconazole	1% in arachis oil	
Itraconazole	1% in arachis oil	
Miconazole	1% in arachis oil	
Natamycin[c]	50 mg/mL	

(Continued)

Table 1 Selected Topical Antimicrobial Drops: Commercially Available and Fortified Hospital-Produced Preparations (*Continued*)

Eye drops	Fortified[a] preparation	Commercial
Antiprotozoal[d]		
Propamidine isethionate (Brolene) (available in UK)		0.1% (1 mg/mL)
Hexamidine isethionate (Desmodine) (available in France)		0.1% (1 mg/mL)
Chlorhexidine digluconate (not commercially available make up in hospital pharmacy)		0.02% (200 mg/mL)
Polyhexamethylene biguanide[e]		0.02% (200 mg/mL)

[a] Produced in hospital pharmacy department.
[b] Use with caution because of theoretical risk of bone marrow aplasia and never prescribe for more than two weeks
[c] Aqueous suspension.
[d] For the treatment of *Acanthamoeba* keratitis.
[e] No commercial preparation available (unlicensed cold chemical sterilizing agent). Can be purchased from Astra Zeneca Biocides as Cosmocil-QC. Indemnity insurance required for use as medical therapy.

Peak aqueous levels are achieved in the first hour and effective levels are maintained for about six hours. Inclusion of adrenaline in the subconjunctival injection prolongs antibiotic activity for 24 hours or more, so that injections may be repeated less frequently. This is contraindicated in patients with cardiac disease, and caution must be exercised in the aged, or when patients are receiving general anesthesia with halothane. Where possible, the injection is delivered close to the site of infection.

A sub-Tenon's injection is delivered in a similar volume but more deeply into the orbit, beneath Tenon's capsule and close to the sclera. Care must be taken to avoid penetration of the globe. It is thought to achieve higher levels in the posterior eye than the more anterior subconjunctival route.

To achieve high levels of chemotherapeutic agents in the posterior segment of the eye, local delivery of antibiotic by periocular or intravitreal injection is employed. Systemic administration of a drug achieves far lower levels in the tissues of the globe than these techniques but is used as an adjunct to other routes in the treatment of endophthalmitis (refer to Chapter 4). It is effective alone in the treatment of adnexal disease. Some of these techniques will be considered further.

Ofloxacin and ciprofloxacin have not yet been used routinely by the intravitreal route to treat endophthalmitis, although ciprofloxacin has been shown to be relatively nontoxic intravitreally in rabbits and to be removed by the active transport route similarly to cephalosporins. Recent studies by systemic routes have shown that insufficient intravitreal levels are achieved to treat serious infective endophthalmitis.

Periocular injections are uncomfortable, requiring local anesthesia, and are not without complications. Conjunctival ischemia and necrosis may occur locally, and orbital hemorrhage and penetration of the globe have been reported. Although high aqueous levels can be achieved, they are not sustained, and the use of frequent topical application of fortified antibiotic preparations is generally preferred to repeated periocular injections in the treatment of microbial keratitis.

Table 2 Selected Antimicrobial Ointments	
Ointment	**Preparation**
Antibacterial, commercial preparations	
Chloramphenicol[a]	1%
Chlortetracycline	1%
Erythromycin	0.5% (not commercially available)
Framycetin	0.5%
Gentamicin	0.3%
Neomycin	0.5%
Rifampicin	2.5% (not commercially available)
Sulphacetamide	2.5–10%
Tetracycline	1%
Antibacterial, Combinations	
Graneodin	
Neomycin	0.25%
Gramicidin	0.025%
Polyfax	
Bacitracin	500 U/g
Polymyxin B	10,000 U/g
Polytrim	
Trimethoprim	0.5%
Polymyxin B	10,000 U/g
Antifungal	
Nystatin	3.3% (not commercially available)
Antiprotozoal	
Dibromopropamidine (Brolene)	0.15%

[a] Use with caution because of theoretical risk of bone marrow aplasia and never prescribe for more than four weeks.

ANTERIOR CHAMBER AND INTRAVITREAL INJECTION

Lavage of the anterior chamber with antibiotic solutions has been employed in the past in the treatment of serious ocular infections but has limited value, since aqueous levels will quickly be reduced by diffusion and by bulk removal of aqueous from the anterior chamber as part of its normal circulation.

Intravitreal injection of antibiotic has a much more important role to play in the treatment of endophthalmitis (Chapter 8). Antibiotic may persist in the vitreous space in effective concentrations for up to 96 hours. Only a small volume is injected (up to 0.2 mL). To avoid excessive elevation of ocular pressure, an equal volume of vitreous fluid is removed prior to injection. Amounts injected are based on studies of toxicity in animals (including primates), since the retina is sensitive in this respect (Chapter 4).

The technique involves entering the vitreous via the pars plana to avoid retinal injury (Chapter 8). This approach is also used to collect a vitreous sample and may be combined with a total vitrectomy to reduce the infective load and facilitate diffusion of injected drugs. The intravitreal injection is given through the same entry site and can be repeated at intervals, depending on the drug(s) selected. Removal of vitreous gel will, however, accelerate loss of antibiotic from the vitreous space by enhancing diffusion across the retina.

Table 3   Selected Intravitreal and Periocular Antibiotics

| Antibiotic | Intravitreal injection | | Periocular dose (subconjunctival) (mg) | Intravenous dose |
	Dose[a] (μg)	Effective duration (hr)		
Amikacin	400	24–48	–	15 mg/kg/24 hr
Ampicillin	500	24	100–125	1 g/4 hr
Amphotericin	5 or 10	24–48	Not used	0.1–1.0 mg/kg/ 24 hr
Carbenicillin	2000	16–24	100	3 g/4 hr
Cefazolin	2000	16	100	1 g/4 hr
Cefuroxime	2000	16–24	125	1.5 g/6 hr
Ceftazidime	2000	16–24	100	1.5 g/6 hr
Clindamycin	1000	16–24	34	0.75 g/8 hr
Erythromycin	500	24	50	0.5 g/6 hr
Flucloxacillin	–	–	100	1 g/6 hr
Gentamicin	200	48	40	5 mg/kg/24 hr
Methicillin	2000	40	100	1 g/4 hr
Miconazole	10	24–48	–	0.6-3.6 g/24 hr
Oxacillin	500	24	100	2 g/4 hr
Penicillin	–	–	300–600	3 MU/4 hr
Vancomycin	1000	72	25	1 g/12 hr

[a] Maximum intravitreal injection volume is usually 0.3 mL, i.e. 0.1 mL of of two antibiotics given in separate syringes and 0.1 mL of unpreserved dexamethasone (inject 400 μ in separate syringe, refer to Chapter 8)

Antibiotic injected into the vitreous is removed by the anterior and posterior routes (Chapter 4). The anterior route for cationic drugs involves diffusion into the posterior chamber and removal by bulk flow in the aqueous humor and, for anionic drugs, by transport across the pigment epithelium of the iris and into the circulation. The posterior route involves diffusion across the retina and, for anionic drugs, active transport by the retinal vessels or retinal pigment epithelium.

For these reasons, cationic drugs such as gentamicin have longer half-lives in the vitreous than anionic drugs such as cephalosporins, which are actively transported out of the vitreous space. This effect persists, though in lesser degree, in the inflamed eye (Chapter 4). Persistence of drug in the vitreous can be prolonged if the same drug is given systemically at the same time since the outward diffusion gradient is decreased. In the case of certain anionic drugs, such as the cephalosporins and ciprofloxacin, levels can be further increased by systemic administration of probenecid. This raises plasma levels by inhibiting renal tubular excretion and also blocks the active transport system out of the eye.

THE SYSTEMIC ROUTE

Systemic medication may be used to treat preseptal and orbital cellulitis, dacryoadenitis, acute dacryocystitis, and the rare condition of ocular erysipelas (necrotizing fasciitis of the eyelids), which may need surgical debridement. In the management of chronic blepharitis, tetracyclines used in low dose (e.g., oxytetracycline 250 mg twice daily) may act by an effect on meibomian oil composition through inhibition of bacterial lipase. In rosacea-associated blepharitis, oral tetracycline 250 mg twice daily or doxycycline 100 mg once daily will be effective in 50% of patients.

Systemic medication may be combined with local therapy in the treatment of ophthalmia neonatorum due to *Neisseria gonorrhoeae*, Chlamydia and, rarely, Pseudomonas to combat systemic manifestations. It will also be used in the treatment of adult patients with chlamydial or gonococcal eye disease.

Systemic chemotherapy has no place in the management of uncomplicated bacterial keratitis. It is usually given in high dosage in the adjunctive treatment of bacterial endophthalmitis and is mandatory for metastatic endophthalmitis when the systemic infection must also be treated. Because of the risks of toxicity, high-dose regimens should be closely monitored.

MYCOLOGICAL PHARMACOPOEIA

This is given in Table 4.

Table 4   Principal Antifungal Drugs Used for Therapy of Ophthalmomycoses[a]

Drug[b] and route of administration	Advantages	Disadvantages
A. Pimaricin (Natamycin) i. Polyene ii. Topical (5%) solution available commercially iii. Used to treat mycotic blepharitis, conjunctivitis, keratitis iv. Used to treat mycotic endophthalmitis in association with other antifungals	a. Broad spectrum of activity including *Fusarium, Aspergillus, Acremonium, Penicillium, Lasiodiplodia, Candida* b. Ophthalmic preparation is well-tolerated, stable and can be sterilized by heat c. Relatively high levels achieved in cornea after topical application d. Treatment of *Fusarium* and *Aspergillus* keratitis yields good results; recommended as first-line treatment in suspected mycotic keratitis	1. Only effective when applied topically 2. Treatment of keratitis due to filamentous fungi other than *Fusarium* and *Aspergillus* and due to yeast-like and related fungi yields variable results 3. Only about 2% of total drug in corneal tissue is bioavailable 4. Allergic conjunctivitis is a side effect
B. Amphotericin B i. Polyene ii. Variably fungistatic, occasionally fungicidal iii. Topical (usually[c] 0.15–0.3%): i.v. and subconjunctival route indicated only in serious cases iv. Intravenous: start with 0.1 mg/kg per day and increase to 1 mg/kg per day: maximum total daily dose is 1.5–2.0 g/day	a. Excellent activity against *Candida, Aspergillus, Lasiodiplodia*; also active against *Cryptococcus, Histoplasma, Blastomyces* and *Fusarium* spp. b. Penetrates deep corneal stroma after topical application: bioavailability sufficient for susceptible fungi c. In addition to direct fungicidal effect, also shows immunoadjuvant properties	1. Less useful than other antifungals against other filamentous fungi 2. Subconjunctival, i.v., intravitreal, and intracameral administration carry risk of toxicity 3. Very toxic when deoxycholate is used as vehicle 4. Some believe this to be inferior to pimaricin in treating clinical mycotic keratitis

(Continued)

Table 4   Principal Antifungal Drugs Used for Therapy of Ophthalmomycoses[a] (*Continued*)

Drug[b] and route of administration	Advantages	Disadvantages
v. Subconjunctival: 2–5 mg in 0.5 mL, every 12–24 hours vi. Intracameral: 20–30 mg in 0.1 mL vii. Intravitreal: 5 mg in 0.1 mL viii. Used to treat conjunctivitis, orbital cellulitis, endophthalmitis, keratitis, etc.	d. Topical preparation (0.15%) from i.v. formulation is well tolerated e. Experimental data suggest that collagen shields soaked in this drug (0.5%) are useful and convenient to treat mycotic keratitis (especially that caused by *Candida albicans*)	5. Monitor renal function when giving i.v. therapy, especially for low serum potassium levels
C. Nystatin i. Polyene ii. Topical solution (25,000–100,000 U/mL) iii. Topical ointment (100,000 U/g)	a. Good in vitro activity against *Candida* sp., *Cryptococcus* sp., *Histoplasma*, *Blastomyces*, and various dermatophytes b. Used clinically to treat superficial keratitis, conjunctivitis and superficial lid lesions due to *Candida*, *Cryptococcus*, and dermatophytes c. Generally nontoxic d. Dermatologic ointments and ophthalmic solutions well tolerated after topical administration	1. Poor in vitro and in vivo activity against many common ocular filamentous fungi. 2. Subconjunctival injection causes local necrosis 3. Intraocular penetration by systemic or topical routes poor 4. Cannot be given intravitreally
D. Flucytosine (5-fluorocytosine) i. Pyrimidine ii. Topical (1%): oral 50–150 mg/kg per day in four divided doses iii. Used to treat *Candida* keratitis and endophthalmitis (metastatic)	a. Active against *Candida*, *Cryptococcus*, and related fungi b. Most useful as adjunctive therapy (topical) for *Candida* keratitis c. Effective for *Candida* endophthalmitis in combination with systemic or intravitreal amphotericin B d. Effective against some strains of *Aspergillus* and *Torula*	1. Cannot be administered alone in treating mycotic keratitis or endophthalmitis due to rapid emergence of resistance 2. Limited spectrum of activity against filamentous fungi 3. Usefulness in treatment of postoperative endophthalmitis not established
E. Clotrimazole i. Imidazole ii. Topical (1% as solution or cream); oral route not used now	a. Broad spectrum of activity especially against *Aspergillus* in vitro and in vivo	1. Poorly soluble in water hence cannot be given parenterally

(*Continued*)

Table 4 Principal Antifungal Drugs Used for Therapy of Ophthalmomycoses[a] (*Continued*)

Drug[b] and route of administration	Advantages	Disadvantages
iii. Used to treat keratitis, non-granulomatous conjunctivitis, superficial lid lesions	b. Believed by some to be the drug of choice for *Aspergillus* keratitis	2. Corneal toxicity associated with long-term use of topical preparation
F. Miconazole i. Imidazole ii. Topical (1%), subconjunctival (10 mg in 0.5 mL), intravenous (600–3600 mg/day), and intravitreal (10 mg in 0.1 mL) iii. Used for treatment of keratitis, non-granulomatous conjunctivitis, endophthalmitis, orbital infection, scleral suppuration, lid lesions, granulomatous conjunctivitis	a. Broad spectrum of activity against many types of fungi including *Aspergillus, Candida, Pseudallescheria boydii* b. Well tolerated by topical and subconjunctival routes to treat keratitis c. Advocated as second-line treatment if no response to pimaricin in keratitis d. Concomitant administration of oral ketoconazole and topical miconazole found useful in clinical fungal keratitis e. Given i.v. for endophthalmitis and for *Lasiodiplodia* keratitis	1. Use of i.v. preparation is occasionally associated with toxicity due to the vehicle used 2. Variable results obtained when treating *Fusarium* keratitis and other ocular *Fusarium* infections
G. Econazole i. Imidazole ii. Topical (1%) in arachis oil	a. Broad spectrum of activity against filamentous fungi, including *Aspergillus, Fusarium, Penicillium*, and many dematiaceous fungi b. Good results obtained by some workers in treating clinical *Fusarium* keratitis; recommended by these workers as first-line treatment for filamentous fungal keratitis	1. Use of topical preparation is associated with ocular irritation 2. Less effective than miconazole against *Candida* species
H. Ketoconazole i. Imidazole ii. Oral (200–600 mg/day). Topical 1% in arachis oil or 2% suspension in artificial tear-drops iii. Used to treat keratitis, endophthalmitis, granulomatous conjunctivitis	a. Broad spectrum of activity against many types of fungi b. Useful in treatment of clinical mycotic keratitis, especially when given orally c. Excellent results obtained in treating keratitis due to	1. Poor in vitro activity versus *Aspergillus fumigatus* and *Fusarium* species 2. Variable results obtained when treating keratitis due to *Aspergillus* or *Fusarium*; poor results in *Lasiodiplodia* keratitis 3. Oral administration may be associated with a transient

(*Continued*)

Table 4 Principal Antifungal Drugs Used for Therapy of Ophthalmomycoses[a] (*Continued*)

Drug[b] and route of administration	Advantages	Disadvantages
	dematiaceous filamentous fungi (not Sphaerosidales)	rise in levels of some serum enzymes 4. Topical therapy is sometimes associated with superficial punctate keratopathy

I. Itraconazole
  i. Triazole
  ii. Oral (200 mg once daily). Topical 1% suspension in artificial tears
  iii. Used for treatment of keratitis, granulomatous conjunctivitis, and endophthalmitis
	a. Broad spectrum of activity against all *Aspergillus* species, *Candida*, many dematiaceous fungi
	b. Excellent results (especially when given orally) in treating keratitis due to *Aspergillus* and dematiaceous fungi (not Sphaerosidales)
	c. Excellent safety profile when given orally
		1. Poor in vitro activity versus *Fusarium* sp. and *Lasiodiplodia*
		2. Variable results obtained in treating *Fusarium* keratitis and poor results in treating *Lasiodiplodia* keratitis
		3. Topical use of the suspension is sometimes associated with superficial punctate keratopathy

J. Fluconazole
  i. Triazole
  ii. Oral (50–200 mg/day). Topical (1%) suspension; intravenous (100 mg/day)
  iii. Used to treat endophthalmitis, keratitis, lid lesions, conjunctivitis, etc.
	a. *Candida* endophthalmitis effectively treated by oral therapy plus vitrectomy or plus intravitreal amphotericin B
	b. Excellent safety profile and good intraocular penetration
	c. Reported to be useful in treating cases of *Candida* keratitis
		1. Spectrum of activity limited to *Candida*, *Cryptococcus*, and related fungi and some strains of *Aspergillus*
		2. Poor activity against *Fusarium*, *Lasiodiplodia*, and other tropical fungi

K. Voriconazole
  i. Azole
  ii. Oral (200 mg twice daily); topical (1%), intravenous, and intravitreal (100 ug/0.1 mL) routes of administration have all been described
  iii. Has been used to successfully treat fungal endophthalmitis and keratitis
	a. Exerts potent activity against a broad spectrum of yeasts and molds
	b. Is more active than amphotericin B against filamentous fungi such as *Aspergillus* species
	c. Achieves good aqueous and vitreous levels with oral administration (53% and 38% of plasma levels, respectively)
	d. Found useful in treating endophthalmitis due to *Candida, Paecilomyces lilacinus, Scedosporium apiospermum,*
		1. Voriconazole monotherapy may sometimes not effect cure; caspofungin may need to be added
		2. Reports on efficacy in ocular mycoses mostly anecdotal; large series of patients with ocular mycoses treated with voriconazole still not studied

(Continued)

Table 4   Principal Antifungal Drugs Used for Therapy of Ophthalmomycoses[a] (*Continued*)

Drug[b] and route of administration	Advantages	Disadvantages
	Lecytophora mutabilis, and *Fusarium* species.	
L. Posaconazole i. Triazole ii. Given orally iii. Has been used to treat keratitis, endophthalmitis and periorbital zygomycosis	a. Has proven efficacious for treating *Fusarium* keratitis and/or endophthalmitis. refractory to other systemic, intracameral, and topical antifungal agents, including systemic and/or topical voriconazole b. Exhibits a good safety profile after oral administration. c. Exhibits broad spectrum of antifungal activity	1. Cannot be given by the intravenous route 2. At present, there is only anecdotal evidence of its efficacy it treating ophthalmomycoses
M. Silver sulphadiazine i. Combination of antimicrobial activity of silver and sulphadiazine ii. 1% ointment iii. Used to treat keratitis iv. Absence of side effects	a. Has a broad spectrum of antifungal activity including *Aspergillus, Fusarium* b. Topical 1% ointment gave better results (80%) than topical 1% miconazole (55%) in one study[d] c. Effective in clinical *Fusarium* and *Aspergillus* keratitis in one study	1. Use of this compound by many workers has not yielded satisfactory results 2. 2% ophthalmic solution was not found effective in superficial or deep keratitis due to *Aspergillus* and *Fusarium*[e]

[a] Compiled from references in Chapter 9.
[b] 1% = 10 mg/mL; 0.1% = 1 mg or 1000 μg/mL.
[c] i.v., intravenous.
[d] Brit J Ophthalmol 1988; 72, 192-195.
[e] PA Thomas, unpublished data.

APPENDIX B

Protozoa, Helminths, and Arachnids: Ocular Disease and Chemotherapy

This appendix summarizes the chemotherapy, the ocular tissues involved (Table 1), and selected features of these infections.

For some infections, several drugs may be available. Consideration should be given to the efficacy and safety of drugs, as well as to the cost and local availability of the drugs or combination of drugs. Thus, it is difficult to define first, second, or third choices. Recommended dosage schedules are provided for adults and pediatric patients. It is important to note that doses are expressed as salt or base. The administered dose may vary substantially between different preparations. Further, special consideration may have to be given to the use of some drugs in pregnancy and in the immunocompromised patient.

PROTOZOA

Acanthamebiasis

Many reports are available for successful medical therapy of *Acanthameba* keratitis. The bisbiguanide (chlorhexidine) or the polymeric biguanide (PHMB), preferably in combination with the aromatic diamidine propamidine (as Brolene), is most effective (Chapter 6). Hexamidine (Desmodine) (0.1%) eye drop has been used four times daily as alternative therapy (Chapter 6).

Microsporidiosis

Several drugs, including propamidine isethionate (0.1% six times daily), Polytrim (polymyxin B sulphate 10,000 U/mL for two months), topical ciprofloxacin (0.3% for two weeks), or itraconazole (200 mg twice daily), have been used but with no or minimum cure. Fumagillin in the form of a water-soluble salt (fumagin bicyclohexylammonium) comes as a reconstitutable powder and has been used successfully for treating microsporidial keratoconjunctivitis at a concentration of about 0.11 mg/mL of active fumagillin in sterile buffered saline. Drops can be instilled every three hours; resolution is usually fairly rapid and seen within a few days.

Toxoplasmosis

Treatment of choice is the combination of the folate inhibitors pyrimethamine at 25 mg/day for three to four weeks for adults or 2 mg/kg per day, then 1 mg/kg per day (maximum 25 mg/day for four weeks for children) and trisulphopyrimidines at 2–6 g/day for three to four weeks for adults and 100–200 mg/kg per day for three to

Table B.1 Involvement of ocular tissues

	Lacrimal sac	Lids	Conjunctiva	Sclera	Cornea	AC	Ciliary body	Iris	Lens	Uvea	Vitreous	Choroid	Retina	Optic nerve	Orbit	Extraocular muscles
Protozoa																
Acanthamoebiasis			[+]	(+)	+			[+]		[+]	au					
Microsporidiosis			[+]		+											
Toxoplasmosis						+			(+)	+pu	+	+	+	(+)		
Pneumocystosis												(+)	+	(+)		
African trypanosomiasis	(+)	+			(+)			(+)		+au		(+)		(+)		
S. American trypanosomiasis		(+)	+							+						
Leishmaniasis[a]	(+)	+	(+)		(+)					+au		(+)	(+)	(+)		
Malaria			(+)					+		+		(+)	+[b]	+		
Babesiosis												+	+	+		
Entamoebiasis		(+)	+							+au		+	(+)	+		
Giardiasis			+				+	+		+au/	pu	+	+	+	+	
Helminths																
Toxocariasis[c]						+	+			+	+	+	+	+		
'Ocular Larva migrans'[c]		(+)	+		(+)	(+)	+	(+)		+	+	+	+	+	+	+
Paragonimiasis[c]		(+)				+			+		+	+	+			
Schistosomiasis[c]			+			+				(+)	+	(+)	(+)			
Alariasis[c]	+						(+)	(+)			+	+	+		(±)	+
Ascariasis[c]	+												+		+	
Baylisascaris[c]							(+)					(+)	(+)	+		
Cysticercosis[d,e]			(+)		(+)	+			(+)		+	+	(+)	+	+	+
Hydatidosis[f]			+								+	(+)	(+)	+	+	
Trichinosis		+	+										(+)	(±)	+	t[g]
Onchocerciasis[d]			+		+	+	+	+	+			+	+	+		

Loiasis

Dirofilariasis

Brugia filariasis

Bancroftian filariasis

Dracunculiasis

Gnathostomiasis

Thelaziasis

Angiostrongyliasis

Philophthalmiasis

Coenurus

Sparagonosis

Arachnida

Ophthalmomyiasis

interna

externa

Phthiariasis (lice)

Demodexiasis

Pentastomids

[a] Depends on type of leishmaniasis.

[b] Retinal haemorrhages, retinopathy after chloroquine therapy.

[c] Can give signs of diffuse unilateral subacute neuroretinitis (DUSN) (Chapter 7).

[d] Posterior, anterior segments involved - variation depends on geographical location.

[e] Cystic stage of *Taenia solium.*

[f] Cystic stage of *Echinococcus granulosus*

[g] In trichinosis, encysted larvae invade the extraocular muscles preferentially, as well as the orbicularis, to produce a characteristic oedema.

[h] Strictly periorbital.

[i] Strictly the globe as well as the orbit.

[j] Strictly eye lashes and lids.

[k] Eyelashes.

Abbreviations: +, Main structure(s) involved. (+), Rare or late effect. [+], Tissue reaction to, but not active infection by, the protozoa or helminth. au, Anterior uveitis. pu, Posterior uveitis.

four weeks for children. For infants treatment should be every two to three days and for newborns with congenital toxoplasmosis treatment should continue for about one year. Pyrimethamine can occasionally cause blood dyscrasias and folic acid deficiency and, rarely, rash, vomiting, convulsions, and shock. To prevent hematological toxicity from pyrimethamine, it is advisable to give leucovorin (folinic acid) at about 10 mg/day, either by injection or orally. (Pyrimethamine alone at 50–75 mg/day has been used to treat CNS toxoplasmosis after sulphonamide sensitivity develops.) Pyrimethamine should not be given in pregnancy since it is teratogenic in animals. In this situation, spiramycin has been recommended at 2–4 g/day for three to four weeks. There are occasional gastrointestinal disturbances and, rarely, allergic reactions to the drug. Spiramycin can be given to children at 50–100 mg/day for three to four weeks. Corticosteroids given systemically may be necessary if active retinochoroiditis is present. Periocular corticosteroid injections are contraindicated. Clindamycin, atovaquone, and azithromycin are alternative therapies, but there is limited data on their use. Further information is given in Chapter 7.

Pneumocystosis

The drug of choice is trimethoprim-sulphamethoxazole combination (trimethoprim at 20 mg/kg per day, sulphamethoxazole at 100 mg/kg per day for both adults and children). AIDS patients will require prolonged therapy. The alternative drug is pentamidine isethionate at 4 mg/kg per day for 14 days for both adults and children. For AIDS patients who develop resistance to trimethoprim-sulphamethoxazole and pentamidine, trimetrexate with leucovorin rescue, or a combination of dapsone and trimethoprim can be effective. Aerosolized pentamidine is available for both prophylaxis and treatment of *Pneumocystis carinii* infection in susceptible hosts. Trimetrexate with leucovorin can cause occasional rash, peripheral neuropathy, bone marrow depression, and increased aminotransferase concentration in some patients.

African Trypanosomiasis

Treatment of choice for early Rhodesian and Gambian disease is suramin sodium. Dosage is 100–200 mg (test dose intravenous), then on days 1, 3, 7, 14, and 21 for adults. For children dosage is 20 mg/kg on days 1, 3, 7, and 21. Frequently there will be vomiting, pruritis, urticaria, parasthesias, hyperesthesia of hands and feet, photophobia, and peripheral neuropathy. Occasional side effects are kidney damage, blood dyscrasias, shock, and optic atrophy. Pentamidine can be used as an alternative in the pre-CNS Gambian disease at a dosage of 4 mg/kg per day for 7 to 10 days. For the late disease with CNS involvement, the drug of choice is melarsoprol at 2–3.6 mg/kg per day intravenous for three doses, then after one week, 3.6 mg/kg per day for three doses for adults. This can be repeated after 10 to 21 days. For children the dosage is 18–25 mg/kg over one month; initial dose of 0.36 mg/kg intravenous, increasing gradually to a maximum of 3.6 mg/kg at intervals of one to five days for a total of 9 to 10 doses. Frequent side effects are myocardial damage, albuminuria, hypertension, colic, Herxheimer-type reaction, encephalopathy, vomiting, and peripheral neuropathy. As a second-line therapy, a combination of melarsoprol and suramin can be used.

South American Trypanosomiasis

The drug of choice is nifurtimox given orally at 8–10 mg/kg per day in four divided doses for 120 days for adults, at 15–20 mg/kg per day in four divided doses for 90 days for children aged 1 to 10 years, and at 12.5–15 mg/kg per day in four divided doses for 90 days for children aged 11 to 16 years. The drug frequently produces anorexia, vomiting, weight loss, loss of memory, sleep disorders, tremor, paresthesia, weakness, and polyneuritis. Rare side effects include convulsions, fever, pulmonary infiltrates and pleural effusion. The alternative treatment is benznidazole at 5–7 mg/kg for 30 to 120 days. There is frequently an allergic rash, dose-dependent polyneuropathy, gastrointestinal disturbances and psychic disturbances.

Leishmaniasis

First-line therapy for all forms of leishmaniasis is antimonials. Sodium stibogluconate is given to adults or children by intramuscular or intravenous administration at 20 mg/kg per day (to a maximum of 800 mg/day) for 20 days. This can be repeated until there is a response. Meglumine antimonate can be used as an alternative antimonial. Stibogluconate can frequently induce muscle pain and joint stiffness as well as bradycardia. There is occasional colic, diarrhea, rash, pruritis, and myocardial damage. Mucocutaneous leishmaniasis can be treated with amphotericin B (as the sodium deoxycholate). The dosage is 0.25–1 mg/kg by small infusion daily, or even every two days, for up to eight weeks for both adults and children.

Malaria

Malaria can be difficult to treat, especially in Southeast Asia where there is an emergence of resistant strains. Therefore, only limited guidelines are provided here, and advice should be sought from a tropical medicine specialist for treatment of infected patients. For treatment of *Plasmodium vivax, P. ovale*, or *P. malariae* infections, chloroquine or mefloquine should be provided. Mefloquine is used for treatment of chloroquine-resistant *P. vivax*. For radical cure of *P. vivax* and *P. ovale* infections, primaquine is used after initial treatment with chloroquine. For treatment of *P. falciparum* infection in chloroquine-sensitive areas, chloroquine can still be used, with quinine being reserved for severe or complicated malaria.

In areas where there is chloroquine-resistance, sulphadoxine-pyrimethamine, halofantrine, or mefloquine may be alternatives to chloroquine, their use being dependent on parasite susceptibility. Quinine or quinine plus sulphadoxine-pyrimethamine are used in severe or complicated malaria. In areas in South-East Asia where there is multi-drug resistance, artemisinin derivatives, halofantrine, or mefloquine may be used. Quinine plus tetracyclines or artemisinin derivatives are used in severe or complicated malaria. For prophylaxis of *P. falciparum*, chloroquine is used in areas where the parasite is sensitive to this drug and where expected transmission is low. In areas where there is chloroquine resistance and significant transmission, chloroquine plus proguanil, or mefloquine, can be used. Sulphadoxine and pyrimethamine can be used for self-treatment in some areas. In areas in Southeast Asia where there is significant transmission and multi-drug resistance, mefloquine or doxycycline should be used.

Malarial Infection and Quinine Toxicity for the Retina

Serious retinal toxicity can occur during routine treatment of malaria with quinine, even if the malarial parasite is not infecting the eye directly. The toxic blood level is close to the therapeutic level; if the patient is below standard weight, then the risk is increased. It is also possible that some people suffer severe ocular damage idiosyncratically even when the dosage is not excessive. The dosage given is standard for intramuscular or intravenous therapy (600 mg twice daily). Blindness develops rapidly on the second to fifth day of therapy and can be reversible, if the quinine is stopped immediately, or permanent. Initially, there is no light perception, with dilated and non-reactive pupils; the fundus may show only attenuated retinal vessels. Pupil responses can return after a few days and normal visual acuity with full fields within 14 days; there may be no apparent permanent damage. However, blindness may be permanent or the patient may be left with good visual acuity but gross field constriction. The fundus can appear normal except for attenuated vessels, which is characteristic of quinine blindness. With greatly increased use of quinine for treating drug-resistant malaria, the potential retinal toxicity of this drug needs documenting.

Babesiosis

The treatment that has been used is a combination of clindamycin and quinine. In adults the dosage is clindamycin 1.2 g twice daily parenterally or 600 mg three times daily orally for seven days, and quinine 650 mg three times daily orally for seven days. For children, the dosage is clindamycin 20–40 mg/kg per day in three doses for seven days and quinine 25 mg/kg per day in three doses for seven days. Adverse side effects of quinine (dihydrochloride and sulphate) can frequently include cinchonism (tinnitus, headache, nausea, abdominal pain, visual disturbance) and can occasionally include hemolytic anemia, other blood dyscrasias, photosensitivity reactions, hypoglycemia, arrhythmias, hypotension, and drug fever. Occasionally, quinine use leads to blindness and sudden death if it is injected too rapidly. Concurrent use of pentamidine and sulphamethoxazole-trimethoprim has been reported to cure a *Babesia divergens* infection.

Entamebiasis

Metronidazole is the drug of choice. It should be given orally as 2 g daily or 800 mg three times daily for five days. There is occasional mild nausea, metallic taste, headache, dizziness, and intolerance to alcohol. An intravenous preparation is available (400 mg, injections, 1 mL equivalent to 5 mg). Pediatric dosage is 35–50 mg/kg per day in three doses for 10 days. It should not be given to pregnant women, especially during the first trimester. Tinidazole may also be used. Diloxanide furoate should follow at 500 mg, three times daily for 10 days. There may be occasional, but mild, gastrointestinal effects and/or urticaria.

Giardiasis

The treatment of choice is metronidazole. For adults it should be 2 g once daily, orally for three days. Pediatric dosage should be 15 mg/kg per day orally in three divided doses for 5 to 10 days. Tinidazole can act as an alternative. Other drugs are furazolidone, given at 100 mg four times daily for 7 to 10 days to adults and 1.25 mg/kg four times daily for 7 to 10 days. There is frequent nausea and vomiting.

HELMINTHS

Toxocariasis

Refer to Chapter 7 for ocular larva migrans and diffuse unilateral subacute neuroretinitis. For chorioretinitis, diethylcarbamazine, thiabendazole, and mebendazole are useful therapies, as well as ivermectin for nematodes and praziquantel for trematodes, but may exacerbate inflammatory lesions by killing the larva. Systemic corticosteroid treatment suppresses the inflammatory response. Retinal detachment secondary to larval granuloma can occur and laser-photocoagulation may be required. Most cases of ocular larval migrans are quiescent and do not require medical or surgical intervention, but if larva are observed in the eye, argon-laser photocoagulation should be used early in the course of the disease to destroy the nematode.

Paragonimiasis

The drug of choice is praziquantel at 25 mg/kg three times daily for two days for both adults and children. There is frequent malaise, headache, and dizziness. Occasionally there may be sedation, abdominal discomfort, fever, sweating, nausea, and eosinophilia. A rare adverse effect is pruritis or rash. The alternative therapy is 10 to 15 doses of bithionol at 30–50 mg/kg on alternate days for both adults and children. Frequent adverse effects are photosensitivity reactions, vomiting, diarrhea, abdominal pain, and urticaria.

Schistosomiasis

In single individuals with *Schistosoma haematobium* infection, and in situations where reinfection is not a problem, praziquantel is favored. This should be given at a dose of 20 mg/kg three times daily for one day for both adults and children. Metrifonate may also be given outside these criteria. An optimal dose for this drug is as yet undetermined. WHO standard regimen is 7.5 mg/kg on three occasions at two-week intervals, but a one-day regimen is required. The same treatment schedule should be used for *S. japonicum, S. intercalatum*, and *S. mekongi* infections. The regimen is also recommended for *S. mansoni* infection, but oxamniquine can also be used at a dosage of 15 mg/kg as a single dose for adults and at 10 mg/kg twice daily as a single dose for children. In East Africa, the dose should be increased to 30 mg/kg per day and in Egypt and South Africa it should be given at the same dosage but for two days. Neuropsychiatric disturbances and seizures have been noted in some patients using this drug, as have convulsions, but these side effects are rare. Occasionally, headache, fever, dizziness, somnolence, nausea, diarrhea, rash, insomnia, hepatic enzyme changes, ECG and EEG changes, and orange-red urine may accompany use of this drug.

Alariasis

Surgical removal of the larva or photocoagulation with laser was used in the two cases that have been recorded (Chapter 6). No antiparasitic drug therapy has been used.

Ascariasis

For systemic infections, albendazole, mebendazole, levamisole, piperazine, or pyrantel can be used. For ectopic infections, surgical removal is the only effective therapy.

Cysticercosis

The treatment recommended is praziquantel at 50 mg/kg per day in three divided doses for 14 days for both adults and children (along with corticosteroids for two to three days before and during therapy), but this does not appear to be particularly useful for ocular (or spinal cord) cysticercosis. Metrifonate at 7.5 mg/kg for five days, repeated six times at two-week intervals is probably more efficacious for ocular disease. Metrifonate causes frequent reversible plasma-cholinesterase inhibition and occasional nausea, vomiting, abdominal pain, headache, and vertigo. Albendazole at 15 mg/kg per day in three divided doses for 28 days for both adults and children is probably the best approach to treatment of this infection. There may be occasional diarrhea, abdominal pain, and other side effects. It may be necessary to remove the intraocular cyst surgically.

Hydatid Disease

Surgical excision of the cyst is recommended. Mebendazole (50 mg/kg/day) or albendazole (15 mg/kg per day) in three divided doses for 28 days is the treatment of choice for neurocysticercosis and is also beneficial as adjunctive therapy.

Trichinosis

Thiabendazole at 25 mg/kg twice daily for five days (maximum 3 g/day) for adults and 25 mg/kg twice daily for five days for children, with steroids if there are severe symptoms, can be given. The efficacy of this compound for trichinosis is difficult to judge, since it appears to be effective when the worms (*Trichinella spiralis*) are located in the intestine, but it may not be effective for migrated worms in the eye. Use of this drug is frequently associated with nausea, vomiting, and vertigo. Occasionally there may be leukopenia, crystalluria, rash, hallucinations, olfactory disturbance, erythema multiforme, and Stevens-Johnson syndrome. Albendazole may be more effective; a single dose of 400 mg can be given to both adults and children. For heavier infections, treatment can be continued for three days. It has been suggested that mebendazole at 200–400 mg three times daily for three days, then 400–500 mg twice daily for 10 days, may be most effective for tissue-phase treatment; there are few adverse effects of use. Ivermectin is a potential new treatment.

Onchocerciasis

The drug of choice is ivermectin (IVM). The dosage is 150 mg/kg orally once, every six months in adults, repeated in patients with heavy ocular infections. Children require only one treatment. More frequent doses will not give increased efficacy. At the dose of 200 mg/kg, it is thought to be as effective as diethylcarbamazine (DEC) in decreasing microfilaria numbers but it produces less ophthalmological reactions. Antihistamines or corticosteroids may be necessary (see Filariasis, this appendix). It may be useful to excise subcutaneous nodules in the head before drug therapy initiation. Ivermectin is a remarkably safe drug; there is occasional fever, pruritis, or tender lymph nodes and, rarely, hypertension. Both IVM and DEC are effective only against microfilaria. In severe cases, suramin should be used to kill the adult worms. The drug is given at 100–200 mg (test dose) intravenously, then 1 g at weekly intervals for five weeks in adults. For children, the dose is 10–20 mg/kg intravenously at weekly intervals for five weeks. Suramin should be used only if ocular microfilaria persist after DEC or IVM therapy and nodulectomy.

Filariasis (Bancroftian and Brugian filariasis, loiasis, and dirofilariasis)

Treatment has been with DEC, which is given orally in a 21-day course: adults—day one, 50 mg after food; day two, 50 mg three times daily; days 4 to 21, 2 mg/kg three times daily; children—day one, 25–50 mg; day two, 25–50 mg; day three, 50–100 mg three times daily; days 4 to 21, 2 mg/kg three times daily. Ivermectin may be useful for these infections, but experience is very limited. DEC induces frequent severe allergic or febrile reactions due to the filarial infection and also gastrointestinal disturbances. Rarely there may be encephalopathy. Mechanical removal of an adult worm from beneath the conjunctiva, after application of topical anesthesia, may be required for lymphatic filariasis, and surgical excision of a lesion or extraction of a worm is usually successful for dirofilariasis of the eye. A subconjunctivally located Loa loa worm requires initial application of topical anesthetic, then suturing of the worm before extraction. Cryoprobe can be used to remove a worm before surgery. DEC should be given with caution in loiasis since it can provoke ocular problems or an encephalopathy. Antihistamines or corticosteroids may be required to decrease allergic reactions due to disintegration of microfilaria in the treatment of all filarial infections, but especially with Loa loa and onchocerciasis.

Dracunculiasis (Guinea Worm)

Drug treatment of this infection is of limited value. Metronidazole has been useful at 250 mg three times daily for 10 days for adults and at 25 mg/kg per day (maximum 750 mg/day) in three doses for 10 days. The drug should not be given to pregnant women, especially in the first trimester. Niridazole has also been used, as has thiabenzadole at a dosage of 25–37.5 mg/kg twice daily for three days for both adults and children. However, this dosage may be toxic and may require reducing. These drugs are thought to reduce inflammation, which facilitates mechanical removal of the emergent worms.

Gnathostomiasis

There is no unequivocal medical therapy for this infection, although mebendazole at 200 mg every three hours for six days has some promise. Albendazole may be useful. The treatment of choice is surgical removal of the parasite. The worm should be frozen before surgery, which should be undertaken early since the patient may die if the nematode enters the central nervous system.

Thelaziasis

Treatment involves mechanical removal of the worms present on the conjunctival surface of the eye, a procedure that should be performed under local anesthesia. Symptoms should disappear immediately, but if they persist the conjunctiva should be re-examined for other worms.

Angiostrongyliasis

Systemic treatment of *Angiostrongyloides cantonensis* and *A. costaricans* is by mebendazole and thiabendazole, respectively. Surgical removal of the worm is required for ocular infection. It may be beneficial to immobilize the worm with a cryoprobe before excision from the posterior segment. Care must be taken in preparing the site of entry such that retinal detachment is prevented. Anterior segment inflammation will require topical atropine sulphate (1%) and topical corticosteroids.

Philophthalmiasis

The worm is removed from the conjunctiva with forceps after gentle pressure.

Coenurus

Surgical removal of the cyst in early infection is the only treatment for the ocular infection. Surgical intervention is of no value if panophthalmitis has developed.

Sparagonosis

Medical treatment has not proven successful for the ocular infection. The worm or nodule must be removed surgically. Injection of 40% ethanol has been used to kill the parasite in situ.

ARACHNIDA

Myiasis

Treatment of external ophthalmomyiasis involves mechanical removal of all offending larvae after application of a topical anesthetic. Steroids or antibiotics should be provided if inflammation or secondary bacterial infection is present. It has been suggested that patching of the eye overnight with anticholinesterase ointment is useful for killing or immobilizing any larvae retained after the initial intervention.

Treatment of internal ophthalmomyiasis is dependent on the extent of the host reaction. If no symptoms are evident, then no treatment is generally necessary, although the patient should be observed during this time. Dead larvae may not require surgical removal in some instances. If an inflammatory reaction is refractory to steroid therapy, then the larvae should be removed from the vitreous, or destroyed by photocoagulation if they are located subretinally.

For orbital ophthalmomyiasis, treatment involves the removal of all invading larvae and control of secondary bacterial infection. Various approaches to killing or narcotizing the larvae have been suggested, including the use of formalin, turpentine, and carbolic acid. These are useful for packing and irrigating infected tissues. If the infestation is severe, exenteration will be necessary.

Phthiariasis

Many treatments have been used for louse infestation of the eyelids. The simplest method involves mechanical removal of lice or nits with forceps, although this may not always be practical. Petroleum applied to the lid margins twice daily for 8 days is often effective against lice but not eggs. Physostigmine (an anticholinesterase), cryotherapy, and argon-laser ablation have been used. Yellow mercuric oxide applied four times daily for 14 days may be effective.

APPENDIX C

Prevention of Hospital-Acquired Ocular Infection

Cross-infection may occur in the accident and emergency room, the outpatient clinic, the ward, and the operating theater.

THE OUTPATIENT CLINIC

Patients receiving treatment within a medical environment are at risk of contracting an infection from another patient or from a health care worker. Such an unexpected outcome can arise from contact with the practitioner's hands or from ophthalmic equipment in contact with the eye. The courteous act of shaking hands with a patient is sufficient to transfer organisms so that hand washing between examining patients, preferably with an antiseptic soap, is considered important.

Adenovirus is the most highly contagious organism and can cross-infect several hundred people with keratoconjunctivitis over a few weeks until a temporary closure of the unit is necessary. This needs urgent investigation using rapid polymerase chain reaction (PCR) identification and typing techniques to establish that it is a hospital-acquired infection outbreak (see Chapter 3). It must be distinguished from the community outbreak that has been the origin of hospital spread. Patients with acutely infected eyes, particularly where adenovirus infection or acute hemorrhagic conjunctivitis (due to enterovirus 70 or coxsackie A24v) is involved, should always be segregated and examined, wearing disposable gloves, in a separate room in the Casualty Department, where equipment in ocular contact is carefully disinfected between patients.

The following disinfection regime is suggested to reduce the chances of hospital-acquired infection:

- Hand washing: perform after handling the lids of any infected eye using an antiseptic impregnated liquid soap preparation from an elbow- or floor-operated container. Such hand wash preparations include 10% povidone iodine (Betadine scrub), which is bactericidal and virucidal, or 4% chlorhexidine (Hibiscrub), which is only bactericidal. For patients suspected of presenting with acute adenovirus infection, disposable gloves should be worn, with hand washing using povidone iodine. Alcoholic hand washes such as Hibiscrub (chlorhexidine gluconate) are ineffective against adenovirus.
- Tonometer heads: these should be wiped after use on a disposable paper towel and then placed into a gallipot of Milton at 1/10 dilution (hypochlorite solution at 1000 ppm) for a minimum of 10 minutes (30 minutes during an outbreak of adenovirus infection). At the end of the clinic, the contents of the gallipot must be discarded and the pot sent for sterilization. A fresh sterile gallipot must be used for each clinic with a new solution of freshly produced hypochlorite

(Milton) solution. All tonometer heads must be soaked for 10 minutes in this new solution before use and then washed with sterile water and dried.

Milton fluid is made of an aqueous solution of sodium hypochlorite and 16.5% sodium chloride and if it has a red label is the original 1% sodium hypochlorite (10,000 ppm) and is only supplied to hospitals and health authorities. Milton found in retail outlets is twice the strength at 2% sodium hypochlorite. Milton can be corrosive to certain items like stainless steel because of the high salt content, unlike ordinary household bleach solution, which does not contain such high salt levels. Milton for hospital use is very stable and nontoxic, as the manufacturing purification process includes the removal of heavy metal ions. Heavy metal ions are responsible for the catalytic breakdown of sodium hypochlorite to sodium chlorate, which is toxic. In the absence of heavy metal ions, Milton breaks down to water and sodium chloride. It is important that dilutions of Milton in hospitals are made up with sterile distilled water.

In addition, there has been concern about the retention of corneal epithelial cells following tonometry and the risk this gives for transferring Creutzfeld Jacob disease (vCJD) (1). Manual cleaning was the most important step in reducing epithelial cell retention. For reducing the risk of vCJD (see below), tonometer heads should be soaked in sodium hypochlorite at 5000 ppm (= 1/2 strength Milton) for one hour or in sodium hydroxide (5%) for one hour.

- **Slit-lamp microscope:** the head rim and chin rest should be wiped with a disposable alcoholic paper towelette.

- **Eye drops:** only single-use should be used in hospital outpatient clinics. Single-use means single use and *not* for use in other patients to empty the container! Microbiological studies have found that if single-use products, e.g., Minims, are used between different patients, then contamination with bacteria, particularly staphylococci, is likely. Essentially, a single-use indication on packaging is much more stringent than "single patient use." Whereas single patient use allows the product, once opened, to be used on the same patient for a specified period of time (three hours under refrigerated conditions is a good guideline), a single-use indication dictates that the product must be discarded immediately after opening and initial use. Further applications of the product would involve a new container being opened. Good examples of single-use are the "Minims range" of products Baush & Lomb, Kingston UK. Since the Minims is available at competitive pricing (approximately £0.30, equivalent to USD 0.15) there is no need for nurses and clinicians to undertake unsafe practices to use up the last drop. Any use outside the licensed use is the clinician's responsibility and is considered off-label.

- **Sampling equipment:** disposable sterile surgical blades are best used to collect corneal scrapings. If the Kimura spatula is used, it must first be sterilized by heat or alcohol. Scrapings should be collected for microscopy and directly cultured onto bacteriological, acanthamebal, and fungal culture agar and into enrichment broth (Chapter 2). Any spillages onto the clinic furniture must be wiped thoroughly with 1% hypochlorite or 10% povidone iodine solution.

- **Dressings:** the infected eye should never be bandaged. All bandages used on non-infected eyes must be sterile and disposable. They should be removed by staff wearing gloves and must not touch surfaces. Use of sterile equipment and dressings is mandatory at all times.

▪ **Sterilizers:** it is better for hospitals to set up a central sterile supply service department (CSSD) than to have each clinic carrying out its own sterilization procedures. With the exception of tonometer heads, the eye clinic should avoid all use of disinfectant solutions for sterilizing equipment. Autoclaving is a common and effective method for sterilizing surgical equipment. Small autoclave units are available for clinics. Heat-sensitive equipment that cannot be autoclaved can be sterilized by a number of alternative processes, including ethylene oxide, gamma irradiation, and gas plasma (Sterrad process).

There has been considerable controversy about defective disinfection and sterilization and the possibility that this gives of transferring vCJD between patients. The American Society of Cataract and Refractive Surgeons has recently issued (2007) Recommended Practices for Cleaning and Sterilizing Intraocular Surgical Instruments and these should be followed (2). In the United Kingdom, the Royal College of Ophthalmologists and the NHS (DHSS) produce ongoing guidelines and these should be consulted on their Web sites.

It has been established that most disinfectants are inadequate for eliminating prion infectivity. A contact time of 15 minutes with a 0.5% chlorinated solution has been found to give the most consistent prion inactivation, but chlorine is corrosive and is thus unsuitable for critical medical devices. Some investigators have shown that a contact time of one hour with 1N sodium hydroxide completely inactivates vCJD.

The combined contribution of cleaning and an effective physical or chemical reprocessing procedure should eliminate the risk of vCJD transmission. As much as a 10^4 reduction in microorganisms is achieved with an effective cleaning regime.

Prions, which are the cause of vCJD, have the following requirements for sterilization:

Autoclaving at 134°C at 30 psi for 18 minutes or greater, at 121°C at 15psi for 4.5 hours, or 132°C for 60 minutes is also now popular.

Prions are resistant to boiling, dry heat, UV, ionizing radiation, ethylene oxide, ethanol, alcoholic iodine, acids, and detergents. In addition, prions are resistant to acetone, formalin, and Lysol®.

THE HOSPITAL WARD

In the past, patients having ophthalmic surgery were admitted to a special eye ward to avoid the risk of cross-infection from general surgical patients and because of specially trained nurses available to look after them. This is not cost effective, particularly when at least 40% of patients have their cataract surgery performed on a day care surgery basis.

Is there a cross infection risk to ophthalmic surgery patients if they share wards with other disciplines? The most likely consequence could be cross-infection via hands, particularly with *Pseudomonas aeruginosa* from a urinary catheterized surgical patient, and this could cause conjunctivitis, particularly in an immediate postoperative setting. However, it is unlikely that such conjunctivitis would progress to endophthalmitis, but keratitis could occur, especially if a bandage contact lens had been placed in the eye or the epithelium had not reformed. It is more sensible for ophthalmic patients to share a ward with a specialty, such as cosmetic surgery, in which patients are not admitted with existing chronic infections and in whom 'clean' as opposed to "clean-contaminated" or "dirty" surgery is to be performed. It is also prudent, although more expensive, to employ separate nurses to look after the eye patients, as opposed to the other patients in a joint ward, to reduce the chances of cross-infection of patients' eyes from handling the non-eye patients, even if they are having "clean" surgery. Such strict criteria may not apply at weekends, particularly if the joint ward is mainly used for day case surgery by other surgical disciplines—the best situation. A separate ward for ophthalmology is ideal.

Drops must never be shared between patients. Each patient must be provided with their own drop bottles, which have an expiry date of no longer than one week, or be given topical drug therapy by single dose preparations. This stringency is necessary to avoid the risk of cross-infection with bacteria from one contaminated, and not yet infected, postoperative eye to another. There is also the need to avoid any possible transfer of highly infectious viruses that can be found in tears, such as hepatitis B and HIV, from one patient to another, when the patient is at risk of rapid virus adsorption from a denuded or damaged corneal epithelium.

There has been concern about possible airborne spread of *Staphylococcus aureus*, particularly multiple-resistant strains, that could colonize or infect eye patients sharing the ward with general surgical patients. This is known to occur with other specialties and therefore patients colonized with multiple-resistant *S. aureus* should not share accommodation in the eye ward. Hospital wards should take general precautions to avoid airborne staphylococcal dispersal by not using woolen blankets and by washing and not sweeping the floor. The ventilation to the ward must be adequate, which will assist in dispersal of shed staphylococci. Patients with chronic ulcers or bedsores should not share the ward with eye patients, since their wounds can become colonized with *Streptococcus pyogenes* that can spread by the airborne route within the ward to infect patients and staff.

THE OPERATING THEATER

Postoperative infection is the most common form of exogenous bacterial endophthalmitis. Sources of organisms include the patient's lid flora, the surgeon (hands, gloves, nose, aspects of technique), contaminated instruments, implants, drugs, irrigations and infusions, and environmental airborne flora. Endogenous endophthalmitis has occurred from contaminated intravenous infusions and blood transfusion during general surgery.

The bacterial flora indigenous to the conjunctival sac and eyelids are important in this setting. The lid margins exhibit both a permanent flora (staphylococci, diphtheroids, and *P. acnes*) and transient pathogens (streptococci and coliforms), which have arisen from the throat flora and the hands, that can adhere to plastic intraocular lenses. Environmental factors are also important.

The lid margins may also be colonized with *Staphylococcus aureus*, especially in atopes. Although similar cultures may be obtained from the two eyes, cultural findings from normal lids vary over 48 hours. For this reason, it is no longer customary to perform preoperative cultures. It is advisable, however, to assume that there is lid colonization with *Staphylococcus aureus* before intraocular surgery in atopes, especially those with active eczema, albeit at a different site. They should be given at least 24 hours of topical antibiotic prophylaxis with fusidic acid (Fucithalmic) or another anti-staphylococcal antibiotic. Preoperative antiseptic preparation with 5% povidone iodine or 0.5% chlorhexidine acetate is mandatory.

Topical antibiotic drops given one day preoperatively can be effective in reducing the bacterial flora of the eye, but are not proven in preventing endophthalmitis. Various preparations are commercially available; fusidic acid as Fucithalmic 1% gel is effective against staphylococci while gentamicin sulphate (0.3%) decreases both lid and conjunctival staphylococcal cultures but is ineffective for streptococci and *P. acnes*. Broad-spectrum cover such as that given by chloramphenicol is useful but should be used with caution, because of its idiosyncratic toxicity for bone marrow—its advantage is penetration of the surface epithelium. Levofloxacin 0.5% is another possibility (refer to Chapter 4).

USE OF POVIDONE IODINE TO REDUCE THE BACTERIAL COUNT ON THE CONJUNCTIVA AND CORNEA BEFORE SURGERY

It is now well established that 5% povidone iodine reduces both the variety and number of bacteria present on the conjunctival and corneal surface by 10–100 fold (3,4). This technique is still in favor, and 5% povidone iodine should be applied onto the ocular surface and allowed to irrigate the cornea, conjunctiva, and palpebral fornices for at least three minutes *before* surgery. In order to achieve this period of time, it is necessary to start the use of 5% povidone iodine drops in the anteroom or waiting area prior to taking the patient into the operating theater. Within the operating theater, the periocular region (lids, brow, and cheeks) should also be treated with 5% povidone iodine before the drape is applied and then a further dose of 5% povidone iodine applied to the ocular surface with the lids retracted. Use of povidone iodine antisepsis in this way has been shown to reduce the rate of postoperative endophthalmitis after cataract surgery but only at a significance level of 5%. However, use of povidone iodine alone at the time of surgery without additional antibiotic prophylaxis has been found to give disappointing prophylaxis against postoperative endophthalmitis in the European multi-center study (reviewed in Chapter 8). 5% povidone iodine is not available commercially in an appropriate form outside the United States and has to be produced elsewhere in hospital pharmacies or by dilution from 10% preparations in the operating theatre. This is not ideal as the 10% preparations are formulated for skin use. Alcoholic preparations of povidone iodine must NOT be applied to the eye. Povidone iodine applied at the end of surgery has no prophylactic effect for preventing endophthalmitis associated with cataract surgery, although there is evidence that the conjunctival bacterial population can be reduced. A useful review by Ciulla et al. (4) of an evidence-based update of bacterial endophthalmitis prophylaxis for cataract surgery highlights the benefits of povidone iodine antisepsis.

A sterile duo procedural pack system specifically designed for invasive ophthalmic procedures, for example, cataract surgery, is being developed by Spectrum Ophthalmics (Vision Pharmaceuticals Ltd. Macclesfield, U.K.). Procedural pack A will contain an aqueous buffered 5% povidone iodine solution packaged in single patient use vials (1 mL and 10 mL) and designed for use in the sterile field in theater. The 10 mL vials are for periocular use (lids, brow, and cheek) prior to the drape being applied and the 1 mL vials are for applying drops onto the ocular surface (cornea, conjunctiva, and palpebral fornices) prior to incision. Both sterile swabs and locking forceps are supplied in a separate pack (B) for application of the povidone iodine to the periocular region.

THEATER HYGIENE DISCIPLINE TO PREVENT HAI

Cleanliness is next to godliness and the eye theater is no exception. Bacteria reside in dirt collected from outside the theater, which contains clostridial spores. In contrast, dust on theater shelves and other horizontal surfaces consists predominantly of shed skin scales from staff and patients, which harbor staphylococci. Hence, a standard of excellence for the theater demands that a minimum of dirt be brought into the theater on shoes, trolley wheels, or equipment. If ward trolleys enter the eye theater, their wheels must be disinfected initially with bleach, but ideally a transfer system should be used in which patients are transferred from a ward trolley to a theater trolley at a transfer station, with the theater trolley never leaving the clean eye theater area. If this standard is maintained and theater floors are properly cleaned, clostridial spores will not be found in the theater. In addition, all modern operating theaters should be serviced by dirty corridors, for removal of used equipment for sterilizing, and be supplied with the services of a central theater sterile supply unit that is maintained by staff experienced in sterilization techniques who clean, disinfect, and sterilize equipment for the next operation. This should no longer be performed on site at the back of the theater by the nurses and orderlies.

All staff undertaking surgery should change into operating theater clothing in designated areas and must not leave the clean theater area wearing operating clothing and shoes. If they leave the theater area, they must discard theater clothing for outside clothing and change into new theater clothing on return—hence the need to provide food for meal breaks within the theater common room. All staff should wear hats and masks, which fully cover nose, mouth, and chin, to reduce staphylococcal shedding and aerosolized mouth droplets as much as possible. Showers should not be taken before surgery because this increases staphylococcal shedding. Beards should not be worn by surgeons when performing cataract surgery, as they are considered a major risk factor for staphylococcal shedding into the surgical wound when using the operating microscope over the patient's eye. All operating staff should perform intensive hand washing with a surgical disinfectant hand scrub such as Betadine (10% povidone iodine) or Hibiscrub (4% chlorhexidine); both these solutions contain detergent to which some staff can become allergic. Before surgery commences, the operating theater doors should be shut, and kept shut, and there should be a minimum movement of staff and onlookers. Surgical drapes are attached to the eyelids, after which the conjunctiva and cornea should be treated for three minutes with a 5% aqueous solution of povidone iodine, which has been shown to reduce, but not abolish, the conjunctival bacterial flora; an alternative disinfecting solution is aqueous chlorhexidine acetate, 0.5%.

THEATER AIR FLOW ENGINEERING TO PREVENT HOSPITAL ACQUIRED INFECTION

While most postoperative infection (conjunctivitis, wound infection, and endophthalmitis) occurs from the patient's own lid and conjunctival flora, the minority (approximately 15%) is probably acquired from the theater staff. This can be reduced, but not abolished, by operating on the patient in the presence of adequate air flow to carry airborne skin squames containing staphylococci, or aerosolized droplets containing beta-hemolytic streptococci (if the surgeons or scrub nurse have a sore throat), away from the surgical wound towards the periphery of the theater and through vents or gaps in the theater doors towards the theater corridors. To achieve this aim, the operating theater requires to be kept at a positive pressure compared to its neighboring rooms and corridors, which can be easily monitored. In addition, the air flow distribution within the theater should be arranged so that the filtered air on entry progresses across the operating field before leaving the theater. Unfortunately, this primary aim of theater air flow is not often achieved in practice! Problems abound but mainly consist of placing air flow entry ports in the wrong site so that most of the air flow crosses the ceiling, missing the surgical wound altogether, and hits the walls opposite instead, drops down to floor level to collect staphylococci shed by staff and then circulates up towards the surgical wound area as dirty air. A modern theater should maintain a minimum of 20 air changes per hour but this is often breached by staff leaving doors open and continually walking in and out—this must not be allowed! Air flow can be easily monitored, and should be on a routine basis, using smoke tests. For more sophisticated analysis, the electronic particle counter has been found particularly useful.

MANAGEMENT OF THE ATOPE

The atopic individual has a symbiotic relationship with *Staphylococcus aureus*, which colonizes the skin, including that of the eyelids and nasal mucosa, in up to 70% of them. *Staphylococcus aureus* also colonizes and may infect eczematous skin. Care is therefore needed when planning surgery, particularly if the patient has an associated blepharitis. The following regime is suggested in addition to others given above:

- whole body bathing, including shampooing hair with Hibiscrub, a detergent skin cleansing solution containing chlorhexidine gluconate, 4%, in an alcoholic base, for 72 hours before surgery
- topical anti-staphylococcal prophylaxis for 72 hours before surgery with fusidic acid gel (Fucithalmic)
- 5% povidone iodine to the conjunctiva for eight minutes before operating
- postoperative oral fusidic acid (750 mg three times daily, enteric coated capsules) or trimethoprim (200 mg twice daily for five days) together with intracameral cefuroxime (1 mg in 0.1 mL) prophylaxis (Chapter 8).

EYE BANKING

The rate of bacterial endophthalmitis complicating penetrating keratoplasty with a corneal graft has been reported as 0.1% to 0.8% (Chapter 6). A higher incidence of postoperative endophthalmitis has been shown in the past from patients receiving corneas from a Sri Lankan eye bank at 1.3% than in those receiving eye bank tissue from western sources (0.14%). This reflects methods of collection and storage of whole eyes for transplantation, as well as the health of the donor. Aminoglycosides (gentamicin, streptomycin) have long been favored for storage solutions but these antibiotics have no effect on streptococci or fungi. Additional anti-streptococcal and antifungal antibiotics should be used in the preservative solution.

Eye banks should set up both quality assurance (QA) as well as quality control (QC) schemes in the management of the removed donor eyes for corneal transplantation. QA involves labeling all drug dilutions properly with the date and time they were made up, together with providing an expiry date. This applies particularly to dilutions of antibiotics held in the refrigerator. Equipment such as the refrigerator and freezer should be fitted with temperature-monitoring graphs to check that they are not failing. It is no longer acceptable to allow infection to occur because of using out-of-date antibiotic solutions that have little potency. QC involves culturing the donor eyes before use to be sure that relative sterility has been obtained and that the donor eye does not harbor *Pseudomonas aeruginosa*, for example. The equipment used within the eye bank laboratory should also be cultured and checked for sterility in an appropriate and recoded manner. The donor individual should have been tested for syphilis, HIV, and hepatitis B and C as a minimum.

Each country should set up its own National Eye Bank and not import whole donor eyes or corneas from other countries. It is not surprising that such practice is associated with an increase in the rate of bacterial endophthalmitis but it also encourages transfer of tissue-associated viruses including herpes (Chapter 6) and vCJD. The export of human donor material from one country to another is wrong ethically, as well as flouting World Health Organization rules. Such practice, especially when performed on a commercial scale, also encourages the acquisition of human donor material in dubious circumstances. This practice must be resisted by ophthalmologists worldwide.

REFERENCES

1. Lim R, Dhillon B, Kurian KM, et al. Retention of corneal epithelial cells following Goldmann tonometry: implications for CJD risk. Brit J Ophthalmol 2003; 87:583–86.
2. Recommended Practices for Cleaning and Sterilizing Intraocular Surgical Instruments. ASCRS - ASORN Special Report. February 16, 2007 (online). Available at: http://www.ascrs.org/9370_1.pdf. Accessed March 29, 2007.
3. Isenberg SJ, Apt L. The Ocular Application of Povidone Iodine. Community Eye Health 2003; 16:30–31.
4. Cuilla AT, Starr BS, Masket, M. Bacterial Endophthalmitis Prophylaxis for Cataract Surgery—An evidence based update. American Academy of Ophthalmol, 2002; 109(1): 13–24.

Contact Lens-Associated Keratitis, *Fusarium*, Disinfecting Solutions, and Hygiene

INTRODUCTION

Contact lenses (CLs) and their storage cases are frequently contaminated with bacteria. This has been estimated in the past for storage cases in the United States and United Kingdom at approximately 40% for bacteria (1,2) and 6% for *Acanthamoeba* (3), especially in the presence of a weak chlorine solution containing 4 ppm of available chlorine; in addition, 11% of sterile saline and other solutions used in CL care were found to be contaminated with bacteria (4). However, these older studies were not based on use of the new multipurpose solutions (see below). The disinfecting solutions did not work well in practice, either because they were difficult for the patient to use or because their chemical activity was limited. There are also patient non-compliance issues to consider with patients often being non-compliants (see below). Disinfectant activity can also be compromised with the presence of organic matter, also biofilms which is a challenge to its effectiveness.

Multipurpose CL cleaning and disinfecting solutions (MPS) were introduced by industry 10 years ago to try to avoid the pitfalls described above. The solution is sterile, marketed to last for one month only, and usually contains both a poloxamer (cleaning agent) and a disinfecting chemical such as polyhexamethylene biguanide (PHMB) 0.0001% viz. ReNu MultiPlus (Bausch & Lomb, Rochester, NY, U.S.A) Complete (Advanced Medical Optics Santa Ana, California, U.S.A) or All-in-One Light (Sauflon, Richmond, London, U.K.) or Polyquad (polyquarternium-1, 0.001%) with an antifungal and anti-amebal agent [myristamidopropyl dimethylamine (Aldox®), 0.0005%] viz. Optifree Express (Alcon, First Worth, Texas, U.S.A.) These are described more fully below. A recent study when the MPS was used properly, without CL tap water rinsing and for only one month, found expected contamination rates of the storage case to be 7% with 78% sterility and without contamination with *Acanthamoeba* (5); the case was air dried while the lens was worn. However, a higher rate of 34% contamination of storage cases and 11% of solution samples was found recently from CL wearers in Hong Kong reflecting the problems of lack of hygiene and compliance in a student population (6,7).

Storage cases should be changed monthly, and then disposed of, to avoid the accumulation of a bacterial biofilm on the inner surface of the case, and on the contact lenses. The biofilm protects all organisms from penetration by the active moieties of the chemical disinfectant. In addition, the storage case should *not* be washed in tap water, as this introduces coliforms and *Acanthamoeba* that colonize the scale at the end of the tap (faucet). Instead the storage case should be washed in boiled-cooled water or with a sterile solution. The pre-used MPS should be discarded daily and the storage case left to air dry, and be stored dry while the contact lenses are worn. Fresh MPS should be placed in the lens storage case each day. Topping up of old, pre-used MPS in the CL storage case should not happen.

PSEUDOMONAS AERUGINOSA KERATITIS

This is the most serious and frequent bacterial infection of the cornea associated with CL wear, which can be devastating with rapid corneal lysis (8,9). The infection occurs in both extended and daily wear of CLs (9), demonstrating that storage case contamination is not a prerequisite requirement for keratitis and that, even in the CL wearer with a storage case, the mode of transmission may be eye to case, rather than the other way around. *P. aeruginosa* is found as a contaminant in only 1.5% of storage cases in the community (1) and thus it more frequently produces keratitis than other species. It is found within the environment of the bathroom, frequently colonizing face flannels and washing machines. For details of pathogenesis, refer to Chapter 1.

Early recognition and treatment is essential to avoid a severe keratitis (Fig. 1). The patient may present to the optometrist or ophthalmologist with a contact lens–associated ulcer, often in the central cornea, that may be 2 mm or greater in size (for other features see Chapter 6). Ideally, the patient is referred to a hospital with a provisional diagnosis where an immediate scraping can be done for a Gram stain and culture. Associated lens storage cases and solutions (including MPS) should also be cultured. The patient should be treated empirically with frequent drop therapy of a quinolone antibiotic using levofloxacin 0.5% or ofloxacin 0.3%, starting with one drop every 15 minutes for the first hour, then every hour for 24 hours, and every other hour for 72 hours and continued once every four hours for at least another week. Because the degree of inflammation can be intense, or become intense during therapy, some clinicians are tempted to also use topical corticosteroids to reduce protease activity by both the bacterium and the polymorphonuclear cell response. This can be disastrous and result in worsening of the infection, with lysis of the cornea, due to inhibition of bactericidal killing by phagocytic peroxidases.

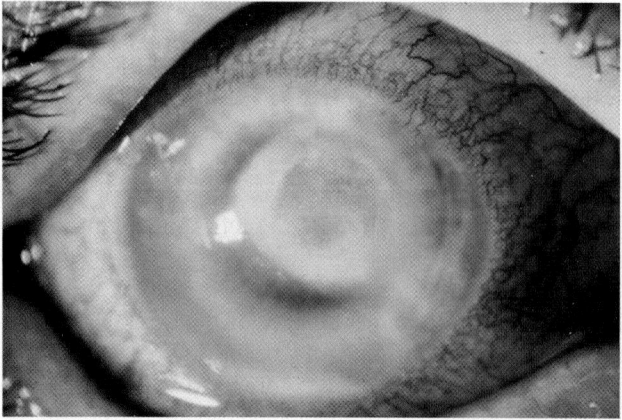

Figure 1   Patient with soft CL-associated *Pseudomonas aeruginosa* keratitis.

Corticosteroids may be safely introduced once it is clear that the bacterial process has stopped advancing but, even then, may be associated with a rapid recurrence so that topical anti-pseudomonal antibiotics must be given in high dosage when topical corticosteroids are introduced and continued during their usage.

P. aeruginosa keratitis is typically found in the central cornea, as with other CL-associated infections, resulting in healing with anterior stromal scarring that can seriously reduce visual acuity within the central axis. Patients should not be encouraged to have graft surgery to remove this scarring, but rather to accept the visual disability that it causes within a normal cornea. CL hygiene and storage case disinfection must aim to eradicate the chance of this infection from occurring.

FUSARIUM KERATITIS ASSOCIATED WITH ReNu WITH MOISTURELOC AND OTHER MULTIPURPOSE SOLUTIONS

In February 2006, an increased number of cases of *Fusarium* keratitis was reported from contact lens wearers initially in Singapore followed by Hong Kong and Malaysia (10); the initial survey was from March 2005 to May 2006. By June 2006, cases were being recognized within the United States (11). Cases were subsequently reported in 33 states and one U.S. Territory. ReNu with MoistureLoc (Bausch & Lomb, Rochester, New York, U.S.A.) was thought to be involved, which was a new multipurpose contact lens solution (MPS) launched in autumn 2004 containing alexidine (0.00045%) as the chemical disinfectant and preservative and a new type of water-retaining polymer that coated the contact lens. This increased comfort for wearing a contact lens due to its water-retaining properties. On April 10, 2006, Bausch & Lomb suspended shipments of ReNu with MoistureLoc and on April 13th removed the product from retailers' shelves in the Far East. On May 16, 2006, following many more cases in the United States, Asia, and Europe, Bausch & Lomb issued a worldwide recall of this product. The company did not recall its existing ReNu MultiPlus and ReNu MultiPurpose solutions, based on 0.0001% polyhexamethylene biguanide without the wetting polymer in MoistureLoc, as there was no evidence to link it with the *Fusarium* epidemic. The final number of cases of *Fusarium*

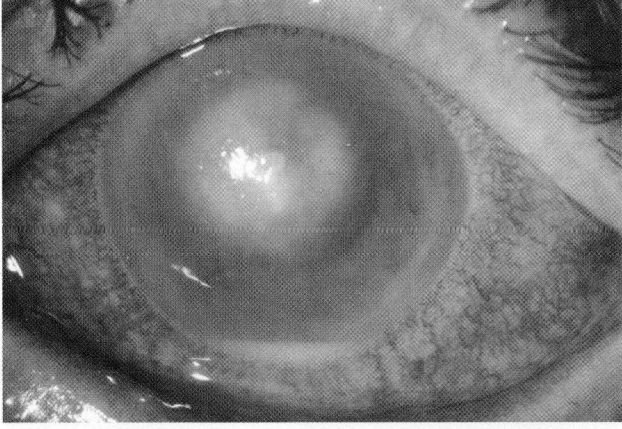

Figure 2 Contact lens wear and ReNu MoistureLoc-associated *Fusarium* keratitis.

keratitis associated with ReNu with MoistureLoc has included 164 confirmed cases in the United States, 68 in Singapore, 12 in Hong Kong, eight in Malaysia, and a few in northern Europe (10,11). Cases still continue to occur in contact lens wearers (CLWs), with 11 cases reported in Spain this last year and one in Portugal (Consuelo Ferrer, personal communication, Alicante) and one seen currently in a CLW in Berlin (Fig. 2). In addition, 14 cases of keratitis have been reported recently from Martinique (Caribbean), of whom 12 were CLWs with 10 using ReNu with MoistureLoc; *F. solani* was cultured from five out of these latter 10 patients (12). There is considerable morbidity, as described by Chang et al. (11).

Bausch & Lomb conducted extensive microbiological tests at its two factories and found no *Fusarium* contamination of its manufacturing lines, equipment, or sterile room environment (11). *Fusarium* spp. were not recovered from any unopened product but were recovered from caps of opened bottles (1 out of 17 tested) of ReNu with MoistureLoc and one of five bottles of ReNu MultiPlus. *Fusarium* was also recovered from 6 of 11 used contact lens storage cases. This led to the conclusion that individual contamination of the bottle cap of ReNu with MoistureLoc occurred with environmental strains of *Fusarium*. This theory is corroborated by finding great genotypic diversity in 39 U.S. isolates and 10 each from Singapore and Hong Kong (11). At least 10 different *Fusarium* species were represented including 19 unique multilocus genotypes, 14 of which were singletons. 77% of the U.S. isolates nested within *Fusarium solani* complex and 18% in the *Fusarium oxysporum* complex. Multilocus genotyping of isolates from Singapore and Hong Kong found all 20 isolates tested were within *Fusarium solani* group 2. *Fusarium solani* complex groups 1 and 2 have previously been identified in sinks and drains (13,14).

Interestingly, *Fusarium* infection was also observed, as part of this epidemic, due to use of other multipurpose solutions, not previously reported and recognized. This identifies that *Fusarium* is a ubiquitous soil fungus that can easily contaminate the cap of the multipurpose solution bottle. These solutions (MPS) have undergone tests to establish a minimum of a 1-log reduction of chemical disinfection for fungi, with *Fusarium solani* ATCC 36031 in an ISO 14729 stand alone test (International Organization for Standardization. 2001. Ophthalmic Optics. CL care products), and a 3-log reduction for designated bacteria, but these

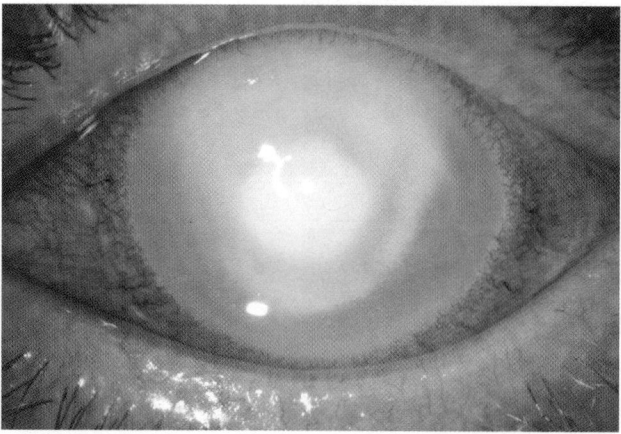

Figure 3   Later presentation of CLW and ReNu MositureLoc-associated *Fusarium* keratitis.

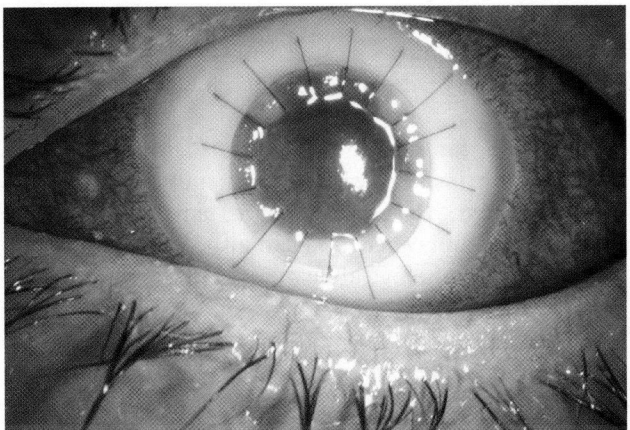

Figure 4 Requirement for a corneal graft for CLW and ReNu MositureLoc-associated *Fusarium* keratitis.

tests take place in ideal conditions in laboratories with selected strains. If a weaker-than-expected chemical disinfectant is used, such as the polymeric biguanide alexidine, then there is insufficient reserve capacity to kill the fungus in field (in-use) conditions. Alexidine was found to be less effective than PHMB or chlorhexidine against *Acanthamoeba* (15).

Fungi belonging to the genus *Fusarium* reside in soil and on plants, primarily in tropical and subtropical climates, and infections are typically seen in agricultural workers of these regions (16). However, this fungus is also found in the soil of northern climates of Europe and the United States when occasional opportunistic infection occurs. It is a mycelial fungus that produces conidiospores. Research analyzing ocular fungal infections in China from 1989 to 2000 indicates a shifting trend in the types of fungus implicated in ocular infection. *Fusarium* infections increased from 54% of cases between 1989 and 1994 to 60% from 1995 to 2000. Concurrently, the percentage of *Aspergillus* infections decreased from 22% to 15% during the same time periods (17). The figure for *Fusarium* infection has now increased to 73% (18). A study of the spectrum of fungi causing ocular infection at Wills Eye Hospital in Philadelphia analyzed data from before the current outbreak (from 1991 to 1999) and found *Fusarium* was the commonest filamentous fungus isolated in this study; *Candida albicans* accounted for 11 out of 24 isolates while *Fusarium* accounted for 6 of 24 isolates. This is a change from prior studies in non-tropical regions (19,20).

Fusarium can cause a variety of infections in humans: keratitis, endophthalmitis, otitis media, onychomycosis, cutaneous infections, pulmonary infections, and endocarditis. Most commonly, *Fusarium solani* is reported, and, unfortunately, this species is relatively intractable to treatment and tends to be resistant to drugs. It penetrates deeper into the corneal stroma than other types of fungi, with a tendency to penetrate Descemet's membrane, and may cause endophthalmitis

The best chance for treatment is early diagnosis and rapid and persistent dosage of an effective antifungal drug. This requires a high index of suspicion. The later the diagnosis, the greater the possibility for the infection to become more difficult to control without sequelae, particularly involvement of the iris and later endophthalmitis. This outbreak had a high degree of morbidity in a relatively young population (median age in United States 41 years, range 12 to 83). Twelve patients had bilateral infection. 23% of the U.S. patients resolved on topical therapy but corneal transplant was required in 55 of 164 patients (34%) in the United States and in 5 of 68 (7%) of patients in Singapore.

A fungal ulcer can sometimes appear similar to bacterial keratitis in initial presentation. Early signs and symptoms include redness and irritation, which can precede a visible infiltrate by several days. Tearing, discharge, pain, photophobia, decrease in visual acuity, and anterior chamber reaction can also occur with fungal keratitis. The infiltrate can vary in morphological features: either a central or peripheral location is possible, as is a deep stromal infiltrate beneath the epithelium. Hypopyon may be present, though typically less so than in bacterial keratitis.

In obtaining a culture, care must be taken to scrape thoroughly and penetrate deeply enough, beyond any superficial epithelial healing, to the location of the infection. Culturing of lenses, lens cases, and the tips and caps of lens solution containers will also provide useful information regarding the presence of fungus and its possible reservoir. One retrospective study found lens case contamination accompanied *Fusarium* keratitis in all patients for whom lens case culture results were available (21).

A topical ophthalmic fungicidal drug, such as natamycin ophthalmic solution (Natacyn, Alcon, Ft. Worth Texas U.S.A.), is the initial treatment of choice, often coupled with repeated corneal scrapings to assure tissue penetration of the drug (22). Sensitivity testing should be performed on *Fusarium* isolates, as they can be resistant to the fungicidal drug natamycin. Use of antifungal imidazoles, which are static and not cidal, may be needed instead (Appendix A) but *Fusarium* isolates may be resistant to them as well. Treatment failure has been reported with voriconazole (23). Further details of *Fusarium* therapy are given in Chapter 9.

Types of CL wear by the U.S. *Fusarium* patients in the outbreak included two-week daily wear (48%), 30-day extended wear (23%), frequent planned replacement (13%), seven-day extended wear (5%), daily disposable (4%), one-year conventional daily wear (3%), and unknown (4%). A case-control study (22 case patients, 32 CLW controls) was mounted that identified only the use of ReNu with MoistureLoc in the month before the onset of symptoms as significantly associated with having *Fusarium* keratitis (odds ratio 22.3; 95% CI 3.1–∞) (11).

It is now thought that the recent *Fusarium* outbreak was the result of several different factors that created an ideal environment for contamination, survival, and multiplication of the fungus. Contamination will have started on the cap of the MPS bottle–a notorious site only exposed to intermittent soaking with the solution and on which microbes and dirt tends to adhere together. Contributing problems included the combination of ingredients in the MoistureLoc lens-care system, with a weak polymeric biguanide disinfectant called alexidine, the polymer polyquarternium 10 moisture retaining polysaccharide that coated the contact lens with a moisture film–and a surfactant, poloxamer 407. This combination worked antagonistically by coating *Fusarium* and protecting it from chemical disinfection (see MPS section below).

In addition, alexidine is a weaker polymeric biguanide than chlorhexidine or PHMB (polyhexamethylene biguanide) although it was used at 0.00045% (4.5 ppm). Alexidine had been shown less effective against *Acanthamoeba* than chlorhexidine or PHMB by a factor of approximately eightfold (15) At a concentration of 0.00045% (4.5 μg/mL), ReNu with MoistureLoc would have had no amebicidal effect on *Acanthamoeba*, as the range for trophociticidal values were 3 to 12 μg/mL (modal value 6 μg/mL), and the cysticidal values were 25 to 100 μg/mL (modal value 25 μg/mL) (15).

Though the combination of a weak disinfectant and a water-retaining polymer appears likely, some questions remain, including why *Fusarium* has also been found with other contact lens multipurpose solutions. Is the present situation one of recognition following the outbreak associated with ReNu with MoistureLoc, which seems most likely, or has there been a real increase in environmental isolates of *Fusarium* due to climate change or other environmental factors, as suggested by the Wills Eye Hospital (Philadelphia, Pennsylvania, U.S.A.) and Chinese studies (18,19)? The figures below, given at ARVO 2006, confirm both the ubiquitous presence of *Fusarium* sp. in the environment of the CLW *before* the introduction of ReNu with MoistureLoc in late 2004, and the increase in incidence due to *Fusarium* sp. in CLWs in south Florida in 2005 and the first quarter of 2006. Why *Fusarium* has emerged as the dominant fungus causing keratitis for the CLW is not known but needs investigating. Is *Fusarium* better able to colonize the particular type of plastic cap used for the bottle of ReNu with MoistureLoc than other fungi?

Epidemiology of Fungus Keratitis in South Florida (ARVO Symposium 2006, Eduardo Alfonso, Miami, Florida, U.S.A.)

1495 corneal cultures in 2004 and 2005 (two per day) presented to Bascom Palmer Institute, University of Miami

    585 (40%) yielded growth on culture
    122 (21%) of culture-positives gave fungal growth

    66 (54%) yielded *Fusarium* sp., of which *F. oxysporum* (69%), *F. solani* (10%)
    56 (46%) yielded *Curvularia, Aspergillus*, and others

    26 of the 122 (21%) fungus culture-positive cases wore CLs and 24 out of 26 yielded *Fusarium* sp.

Incidence of *Fusarium* Infection in CLW in South Florida (24)

    2004 – 8/30 (27%)
    2005 – 18/36 (50%)
    2006 (Jan. 1st to March 31st) – 70%
    No fungal cases occurred with Daily Wear disposable CLs
    This data from Miami is interesting because 77% of *Fusarium* isolates overall in the United States causing infection were of the *Fusarium solani* complex rather than *F. oxysporum*, which is more commonly found in the south Florida patient population.

The list below summarizes some of the main predisposing factors that can compromise the overall success of a multipurpose disinfecting product, particularly for a relatively hard-to-kill fungus such as *Fusarium*:

▪ Biguanides viz. chlorhexidine, alexidine, and PHMB are cationic with a strong net positive charge so bind to negatively charged surfaces such as CLs. The pharmacokinetics of lens/preservative interaction studies (uptake/release) demonstrates this.

▪ The presence of no-rub multipurpose solutions, which have crept into the care regime over the last year. This idea primarily came about because it was well recognized that patients were generally non-compliant with instructions relating to the rub-and-rinse step, with many merely removing the lens from their eye and dropping it into their storage case overnight. These no-rub products were developed to work optimally on the large amount of loosely bound, non-denatured protein that is found on many conventional lens materials. To maintain optimal performance there is an argument to reinstate the rub regimen, particularly for the new silicone hydrogel lenses.

- Failure to use fresh multipurpose solution each day in the lens storage case, thus storing contact lenses in pre-used solution. This 'topping-up' technique was identified by Chang et al. (11) in their case-control study and was significant on univariate analysis for the outbreak patients. Exploring why re-use of this multipurpose solution (ReNu with MoistureLoc) can promote *Fusarium* growth in lens cases needs investigation.

- Preservative (disinfecting chemical) interaction with plastic containers giving rise to a reduced level of biological activity. This has been a past problem but is not happening now.

- Loss of preservative (disinfecting chemical) through the container wall due to moisture loss. This is more of a consideration when greater surface area to volume ratios of the container provides a less stable product exacerbated by storage in hot temperatures.

- General hygiene with concomitant compliance is a major prerequisite.

- Repeated exposure to low concentrations of antimicrobial contributing to the development of resistance.

- Unreasonable expectation of disinfectants that are challenged beyond their capability should circumstances dictate, viz. presence of organic soil, development of biofilms, protection by water-retaining polymers. In addition, a balance has to exist between good antimicrobial effectivity and low toxicity.

DISINFECTION SOLUTIONS

Hydrogen Peroxide

This is the most effective chemical disinfectant against bacteria and *Acanthamoeba*, including trophozoites and cysts, and acts by oxidizing the organism (25). It does not remove protein from the CL, which requires a cleaning process (rinse and rub) with a separate cleaning solution. Hydrogen peroxide that has not been neutralized and is carried onto the cornea with the lens causes an acute red painful eye, with sterile inflammatory corneal infiltrates occurring due to oxidative damage to the epithelial surface. Neutralization is best performed after overnight storage in a vented storage case to release liberated oxygen; use of a non-vented case has resulted in serious ocular trauma from an explosive propulsion of the case lid into the eye. This type of system is called a 'two-step' procedure. Because some CL wearers forget to neutralize the storage case solution in the morning, a "one-step" product has been produced, based on adding a neutralizing tablet to the storage case when the CL are placed in the case for disinfection. The problem with products so far has been the rapid neutralization of the hydrogen peroxide after approximately 10 minutes. This is insufficient time to kill contaminating microbes adherent to the CL. We have found a high bacterial contamination rate of storage cases containing neutralized hydrogen peroxide from the 'one-step' product, which can in addition be contaminated through their venting holes in their lids. These venting holes also leak fluid contents out of the case. Careful use of the two-step hydrogen peroxide procedure remains the gold standard for CL disinfection, which will prevent microbial keratitis associated with CL wear, but is unlikely to be performed properly by most CL wearers. Finally, the peroxide distorts high water content ionic CL [U.S. Food and Drug Administration, (FDA) group 4] and it can take up to one hour after insertion before full visual acuity is restored.

Multipurpose Solutions

Due to compliance problems with hydrogen peroxide, multipurpose solutions have been produced by industry to clean and store CL with a single solution without the need for neutralization. This is usually achieved by combining a poloxamer (detergent) with a chemical disinfectant [polyhexamethylene biguanide (PHMB) or polyquad (poly-quaternium-1)] and Aldox® with appropriate buffers and EDTA. It is provided as a sterile solution in sufficient quantity for rub–and-rinse cleaning and storing of the CL, and washing out of the case. Products may contain from 5 ppm (0.0005%) PHMB, which is most effective against bacteria and fungi and is also acanthamebicidal for 10^2 cysts, to those containing 0.5 ppm (0.00005%) PHMB, which is less effective against bacteria and has no activity against *Acanthamoeba*. At this low concentration, reliance for eradicating *Acanthamoeba* depends solely on ·cleaning by the rub–and-rinse technique. Polyquad is used at low concentrations that have poor bactericidal activity and no acanthamebicidal activity, but the addition of Aldox, as a combination product, has given additional anti-amebic and antifungal activity.

Multipurpose solutions provide the easiest technique for the CL wearer to clean and disinfect the CL and, therefore, compliance can be expected to improve (7). The main advantage of these solutions is that the product is sterile, and there is no need to wash the storage case with tap water. The poloxamers used have a good surfactant action for removal of microbes from adhering to the CL. Provided the storage case is changed monthly, and that tap water contamination is avoided, these solutions represent the most compliance-friendly method with a reasonable, but not the best, bactericidal activity.

Use of solutions with PHMB as the disinfectant at a concentration of 5 ppm (0.0005%) gives an enhanced microbicidal effect, including activity against *Acanthamoeba*, but most solutions use a concentration of 1 ppm (0.0001%), because of punctate keratopathy that can occur with FDA group 2 lenses when the PHMB is apparently concentrated into lipid deposits attached to their surface. It has been found experimentally that ReNu Multiplus (Bausch & Lomb) MPS, with 1 ppm of PHMB, is more effective against *Acanthamoeba* than other MPS with 1 ppm PHMB (26).

In order for a disinfectant like PHMB to be cysticidal, it has to gain entry to the trophozoite internalized within the cyst. The most obvious route is via the ostioles or the pores in the double cell wall that connect the outer exocyst and inner endocyst, thus allowing the internalized ameba to communicate with its outside environment. It is believed that the ostioles are plugged with mucopolysaccharide, which has to be destabilized to allow penetration of the disinfectant. The effectiveness of the higher concentration of PHMB (5 ppm), found in All-in-One, is thought to be due to the binding of this highly positively charged molecule to the mucopolysaccharide, resulting in penetration and irreversible damage to the cell membrane and contents. Cell damage by PHMB is associated first of all with leakage of calcium ions from the plasma membrane of the trophozoite or the internalized ameba within the cyst or a bacterial or fungal cell. This factor may be potentiated by the hydranate chelating agent in ReNu Multiplus, which would explain the enhanced trophozoiticidal and cysticidal action found with this MPS as compared to the other solutions containing 1 ppm PHMB. Rapid chelation of calcium ions outside the cell will hasten further leakage from within the cell, leading to protein disruption and ultimately irreversible cellular damage. An alternative hypothesis is the inclusion of sodium borate and boric acid in the formulation of ReNu Multiplus, which has been shown to potentiate the activity of PHMB twofold, although EDTA at 0.1% w/v, also present in ReNu Multiplus, has been found to inhibit activity fourfold. Since the PHMB concentration used at 1ppm (μg/mL) is at its minimum effective level (MTAC), the effects of the formulatory adjuvants will be crucial and explain the differences found in activity of the various commercial MPSs all containing the same concentration of PHMB at 1 ppm.

The withdrawal of Complete MoisturePlus (AMO), because of an associated use with *Acanthamoeba* keratitis, is describedin the 'Addendum' below.

Chlorine

This was used mainly in the United Kingdom but has now ceased. Two organochlorine compounds were manufactured that liberated chlorine at approximately 4 ppm (sodium dichloroisocyanurate) or 8 ppm (halazone). Both were produced as tablets that were dissolved in the storage case in sterile saline; unfortunately, tap water was often used instead. Neither compound cleaned the CL, which required a separate reagent for the rinse-and-rub technique. Both were bactericidal, halazone being more active, but could be readily inactivated by extraneous organic material such as heavy bacterial contamination or tap water scale. Neither was active against *Acanthamoeba* cysts but halazone was active against trophozoites. The weaker compound, marketed as 'Softab,' was found to be used often by patients with *Acanthamoeba* keratitis (15) and the use of this product for CL disinfection was described as an increased risk factor for this infection. The use of chlorine tablet disinfection was also found to occur more frequently in patients with microbial keratitis, probably because wearers were dissolving them in tap water instead of sterile saline.

HYGIENE REQUIRED TO PREVENT INFECTION (MICROBIAL KERATITIS)

The ideal way to prevent microbial keratitis is the daily-wear use of disposable CLs, which are thrown away each day instead of being disinfected in a storage case. Wearing soft (hydrogel) CLs at night as extended wear CLs has been shown to increase the risk for keratitis five times, from 1:2500 to 1:500 CL wearers annually (27). The use of the new silicone hydrogel lenses is discussed in Chapter 6 with the latest results for causing keratitis, showing that their use for extended (overnight) wear has not reduced the incidence of microbial keratitis compared with the older hydrogels.

Unfortunately, the CL industry is guilty of misusing the word "disposable" to market CLs that have a suggested disposal date after a period of daily wear of two weeks or one month, in comparison to the non-disposable soft CL, which is changed yearly. This has led to non-compliance by CL wearers using the disposable CL that is not disposed of for one month! Such wearers must be as vigorous with hygiene and disinfection as users of non-disposable CLs.

The CL must be rinsed with a cleaning chemical after each day of use to remove bacteria and *Acanthamoeba* that have become adherent to the CL while worn. This is achieved by a rinse-and-rub technique in the hand with, for example, a poloxamer surfactant. If most microbes are removed by this technique, there is greater chance that the remaining microbes will be killed by the chemical disinfectant. This depends on the CL being placed in a clean, dry storage case for disinfection. If the case has been kept wet while the CLs have been worn, it may become heavily contaminated with bacteria and possibly ameba. If there is gross bacterial contamination of the case, this will result in the fresh chemical disinfectant being inactivated. Bacteria will then form a biofilm on the surface of the CL, to which *Acanthamoeba* can readily adhere and in greater numbers than without biofilm being present. The CL carries these microbial contaminants to the surface of the eye and infection can proceed.

The storage case can become contaminated from various domestic sources. In a recent intensive study of the domestic kitchen, sink taps and cloths were found colonized with *P. aeruginosa* from 3% of samples collected, while up to 70% of samples from domestic washing machines were similarly contaminated (Malcolm, 1996, personal communication). These sites should be sampled in epidemiological studies of *P. aeruginosa* infection.

It is therefore essential that the storage case should be emptied of fluid contents when the CL is worn. The case should be washed each day, after the CLs have been placed in the eye, with boiled-cooled water at 75°C, to pasteurize the case, but if this is not possible then a multipurpose solution can be used instead. The case should then be emptied and shaken dry; it can also be wiped dry with disposable paper tissues. When kept dry, the coliform (Gram-negative) bacteria are killed after several hours and ameba will form cysts instead of multiplying. When the CLs and chemical disinfectant are added, at the end of the day (or night for night-workers), there will be a greater chance of killing most bacteria because smaller numbers are present and the chance of inactivation is reduced. The rate of kill of bacteria is governed by the time required to gain a 10-fold reduction in the logarithmic (base 10) count, the so-called chemical D value, and therefore the chance for sterility of the CL within the storage case, depends on the initial numbers present. *Acanthamoeba* will only be killed if the chemical disinfectant used has activity against it, both trophozoites and cysts. It is our opinion that all commercial products used to disinfect CLs should have activity against *Acanthamoeba* to prevent this painful infection associated with CL wear.

Tap water as supplied in the domestic bathroom wash basin has been proven by molecular techniques to contain the same strain of *Acanthamoeba*, in this case *A. griffini*, as that found in the storage case and corneal infection (epithelial and anterior stromal disease); the CL wearer had been storing the soft (hydrogel) CLs in tap water (15). Tap water has also been shown to contain *Acanthamoeba* in up to 30% of homes in the United Kingdom (28,29) but only 3% of homes in the United States due to the lack of a roof storage tank (30). Tap water should never be used to store soft (hydrogel) CLs. Tap water should also not be used to rinse or wash out the case but unfortunately this happens frequently, and is used to wash the case by approximately 40% of U.K. soft (hydrogel) CL wearers. Tap water has been used for many years to wash hard or rigid gas permeable CLs, with which the risk of microbial keratitis is much less (approximately 10-fold), but cannot be recommended. Showering while wearing CLs has not yet been proven as a risk factor for microbial or *Acanthamoeba* keratitis, but is best avoided if the wearer has sufficient visual acuity to manage without CL use. If CLs are worn while showering, then the CLs should be placed afterwards in a dry case with fresh chemical disinfectant. Health care workers who wear contact lenses are not considered to be at greater risk of contracting microbial keratitis than others, provided their contact lens care hygiene is compliant as described above.

The prospective cohort Hong Kong Study (27) confirmed a previous finding that smoking is a risk factor for CL-associated microbial keratitis. The reason for this is not entirely clear, except that heavy smoking may limit the chances for good hygiene practice; nicotine-stained fingers per se should not matter. In addition, could there be an interaction between the CL being worn and nicotine from inhaled/ exhaled smoke compromising the corneal surface epithelium by a toxic reaction to allow bacterial invasion?

COMPLIANCE VERSUS NON-COMPLIANCE

It is easy to blame the CL wearer for not complying with inadequate or complicated instructions provided by industry labels or leaflets that often differ between products. It is therefore important that all practitioners provide good advice on CL hygiene to their CL wearers, including clear printed instructions that do not conflict with those provided on the disinfecting products sold. CL wearers are non-compliant by nature and therefore require constant education of the dangers of infection and how to avoid it. Labeling of different products should be coordinated by national governments and needs to stress the importance of using sterile solutions instead of tap water in the storage case. Compliance can be usefully monitored in a practice by use of a "health belief questionnaire" (7).

CLASSIFICATION OF CONTACT LENSES FOR STUDIES OF ASSOCIATED INFECTION

The FDA classification of soft hydrogel contact lenses has been used as follows:
FDA group 1: non-ionic, water content < 50%
FDA group 2: non-ionic, water content > 50%
FDA group 3: ionic, water content < 50%
FDA group 4: ionic, water content > 50%
Silicone hydrogel lenses are discussed in detail in Chapter 8 sections on Keratitis and *Acanthamoeba* contact lens binding studies.

HEALTH WARNING

All CL wearers should be warned of the risk of infection and of the consequent visual disability that can occur. They should be warned that extended wear has an increased risk of infection over daily wear of five to 10 times. They should be instructed carefully in hygiene and disinfection principles, with the advantages and disadvantages of different types of disinfection clearly explained. All CL wearers should be warned not to place domestic tap water in their storage cases, either for cleaning or for storing their CLs. Young teenagers with insufficient money to care for their CL hygiene should not be offered CLs; similarly, other members of the community thought to be non-compliant should be fitted instead with spectacles for their visual correction.

ADDENDUM – AUGUST 2007

The Department of Ophthalmology at University of Illinois, Chicago has experienced a considerable increase in the number of cases of *Acanthamoeba* keratitis (AK) between June 1st 2003 and November 30th 2005, when 40 cases of AK were confirmed, 38 in CLW of whom 35 were soft CLs (31). In addition, in January 2007, center for disease control (CDC) (Atlanta) conducted a multiple state analysis in 13 centres and established an increase in culture confirmed AK cases, which began in 2004 with wide geographic distribution. An investigation has been conducted among 39 soft CLW, with 36 using MPS of which 21 (58%) reported use of Complete MoisturePlus (AMOCMP) in the month before symptom onset; 20/36 (56%) used AMOCMP as their primary MPS and 14/36 (39%) as their exclusive MPS (32). The 36 soft CLW AK patients with symptom onset before May 24, 2007, using one or more MPS, were compared with 124 *Fusarium* controls using one or more MPS (11). Results showed that 21/36 (58%) of AK patients had used AMOCMP compared with 8/124 (6%) of *Fusarium* controls (odds ratio: 20.3; [CI = 7.6–53.9]). AMOCMP lot numbers were available for 10 patients when no single log number was repeated suggesting no intrinsic contamination. Because of this preliminary association between AMOCMP and the development of AK with its use, AMO undertook a voluntary recall of Complete® MoisturePlus worldwide in March 2007. It would seem that the problem with AMOCMP mirrors that of ReNu MoistureLoc with the wetting agent polymer interfering with the disinfecting effect of the MPS. The source of the Acanthamoeba was probably use of tap water in CL storage case hygiene (Chapter 8).

REFERENCES

1. Devonshire P, Munro FA, Abernethy C, et al. Microbial contamination of contact lens cases in the West of Scotland. Brit J Ophthalmol 1993; 77:41–45.
2. Gray TB, Cursons TM, Sherwan JF, et al. *Acanthamoeba*, fungal and bacterial contamination of contact lens storage cases. Brit J Ophthalmol 1995; 79:601–605.
3. Larkin DF, Kilvington S, Easty DL. Contamination of contact lens storage cases by Acanthamoeba and bacteria. Brit J Ophthalmol 1990; 74:133–135.
4. Seal DV, Stapleton F, Dart J. Possible environmental sources of *Acanthamoeba* sp. in contact lens wearers. Brit J Ophthalmol 1992; 76:424–427.
5. Seal DV, Dalton A, Doris D. Disinfection of contact lenses without tap water rinsing-is it effective? Eye 1999:13;226–230.
6. Yung MS, Boost M, Cho P, et al. Microbial contamination of contact lenses and lens care accessories of soft contact lens wearers (university students). in Hong Kong. Ophthalmic Physiol Opt 2007; 27:11–21.
7. Fan D, Lam D, Houang E, Wong E, Seal D. Health Belief and Health Practice in Contact Lens Wear a dichotomy? CLAO Journal 2002; 28:36 39.
8. Fleiszig SM, Evans DJ. The pathogenesis of bacterial keratitis: studies with *Pseudomonas aeruginosa*. Clin Exp Optom 2002; 85:271–278.
9. Stapleton F, Dart JKG, Seal DV. Epidemiology of *Pseudomonas aeruginosa* in contact lens wearers. Epidemiol Infect 1995; 114:395–402.
10. Khor WB, Aung T, Saw SM, et al. An outbreak of *Fusarium* keratitis associated with contact lens wear in Singapore. J Am Med Assoc (JAMA). 2006; 295:2867–2873.
11. Chang DC, Grant GB, O'Donnell KO, et al. Multistate outbreak of *Fusarium* keratitis associated with use of a contact lens solution. JAMA 2006; 296:953–963.

12.   Donnio A, van Nuoi DN, Catanese M, et al. Outbreak of keratomycosis attributable to *Fusarium solani* in the French West Indes. Am J Ophthalmol 2007; 143:356–358.

13.   Zhang N, O'Donnell K, Sutton DA, et al. Members of the *Fusarium solani* species complex that cause infections in both humans and plants are common in the environment. J Clin Microbiol 2006; 44:2186–2190.

14.   O'Donnell K, Sutton DA, Rinaldi MG, et al. Genetic diversity of human pathogenic members of the *Fusarium oxysporum* complex inferred from multilocus DNA sequence data and amplified fragment length polymorphism analyses: evidence for the recent dispersion of a geographically widespread clonal lineage and nosocomial origin. J Clin Microbiol 2004; 42:5109–5120.

15.   Seal, DV. *Acanthamoeba* keratitis update—incidence, molecular epidemiology and new drugs for treatment. Eye 2003; 17:893–905.

16.   Doczi I, Gyetvai T, Kredics L, et al. Involvement of *Fusarium* spp. in fungal keratitis. Clin Microbiol Infect 2004; 10:9:773–6.

17.   Sun XG, Zhang Y, Li R, Wang ZQ, et al. Etiological analysis on ocular fungal infection in the period of 1989–2000. Chin Med J 2004; 117:4:598–600.

18.   Xie L, Zhong W, Shi W, et al. Spectrum of fungal keratitis in north China. Ophthalmol 2006; 113:1943–1948.

19.   Tanure MA, Cohen EJ, Sudesh S, et al. Spectrum of fungal keratitis at Wills Eye hospital, Philadelphia Pennsylvania. Cornea 2000; 19:3:307–12.

20.   Houang E, Lam D, Seal D, et al. Microbial Keratitis in Hong Kong—relationship to climate, environment and contact lens disinfection. Trans Roy Soc Trop Med & Hyg, 2001; 95:361–67.

21.   Alfonso E, Cantu-Dibildox J, Munir W, et al. Insurgence of *Fusarium* keratitis associated with contact lens wear. Arch Ophthalmol 2006; 124:7:941–947.

22.   Thomas PA. Fungal infections of the cornea. Eye 2003; 17:8:852–62.

23.   Giaconi JA, Marangon FB, Miller D, et al. Voriconazole and fungal keratitis: a report of two treatment failures. J Ocul Pharmacol Ther 2006; 22:437–439.

24.   Alfonso EC, Miller D, Cantu-Dibildox J, et al. Fungal keratitis associated with non-therapeutic soft contact lenses. Am J Ophthalmol 2006; 142:154–155.

25.   Zennetti S, Siori PL, Pinna A, et al. Susceptibility of *Acanthamoeba castellanii* to contact lens disinfectant solutions. Antimicrob Ag Chemother 1995; 39:1596–1598.

26.   Beattie TK, Seal DV, Tomalinson A, McFayden AK, Grimason A. Determination of Amoebicidal Activities of Multipurpose Contact Lens Solutions by Using a Most Probable Number Enumeration Technique. J Clin Microbiol 2003; 41:2992–3000.

27.   Lam, D., E. Houang, D. Fan, D. Lyon, D. V. Seal, E. Wong, and Hong Kong Microbial Keratitis Study Group. Incidences and risk factors for Microbial Keratitis in Hong Kong: comparison with Europe and North America. Eye 2002: 16:608–618.

28.   Seal DV, Kirkness CM, Bennett HGB, Peterson M & Keratitis Study Group. *Acanthamoeba* keratitis in Scotland: risk factors for contact lens wearers. Contact Lens & Ant Eye 1999; 22:58–68.

29.   Kilvington S, Gray T, Dart JKG, et al. Acanthamoeba keratitis: the role of domestic tap water contamination in the United Kingdom. Invest Ophthalmol 2004; 45:165–169.

30.   Shoff M, Rogerson A, Seal DV, et al. Prevalence of *Acanthamoeba* and other naked amoebae in South Florida domestic water. J Water Health 2008; in press.

31.   Joslin CE, Tu EY, McMahon TT, et al. Epidemiological characteristics of a Chicago-area *Acanthamoeba* keratitis outbreak. Am J Ophthalmol 2006; 142:212–17.

32.   Bryant K, Bugante J, Chang T, et al. *Acanthamoeba* keratitis multiple states, 2005–2007. Morbid Mortal Weekly Report (June 1) 2007; 56:532–34 (http://www.cdc.gov/mmwr®).

Index